STOCHASTIC FINANCE

STOCHASTIC FINANCE

Edited by

A.N. SHIRYAEV
Steklov Mathematical Institute, Moscow, Russia

M.R. GROSSINHO
ISEG, Universidade Técnica de Lisboa, Portugal

P.E. OLIVEIRA
Universidade de Coimbra, Portugal

M.L. ESQUÍVEL
FCT, Universidade Nova de Lisboa, Monte de Caparica, Portugal

Springer

Library of Congress Cataloging-in-Publication Data

Stochastic finance / edited by A.N. Shiryaev ... [et al.]/
p. cm.
"From 26–30 September 2004, the 'International Conference on Stochastic Finance 2004' took place at Instituto Superior de Economia e Gestão (ISEG) da Universidade Técnica de Lisboa, in Portugal"—Preface.
Includes bibliographical references.
ISBN-10: 0-387-28262-9 (alk. paper)
ISBN-10: 0-387-28359-5 (e-book)
ISBN-13: 978-0387-28262-6 (ISBN-13)
ISBN-13: 978-0387-28359-3 (e-ISBN-13)
1. Business mathematics—Congresses. 2. Finance—Congresses. 3. Stochastic analysis—Congresses. I. Shiriaev, Albert Nikolaevich.

HF5691.S68 2006
332'.01'51922—dc22

2005051577

AMS Subject Classifications: 91B70, 91B28, 60G70

Printed in the United States of America.

9 8 7 6 5 4 3 2 1 SPIN 11367710

springeronline.com

Preface

From 26–30 September 2004, the "International Conference on *Stochastic Finance 2004*" took place at INSTITUTO SUPERIOR DE ECONOMIA E GESTÃO (ISEG) da Universidade Técnica de Lisboa, in Portugal. The conference was one of the biggest international forums for scientists and practitioners working in financial mathematics and financial engineering.

Taking place just before the conference, on 20–24 September 2004 was the "Autumn School on Stochastic Finance 2004" hosted by the Universidade de Coimbra. The goal of this event was to present instances of the interaction of finance and mathematics by means of a coherent combination of five courses of introductory lectures, delivered by specialists, in order to stimulate and reinforce the understanding of the subject and to provide an opportunity for graduate students and researchers to develop some competence in financial mathematics and thereby simplify their participation in the conference.

At both meetings the organizing and scientific committees worked in close contact, which was crucial for inviting many leading specialists in financial mathematics and financial engineering — eleven plenary lecturers and eleven invited speakers. Besides these presentations, the conference included more than eighty contributed talks distributed among eight thematic sessions: *Mathematical Finance–Stochastic Models, Derivative Pricing, Interest Rate Term Structure Modelling, Portfolio Management, Integrated Risk Management, Mathematical Economics, Finance, and Quantitative and Computational Models and Methods.*

Stochastic financial mathematics is now one of the most rapidly developing fields of mathematics and applied mathematics. It has very close ties with economics and is oriented to the solution of problems appearing every day in real financial markets. We recall here an extract from the "Editorial" note presented in volume 1, issue 1 of the journal *Finance and Stochastics* that Springer-Verlag began publishing in 1997:

> "Nearly a century ago, Louis Bachelier published his thesis "Théorie de la speculation", Ann. Sci. École Norm. Sup. 3 (1900), in which he invented Brownian motion as a tool for the analysis of financial markets. A.N. Kolmogorov, in his own landmark work "Über die analytischen Methoden in der Wahrscheinlichkeitsrechnung", Math. Annalen 104 (1931), pp.415-458, credits Bachelier with the first systematic study of stochastic processes in continuous time. But in addition, Bachelier's thesis marks the beginning of the theory of option pricing, now an integral part of modern finance. Thus the year 1900 may be considered as birth date of both Finance and Stochastics. For the first seven decades following Bachelier, finance and stochastics followed more or less independently. The theory of stochastic processes grew fast and incorporating classical calculus became a powerful mathematical tool — called stochastic calculus. Finance lay dormant until the middle of the twentieth century, and then was resurrected as an offshoot of general equilibrium theory in economics. With the work in the late 1960s and early 1970s of Black, Merton, Samuelson and Scholes, modelling stock prices as geometric Brownian motion and using this model to study equilibrium and arbitrage pricing, the two disciplines were reunited. Soon it was discovered how well suited stochastic calculus with its rich mathematical structure — martingale theory, Itô calculus, stochastic integration and PDE's — was for a rigorous analysis of contemporary finance, which would lead one to believe (erroneously) that also these tools were invented with the application to finance in mind. Since then the interplay of these two disciplines has become an ever growing research field with great impact both on the theory and practice of financial markets".

The aims formulated in this text were the leading ideas for our conference. Indeed, all talks had, first of all, financial meanings and interpretations. All talks used and developed stochastic methods or solutions for real problems. Such joint mutual collaboration was useful both for financial economics and stochastic theory, and it could bring the mathematical and financial communities together.

In the present volume the reader can find some papers based on the plenary and invited lectures and on some contributed talks selected for publication.

The editorial committee of these proceedings expresses its deep gratitude to those who contributed their work to this volume and those who kindly helped us in refereeing them.

It is our pleasure to express our thanks to the scientific committee of the conference, as well as to plenary and invited lecturers and all the participants of Stochastic Finance 2004; their presence and their work formed the main contribution to the success of the conference.

A special acknowledgement is due to the Governador do Banco de Portugal (Governor of the Portuguese Central Bank) for his sharp advice and sponsorship of the event.

We thank the financial supporters:

Arkimed Innovative Technologies,
Associacão Portuguesa de Seguradores,
Caixa Geral de Depósitos,
Comissão do Mercado de Valores Mobiliários,
Delta Cafés,
Fundação Luso-Americana,
Fundação Oriente,
Império Bonança,
Portugalia Airlines
Grupo Sumol.

Our gratitude goes to CIM (Centro Internacional de Matemática) for suggesting the organization of this event within its annual scientific planning, and to the academic institutions

CMUC (Centro de Matemática da Universidade de Coimbra),
FCT-UNL (Fac. de Ciências e Tecnologia da Univ. Nova de Lisboa),
UECE (Unidade de Estudos sobre a Complexidade na Economia),
FCT (Fundação para a Ciência e a Tecnologia),

for their support.

A special word of thanks is due to CEMAPRE (Centro de Matemática Aplicada à Previsão e Decisão Económica) whose support has been crucial for the viability of the event. We thank the staff that at different moments and in diverse tasks were key collaborators to the organizing procedure: Ana Sofia Nunes (computer support), Maria do Rosário Pato (secretary) and Maria Júlia Marmelada (public relations).

Thanks are due to Béatrice Huberty, the editorial secretary who prepared this volume, for her proficiency and dedicated work.

Our appreciation goes John Martindale and Robert Saley, editor and assistant editor of Springer, respectively, for their continuous support and active interest in the development of this project.

We sincerely hope that this volume will be an essential contribution to the literature in financial mathematics and financial engineering.

Albert Shiryaev
Maria do Rosário Grossinho
Lisbon, *Paulo Eduardo Oliveira*
Manuel Leote Esquível

International Conference

Stochastic Finance 2004

Lisbon, 26-30 September

Chairpersons

- *Albert N. Shiryaev* - Scientific Committee
  Steklov Mathematical Institute, Moscow, Russia
- *Maria do Rosário Grossinho* - Organizing Committee
  ISEG, Universidade Técnica de Lisboa, Portugal

Scientific Committee

- *Ole Barndorff-Nielsen*, University of Aarhus, Denmark
- *Ernst Eberlein,* University of Freiburg, Germany
- *Albert N. Shiryaev,* Steklov Mathematical Institute, Moscow, Russia

Organizing Committee

- *Paulo Brito,* ISEG, Universidade Técnica de Lisboa, Portugal
- *Manuel L. Esquível*, FCT, Universidade Nova de Lisboa, Portugal
- *Maria do Rosário Grossinho,* ISEG, Univ. Técnica de Lisboa, Portugal
- *João Nicolau*, ISEG, Universidade Técnica de Lisboa, Portugal
- *Paulo Eduardo Oliveira,* Universidade de Coimbra, Portugal

Plenary Lecturers

- *Yacine Aït-Sahalia,* Princeton University, USA
- *Ole E. Barndorff-Nielsen*, University of Aarhus, Denmark
- *Tomas Björk,* Stockholm School of Economics, Sweden
- *Hans Föllmer,* Humboldt Universität zu Berlin, Germany
- *Paul Embrechts,* ETH Zurich, Switzerland
- *Ioannis Karatzas,* Columbia University, USA
- *Nicole El Karoui*, Ecole Polytechnique, Palaiseau, France
- *Stanley Pliska,* University of Illinois at Chicago, USA
- *Marek Rutkowski*, Warsaw University of Technology, Poland
- *Eckhard Platen,* University of Technology, Sydney, Australia
- *Albert Shiryaev,* Steklov Mathematical Institute, Russia

Invited Speakers

- *Bent Jesper Christensen*, University of Aarhus, Denmark
- *Ernst Eberlein,* University of Freiburg, Germany
- *Monique Jeanblanc,* Université d'Evry Val d'Essonne, France
- *Youri Kabanov,* Université de Franche-Comté, France
- *Claudia Klüeppelberg*, München University of Technology, Germany
- *Dmitry Kramkov,* Carnegie Mellon University, USA
- *Yoshio Miyahara,* Nagoya City University, Japan
- *Marek Musiela,* BNP Paribas Bank
- *Wolfgang Runggaldier,* Universitá di Padova, Italy
- *Walter Schachermayer,* Vienna University of Technology, Austria
- *Michael Søerensen*, University of Copenhagen, Denmark
- *Nizar Touzi,* CREST, Laboratoire de Finance et Assurance, France
- *Xun-Yu Zhou,* The Chinese University of Hong Kong

Contents

Part II Contributed Talks

List of Contributors

Albrecht Irle
Mathematisches Seminar
Universität Kiel
D-24098 Kiel, Germany
irle@math.uni-kiel.de

Alexander Lindner
Center for Mathematical Sciences
Munich University of Technology
D-85747 Garching, Germany
lindner@ma.tum.de

Claudia Klüppelberg
Center for Mathematical Sciences
Munich University of Technology
D-85747 Garching, Germany
cklu@ma.tum.de

Eckhard Platen
University of Technology Sydney,
School of Finance & Economics
and Department of Mathematical Sciences
PO Box 123
Broadway, NSW, 2007, Australia
eckhard.platen@uts.edu.au

Ferry Jaya Permana
Delft University of Technology
Mekelweg 4
2628 CD Delft, The Netherlands
f.j.permana@ewi.tudelft.nl

Jeannette H.C. Woerner
Institut für Mathematische Stochastik, Universität Göttingen
Maschmühlenweg 8-10
D-37073 Göttingen, Germany
woerner@math.uni-goettingen.de

Jörn Sass
RICAM, Austrian Academy of Sciences
Altenberger Str. 69
A-4040 Linz, Austria
joern.sass@oeaw.ac.at

Lan Zhang
Department of Statistics
Carnegie Mellon University
Pittsburgh, PA 15213, USA
lzhang@stat.cmu.edu

Lian Yu
Department of Industrial Engineering and Operations Research
University of California
Berkeley, CA 94720
liany@ieor.berkeley.edu

Marek Rutkowski
School of Mathematics
University of New South Wales
Sydney, NSW 2052, Australia
marekr@maths.unsw.edu.au

Monique Jeanblanc
Département de Mathématiques
Université d'Évry Val d'Essonne
91025 Évry Cedex, France
monique.jeanblanc@univ-evry.fr

Neil Shephard
Nuffield College
Oxford OX1 1NF, U.K.
neil.shephard@nuf.ox.ac.uk

Ole E. Barndorff-Nielsen
Department of Mathematical
Sciences, University of Aarhus
Ny Munkegade
DK-8000 Aarhus C, Denmark
oebn@imf.au.dk

Per A. Mykland
Department of Statistics
The University of Chicago
Chicago, IL 60637, USA
mykland@uchicago.edu

Raquel M. Gaspar
Department of Finance, Stockholm
School of Economics
Dept. Finance, P.O. Box 6501
113 83 Stockholm, Sweden
raquel.gaspar@hhs.se

Sara Di Emidio
Dipartimento di Matematica Pura
ed Applicata
Universitá di Padova
Padova, Italy
sara.diemidio@gmail.com

Shuzhong Zhang
Department of Systems Engineering
and Engineering Management
The Chinese University of Hong
Kong Shatin
Hong Kong
zhang@se.cuhk.edu.hk

Stanley R. Pliska
Department of Finance, University
of Illinois at Chicago
601 S. Morgan Street
Chicago, IL 60607-7124 USA
srpliska@uic.edu

Svetlana Borovkova
Delft University of Technology
Mekelweg 4
2628 CD Delft, The Netherlands
s.a.borovkova@ewi.tudelft.nl

Tomasz R. Bielecki
Department of Applied Mathematics
Illinois Institute of Technology
Chicago, IL 60616, USA
bielecki@iit.edu

Vicky Fasen
Center for Mathematical Sciences
Munich University of Technology
D-85747 Garching, Germany
fasen@ma.tum.de

Wolfgang J.Runggaldier
Dipartimento di Matematica Pura
ed Applicata
Universitá di Padova
Padova, Italy
runggal@math.unipd.it

Xun Yu Zhou
Department of Systems Engineering
and Engineering Management
The Chinese University of Hong
Kong Shatin
Hong Kong
xyzhou@se.cuhk.edu.hk

Yacine Aït-Sahalia
Bendheim Center for Finance,
Princeton University,
Princeton, NJ 08540, USA
yacine@princeton.edu

Part I

Plenary and Invited Lectures

1

How Often to Sample a Continuous-Time Process in the Presence of Market Microstructure Noise*

Yacine Aït-Sahalia, Per A. Mykland, and Lan Zhang

[1] Bendheim Center for Finance, Princeton University, Princeton, NJ 08540 and NBER
yacine@princeton.edu
[2] Department of Statistics, The University of Chicago, Chicago, IL 60637
mykland@uchicago.edu
[3] Department of Statistics, Carnegie Mellon University, Pittsburgh, PA 15213
lzhang@stat.cmu.edu

Note: Reprint of: Yacine Aït-Sahalia, Per A. Mykland and Lan Zhang, *How Often to Sample a Continuous-Time Process in the Presence of Market Microstructure Noise*, Review of Financial Studies, published in Volume 18 **2** (2005), pages 351–416, Oxford University Press (by permission of Oxford University Press).

Summary. In theory, the sum of squares of log returns sampled at high frequency estimates their variance. When market microstructure noise is present but unaccounted for, however, we show that the optimal sampling frequency is finite and derive its closed-form expression. But even with optimal sampling, using say five minute returns when transactions are recorded every second, a vast amount of data is discarded, in contradiction to basic statistical principles. We demonstrate that modelling the noise and using all the data is a better solution, even if one misspecifies the noise distribution. So the answer is: sample as often as possible.

Over the past few years, price data sampled at very high frequency have become increasingly available, in the form of the Olsen dataset of currency exchange rates or the TAQ database of NYSE stocks. If such data were not affected by market microstructure noise, the realized volatility of the process

* We are grateful for comments and suggestions from the editor, Maureen O'Hara, and two anonymous referees, as well as seminar participants at Berkeley, Harvard, NYU, MIT, Stanford, the Econometric Society and the Joint Statistical Meetings. Financial support from the NSF under grants SBR-0111140 (Aït-Sahalia), DMS-0204639 (Mykland and Zhang) and the NIH under grant RO1 AG023141-01 (Zhang) is also gratefully acknowledged.

(i.e., the average sum of squares of log-returns sampled at high frequency) would estimate the returns' variance, as is well known. In fact, sampling as often as possible would theoretically produce in the limit a perfect estimate of that variance.

We start by asking whether it remains optimal to sample the price process at very high frequency in the presence of market microstructure noise, consistently with the basic statistical principle that, ceteris paribus, more data is preferred to less. We first show that, if noise is present but unaccounted for, then the optimal sampling frequency is finite, and we derive a closed-form formula for it. The intuition for this result is as follows. The volatility of the underlying efficient price process and the market microstructure noise tend to behave differently at different frequencies. Thinking in terms of signal-to-noise ratio, a log-return observed from transaction prices over a tiny time interval is mostly composed of market microstructure noise and brings little information regarding the volatility of the price process since the latter is (at least in the Brownian case) proportional to the time interval separating successive observations. As the time interval separating the two prices in the log-return increases, the amount of market microstructure noise remains constant, since each price is measured with error, while the informational content of volatility increases. Hence very high frequency data are mostly composed of market microstructure noise, while the volatility of the price process is more apparent in longer horizon returns. Running counter to this effect is the basic statistical principle mentioned above: in an idealized setting where the data are observed without error, sampling more frequently cannot hurt. What is the right balance to strike? What we show is that these two effects compensate each other and result in a finite optimal sampling frequency (in the root mean squared error sense) so that some time aggregation of the returns data is advisable.

By providing a quantitative answer to the question of how often one should sample, we hope to reduce the arbitrariness of the choices that have been made in the empirical literature using high frequency data: for example, using essentially the same Olsen exchange rate series, these somewhat ad hoc choices range from 5 minute intervals (e.g., [5], [8] and [19]) to as long as 30 minutes (e.g., [6]). When calibrating our analysis to the amount of microstructure noise that has been reported in the literature, we demonstrate how the optimal sampling interval should be determined: for instance, depending upon the amount of microstructure noise relative to the variance of the underlying returns, the optimal sampling frequency varies from 4 minutes to 3 hours, if 1 day's worth of data is used at a time. If a longer time period is used in the analysis, then the optimal sampling frequency can be considerably longer than these values.

But even if one determines the sampling frequency optimally, it remains the case that the empirical researcher is not making use of the full data at his/her disposal. For instance, suppose that we have available transaction records on a liquid stock, traded once every second. Over a typical 6.5 hour day, we therefore start with $23,400$ observations. If one decides to sample once every 5 minutes, then – whether or not this is the optimal sampling frequency – this amounts to retaining only 78 observations. Said differently, one is throwing away 299 out of every 300 transactions. From a statistical perspective, this is unlikely to be the optimal solution, even though it is undoubtedly better than computing a volatility estimate using noisy squared log-returns sampled every second. Somehow, an optimal solution should make use of all the data, and this is where our analysis goes next.

So, if one decides to account for the presence of the noise, how should one go about doing it? We show that modelling the noise term explicitly restores the first order statistical effect that sampling as often as possible is optimal. This will involve an estimator different from the simple sum of squared log-returns. Since we work within a fully *parametric* framework, *likelihood* is the key word. Hence we construct the likelihood function for the observed log-returns, which include microstructure noise. To do so, we must postulate a model for the noise term. We assume that the noise is Gaussian. In light of what we know from the sophisticated theoretical microstructure literature, this is likely to be overly simplistic and one may well be concerned about the effect(s) of this assumption. Could it do more harm than good? Surprisingly, we demonstrate that our likelihood correction, based on Gaussianity of the noise, works *even if* one misspecifies the assumed distribution of the noise term. Specifically, if the econometrician assumes that the noise terms are normally distributed when in fact they are not, not only is it still optimal to sample as often as possible (unlike the result when no allowance is made for the presence of noise), but the estimator has the *same variance* as if the noise distribution had been correctly specified. This robustness result is, we think, a major argument in favor of incorporating the presence of the noise when estimating continuous time models with high frequency financial data, even if one is unsure about what is the true distribution of the noise term.

In other words, the answer to the question we pose in our title is "as often as possible", provided one accounts for the presence of the noise when designing the estimator (and we suggest maximum likelihood as a means of doing so). If one is unwilling to account for the noise, then the answer is to rely on the finite optimal sampling frequency we start our analysis with, but we stress that while it is optimal if one insists upon using sums of squares of log-returns, this is not the best possible approach to estimate volatility given the complete high frequency dataset at hand.

In a companion paper ([43]), we study the corresponding *nonparametric* problem, where the volatility of the underlying price is a stochastic process, and nothing else is known about it, in particular no parametric structure. In that case, the object of interest is the integrated volatility of the process over a fixed time interval, such as a day, and we show how to estimate it using again all the data available (instead of sparse sampling at an arbitrarily lower frequency of, say, 5 minutes). Since the model is nonparametric, we no longer use a likelihood approach but instead propose a solution based on subsampling and averaging, which involves estimators constructed on two different time scales, and demonstrate that this again dominates sampling at a lower frequency, whether arbitrary or optimally determined.

This paper is organized as follows. We start by describing in Section 1.1 our reduced form setup and the underlying structural models that support it. We then review in Section 1.2 the base case where no noise is present, before analyzing in Section 1.3 the situation where the presence of the noise is ignored. In Section 1.4, we examine the concrete implications of this result for empirical work with high frequency data. Next, we show in Section 1.5 that accounting for the presence of the noise through the likelihood restores the optimality of high frequency sampling. Our robustness results are presented in Section 1.6 and interpreted in Section 1.7. We study the same questions when the observations are sampled at random time intervals, which are an essential feature of transaction-level data, in Section 1.8. We then turn to various extensions and relaxation of our assumptions in Section 1.9: we add a drift term, then serially correlated and cross-correlated noise respectively. Section 1.10 concludes. All proofs are in the Appendix.

1.1 Setup

Our basic setup is as follows. We assume that the underlying process of interest, typically the log-price of a security, is a time-homogenous diffusion on the real line

$$dX_t = \mu(X_t; \theta)dt + \sigma dW_t \, , \tag{1.1}$$

where $X_0 = 0$, W_t is a Brownian motion, $\mu(.,.)$ is the drift function, σ^2 the diffusion coefficient and θ the drift parameters, $\theta \in \Theta$ and $\sigma > 0$. The parameter space is an open and bounded set. As usual, the restriction that σ is constant is without loss of generality since in the univariate case a one-to-one transformation can always reduce a known specification $\sigma(X_t)$ to that case. Also, as discussed in [4], the properties of parametric estimators in this model are quite different depending upon whether we estimate θ alone, σ^2 alone, or both parameters together. When the data are noisy, the main effects that we describe are already present in the simpler of these three cases, where σ^2 alone is estimated, and so we focus on that case. Moreover, in the high frequency context we have in mind, the diffusive component of (1.1) is of order $(dt)^{1/2}$ while the drift component is of order dt only, so the drift component is mathematically negligible at high frequencies. This is validated empirically: including a drift actually deteriorates the performance of variance estimates from high frequency data since the drift is estimated with a large standard error. Not centering the log returns for the purpose of variance estimation produces more accurate results (see [38]). So we simplify the analysis one step further by setting $\mu = 0$, which we do until Section 1.9.1, where we then show that adding a drift term does not alter our results. In Section 1.9.4, we discuss the situation where the instantaneous volatility σ is stochastic.

But for now,

$$X_t = \sigma W_t \,. \tag{1.2}$$

Until Section 1.8, we treat the case where the observations occur at equidistant time intervals Δ, in which case the parameter σ^2 is therefore estimated at time T on the basis of $N+1$ discrete observations recorded at times $\tau_0 = 0$, $\tau_1 = \Delta$,..., $\tau_N = N\Delta = T$. In Section 1.8, we let the sampling intervals be themselves random variables, since this feature is an essential characteristic of high frequency transaction data.

The notion that the observed transaction price in high frequency financial data is the unobservable efficient price plus some noise component due to the imperfections of the trading process is a well established concept in the market microstructure literature (see for instance [10]). So, where we depart from the inference setup previously studied in [4] is that we now assume that, instead of observing the process X at dates τ_i, we observe X with error:

$$\tilde{X}_{\tau_i} = X_{\tau_i} + U_{\tau_i} \,, \tag{1.3}$$

where the $U_{\tau_i}'s$ are i.i.d. noise with mean zero and variance a^2 and are independent of the W process. In that context, we view X as the efficient log-price, while the observed $\tilde{X}$ is the transaction log-price. In an efficient market, X_t is the log of the expectation of the final value of the security conditional on all publicly available information at time t. It corresponds to the log-price that would be in effect in a perfect market with no trading imperfections, frictions, or informational effects. The Brownian motion W is the process representing the arrival of new information, which in this idealized setting is immediately impounded in X.

By contrast, U_t summarizes the noise generated by the mechanics of the trading process. What we have in mind as the source of noise is a diverse array of market microstructure effects, either information or non-information related, such as the presence of a bid-ask spread and the corresponding bounces, the differences in trade sizes and the corresponding differences in representativeness of the prices, the different informational content of price changes due to informational asymmetries of traders, the gradual response of prices to a block trade, the strategic component of the order flow, inventory control effects, the discreteness of price changes in markets that are not decimalized, etc., all summarized into the term U. That these phenomena are real are important is an accepted fact in the market microstructure literature, both theoretical and empirical. One can in fact argue that these phenomena justify this literature.

We view (1.3) as the simplest possible reduced form of structural market microstructure models. The efficient price process X is typically modelled as a random walk, i.e., the discrete time equivalent of (1.2). Our specification coincides with that of [29], who discusses the theoretical market microstructure underpinnings of such a model and argues that the parameter a is a summary measure of market quality. Structural market microstructure models do generate (1.3). For instance, [39] proposes a model where U is due entirely to the bid-ask spread. [28] notes that in practice there are sources of noise other than just the bid-ask spread, and studies their effect on the Roll model and its estimators.

Indeed, a disturbance U can also be generated by adverse selection effects as in [20] and [21], where the spread has two components: one that is due to monopoly power, clearing costs, inventory carrying costs, etc., as previously, and a second one that arises because of adverse selection whereby the specialist is concerned that the investor on the other side of the transaction has superior information. When asymmetric information is involved, the disturbance U would typically no longer be uncorrelated with the W process and would exhibit autocorrelation at the first order, which would complicate our analysis without fundamentally altering it: see Sections 1.9.2 and 1.9.3 where we relax the assumptions that the $U's$ are serially uncorrelated and independent of the W process, respectively.

The situation where the measurement error is primarily due to the fact that transaction prices are multiples of a tick size (i.e., $\tilde{X}_{\tau_i} = m_i \kappa$ where κ is the tick size and m_i is the integer closest to X_{τ_i}/κ) can be modelled as a rounding off problem (see [14], [23] and [31]). The specification of the model in [27] combines both the rounding and bid-ask effects as the dual sources of the noise term U. Finally, structural models, such as that of [35], also give rise to reduced forms where the observed transaction price $\tilde{X}$ takes the form of an unobserved fundamental value plus error.

With (1.3) as our basic data generating process, we now turn to the questions we address in this paper: how often should one sample a continuous-time process when the data are subject to market microstructure noise, what are the implications of the noise for the estimation of the parameters of the X process, and how should one correct for the presence of the noise, allowing for the possibility that the econometrician misspecifies the assumed distribution of the noise term, and finally allowing for the sampling to occur at random points in time? We proceed from the simplest to the most complex situation by adding one extra layer of complexity at a time: Figure 1.1 shows the three sampling schemes we consider, starting with fixed sampling without market microstructure noise, then moving to fixed sampling with noise and concluding with an analysis of the situation where transaction prices are not only subject to microstructure noise but are also recorded at random time intervals.

1.2 The Baseline Case: No Microstructure Noise

We start by briefly reviewing what would happen in the absence of market microstructure noise, that is when $a = 0$. With X denoting the log-price, the first differences of the observations are the log-returns $Y_i = \tilde{X}_{\tau_i} - \tilde{X}_{\tau_{i-1}}$, $i = 1, ..., N$. The observations $Y_i = \sigma\left(W_{\tau_{i+1}} - W_{\tau_i}\right)$ are then i.i.d. $N(0, \sigma^2 \Delta)$ so the likelihood function is

$$l(\sigma^2) = -N \ln(2\pi\sigma^2\Delta)/2 - (2\sigma^2\Delta)^{-1} Y'Y \,, \tag{1.4}$$

where $Y = (Y_1, ..., Y_N)'$.. The maximum-likelihood estimator of σ^2 coincides with the discrete approximation to the quadratic variation of the process

$$\hat{\sigma}^2 = \frac{1}{T}\sum_{i=1}^{N} Y_i^2 \tag{1.5}$$

which has the following exact small sample moments:

$$E\left[\hat{\sigma}^2\right] = \frac{1}{T}\sum_{i=1}^{N} E\left[Y_i^2\right] = \frac{N\left(\sigma^2 \Delta\right)}{T} = \sigma^2,$$

$$\mathrm{Var}\left[\hat{\sigma}^2\right] = \frac{1}{T^2}\,\mathrm{Var}\left[\sum_{i=1}^{N} Y_i^2\right] = \frac{1}{T^2}\left(\sum_{i=1}^{N}\mathrm{Var}\left[Y_i^2\right]\right) = \frac{N}{T^2}\left(2\sigma^4\Delta^2\right) = \frac{2\sigma^4\Delta}{T}$$

and the following asymptotic distribution

$$T^{1/2}\left(\hat{\sigma}^2 - \sigma^2\right) \underset{T\longrightarrow\infty}{\longrightarrow} N(0, \omega), \tag{1.6}$$

where

$$\omega = \mathrm{AVAR}(\hat{\sigma}^2) = \Delta E\left[-\ddot{l}(\sigma^2)\right]^{-1} = 2\sigma^4\Delta. \tag{1.7}$$

Thus selecting Δ as small as possible is optimal for the purpose of estimating σ^2.

1.3 When the Observations Are Noisy But the Noise Is Ignored

Suppose now that market microstructure noise is present but the presence of the $U's$ is ignored when estimating σ^2. In other words, we use the log-likelihood (1.4) even though the true structure of the observed log-returns $Y_i's$ is given by an MA(1) process since

$$\begin{aligned} Y_i &= \tilde{X}_{\tau_i} - \tilde{X}_{\tau_{i-1}} \\ &= X_{\tau_i} - X_{\tau_{i-1}} + U_{\tau_i} - U_{\tau_{i-1}} \\ &= \sigma\left(W_{\tau_i} - W_{\tau_{i-1}}\right) + U_{\tau_i} - U_{\tau_{i-1}} \\ &\equiv \varepsilon_i + \eta\varepsilon_{i-1}, \end{aligned} \tag{1.8}$$

where the $\varepsilon_i's$ are uncorrelated with mean zero and variance γ^2 (if the $U's$ are normally distributed, then the $\varepsilon_i's$ are i.i.d.). The relationship to the original parametrization (σ^2, a^2) is given by

$$\gamma^2(1+\eta^2) = \mathrm{Var}[Y_i] = \sigma^2\Delta + 2a^2 \tag{1.9}$$

$$\gamma^2\eta = \mathrm{Cov}(Y_i, Y_{i-1}) = -a^2. \tag{1.10}$$

Equivalently, the inverse change of variable is given by

$$\gamma^2 = \frac{1}{2}\left\{2a^2 + \sigma^2\Delta + \sqrt{\sigma^2\Delta\left(4a^2 + \sigma^2\Delta\right)}\right\} \tag{1.11}$$

$$\eta = \frac{1}{2a^2}\left\{-2a^2 - \sigma^2\Delta + \sqrt{\sigma^2\Delta\left(4a^2 + \sigma^2\Delta\right)}\right\}. \tag{1.12}$$

Two important properties of the log-returns $Y_i's$ emerge from the two equations (1.9)-(1.10). First, it is clear from (1.9) that microstructure noise leads to spurious variance in observed log-returns, $\sigma^2\Delta + 2a^2$ vs. $\sigma^2\Delta$. This is consistent with the predictions of theoretical microstructure models. For instance, [16] develop a model linking the arrival of information, the timing of trades, and the resulting price process. In their model, the transaction price will be a biased representation of the efficient price process, with a variance that is both overstated and heteroskedastic as a result of the fact that transactions (hence the recording of an observation on the process $\tilde{X}$) occur at intervals that are time-varying. While our specification is too simple to capture the rich joint dynamics of price and sampling times predicted by their model, heteroskedasticity of the observed variance will also arise in our case once we allow for time variation of the sampling intervals (see Section 1.8 below).

In our model, the proportion of the total return variance that is market microstructure-induced is

$$\pi = \frac{2a^2}{\sigma^2\Delta + 2a^2} \tag{1.13}$$

at observation interval Δ. As Δ gets smaller, π gets closer to 1, so that a larger proportion of the variance in the observed log-return is driven by market microstructure frictions, and correspondingly a lesser fraction reflects the volatility of the underlying price process X.

Second, (1.10) implies that $-1 < \eta < 0$, so that log-returns are (negatively) autocorrelated with first order autocorrelation $-a^2/(\sigma^2\Delta + 2a^2) = -\pi/2$. It has been noted that market microstructure noise has the potential to explain the empirical autocorrelation of returns. For instance, in the simple Roll model, $U_t = (s/2)Q_t$ where s is the bid/ask spread and Q_t, the order flow indicator, is a binomial variable that takes the values $+1$ and -1 with equal probability. Therefore $\mathrm{Var}[U_t] = a^2 = s^2/4$. Since $\mathrm{Cov}(Y_i, Y_{i-1}) = -a^2$, the bid/ask spread can be recovered in this model as $s = 2\sqrt{-\rho}$ where $\rho = \gamma^2\eta$ is the first order autocorrelation of returns. [18] proposed to adjust variance estimates to control for such autocorrelation and [28] studied the resulting estimators. In [41], U arises because of the strategic trading of institutional investors which is then put forward as an explanation for the observed serial correlation of returns. [33] show that infrequent trading has implications for the variance and autocorrelations of returns. Other empirical patterns in high frequency financial data have been documented: leptokurtosis, deterministic patterns and volatility clustering.

Our first result shows that the optimal sampling frequency is finite when noise is present but unaccounted for. The estimator $\hat{\sigma}^2$ obtained from maximizing the misspecified log-likelihood (1.4) is quadratic in the $Y_i's$: see (1.5). In order to obtain its exact (i.e., small sample) variance, we therefore need to calculate the fourth order cumulants of the $Y_i's$ since

$$\mathrm{Cov}\left(Y_i^2, Y_j^2\right) = 2\,\mathrm{Cov}\left(Y_i, Y_j\right)^2 + \mathrm{Cum}(Y_i, Y_i, Y_j, Y_j) \tag{1.14}$$

(see e.g., Section 2.3 of [36] for definitions and properties of the cumulants). We have:

Lemma 1. *The fourth cumulants of the log-returns are given by*

$$\mathrm{Cum}(Y_i, Y_j, Y_k, Y_l) = \begin{cases} 2\ \mathrm{Cum}_4\left[U\right] & \text{if } i = j = k = l\,, \\ (-1)^{s(i,j,k,l)}\, \mathrm{Cum}_4\left[U\right], & \text{if } \max(i,j,k,l) = \min(i,j,k,l) + 1, \\ 0 & \text{otherwise}, \end{cases} \tag{1.15}$$

where $s(i,j,k,l)$ denotes the number of indices among (i,j,k,l) that are equal to $\min(i,j,k,l)$ and U denotes a generic random variable with the common distribution of the $U_{\tau_i}'s$. Its fourth cumulant is denoted $\mathrm{Cum}_4\left[U\right]$.

Now U has mean zero, so in terms of its moments

$$\mathrm{Cum}_4\left[U\right] = E\left[U^4\right] - 3\left(E\left[U^2\right]\right)^2. \tag{1.16}$$

In the special case where U is normally distributed, $\mathrm{Cum}_4\left[U\right] = 0$ and as a result of (1.14) the fourth cumulants of the log-returns are all 0 (since W is normal, the log-returns are also normal in that case). If the distribution of U is binomial as in the simple bid/ask model described above, then $\mathrm{Cum}_4\left[U\right] = -s^4/8$; since in general s will be a tiny percentage of the asset price, say $s = 0.05\%$, the resulting $\mathrm{Cum}_4\left[U\right]$ will be very small.

We can now characterize the root mean squared error

$$RMSE\left[\hat{\sigma}^2\right] = \left(\left(E\left[\hat{\sigma}^2\right] - \sigma^2\right)^2 + \mathrm{Var}\left[\hat{\sigma}^2\right]\right)^{1/2}$$

of the estimator:

Theorem 1. *In small samples (finite T), the bias and variance of the estimator $\hat{\sigma}^2$ are given by*

$$E\left[\hat{\sigma}^2\right] - \sigma^2 = \frac{2a^2}{\Delta}, \tag{1.17}$$

$$\operatorname{Var}\left[\hat{\sigma}^2\right] = \frac{2\left(\sigma^4\Delta^2 + 4\sigma^2\Delta a^2 + 6a^4 + 2\operatorname{Cum}_4[U]\right)}{T\Delta} - \\ -\frac{2\left(2a^4 + \operatorname{Cum}_4[U]\right)}{T^2}. \tag{1.18}$$

Its RMSE has a unique minimum in Δ which is reached at the optimal sampling interval

$$\Delta^* = \left(\frac{2a^4T}{\sigma^4}\right)^{1/3}\left(\left(1-\sqrt{1-\frac{2\left(3a^4+\operatorname{Cum}_4[U]\right)^3}{27\sigma^4a^8T^2}}\right)^{1/3} \right. \\ \left. + \left(1+\sqrt{1-\frac{2\left(3a^4+\operatorname{Cum}_4[U]\right)^3}{27\sigma^4a^8T^2}}\right)^{1/3}\right). \tag{1.19}$$

As T grows, we have

$$\Delta^* = \frac{2^{2/3}a^{4/3}}{\sigma^{4/3}}T^{1/3} + O\left(\frac{1}{T^{1/3}}\right). \tag{1.20}$$

The trade-off between bias and variance made explicit in (1.17)-(1.19) is not unlike the situation in nonparametric estimation with Δ^{-1} playing the role of the bandwidth h. A lower h reduces the bias but increases the variance, and the optimal choice of h balances the two effects.

Note that these are exact small sample expressions, valid for all T. Asymptotically in T, $\operatorname{Var}\left[\hat{\sigma}^2\right] \to 0$, and hence the RMSE of the estimator is dominated by the bias term which is independent of T. And given the form of the bias (1.17), one would in fact want to select the largest Δ possible to minimize the bias (as opposed to the smallest one as in the no-noise case of Section 1.2). The rate at which Δ^* should increase with T is given by (1.20). Also, in the limit where the noise disappears ($a \to 0$ and $\operatorname{Cum}_4[U] \to 0$), the optimal sampling interval Δ^* tends to 0.

How does a small departure from a normal distribution of the microstructure noise affect the optimal sampling frequency? The answer is that a small positive (resp. negative) departure of Cum 4 $[U]$ starting from the normal value of 0 leads to an increase (resp. decrease) in Δ^*, since

$$\Delta^* = \Delta^*_{\text{normal}} + \\ + \frac{\left(\left(1+\sqrt{1-\frac{2a^4}{T^2\sigma^4}}\right)^{2/3} - \left(1-\sqrt{1-\frac{2a^4}{T^2\sigma^4}}\right)^{2/3}\right)}{3\;2^{1/3}a^{4/3}T^{1/3}\sqrt{1-\frac{2a^4}{T^2\sigma^4}}\,\sigma^{8/3}}\operatorname{Cum}4[U] + \\ + O\left(\operatorname{Cum}4[U]^2\right), \tag{1.21}$$

where Δ^*_{normal} is the value of Δ^* corresponding to $\text{Cum}_4[U] = 0$. And of course the full formula (1.20) can be used to get the exact answer for any departure from normality instead of the comparative static one.

Another interesting asymptotic situation occurs if one attempts to use higher and higher frequency data ($\Delta \to 0$, say sampled every minute) over a fixed time period (T fixed, say a day). Since the expressions in Theorem 1 are exact small sample ones, they can in particular be specialized to analyze this situation. With $n = T/\Delta$, it follows from (1.17)-(1.19) that

$$E\left[\hat{\sigma}^2\right] = \frac{2na^2}{T} + o(n) = \frac{2nE\left[U^2\right]}{T} + o(n), \tag{1.22}$$

$$\text{Var}\left[\hat{\sigma}^2\right] = \frac{2n\left(6a^4 + 2\,\text{Cum}_4[U]\right)}{T^2} + o(n) = \frac{4nE\left[U^4\right]}{T^2} + o(n), \tag{1.23}$$

so $(T/2n)\hat{\sigma}^2$ becomes an estimator of $E\left[U^2\right] = a^2$ whose asymptotic variance is $E\left[U^4\right]$. Note in particular that $\hat{\sigma}^2$ estimates the variance of the noise, which is essentially unrelated to the object of interest σ^2. This type of asymptotics is relevant in the stochastic volatility case we analyze in our companion paper [43].

Our results also have implications for the two parallel tracks that have developed in the recent financial econometrics literature dealing with discretely observed continuous-time processes. One strand of the literature has argued that estimation methods should be robust to the potential issues arising in the presence of high frequency data and, consequently, be asymptotically valid without requiring that the sampling interval Δ separating successive observations tend to zero (see, e.g., [2], [3] and [26]). Another strand of the literature has dispensed with that constraint, and the asymptotic validity of these methods requires that Δ tend to zero instead of or in addition to, an increasing length of time T over which these observations are recorded (see, e.g., [6], [7] and [8]).

The first strand of literature has been informally warning about the potential dangers of using high frequency financial data without accounting for their inherent noise (see e.g., page 529 of [2]), and we propose a formal modelization of that phenomenon. The implications of our analysis are most salient for the second strand of the literature, which is predicated on the use of high frequency data but does not account for the presence of market microstructure noise. Our results show that the properties of estimators based on the local sample path properties of the process (such as the quadratic variation to estimate σ^2) change dramatically in the presence of noise. Complementary to this are the results of [22] which show that the presence of even increasingly negligible noise is sufficient to adversely affect the identification of σ^2.

1.4 Concrete Implications for Empirical Work with High Frequency Data

The clear message of Theorem 1 for empirical researchers working with high frequency financial data is that it may be optimal to sample less frequently. As discussed in the Introduction, authors have reduced their sampling frequency below that of the actual record of observations in a somewhat ad hoc fashion, with typical choices 5 minutes and up. Our analysis provides not only a theoretical rationale for sampling less frequently, but also delivers a precise answer to the question of "how often one should sample?" For that purpose, we need to calibrate the parameters appearing in Theorem 1, namely σ, α, $\mathrm{Cum}_4[U]$, Δ and T. We assume in this calibration exercise that the noise is Gaussian, in which case $\mathrm{Cum}_4[U] = 0$.

1.4.1 Stocks

We use existing studies in empirical market microstructure to calibrate the parameters. One such study is [35], who estimated on the basis of a sample of 274 NYSE stocks that approximately 60% of the total variance of price changes is attributable to market microstructure effects (they report a range of values for π from 54% in the first half hour of trading to 65% in the last half hour, see their Table 4; they also decompose this total variance into components due to discreteness, asymmetric information, transaction costs and the interaction between these effects). Given that their sample contains an average of 15 transactions per hour (their Table 1), we have in our framework

$$\pi = 60\%, \; \Delta = 1/(15 \times 7 \times 252)\,. \tag{1.24}$$

These values imply from (1.13) that $a = 0.16\%$ if we assume a realistic value of $\sigma = 30\%$ per year. (We do not use their reported volatility number since they apparently averaged the variance of price changes over the 274 stocks instead of the variance of the returns. Since different stocks have different price levels, the price variances across stocks are not directly comparable. This does not affect the estimated fraction π however, since the price level scaling factor cancels out between the numerator and the denominator).

The magnitude of the effect is bound to vary by type of security, market and time period. [29] estimates the value of a to be 0.33%. Some authors have reported even larger effects. Using a sample of NASDAQ stocks, [32] estimate that about 50% of the daily variance of returns in due to the bid-ask effect. With $\sigma = 40\%$ (NASDAQ stocks have higher volatility), the values

$$\pi = 50\%, \; \Delta = 1/252$$

yield the value $a = 1.8\%$. Also on NASDAQ, [12] estimate that 11% of the variance of weekly returns (see their Table 4, middle portfolio) is due to bid-ask effects. The values

$$\pi = 11\%,\ \Delta = 1/52$$

imply that $a = 1.4\%$.

In Table 1.1, we compute the value of the optimal sampling interval Δ^* implied by different combinations of sample length (T) and noise magnitude (a). The volatility of the efficient price process is held fixed at $\sigma = 30\%$ in Panel A, which is a realistic value for stocks. The numbers in the table show that the optimal sampling frequency can be substantially affected by even relatively small quantities of microstructure noise. For instance, using the value $a = 0.15\%$ calibrated from [35], we find an optimal sampling interval of 22 minutes if the sampling length is 1 day; longer sample lengths lead to higher optimal sampling intervals. With the higher value of $a = 0.3\%$, approximating the estimate from [29], the optimal sampling interval is 57 minutes. A lower value of the magnitude of the noise translates into a higher frequency: for instance, $\Delta^* = 5$ minutes if $a = 0.05\%$ and $T = 1$ day. Figure 1.2 displays the RMSE of the estimator as a function of Δ and T, using parameter values $\sigma = 30\%$ and $a = 0.15\%$. The figure illustrates the fact that deviations from the optimal choice of Δ lead to a substantial increase in the RMSE: for example, with $T = 1$ month, the RMSE more than doubles if, instead of the optimal $\Delta^* = 1$ hour, one uses $\Delta = 15$ minutes.

1.4.2 Currencies

Looking now at foreign exchange markets, empirical market microstructure studies have quantified the magnitude of the bid-ask spread. For example, [9] computes the average bid/ask spread s in the wholesale market for different currencies and reports values of $s = 0.05\%$ for the German mark, and 0.06% for the Japanese yen (see Panel B of his Table 2). We calculated the corresponding numbers for the 1996-2002 period to be 0.04% for the mark (followed by the euro) and 0.06% for the yen. Emerging market currencies have higher spreads: for instance, $s = 0.12\%$ for Korea and 0.10% for Brazil. During the same period, the volatility of the exchange rate was $\sigma = 10\%$ for the German mark, 12% for the Japanese yen, 17% for Brazil and 18% for Korea. In Panel B of Table 1.1, we compute Δ^* with $\sigma = 10\%$, a realistic value for the euro and yen. As we noted above, if the sole source of the noise were a bid/ask spread of size s, then a should be set to $s/2$. Therefore Panel B reports the values of Δ^* for values of a ranging from 0.02% to 0.1%. For example, the dollar/euro or dollar/yen exchange rates (calibrated to $\sigma = 10\%$, $a = 0.02\%$) should be sampled every $\Delta^* = 23$ minutes if the overall sample length is $T = 1$ day, and every 1.1 hours if $T = 1$ year.

Furthermore, using the bid/ask spread alone as a proxy for all microstructure frictions will lead, except in unusual circumstances, to an understatement of the parameter a, since variances are additive. Thus, since Δ^* is increasing in a, one should interpret the value of Δ^* read off 1.1 on the row corresponding to $a = s/2$ as a lower bound for the optimal sampling interval.

1.4.3 Monte Carlo Evidence

To validate empirically these results, we perform Monte Carlo simulations. We simulate $M = 10,000$ samples of length $T = 1$ year of the process X, add microstructure noise U to generate the observations $\tilde{X}$ and then the log returns Y. We sample the log-returns at various intervals Δ ranging from 5 minutes to 1 week and calculate the bias and variance of the estimator $\hat{\sigma}^2$ over the M simulated paths. We then compare the results to the theoretical values given in (1.17)-(1.19) of Theorem 1. The noise distribution is Gaussian, $\sigma = 30\%$ and $a = 0.15\%$ – the values we calibrated to stock returns data above. Table 1.2 shows that the theoretical values are in close agreement with the results of the Monte Carlo simulations.

The table also illustrates the magnitude of the bias inherent in sampling at too high a frequency. While the value of σ^2 used to generate the data is 0.09, the expected value of the estimator when sampling every 5 minutes is 0.18, so on average the estimated quadratic variation is twice as big as it should be in this case.

1.5 Incorporating Market Microstructure Noise Explicitly

So far we have stuck to the sum of squares of log-returns as our estimator of volatility. We then showed that, for this estimator, the optimal sampling frequency is finite. But this implies that one is discarding a large proportion of the high frequency sample (299 out of every 300 observations in the example described in the Introduction), in order to mitigate the bias induced by market microstructure noise. Next, we show that if we explicitly incorporate the $U's$ into the likelihood function, then we are back in the situation where the optimal sampling scheme consists in sampling as often as possible – i.e., using all the data available.

Specifying the likelihood function of the log-returns, while recognizing that they incorporate noise, requires that we take a stand on the distribution of the noise term. Suppose for now that the microstructure noise is normally distributed, an assumption whose effect we will investigate below in Section 1.6. Under this assumption, the likelihood function for the $Y's$ is given by

$$l(\eta, \gamma^2) = -\ln \det(V)/2 - N \ln(2\pi\gamma^2)/2 - (2\gamma^2)^{-1} Y' V^{-1} Y , \qquad (1.25)$$

where the covariance matrix for the vector $Y = (Y_1, ..., Y_N)'$ is given by $\gamma^2 V$, where

$$V = [v_{ij}]_{i,j=1,\ldots,N} = \begin{pmatrix} 1+\eta^2 & \eta & 0 & \cdots & 0 \\ \eta & 1+\eta^2 & \eta & \ddots & \vdots \\ 0 & \eta & 1+\eta^2 & \ddots & 0 \\ \vdots & \ddots & \ddots & \ddots & \eta \\ 0 & \cdots & 0 & \eta & 1+\eta^2 \end{pmatrix}. \tag{1.26}$$

Further,

$$\det(V) = \frac{1-\eta^{2N+2}}{1-\eta^2}, \tag{1.27}$$

and, neglecting the end effects, an approximate inverse of V is the matrix $\Omega = [\omega_{ij}]_{i,j=1,\ldots,N}$ where

$$\omega_{ij} = \left(1-\eta^2\right)^{-1}(-\eta)^{|i-j|}$$

(see [15]). The product $V\Omega$ differs from the identity matrix only on the first and last rows. The exact inverse is $V^{-1} = \left[v^{ij}\right]_{i,j=1,\ldots,N}$ where

$$\begin{aligned} v^{ij} = {} & \left(1-\eta^2\right)^{-1}\left(1-\eta^{2N+2}\right)^{-1}\Big\{(-\eta)^{|i-j|} - (-\eta)^{i+j} - (-\eta)^{2N-i-j+2} - \\ & - (-\eta)^{2N+|i-j|+2} + (-\eta)^{2N+i-j+2} + (-\eta)^{2N-i+j+2}\Big\} \end{aligned} \tag{1.28}$$

(see [24] and [40]).

From the perspective of practical implementation, this estimator is nothing else than the MLE estimator of an MA(1) process with Gaussian errors: any existing computer routines for the MA(1) situation can therefore be applied (see e.g., Section 5.4 in [25]). In particular, the likelihood function can be expressed in a computationally efficient form by triangularizing the matrix V, yielding the equivalent expression:

$$l(\eta, \gamma^2) = -\frac{1}{2}\sum_{i=1}^{N} \ln(2\pi d_i) - \frac{1}{2}\sum_{i=1}^{N} \frac{\tilde{Y}_i^2}{d_i}, \tag{1.29}$$

where

$$d_i = \gamma^2 \frac{1+\eta^2+\ldots+\eta^{2i}}{1+\eta^2+\ldots+\eta^{2(i-1)}},$$

and the $\tilde{Y}_i's$ are obtained recursively as $\tilde{Y}_1 = Y_1$ and for $i = 2, \ldots, N$:

$$\tilde{Y}_i = Y_i - \frac{\eta\left(1+\eta^2+\ldots+\eta^{2(i-2)}\right)}{1+\eta^2+\ldots+\eta^{2(i-1)}}\tilde{Y}_{i-1}.$$

This latter form of the log-likelihood function involves only single sums as opposed to double sums if one were to compute $Y'V^{-1}Y$ by brute force using the expression of V^{-1} given above.

We now compute the distribution of the MLE estimators of σ^2 and a^2, which follows by the delta method from the classical result for the MA(1) estimators of γ and η :

Proposition 1. *When U is normally distributed, the MLE $(\hat{\sigma}^2, \hat{a}^2)$ is consistent and its asymptotic variance is given by*

$$\text{AVAR}_{normal}(\hat{\sigma}^2, \hat{a}^2) = \begin{pmatrix} 4\sqrt{\sigma^6 \Delta \left(4a^2 + \sigma^2 \Delta\right)} + 2\sigma^4 \Delta & -\sigma^2 \Delta h(\Delta, \sigma^2, a^2) \\ \bullet & \frac{\Delta}{2}\left(2a^2 + \sigma^2 \Delta\right) h(\Delta, \sigma^2, a^2) \end{pmatrix}$$

with

$$h(\Delta, \sigma^2, a^2) \equiv 2a^2 + \sqrt{\sigma^2 \Delta \left(4a^2 + \sigma^2 \Delta\right)} + \sigma^2 \Delta . \tag{1.30}$$

Since $\text{AVAR}_{\text{normal}}(\hat{\sigma}^2)$ is increasing in Δ, it is optimal to sample as often as possible. Further, since

$$\text{AVAR}_{\text{normal}}(\hat{\sigma}^2) = 8\sigma^3 a \Delta^{1/2} + 2\sigma^4 \Delta + o(\Delta) , \tag{1.31}$$

the loss of efficiency relative to the case where no market microstructure noise is present (and $\text{AVAR}(\hat{\sigma}^2) = 2\sigma^4 \Delta$ as given in (1.7) if $a^2 = 0$ is not estimated, or $\text{AVAR}(\hat{\sigma}^2) = 6\sigma^4 \Delta$ if $a^2 = 0$ is estimated) is at order $\Delta^{1/2}$. Figure 1.3 plots the asymptotic variances of $\hat{\sigma}^2$ as functions of Δ with and without noise (the parameter values are again $\sigma = 30\%$ and $a = 0.15\%$). Figure 1.4 reports histograms of the distributions of $\hat{\sigma}^2$ and $\hat{a}^2$ from 10,000 Monte Carlo simulations with the solid curve plotting the asymptotic distribution of the estimator from Proposition 1. The sample path is of length $T = 1$ year, the parameter values the same as above, and the process is sampled every 5 minutes – since we are now accounting explicitly for the presence of noise, there is no longer a reason to sample at lower frequencies. Indeed, the figure documents the absence of bias and the good agreement of the asymptotic distribution with the small sample one.

1.6 The Effect of Misspecifying the Distribution of the Microstructure Noise

We now study the situation where one attempts to incorporate the presence of the $U's$ into the analysis, as in Section 1.5, but mistakenly assumes a misspecified model for them. Specifically, we consider the case where the $U's$ are assumed to be normally distributed when in reality they have a different distribution. We still suppose that the $U's$ are i.i.d. with mean zero and variance a^2.

Since the econometrician assumes the $U's$ to have a normal distribution, inference is still done with the log-likelihood $l(\sigma^2, a^2)$, or equivalently $l(\eta, \gamma^2)$

given in (1.25), using (1.9)-(1.10). This means that the scores $\dot{l}_{\sigma^2}$ and $\dot{l}_{a^2}$, or equivalently (C.1) and (C.2), are used as moment functions (or "estimating equations"). Since the first order moments of the moment functions only depend on the second order moment structure of the log-returns $(Y_1, ..., Y_N)$, which is unchanged by the absence of normality, the moment functions are unbiased under the true distribution of the $U's$:

$$E_{\text{true}}[\dot{l}_\eta] = E_{\text{true}}[\dot{l}_{\gamma^2}] = 0\,, \tag{1.32}$$

and similarly for $\dot{l}_{\sigma^2}$ and $\dot{l}_{a^2}$. Hence the estimator $(\hat{\sigma}^2, \hat{a}^2)$ based on these moment functions is consistent and asymptotically unbiased (even though the likelihood function is misspecified.)

The effect of misspecification therefore lies in the asymptotic variance matrix. By using the cumulants of the distribution of U, we express the asymptotic variance of these estimators in terms of deviations from normality. But as far as computing the actual estimator, nothing has changed relative to Section 1.5: we are still calculating the MLE for an MA(1) process with Gaussian errors and can apply exactly the same computational routine.

However, since the error distribution is potentially misspecified, one could expect the asymptotic distribution of the estimator to be altered. This turns out not be the case, as far as $\hat{\sigma}^2$ is concerned:

Theorem 2. *The estimators $(\hat{\sigma}^2, \hat{a}^2)$ obtained by maximizing the possibly misspecified log-likelihood (1.25) are consistent and their asymptotic variance is given by*

$$\text{AVAR}_{true}(\hat{\sigma}^2, \hat{a}^2) = \text{AVAR}_{normal}(\hat{\sigma}^2, \hat{a}^2) + \text{Cum}_4\,[U] \begin{pmatrix} 0 & 0 \\ 0 & \Delta \end{pmatrix}, \tag{1.33}$$

where $\text{AVAR}_{normal}(\hat{\sigma}^2, \hat{a}^2)$ is the asymptotic variance in the case where the distribution of U is normal, that is, the expression given in Proposition 1.

In other words, the asymptotic variance of $\hat{\sigma}^2$ is identical to its expression if the $U's$ had been normal. Therefore the correction we proposed for the presence of market microstructure noise relying on the assumption that the noise is Gaussian is robust to misspecification of the error distribution.

Documenting the presence of the correction term through simulations presents a challenge. At the parameter values calibrated to be realistic, the order of magnitude of a is a few basis points, say $a = 0.10\% = 10^{-3}$. But if U is of order 10^{-3}, $\text{Cum}_4[U]$ which is of the same order as U^4, is of order 10^{-12}. In other words, with a typical noise distribution, the correction term in (1.33) will not be visible.

To nevertheless make it discernible, we use a distribution for U with the same calibrated standard deviation a as before, but a disproportionately large fourth cumulant. Such a distribution can be constructed by letting $U = \omega T_\nu$

where $\omega > 0$ is constant and T_ν is a Student t distribution with v degrees of freedom. T_ν has mean zero, finite variance as long as $v > 2$ and finite fourth moment (hence finite fourth cumulant) as long as $v > 4$. But as v approaches 4 from above, $E[T_\nu^4]$ tends to infinity. This allows us to produce an arbitrarily high value of $\text{Cum}_4[U]$ while controlling for the magnitude of the variance. The specific expressions of a^2 and $\text{Cum}_4[U]$ for this choice of U are given by

$$a^2 = \text{Var}[U] = \frac{\omega^2\nu}{\nu - 2}, \tag{1.34}$$

$$\text{Cum}_4\,[U] = \frac{6\omega^4\nu^2}{(\nu - 4)\,(\nu - 2)^2}. \tag{1.35}$$

Thus we can select the two parameters (ω, ν) to produce desired values of $(a^2, \text{Cum}_4\,[U])$. As before, we set $a = 0.15\%$. Then, given the form of the asymptotic variance matrix (1.33), we set $\text{Cum}_4\,[U]$ so that $\text{Cum}_4\,[U]\,\Delta = \text{AVAR}_{\text{normal}}(\hat{a}^2)/2$. This makes $\text{AVAR}_{\text{true}}(\hat{a}^2)$ by construction 50% larger than $\text{AVAR}_{\text{normal}}(\hat{a}^2)$. The resulting values of (ω, ν) from solving (1.34)-(1.35) are $\omega = 0.00115$ and $v = 4.854$. As above, we set the other parameters to $\sigma = 30\%$, $T = 1$ year, and $\Delta = 5$ minutes. Figure 1.5 reports histograms of the distributions of $\hat{\sigma}^2$ and $\hat{a}^2$ from 10,000 Monte Carlo simulations. The solid curve plots the asymptotic distribution of the estimator, given now by (1.33). There is again good adequacy between the asymptotic and small sample distributions. In particular, we note that as predicted by Theorem 2, the asymptotic variance of $\hat{\sigma}^2$ is unchanged relative to Figure 1.4 while that of $\hat{a}^2$ is 50% larger. The small sample distribution of $\hat{\sigma}^2$ appears unaffected by the non-Gaussianity of the noise; with a skewness of 0.07 and a kurtosis of 2.95, it is closely approximated by its asymptotic Gaussian limit. The small sample distribution of $\hat{a}^2$ does exhibit some kurtosis (4.83), although not large relative to that of the underlying noise distribution (the values of ω and ν imply a kurtosis for U of $3 + 6/(\nu - 4) = 10$). Similar simulations but with a longer time span of $T = 5$ years are even closer to the Gaussian asymptotic limit: the kurtosis of the small sample distribution of $\hat{a}^2$ goes down to 2.99.

1.7 Robustness to Misspecification of the Noise Distribution

Going back to the theoretical aspects, the above Theorem 2 has implications for the use of the Gaussian likelihood l that go beyond consistency, namely that this likelihood can also be used to estimate the distribution of $\hat{\sigma}^2$ under misspecification. With l denoting the log-likelihood assuming that the $U's$ are Gaussian, given in (1.25), $-\ddot{l}(\hat{\sigma}^2, \hat{a}^2)$ denote the observed information matrix in the original parameters σ^2 and a^2. Then

$$\hat{V} = \widehat{\text{AVAR}}_{\text{normal}} = \left(-\frac{1}{T}\ddot{l}(\hat{\sigma}^2, \hat{a}^2)\right)^{-1}$$

is the usual estimate of asymptotic variance when the distribution is correctly specified as Gaussian. Also note, however, that otherwise, so long as $(\hat{\sigma}^2, \hat{a}^2)$ is consistent, $\hat{V}$ is also a consistent estimate of the matrix $\text{AVAR}_{\text{normal}}(\hat{\sigma}^2, \hat{a}^2)$. Since this matrix coincides with $\text{AVAR}_{\text{true}}(\hat{\sigma}^2, \hat{a}^2)$ for all but the (a^2, a^2) term (see (1.33)), the asymptotic variance of $T^{1/2}(\hat{\sigma}^2 - \sigma^2)$ is consistently estimated by $\hat{V}_{\sigma^2\sigma^2}$. The similar statement is true for the covariances, but not, obviously, for the asymptotic variance of $T^{1/2}(\hat{a}^2 - a^2)$.

In the likelihood context, the possibility of estimating the asymptotic variance by the observed information is due to the second Bartlett identity. For a general log likelihood l, if $S \equiv E_{\text{true}}[\dot{l}\dot{l}']/N$ and $D \equiv -E_{\text{true}}[\ddot{l}]/N$ (differentiation refers to the original parameters (σ^2, a^2), not the transformed parameters (γ^2, η)) this identity says that

$$S - D = 0\,. \tag{1.36}$$

It implies that the asymptotic variance takes the form

$$\text{AVAR} = \Delta(DS^{-1}D)^{-1} = \Delta D^{-1}. \tag{1.37}$$

It is clear that (1.37) remains valid if the second Bartlett identity holds only to first order, i.e.,

$$S - D = o(1) \tag{1.38}$$

as $N \to \infty$, for a general criterion function l which satisfies $E_{\text{true}}[\dot{l}] = o(N)$.

However, in view of Theorem 2, equation (1.38) cannot be satisfied. In fact, we show in Appendix E that

$$S - D = \text{Cum}_4\,[U]\, gg' + o(1)\,, \tag{1.39}$$

where

$$g = \begin{pmatrix} g_{\sigma^2} \\ g_{a^2} \end{pmatrix} = \begin{pmatrix} \frac{\Delta^{1/2}}{\sigma(4a^2+\sigma^2\Delta)^{3/2}} \\ \frac{1}{2a^4}\left(1 - \frac{\Delta^{1/2}\sigma(6a^2+\sigma^2\Delta)}{(4a^2+\sigma^2\Delta)^{3/2}}\right) \end{pmatrix}. \tag{1.40}$$

From (1.40), we see that $g \neq 0$ whenever $\sigma^2 > 0$. This is consistent with the result in Theorem 2 that the true asymptotic variance matrix, $\text{AVAR}_{\text{true}}(\hat{\sigma}^2, \hat{a}^2)$, does not coincide with the one for Gaussian noise, $\text{AVAR}_{\text{normal}}(\hat{\sigma}^2, \hat{a}^2)$. On the other hand, the 2×2 matrix gg' is of rank 1, signaling that there exist linear combinations that will cancel out the first column of $S - D$. From what we already know of the form of the correction matrix, D^{-1} gives such a combination, so that the asymptotic variance of the original parameters (σ^2, a^2) will have the property that its first column is not subject to correction in the absence of normality.

A curious consequence of (1.39) is that while the observed information can be used to estimate the asymptotic variance of $\hat{\sigma}^2$ when a^2 is not known, this is not the case when a^2 is known. This is because the second Bartlett identity also fails to first order when considering a^2 to be known, i.e., when differentiating with respect to σ^2 only. Indeed, in that case we have from the upper left component in the matrix equation (1.39)

$$\begin{aligned} S_{\sigma^2\sigma^2} - D_{\sigma^2\sigma^2} &= N^{-1}E_{\text{true}}\left[\dot{l}_{\sigma^2\sigma^2}(\sigma^2,a^2)^2\right] + N^{-1}E_{\text{true}}\left[\ddot{l}_{\sigma^2\sigma^2}(\sigma^2,a^2)\right] \\ &= \text{Cum}_4\left[U\right](g_{\sigma^2})^2 + o(1) \end{aligned}$$

which is not $o(1)$ unless $\text{Cum}_4\left[U\right] = 0$.

To make the connection between Theorem 2 and the second Bartlett identity, one needs to go to the log profile likelihood

$$\lambda(\sigma^2) \quad \equiv \quad \sup_{a^2} \, l(\sigma^2, a^2) \, . \tag{1.41}$$

Obviously, maximizing the likelihood $l(\sigma^2, a^2)$ is the same as maximizing $\lambda(\sigma^2)$. Thus one can think of σ^2 as being estimated (when α^2 is unknown) by maximizing the criterion function $\lambda(\sigma^2)$, or by solving $\dot{\lambda}(\hat{\sigma}^2) = 0$. Also, the observed profile information is related to the original observed information by

$$\ddot{\lambda}(\hat{\sigma}^2)^{-1} = \left[\ddot{l}(\hat{\sigma}^2,\hat{a}^2)^{-1}\right]_{\sigma^2\sigma^2}, \tag{1.42}$$

i.e., the first (upper left hand corner) component of the inverse observed information in the original problem. We recall the rationale for equation (1.42) in Appendix E, where we also show that $E_{\text{true}}[\dot{\lambda}] = o(N)$. In view of Theorem 2, $\ddot{\lambda}(\hat{\sigma}^2)$ can be used to estimate the asymptotic variance of $\hat{\sigma}^2$ under the true (possibly non-Gaussian) distribution of the $U's$, and so it must be that the criterion function λ satisfies (1.38), that is

$$N^{-1}E_{\text{true}}[\dot{\lambda}(\sigma^2)^2] + N^{-1}E_{\text{true}}[\ddot{\lambda}(\sigma^2)] = o(1) \, . \tag{1.43}$$

This is indeed the case, as shown in Appendix E.

This phenomenon is related, although not identical, to what occurs in the context of quasi-likelihood (for comprehensive treatments of quasi-likelihood theory, see the books by [30] and [37], and the references therein, and for early econometrics examples see [34] and [42]). In quasi-likelihood situations, one uses a possibly incorrectly specified score vector which is nevertheless required to satisfy the second Bartlett identity. What makes our situation unusual relative to quasi-likelihood is that the interest parameter σ^2 and the nuisance parameter a^2 are entangled in the same estimating equations ($\dot{l}_{\sigma^2}$ and $\dot{l}_{a^2}$ from the Gaussian likelihood) in such a way that the estimate of σ^2 depends, to first order, on whether a^2 is known or not. This is unlike the typical development of quasi-likelihood, where the nuisance parameter separates out (see, e.g., Table 9.1, page 326 of [37]). Thus only by going to the profile likelihood λ can one make the usual comparison to quasi-likelihood.

1.8 Randomly Spaced Sampling Intervals

One essential feature of transaction data in finance is that the time that separates successive observations is random, or at least time-varying. So, as in

[4], we are led to consider the case where $\Delta_i = \tau_i - \tau_{i-1}$ are either deterministic and time-varying, or random in which case we assume for simplicity that they are i.i.d., independent of the W process. This assumption, while not completely realistic (see [17] for a discrete time analysis of the autoregressive dependence of the times between trades) allows us to make explicit calculations at the interface between the continuous and discrete time scales. We denote by N_T the number of observations recorded by time T. N_T is random if the $\Delta' s$ are. We also suppose that U_{τ_i} can be written U_i, where the U_i are i.i.d. and independent of the W process and the $\Delta_i' s$. Thus, the observation noise is the same at all observation times, whether random or nonrandom. If we define the Y_is as before, in the first two lines of (1.8), though the MA(1) representation is not valid in the same form.

We can do inference conditionally on the observed sampling times, in light of the fact that the likelihood function using all the available information is

$$L\left(Y_N, \Delta_N, ..., Y_1, \Delta_1; \beta, \psi\right) = L\left(Y_N, ..., Y_1 | \Delta_N, ..., \Delta_1; \beta\right) \times L\left(\Delta_N, ..., \Delta_1; \psi\right)$$

where β are the parameters of the state process, that is (σ^2, a^2), and ψ are the parameters of the sampling process, if any (the density of the sampling intervals density $L\left(\Delta_{N_T}, ..., \Delta_1; \psi\right)$ may have its own nuisance parameters ψ, such as an unknown arrival rate, but we assume that it does not depend on the parameters β of the state process.) The corresponding log-likelihood function is

$$\sum_{n=1}^{N} \ln L\left(Y_N, ..., Y_1 | \Delta_N, ..., \Delta_1; \beta\right) + \sum_{n=1}^{N-1} \ln L\left(\Delta_N, ..., \Delta_1; \psi\right) , \tag{1.44}$$

and since we only care about β, we only need to maximize the first term in that sum.

We operate on the covariance matrix Σ of the log-returns $Y's$, now given by

$$\Sigma = \begin{pmatrix} \sigma^2\Delta_1 + 2a^2 & -a^2 & 0 & \cdots & 0 \\ -a^2 & \sigma^2\Delta_2 + 2a^2 & -a^2 & \ddots & \vdots \\ 0 & -a^2 & \sigma^2\Delta_3 + 2a^2 & \ddots & 0 \\ \vdots & \ddots & \ddots & \ddots & -a^2 \\ 0 & \cdots & 0 & -a^2 & \sigma^2\Delta_n + 2a^2 \end{pmatrix} . \tag{1.45}$$

Note that in the equally spaced case, $\Sigma = \gamma^2 V$. But now Y no longer follows an MA(1) process in general. Furthermore, the time variation in $\Delta_i' s$ gives rise to heteroskedasticity as is clear from the diagonal elements of Σ. This is consistent with the predictions of the model of [16] where the variance of the transaction price process $\tilde{X}$ is heteroskedastic as a result of the influence of the sampling times. In their model, the sampling times are autocorrelated and correlated with the evolution of the price process, factors we have assumed away here. However, [4] show how to conduct likelihood inference in such a situation.

The log-likelihood function is given by

$$\ln L\left(Y_N, ..., Y_1 | \Delta_N, ..., \Delta_1; \beta\right) \equiv l(\sigma^2, a^2) \qquad (1.46)$$
$$= -\ln\det(\Sigma)/2 - N\ln(2\pi)/2 - Y'\Sigma^{-1}Y/2\,.$$

In order to calculate this log-likelihood function in a computationally efficient manner, it is desirable to avoid the "brute force" inversion of the $N \times N$ matrix Σ. We extend the method used in the MA(1) case (see (1.29)) as follows. By Theorem 5.3.1 in [13], and the development in the proof of their Theorem 5.4.3, we can decompose Σ in the form $\Sigma = LDL^T$, where L is a lower triangular matrix whose diagonals are all 1 and D is diagonal. To compute the relevant quantities, their Example 5.4.3 shows that if one writes $D = \text{diag}(g_1, ..., g_n)$ and

$$L = \begin{pmatrix} 1 & 0 & 0 & \cdots & 0 \\ \kappa_2 & 1 & 0 & \ddots & \vdots \\ 0 & \kappa_3 & 1 & \ddots & 0 \\ \vdots & \ddots & \ddots & \ddots & 0 \\ 0 & \cdots & 0 & \kappa_n & 1 \end{pmatrix}, \qquad (1.47)$$

then the $g_k' s$ and $\kappa_k' s$ follow the recursion equation $g_1 = \sigma^2\Delta_1 + 2a^2$ and for $i = 2, ..., N$:

$$\kappa_i = -a^2/g_{i-1} \qquad \text{and} \qquad g_i = \sigma^2\Delta_i + 2a^2 + \kappa_i a^2. \qquad (1.48)$$

Then, define $\tilde{Y} = L^{-1}Y$ so that $Y'\Sigma^{-1}Y = \tilde{Y}'D^{-1}\tilde{Y}$. From $Y = L\tilde{Y}$, it follows that $\tilde{Y}_1 = Y_1$ and, for $i = 2, ..., N$:

$$\tilde{Y}_i = Y_i - \kappa_i\tilde{Y}_{i-1}\,.$$

And $\det(\Sigma) = \det(D)$ since $\det(L) = 1$. Thus we have obtained a computationally simple form for (1.46) that generalizes the MA(1) form (1.29) to the case of non-identical sampling intervals:

$$l(\sigma^2, a^2) = -\frac{1}{2}\sum_{i=1}^{N}\ln\left(2\pi g_i\right) - \frac{1}{2}\sum_{i=1}^{N}\frac{\tilde{Y}_i^2}{g_i}\,. \qquad (1.49)$$

We can now turn to statistical inference using this likelihood function. As usual, the asymptotic variance of $T^{1/2}(\hat{\sigma}^2 - \sigma^2, \hat{a}^2 - a^2)$ is of the form

$$\text{AVAR}(\hat{\sigma}^2, \hat{a}^2) = \lim_{T\ ->\ \infty}\begin{pmatrix} \frac{1}{T}E\left[-\ddot{l}_{\sigma^2\sigma^2}\right] & \frac{1}{T}E\left[-\ddot{l}_{\sigma^2a^2}\right] \\ \bullet & \frac{1}{T}E\left[-\ddot{l}_{a^2a^2}\right] \end{pmatrix}^{-1}. \qquad (1.50)$$

To compute this quantity, suppose in the following that β_1 and β_2 can represent either σ^2 or a^2. We start with:

Lemma 2. *Fisher's Conditional Information is given by*

$$E\left[-\ddot{l}_{\beta_2\beta_1}\middle|\Delta\right]=-\frac{1}{2}\frac{\partial^2\ln\det\Sigma}{\partial\beta_2\beta_1}. \tag{1.51}$$

To compute the asymptotic distribution of the MLE of (β_1,β_2), one would then need to compute the inverse of $E\left[-\ddot{l}_{\beta_2\beta_1}\right]=E_\Delta\left[E\left[-\ddot{l}_{\beta_2\beta_1}\middle|\Delta\right]\right]$ where E_Δ denotes expectation taken over the law of the sampling intervals. From (1.51), and since the order of E_Δ and $\partial^2/\partial\beta_2\beta_1$ can be interchanged, this requires the computation of

$$E_\Delta\left[\ln\det\Sigma\right]=E_\Delta\left[\ln\det D\right]=\sum_{i=1}^{N}E_\Delta\left[\ln\left(g_i\right)\right]$$

where from (1.48) the $g_i's$ are given by the continuous fraction

$$\begin{aligned}
g_1&=\sigma^2\Delta_1+2a^2,\\
g_2&=\sigma^2\Delta_2+2a^2-\frac{a^4}{\sigma^2\Delta_1+2a^2},\\
g_3&=\sigma^2\Delta_3+2a^2-\frac{a^4}{\sigma^2\Delta_2+2a^2-\frac{a^4}{\sigma^2\Delta_1+2a^2}},
\end{aligned}$$

and in general

$$g_i=\sigma^2\Delta_i+2a^2-\frac{a^4}{\sigma^2\Delta_{i-1}+2a^2-\frac{a^4}{\ddots}}.$$

It therefore appears that computing the expected value of $\ln\left(g_i\right)$ over the law of $(\Delta_1,\Delta_2,...,\Delta_i)$ will be impractical.

1.8.1 Expansion Around a Fixed Value of Δ

To continue further with the calculations, we propose to expand around a fixed value of Δ, namely $\Delta_0=E\left[\Delta\right]$. Specifically, suppose now that

$$\Delta_i=\Delta_0\left(1+\varepsilon\xi_i\right), \tag{1.52}$$

where ε and Δ_0 are nonrandom, the $\xi_i's$ are i.i.d. random variables with mean zero and finite distribution. We will Taylor-expand the expressions above around $\varepsilon=0$, i.e., around the non-random sampling case we have just finished dealing with. Our expansion is one that is valid when the randomness of the sampling intervals remains small, i.e., when $\mathrm{Var}\left[\Delta_i\right]$ is small, or $o(1)$.

Then we have $\Delta_0 = E[\Delta] = O(1)$ and $\mathrm{Var}[\Delta_i] = \Delta_0^2 \varepsilon^2 \mathrm{Var}[\xi_i]$. The natural scaling is to make the distribution of ξ_i finite, i.e., $\mathrm{Var}[\xi_i] = O(1)$, so that $\varepsilon^2 = O(\mathrm{Var}[\Delta_i]) = o(1)$. But any other choice would have no impact on the result since $\mathrm{Var}[\Delta_i] = o(1)$ implies that the product $\varepsilon^2 \mathrm{Var}[\xi_i]$ is $o(1)$ and whenever we write reminder terms below they can be expressed as $O_p(\varepsilon^3 \xi^3)$ instead of just $O(\varepsilon^3)$. We keep the latter notation for clarity given that we set $\xi_i = O_p(1)$. Furthermore, for simplicity, we take the $\xi_i' s$ to be bounded.

We emphasize that the time increments or durations Δ_i do not tend to zero length as $\varepsilon \to 0$. It is only the variability of the Δ_i's that goes to zero.

Denote by Σ_0 the value of Σ when Δ is replaced by Δ_0, and let Ξ denote the matrix whose diagonal elements are the terms $\Delta_0 \xi_i$, and whose off-diagonal elements are zero. We obtain:

Theorem 3. *The MLE $(\hat{\sigma}^2, \hat{a}^2)$ is again consistent, this time with asymptotic variance*

$$\mathrm{AVAR}(\hat{\sigma}^2, \hat{a}^2) = A^{(0)} + \varepsilon^2 A^{(2)} + O(\varepsilon^3), \tag{1.53}$$

where

$$A^{(0)} = \begin{pmatrix} 4\sqrt{\sigma^6 \Delta_0 (4a^2 + \sigma^2 \Delta_0)} + 2\sigma^4 \Delta_0 & -\sigma^2 \Delta_0 h(\Delta_0, \sigma^2, a^2) \\ \bullet & \frac{\Delta_0}{2}(2a^2 + \sigma^2 \Delta_0) h(\Delta_0, \sigma^2, a^2) \end{pmatrix},$$

and

$$A^{(2)} = \frac{\mathrm{Var}[\xi]}{(4a^2 + \Delta_0 \sigma^2)} \begin{pmatrix} A^{(2)}_{\sigma^2\sigma^2} & A^{(2)}_{\sigma^2 a^2} \\ \bullet & A^{(2)}_{a^2 a^2} \end{pmatrix},$$

with

$$\begin{aligned} A^{(2)}_{\sigma^2\sigma^2} &= -4\left(\Delta_0^2 \sigma^6 + \Delta_0^{3/2} \sigma^5 \sqrt{4a^2 + \Delta_0 \sigma^2}\right), \\ A^{(2)}_{\sigma^2 a^2} &= \Delta_0^{3/2} \sigma^3 \sqrt{4a^2 + \Delta_0 \sigma^2}\,(2a^2 + 3\Delta_0 \sigma^2) + \Delta_0^2 \sigma^4 (8a^2 + 3\Delta_0 \sigma^2), \\ A^{(2)}_{a^2 a^2} &= -\Delta_0^2 \sigma^2 \left(2a^2 + \sigma\sqrt{\Delta_0}\sqrt{4a^2 + \Delta_0 \sigma^2} + \Delta_0 \sigma^2\right)^2. \end{aligned}$$

In connection with the preceding result, we underline that the quantity $\mathrm{AVAR}(\hat{\sigma}^2, \hat{a}^2)$ is a limit as $T \to \infty$, as in (1.50). The equation (1.53), therefore, is an expansion in ε *after* $T \to \infty$.

Note that $A^{(0)}$ is the asymptotic variance matrix already present in Proposition 1, except that it is evaluated at $\Delta_0 = E[\Delta]$. Note also that the second order correction term is proportional to $\mathrm{Var}[\xi]$, and is therefore zero in the absence of sampling randomness. When that happens, $\Delta = \Delta_0$ with probability one and the asymptotic variance of the estimator reduces to the leading term $A^{(0)}$, i.e., to the result in the fixed sampling case given in Proposition 1.

1.8.2 Randomly Spaced Sampling Intervals and Misspecified Microstructure Noise

Suppose now, as in Section 1.6, that the $U's$ are i.i.d., have mean zero and variance a^2, but are otherwise not necessarily Gaussian. We adopt the same approach as in Section 1.6, namely to express the estimator's properties in terms of deviations from the deterministic and Gaussian case. The additional correction terms in the asymptotic variance are given in the following result.

Theorem 4. *The asymptotic variance is given by*

$$\begin{aligned} \mathrm{AVAR}_{true}(\hat{\sigma}^2, \hat{a}^2) = & \left(A^{(0)} + \mathrm{Cum}_4\left[U\right] B^{(0)}\right) + \\ & + \varepsilon^2 \left(A^{(2)} + \mathrm{Cum}_4\left[U\right] B^{(2)}\right) + O(\varepsilon^3), \end{aligned} \tag{1.54}$$

where $A^{(0)}$ and $A^{(2)}$ are given in the statement of Theorem 3 and

$$B^{(0)} = \begin{pmatrix} 0 & 0 \\ 0 & \Delta_0 \end{pmatrix},$$

while

$$B^{(2)} = \mathrm{Var}[\xi] \begin{pmatrix} B^{(2)}_{\sigma^2\sigma^2} & B^{(2)}_{\sigma^2 a^2} \\ \bullet & B^{(2)}_{a^2a^2} \end{pmatrix},$$

with

$$B^{(2)}_{\sigma^2\sigma^2} = \frac{10\Delta_0^{3/2}\sigma^5}{(4a^2+\Delta_0\sigma^2)^{5/2}} + \frac{4\Delta_0^2\sigma^6\left(16a^4+11a^2\Delta_0\sigma^2+2\Delta_0^2\sigma^4\right)}{(2a^2+\Delta_0\sigma^2)^3(4a^2+\Delta_0\sigma^2)^2},$$

$$\begin{aligned} B^{(2)}_{\sigma^2 a^2} = & \frac{-\Delta_0^2\sigma^4}{(2a^2+\Delta_0\sigma^2)^3(4a^2+\Delta_0\sigma^2)^{5/2}} \times \\ & \times \left(\sqrt{4a^2+\Delta_0\sigma^2}\left(32a^6+64a^4\Delta_0\sigma^2+35a^2\Delta_0^2\sigma^4+6\Delta_0^3\sigma^6\right) + \right. \\ & \left. + \Delta_0^{1/2}\sigma\left(116a^6+126a^4\Delta_0\sigma^2+47a^2\Delta_0^2\sigma^4+6\Delta_0^3\sigma^6\right)\right), \end{aligned}$$

$$B^{(2)}_{a^2a^2} = \frac{16a^8\Delta_0^{5/2}\sigma^3\left(13a^4+10a^2\Delta_0\sigma^2+2\Delta_0^2\sigma^4\right)}{(2a^2+\Delta_0\sigma^2)^3(4a^2+\Delta_0\sigma^2)^{5/2}\left(2a^2+\sigma^2\Delta-\sqrt{\sigma^2\Delta\left(4a^2+\sigma^2\Delta\right)}\right)^2}.$$

The term $A^{(0)}$ is the base asymptotic variance of the estimator, already present with fixed sampling and Gaussian noise. The term $\mathrm{Cum}_4\left[U\right] B^{(0)}$ is the correction due to the misspecification of the error distribution. These two terms are identical to those present in Theorem 2. The terms proportional to ε^2 are the further correction terms introduced by the randomness of the sampling. $A^{(2)}$ is the base correction term present even with Gaussian noise in Theorem 3, and $\mathrm{Cum}_4\left[U\right] B^{(2)}$ is the further correction due to the sampling randomness. Both $A^{(2)}$ and $B^{(2)}$ are proportional to $\mathrm{Var}[\xi]$ and hence vanish in the absence of sampling randomness.

1.9 Extensions

In this section, we briefly sketch four extensions of our basic model. First, we show that the introduction of a drift term does not alter our conclusions. Then we examine the situation where market microstructure noise is serially correlated; there, we show that the insight of Theorem 1 remains valid, namely that the optimal sampling frequency is finite. Third, we then turn to the case where the noise is correlated with the efficient price signal. Fourth, we discuss what happens if volatility is stochastic.

In a nutshell, each one of these assumptions can be relaxed without affecting our main conclusion, namely that the presence of the noise gives rise to a finite optimal sampling frequency. The second part of our analysis, dealing with likelihood corrections for microstructure noise, will not necessarily carry through unchanged if the assumptions are relaxed (for instance, there is not even a known likelihood function if volatility is stochastic, and the likelihood must be modified if the assumed variance-covariance structure of the noise is modified).

1.9.1 Presence of a Drift Coefficient

What happens to our conclusions when the underlying X process has a drift? We shall see in this case that the presence of the drift does not alter our earlier conclusions. As a simple example, consider linear drift, i.e., replace (1.2) with

$$X_t = \mu t + \sigma W_t \,. \tag{1.55}$$

The contamination by market microstructure noise is as before: the observed process is given by (1.3).

As before, we first-difference to get the log-returns $Y_i = \tilde{X}_{\tau_i} - \tilde{X}_{\tau_{i-1}} + U_{\tau_i} - U_{\tau_{i-1}}$. The likelihood function is now

$$\begin{aligned} &\ln L\,(Y_N, ..., Y_1 | \Delta_N, ..., \Delta_1; \beta) \equiv l(\sigma^2, a^2, \mu) \\ &= -\ln\det(\Sigma)/2 - N\ln(2\pi)/2 - (Y - \mu\Delta)'\Sigma^{-1}(Y - \mu\Delta)/2 \,, \end{aligned}$$

where the covariance matrix is given in (1.45), and where $\Delta = (\Delta_1, ..., \Delta_N)'$. If β denotes either σ^2 or a^2, one obtains

$$\ddot{l}_{\mu\beta} = \Delta' \frac{\partial \Sigma^{-1}}{\partial \beta}(Y - \mu\Delta) \,,$$

so that $E[\ddot{l}_{\mu\beta}|\Delta] = 0$ no matter whether the $U's$ are normally distributed or have another distribution with mean 0 and variance a^2. In particular,

$$E[\ddot{l}_{\mu\beta}] = 0 \,. \tag{1.56}$$

Now let $E[\ddot{l}]$ be the 3×3 matrix of expected second likelihood derivatives. Let $E[\ddot{l}] = -TE[\Delta]D + o(T)$. Similarly define $\mathrm{Cov}(\dot{l}, \dot{l}) = TE[\Delta]S + o(T)$. As before, when the $U's$ have a normal distribution, $S = D$, and otherwise that is not the case. The asymptotic variance matrix of the estimators is of the form $\text{AVAR} = E[\Delta]D^{-1}SD^{-1}$.

Let D_{σ^2,a^2} be the corresponding 2×2 matrix when estimation is carried out on σ^2 and a^2 for known μ, and D_μ is the asymptotic information on μ for known σ^2 and a^2. Similarly define S_{σ^2,a^2} and $\text{AVAR}_{\sigma^2,a^2}$. Since D is block diagonal by (1.56),

$$D = \begin{pmatrix} D_{\sigma^2,a^2} & 0 \\ 0' & D_\mu \end{pmatrix},$$

it follows that

$$D^{-1} = \begin{pmatrix} D_{\sigma^2,a^2}^{-1} & 0 \\ 0' & D_\mu^{-1} \end{pmatrix}.$$

Hence

$$\text{AVAR}(\hat{\sigma}^2, \hat{a}^2) = E[\Delta]D_{\sigma^2,a^2}^{-1}S_{\sigma^2,a^2}D_{\sigma^2,a^2}^{-1}. \tag{1.57}$$

The asymptotic variance of $(\hat{\sigma}^2, \hat{a}^2)$ is thus the same as if μ were known, in other words, as if $\mu = 0$, which is the case that we focused on in all the previous sections.

1.9.2 Serially Correlated Noise

We now examine what happens if we relax the assumption that the market microstructure noise is serially independent. Suppose that, instead of being i.i.d. with mean 0 and variance a^2, the market microstructure noise follows

$$dU_t = -bU_t dt + c dZ_t \tag{1.58}$$

where $b > 0$, $c > 0$ and Z is a Brownian motion independent of W. $U_\Delta | U_0$ has a Gaussian distribution with mean $e^{-b\Delta}U_0$ and variance $\frac{c^2}{2b}\left(1 - e^{-2b\Delta}\right)$. The unconditional mean and variance of U are 0 and $a^2 = \frac{c^2}{2b}$. The main consequence of this model is that the variance contributed by the noise to a log-return observed over an interval of time Δ is now of order $O(\Delta)$, that is of the same order as the variance of the efficient price process $\sigma^2\Delta$, instead of being of order $O(1)$ as previously. In other words, log-prices observed close together have very highly correlated noise terms. Because of this feature, this model for the microstructure noise would be less appropriate if the primary source of the noise consists of bid-ask bounces. In such a situation, the fact that a transaction is on the bid or ask side has little predictive power for the next transaction, or at least not enough to predict that two successive transactions are on the same side with very high probability (although [11] have argued that serial correlation in the transaction type can be a component of the bid-ask spread, and extended the model of [39] to allow for it). On the other hand, the model (1.58) can better capture effects such as the gradual adjustment of prices in response to a shock such as a large trade. In practice, the noise term probably encompasses both of these examples, resulting in a situation where the variance contributed by the noise has both types of components, some of order $O(1)$, some of lower orders in Δ.

The observed log-returns take the form

$$\begin{aligned} Y_i &= \tilde{X}_{\tau_i} - \tilde{X}_{\tau_{i-1}} + U_{\tau_i} - U_{\tau_{i-1}}, \\ &= \sigma\left(W_{\tau_i} - W_{\tau_{i-1}}\right) + U_{\tau_i} - U_{\tau_{i-1}}, \\ &\equiv w_i + u_i, \end{aligned}$$

where the $w_i's$ are i.i.d. $N(0,\sigma^2\Delta)$, the $u_i's$ are independent of the $w_i's$, so we have $\mathrm{Var}\left[Y_i\right] = \sigma^2\Delta + E\left[u_i^2\right]$, and they are Gaussian with mean zero and variance

$$E\left[u_i^2\right] = E\left[\left(U_{\tau_i} - U_{\tau_{i-1}}\right)^2\right] = \frac{c^2\left(1-e^{-b\Delta}\right)}{b} = c^2\Delta + o(\Delta), \tag{1.59}$$

instead of $2a^2$.

In addition, the $u_i's$ are now serially correlated at all lags since

$$E\left[U_{\tau_i}U_{\tau_k}\right] = \frac{c^2\left(1-e^{-b\Delta(i-k)}\right)}{2b},$$

for $i \geq k$. The first order correlation of the log-returns is now

$$\mathrm{Cov}\left(Y_i, Y_{i-1}\right) = -\frac{c^2\left(1-e^{-b\Delta}\right)^2}{2b} = -\frac{c^2 b}{2}\Delta^2 + o\left(\Delta^2\right),$$

instead of η.

The result analogous to Theorem 1 is as follows. If one ignores the presence of this type of serially correlated noise when estimating σ^2, then:

Theorem 5. *In small samples (finite T), the RMSE of the estimator $\hat{\sigma}^2$ is given by*

$$\begin{aligned} RMSE\left[\hat{\sigma}^2\right] &= \left(\frac{c^4\left(1-e^{-b\Delta}\right)^2}{b^2\Delta^2} + \frac{c^4\left(1-e^{-b\Delta}\right)^2\left(\frac{T}{\Delta}e^{-2b\Delta} - 1 + e^{-2Tb}\right)}{T^2b^2\left(1+e^{-b\Delta}\right)^2} + \right. \\ &\quad \left. + \frac{2}{T\Delta}\left(\sigma^2\Delta + \frac{c^2\left(1-e^{-b\Delta}\right)}{b}\right)^2\right)^{1/2} \qquad (1.60) \\ &= c^2 - \frac{bc^2}{2}\Delta + \frac{\left(\sigma^2+c^2\right)^2\Delta}{c^2T} + O(\Delta^2) + O\left(\frac{1}{T^2}\right), \end{aligned}$$

so that for large T, starting from a value of c^2 in the limit where $\Delta \to 0$, increasing Δ first reduces $RMSE\left[\hat{\sigma}^2\right]$. Hence the optimal sampling frequency is finite.

One would expect this type of noise to be not nearly as bad as i.i.d. noise for the purpose of inferring σ^2 from high frequency data. Indeed, the variance of the noise is of the same order $O(\Delta)$ as the variance of the efficient price process. Thus log returns computed from transaction prices sampled close together are not subject to as much noise as previously ($O(\Delta)$ vs. $O(1)$) and the squared bias β^2 of the estimator $\hat{\sigma}^2$ no longer diverges to infinity as $\Delta \to 0$: it has the finite limit c^4. Nevertheless, β^2 first decreases as Δ increases from 0, since

$$\beta^2 = \left(E\left[\hat{\sigma}^2\right] - \sigma^2\right)^2 = \frac{c^4\left(1 - e^{b\Delta}\right)^2}{b^2 \Delta^2 e^{2b\Delta}} ,$$

and $\partial b_2/\partial\Delta \to -bc^4 < 0$ as $\Delta \to 0$. For large enough T, this is sufficient to generate a finite optimal sampling frequency.

To calibrate the parameter values b and c, we refer to the same empirical microstructure studies we mentioned in Section 1.4. We now have $\pi = E\left[u_i^2\right] / (\sigma^2\Delta + E\left[u_i^2\right])$ as the proportion of total variance that is microstructure-induced; we match it to the numbers in (1.24) from [35]. In their Table 5, they report the first order correlation of price changes (hence returns) to be approximately $\rho = -0.2$ at their frequency of observation. Here $\rho = \text{Cov}\left(Y_i, Y_{i-1}\right) / \text{Var}\left[Y_i\right]$. If we match $\pi = 0.6$ and $\rho = -0.2$, with $\sigma = 30\%$ as before, we obtain (after rounding) $c = 0.5$ and $b = 3 \times 10^4$. Figure 1.6 displays the resulting RMSE of the estimator as a function of Δ and T. The overall picture is comparable to Figure 1.2.

As for the rest of the analysis of the paper, dealing with likelihood corrections for microstructure noise, the covariance matrix of the log-returns, $\gamma^2 V$ in (1.26), should be replaced by the matrix whose diagonal elements are

$$\text{Var}\left[Y_i^2\right] = E\left[w_i^2\right] + E\left[u_i^2\right] = \sigma^2\Delta + \frac{c^2\left(1 - e^{-b\Delta}\right)}{b} ,$$

and off-diagonal elements $i > j$ are:

$$\begin{aligned}
\text{Cov}\left(Y_i, Y_j\right) &= E\left[Y_i Y_j\right] = E\left[\left(w_i + u_i\right)\left(w_j + u_j\right)\right] , \\
&= E\left[u_i u_j\right] = E\left[\left(U_{\tau_i} - U_{\tau_{i-1}}\right)\left(U_{\tau_j} - U_{\tau_{j-1}}\right)\right] , \\
&= E\left[U_{\tau_i} U_{\tau_j}\right] - E\left[U_{\tau_i} U_{\tau_{j-1}}\right] - E\left[U_{\tau_{i-1}} U_{\tau_j}\right] + E\left[U_{\tau_{i-1}} U_{\tau_{j-1}}\right] , \\
&= -\frac{c^2\left(1 - e^{-b\Delta}\right)^2 e^{-b\Delta(i-j-1)}}{2b} .
\end{aligned}$$

Having modified the matrix $\gamma^2 V$, the artificial "normal" distribution that assumes i.i.d. $U's$ that are $N(0, \alpha^2)$ would no longer use the correct second moment structure of the data. Thus we cannot relate a priori the asymptotic variance of the estimator of the estimator $\hat{\sigma}^2$ to that of the i.i.d. Normal case, as we did in Theorem 2.

1.9.3 Noise Correlated with the Price Process

We have assumed so far that the U process was uncorrelated with the W process. Microstructure noise attributable to informational effects is likely to be correlated with the efficient price process, since it is generated by the response of market participants to information signals (i.e., to the efficient price process). This would be the case for instance in the bid-ask model with adverse selection of [20]. When the U process is no longer uncorrelated from the W process, the form of the variance matrix of the observed log-returns Y must be altered, replacing $\gamma^2 v_{ij}$ in (1.26) with

$$\begin{aligned}\mathrm{Cov}(Y_i, Y_j) &= \mathrm{Cov}\left(\sigma(W_{\tau_i} - W_{\tau_{i-1}}) + U_{\tau_i} - U_{\tau_{i-1}}, \sigma(W_{\tau_j} - W_{\tau_{j-1}}) + \right.\\ &\quad \left. + U_{\tau_j} - U_{\tau_{j-1}}\right),\\ &= \sigma^2 \Delta \delta_{ij} + \mathrm{Cov}(\sigma(W_{\tau_i} - W_{\tau_{i-1}}), U_{\tau_j} - U_{\tau_{j-1}}) + \\ &\quad + \mathrm{Cov}(\sigma(W_{\tau_j} - W_{\tau_{j-1}}), U_{\tau_i} - U_{\tau_{i-1}}) + \\ &\quad + \mathrm{Cov}(U_{\tau_i} - U_{\tau_{i-1}}, U_{\tau_j} - U_{\tau_{j-1}}),\end{aligned}$$

where δ_{ij} is the Kronecker symbol.

The small sample properties of the misspecified MLE for σ^2 analogous to those computed in the independent case, including its RMSE, can be obtained from

$$E\left[\hat{\sigma}^2\right] = \frac{1}{T}\sum_{i=1}^{N} E\left[Y_i^2\right],$$

$$\mathrm{Var}\left[\hat{\sigma}^2\right] = \frac{1}{T^2}\sum_{i=1}^{N} \mathrm{Var}\left[Y_i^2\right] + \frac{2}{T^2}\sum_{i=1}^{N}\sum_{j=1}^{i-1} \mathrm{Cov}\left(Y_i^2, Y_j^2\right).$$

Specific expressions for all these quantities depend upon the assumptions of the particular structural model under consideration: for instance, in the [20] model (see his Proposition 6), the $U's$ remain stationary, the transaction noise U_{τ_i} is uncorrelated with the return noise during the previous observation period, i.e., $U_{\tau_{i-1}} - U_{\tau_{i-2}}$, and the efficient return $\sigma\left(W_{\tau_i} - W_{\tau_{i-1}}\right)$ is also uncorrelated with the transaction noises $U_{\tau_{i+1}}$ and $U_{\tau_{i-2}}$. With these in hand, the analysis of the RMSE and its minimum can then proceed as above. As for the likelihood corrections for microstructure noise, the same caveat as in serially correlated U case applies: having modified the matrix $\gamma^2 V$, the artificial "normal" distribution would no longer use the correct second moment structure of the data and the likelihood must be modified accordingly.

1.9.4 Stochastic Volatility

One important departure from our basic model is the case where volatility is stochastic. The observed log-returns are still generated by equation (1.3). Now, however, the constant volatility assumption (1.2) is replaced by

$$dX_t = \sigma_t dW_t \,. \tag{1.61}$$

The object of interest in much of the literature on high frequency volatility estimation (see e.g., [8] and [6]) is then the integral

$$\int_0^T \sigma_t^2 dt \tag{1.62}$$

over a fixed time period $[0, T]$, or possibly several such time periods. The estimation is based on observations $0 = t_0 < t_1 < \ldots < t_n = T$, and asymptotic results are obtained when $\max \Delta t_i \to 0$. The usual estimator for (1.62) is the "realized variance"

$$\sum_{i=1}^{n} (\tilde{X}_{t_{i+1}} - \tilde{X}_{t_i})^2 . \tag{1.63}$$

In the context of stochastic volatility, ignoring market microstructure noise leads to an even more dangerous situation than when σ is constant and $T \to \infty$. We show in the companion paper [43] that, after suitable scaling, the realized variance is a consistent and asymptotically normal estimator – but of the quantity $2a^2$. This quantity has, in general, nothing to do with the object of interest (1.62). Said differently, market microstructure noise totally swamps the variance of the price signal at the level of the realized variance. To obtain a finite optimal sampling interval, one needs that $a^2 \to 0$ as $n \to \infty$, that is the amount of noise must disappear asymptotically. For further developments on this topic, we refer to [43].

1.10 Conclusions

We showed that the presence of market microstructure noise makes it optimal to sample less often than would otherwise be the case in the absence of noise, and we determined accordingly the optimal sampling frequency in closed-form.

We then addressed the issue of what to do about it, and showed that modelling the noise term explicitly restores the first order statistical effect that sampling as often as possible is optimal. We also demonstrated that this remains the case if one misspecifies the assumed distribution of the noise term. If the econometrician assumes that the noise terms are normally distributed when in fact they are not, not only is it still optimal to sample as often as possible, but the estimator has the same asymptotic variance as if the noise distribution had been correctly specified. This robustness result is, we think, a major argument in favor of incorporating the presence of the noise when estimating continuous time models with high frequency financial data, even if one is unsure about what is the true distribution of the noise term. Hence, the answer to the question we pose in our title is "as often as possible," provided one accounts for the presence of the noise when designing the estimator.

Appendix A – Proof of Lemma 1

To calculate the fourth cumulant $\mathrm{Cum}(Y_i, Y_j, Y_k, Y_l)$, recall from (1.8) that the observed log-returns are

$$Y_i = \sigma\left(W_{\tau_i} - W_{\tau_{i-1}}\right) + U_{\tau_i} - U_{\tau_{i-1}} \,.$$

First, note that the τ_i are nonrandom, and W is independent of the $U's$, and has Gaussian increments. Second, the cumulants are multilinear, so

$$\begin{aligned}
&\mathrm{Cum}(Y_i, Y_j, Y_k, Y_l) \\
&= \mathrm{Cum}\left(\sigma\left(W_{\tau_i} - W_{\tau_{i-1}}\right) + U_{\tau_i} - U_{\tau_{i-1}}, \sigma\left(W_{\tau_j} - W_{\tau_{j-1}}\right) + U_{\tau_j} - U_{\tau_{j-1}},\right. \\
&\quad \left.\sigma\left(W_{\tau_k} - W_{\tau_{k-1}}\right) + U_{\tau_k} - U_{\tau_{k-1}}, \sigma\left(W_{\tau_l} - W_{\tau_{l-1}}\right) + U_{\tau l} - U_{\tau_{l-1}}\right), \\
&= \sigma^4 \,\mathrm{Cum}(W_{\tau_i} - W_{\tau_{i-1}}, W_{\tau_j} - W_{\tau_{j-1}}, W_{\tau_k} - W_{\tau_{k-1}}, W_{\tau_l} - W_{\tau_{l-1}}) + \\
&\quad + \sigma^3 \,\mathrm{Cum}(W_{\tau_i} - W_{\tau_{i-1}}, W_{\tau_j} - W_{\tau_{j-1}}, W_{\tau_k} - W_{\tau_{k-1}}, U_{\tau_l} - U_{\tau_{l-1}})[4] + \\
&\quad + \sigma^2 \,\mathrm{Cum}(W_{\tau_i} - W_{\tau_{i-1}}, W_{\tau_j} - W_{\tau_{j-1}}, U_{\tau_k} - U_{\tau_{k-1}}, U_{\tau_l} - U_{\tau_{l-1}})[6] + \\
&\quad + \sigma \,\mathrm{Cum}(W_{\tau_i} - W_{\tau_{i-1}}, U_{\tau_j} - U_{\tau_{j-1}}, U_{\tau_k} - U_{\tau_{k-1}}, U_{\tau_l} - U_{\tau_{l-1}})[4] + \\
&\quad + \mathrm{Cum}(U_{\tau_i} - U_{\tau_{i-1}}, U_{\tau_j} - U_{\tau_{j-1}}, U_{\tau_k} - U_{\tau_{k-1}}, U_{\tau_l} - U_{\tau_{l-1}}) \,.
\end{aligned}$$

Out of these terms, only the last is nonzero because W has Gaussian increments (so all cumulants of its increments of order greater than two are zero), and is independent of the $U's$ (so all cumulants involving increments of both W and U are also zero.) Therefore,

$$\mathrm{Cum}(Y_i, Y_j, Y_k, Y_l) = \mathrm{Cum}(U_{\tau_i} - U_{\tau_{i-1}}, U_{\tau_j} - U_{\tau_{j-1}}, U_{\tau_k} - U_{\tau_{k-1}}, U_{\tau_l} - U_{\tau_{l-1}}) \,.$$

If $i = j = k = l$, we have:

$$\begin{aligned}
&\mathrm{Cum}(U_{\tau_i} - U_{\tau_{i-1}}, U_{\tau_i} - U_{\tau_{i-1}}, U_{\tau_i} - U_{\tau_{i-1}}, U_{\tau_i} - U_{\tau_{i-1}}) \\
&= \mathrm{Cum}_4(U_{\tau_i} - U_{\tau_{i-1}}) \\
&= \mathrm{Cum}_4(U_{\tau_i}) + \mathrm{Cum}_4(-U_{\tau_{i-1}}) \\
&= 2 \;\mathrm{Cum}_4\left[U\right] ,
\end{aligned}$$

with the second equality following from the independence of U_{τ_i} and $U_{\tau_{i-1}}$, and the third from the fact that the cumulant is of even order.

If $\max(i,j,k,l) = \min(i,j,k,l) + 1$, two situations arise. Set $m = \min(i,j,k,l)$ and $M = \max(i,j,k,l)$. Also set $s = s(i,j,k,l) = \#\{i,j,k,l = m\}$. If s is odd, say $s = 1$ with $i = m$, and $j,k,l = M = m+1$, we get a term of the form

$$\mathrm{Cum}(U_{\tau_m} - U_{\tau_{m-1}}, U_{\tau_{m+1}} - U_{\tau_m}, U_{\tau_{m+1}} - U_{\tau_m}, U_{\tau_{m+1}} - U_{\tau_m}) = -\,\mathrm{Cum}_4(U_{\tau_m}) \,.$$

By permutation, the same situation arises if $s = 3$. If instead s is even, i.e., $s = 2$, then we have terms of the form

$$\mathrm{Cum}(U_{\tau_m} - U_{\tau_{m-1}}, U_{\tau_m} - U_{\tau_{m-1}}, U_{\tau_{m+1}} - U_{\tau_m}, U_{\tau_{m+1}} - U_{\tau_m}) = \mathrm{Cum}_4(U_{\tau_m}) \,.$$

Finally, if at least one pair of indices in the quadruple (i,j,k,l) is more than one integer apart, then

$$\mathrm{Cum}(U_{\tau_i} - U_{\tau_{i-1}}, U_{\tau_j} - U_{\tau_{j-1}}, U_{\tau_k} - U_{\tau_{k-1}}, U_{\tau_l} - U_{\tau_{l-1}}) = 0$$

by independence of the $U's$.

Appendix B –Proof of Theorem 1

Given The estimator (1.5) has the following expected value

$$E\left[\hat{\sigma}^2\right] = \frac{1}{T}\sum_{i=1}^{N} E\left[Y_i^2\right] = \frac{N\left(\sigma^2\Delta + 2a^2\right)}{T} = \sigma^2 + \frac{2a^2}{\Delta}.$$

The estimator's variance is

$$\operatorname{Var}\left[\hat{\sigma}^2\right] = \frac{1}{T^2}\operatorname{Var}\left[\sum_{i=1}^{N} Y_i^2\right] = \frac{1}{T^2}\sum_{i,j=1}^{N}\operatorname{Cov}\left(Y_i^2, Y_j^2\right).$$

Applying Lemma 1 in the special case where the first two indices and the last two respectively are identical yields

$$\operatorname{Cum}(Y_i, Y_i, Y_j, Y_j) = \begin{cases} 2\ \operatorname{Cum}_4[U] & \text{if } j = i\,, \\ \operatorname{Cum}_4[U] & \text{if } j = i+1 \text{ or } j = i-1\,, \\ 0 \text{ otherwise}\,. \end{cases} \tag{B.1}$$

In the middle case, i.e., whenever $j = i+1$ or $j = i-1$, the number s of indices that are equal to the minimum index is always 2. Combining (B.1) with (1.14), we have

$$\begin{aligned}
\operatorname{Var}\left[\hat{\sigma}^2\right] &= \frac{1}{T^2}\sum_{i=1}^{N}\operatorname{Cov}\left(Y_i^2, Y_i^2\right) + \frac{1}{T^2}\sum_{i=1}^{N-1}\operatorname{Cov}\left(Y_i^2, Y_{i+1}^2\right) + \frac{1}{T^2}\sum_{i=2}^{N}\operatorname{Cov}\left(Y_i^2, Y_{i-1}^2\right) \\
&= \frac{1}{T^2}\sum_{i=1}^{N}\left\{2\operatorname{Cov}\left(Y_i, Y_i\right)^2 + 2\ \operatorname{Cum}_4[U]\right\} + \\
&+ \frac{1}{T^2}\sum_{i=1}^{N-1}\left\{2\operatorname{Cov}\left(Y_i, Y_{i+1}\right)^2 + \operatorname{Cum}_4[U]\right\} + \\
&+ \frac{1}{T^2}\sum_{i=2}^{N}\left\{2\operatorname{Cov}\left(Y_i, Y_{i-1}\right)^2 + \operatorname{Cum}_4[U]\right\} \\
&= \frac{2N}{T^2}\left\{\operatorname{Var}[Y_i]^2 + \operatorname{Cum}_4[U]\right\} + \frac{2(N-1)}{T^2}\left\{2\operatorname{Cov}\left(Y_i, Y_{i-1}\right)^2 + \operatorname{Cum}_4[U]\right\},
\end{aligned}$$

with $\operatorname{Var}[Y_i]$ and $\operatorname{Cov}(Y_i, Y_{i-1}) = \operatorname{Cov}(Y_i, Y_{i+1})$ given in (1.9)-(1.10), so that

$$\begin{aligned}
\operatorname{Var}\left[\hat{\sigma}^2\right] &= \frac{2N}{T^2}\left\{\left(\sigma^2\Delta + 2a^2\right)^2 + \operatorname{Cum}_4[U]\right\} + \frac{2(N-1)}{T^2}\left\{2a^4 + \operatorname{Cum}_4[U]\right\}, \\
&= \frac{2\left(\sigma^4\Delta^2 + 4\sigma^2\Delta a^2 + 6a^4 + 2\operatorname{Cum}_4[U]\right)}{T\Delta} - \frac{2\left(2a^4 + \operatorname{Cum}_4[U]\right)}{T^2},
\end{aligned}$$

since $N = T/\Delta$. The expression for the RMSE follows from those for the expected value and variance given in (1.17) and (1.19):

$$\begin{aligned}
RMSE\left[\hat{\sigma}^2\right] = \Bigg(&\frac{4a^4}{\Delta^2} + \frac{2\left(\sigma^4\Delta^2 + 4\sigma^2\Delta a^2 + 6a^4 + 2\operatorname{Cum}_4[U]\right)}{T\Delta} - \\
&- \frac{2\left(2a^4 + \operatorname{Cum}_4[U]\right)}{T^2}\Bigg)^{1/2}.
\end{aligned} \tag{B.2}$$

The optimal value Δ^* of the sampling interval given in (1.20) is obtained by minimizing $RMSE\left[\hat{\sigma}^2\right]$ over Δ. The first order condition that arises from setting $\partial \, \mathrm{RMSE}\left[\hat{\sigma}^2\right]/\partial\Delta$ to 0 is the cubic equation in Δ :

$$\Delta^3 - \frac{2\left(3a^4 + \mathrm{Cum}_4\left[U\right]\right)}{\sigma^4}\Delta - \frac{4a^4T}{\sigma^4} = 0 \,. \tag{B.3}$$

We now show that (B.3) has a unique positive root, and that it corresponds to a minimum of $\mathrm{RMSE}\left[\hat{\sigma}^2\right]$. We are therefore looking for a real positive root in $\Delta = z$ to the cubic equation

$$z^3 + pz - q = 0 \,, \tag{B.4}$$

where $q > 0$ and $p < 0$ since from (1.16):

$$3a^4 + \mathrm{Cum}_4\left[U\right] = 3a^4 + E\left[U^4\right] - 3E\left[U^2\right]^2 = E\left[U^4\right] > 0 \,.$$

Using Vièta's change of variable from z to w given by $z = w - p/(3w)$ reduces, after multiplication by w^3, the cubic to the quadratic equation

$$y^2 - qy - \frac{p^3}{27} = 0 \tag{B.5}$$

in the variable $y \equiv w^3$.

Define the discriminant

$$D = \left(\frac{p}{3}\right)^3 + \left(\frac{q}{2}\right)^2 .$$

The two roots of (B.5)

$$y_1 = \frac{q}{2} + D^{1/2} \quad \text{and} \quad y_2 = \frac{q}{2} - D^{1/2}$$

are real if $D \geq 0$ (and distinct if $D > 0$) and complex conjugates if $D < 0$. Then the three roots of (B.4) are

$$\begin{aligned}
z_1 &= y_1^{1/3} + y_2^{1/3} \,, \\
z_2 &= -\frac{1}{2}\left(y_1^{1/3} + y_2^{1/3}\right) + \mathrm{i}\frac{3^{1/2}}{2}\left(y_1^{1/3} - y_2^{1/3}\right), \\
z_3 &= -\frac{1}{2}\left(y_1^{1/3} + y_2^{1/3}\right) - \mathrm{i}\frac{3^{1/2}}{2}\left(y_1^{1/3} - y_2^{1/3}\right),
\end{aligned} \tag{B.6}$$

(see e.g., Section 3.8.2 in [1]). If $D > 0$, the two roots in y are both real and positive because $p < 0$ and $q > 0$ imply

$$y_1 > y_2 > 0$$

and hence of the three roots given in (B.6), z_1 is real and positive and z_2 and z_3 are complex conjugates. If $D = 0$, then $y_1 = y_2 = q/2 > 0$ and the three roots are real (two of which are identical) and given by

$$\begin{aligned}
z_1 &= y_1^{1/3} + y_2^{1/3} = 2^{2/3}q^{1/3} \,, \\
z_2 = z_3 &= -\frac{1}{2}\left(y_1^{1/3} + y_2^{1/3}\right) = -\frac{1}{2}z_1 \,.
\end{aligned}$$

Of these, $z_1 > 0$ and $z_2 = z_3 < 0$. If $D < 0$, the three roots are distinct and real because

$$y_1 = \frac{q}{2} + \mathrm{i}(-D)^{1/2} \equiv re^{\mathrm{i}\theta}, \quad y_2 = \frac{q}{2} - \mathrm{i}(-D)^{1/2} \equiv re^{-\mathrm{i}\theta},$$

so

$$y_1^{1/3} = r^{1/3}e^{\mathrm{i}\theta/3}, \quad y_2^{1/3} = r^{1/3}e^{-\mathrm{i}\theta/3},$$

and therefore

$$y_1^{1/3} + y_2^{1/3} = 2r^{1/3}\cos(\theta/3), \quad y_1^{1/3} - y_2^{1/3} = 2\mathrm{i}r^{1/3}\sin(\theta/3),$$

so that

$$\begin{aligned} z_1 &= 2r^{1/3}\cos(\theta/3) \\ z_2 &= -r^{1/3}\cos(\theta/3) + 3^{1/2}r^{1/3}\sin(\theta/3) \\ z_3 &= -r^{1/3}\cos(\theta/3) - 3^{1/2}r^{1/3}\sin(\theta/3) . \end{aligned}$$

Only z_1 is positive because $q > 0$ and $(-D)^{1/2} > 0$ imply that $0 < \theta < \pi/2$. Therefore $\cos(\theta/3) > 0$, so $z_1 > 0$; $\sin(\theta/3) > 0$, so $z_3 < 0$; and

$$\cos(\theta/3) > \cos(\pi/6) = \frac{3^{1/2}}{2} = 3^{1/2}\sin(\pi/6) > 3^{1/2}\sin(\theta/3),$$

so $z_2 < 0$.

Thus the equation (B.4) has exactly one root that is positive, and it is given by z_1 in (B.6). Since $\mathrm{RMSE}\left[\hat{\sigma}^2\right]$ is of the form

$$\begin{aligned} &\mathrm{RMSE}\left[\hat{\sigma}^2\right] \\ &= \left(\frac{2T\Delta^3\sigma^4 - 2\Delta^2\left(2a^4 - 4a^2T\sigma^2 + \mathrm{Cum}_4[U]\right) + 2\Delta\left(6a^4T + 2T\,\mathrm{Cum}_4[U]\right) + 4a^4T^2}{T^2\Delta^2}\right)^{1/2} \\ &= \left(\frac{a_3\Delta^3 + a_2\Delta^2 + a_1\Delta + a_0}{T^2\Delta^2}\right)^{1/2}, \end{aligned}$$

with $a_3 > 0$, it tends to $+\infty$ when Δ tends to $+\infty$. Therefore that single positive root corresponds to a minimum of $\mathrm{RMSE}\left[\hat{\sigma}^2\right]$ which is reached at

$$\begin{aligned} \Delta^* &= y_1^{1/3} + y_2^{1/3} \\ &= \left(\frac{q}{2} + D^{1/2}\right)^{1/3} + \left(\frac{q}{2} - D^{1/2}\right)^{1/3}. \end{aligned}$$

Replacing q and p by their values in the expression above yields (1.20). As shown above, if the expression inside the square root in formula (1.20) is negative, the resulting Δ^* is still a positive real number.

Appendix C –Proof of Proposition 1

The result follows from an application of the delta method to the known properties of the MLE estimator of an MA(1) process (Section 5.4 in [25]), as follows. Because we re-use these calculations below in the proof of Theorem 2 (whose result cannot be inferred from known MA(1) properties), we recall some of the expressions of the score vector of the MA(1) likelihood. The partial derivatives of the log-likelihood function (1.25) have the form

$$\dot{l}_{\eta} = -\frac{1}{2}\frac{\partial \ln \det(V)}{\partial \eta} - \frac{1}{2\gamma^2} Y' \frac{\partial V^{-1}}{\partial \eta} Y \,, \tag{C.1}$$

and

$$\dot{l}_{\gamma^2} = -\frac{N}{2\gamma^2} + \frac{1}{2\gamma^4} Y' V^{-1} Y \,. \tag{C.2}$$

so that the MLE for γ^2 is

$$\hat{\gamma}^2 = \frac{1}{N} Y' V^{-1} Y \,. \tag{C.3}$$

At the true parameters, the expected value of the score vector is zero: $E\left[\dot{l}_{\eta}\right] = E\left[\dot{l}_{\gamma^2}\right] = 0$. Hence it follows from (C.1) that

$$E\left[Y' \frac{\partial V^{-1}}{\partial \eta} Y\right] = -\gamma^2 \frac{\partial \ln \det(V)}{\partial \eta} = -\gamma^2 \frac{2\eta\left(1 - (1+N)\,\eta^{2N} + N\eta^{2(1+N)}\right)}{(1-\eta^2)\left(1-\eta^{2(1+N)}\right)} \,,$$

thus as $N \to \infty$

$$E\left[Y' \frac{\partial V^{-1}}{\partial \eta} Y\right] = -\frac{2\eta\gamma^2}{(1-\eta^2)} + o(1) \,.$$

Similarly, it follows from (C.2) that

$$E\left[Y' V^{-1} Y\right] = N\gamma^2.$$

Turning now to Fisher's information, we have

$$E\left[-\ddot{l}_{\gamma^2\gamma^2}\right] = -\frac{N}{2\gamma^4} + \frac{1}{\gamma^6} E\left[Y' V^{-1} Y\right] = \frac{N}{2\gamma^4} \,, \tag{C.4}$$

whence the asymptotic variance of $T^{1/2}(\hat{\gamma}^2 - \gamma^2)$ is $2\gamma^4 \Delta$. We also have that

$$E\left[-\ddot{l}_{\gamma^2\eta}\right] = \frac{1}{2\gamma^4} E\left[Y' \frac{\partial V^{-1}}{\partial \eta} Y\right] = -\frac{\eta}{\gamma^2(1-\eta^2)} + o(1) \,, \tag{C.5}$$

whence the asymptotic covariance of $T^{1/2}(\hat{\gamma}^2 - \gamma^2)$ and $T^{1/2}(\hat{\eta} - \eta)$ is zero.

To evaluate $E\left[-\ddot{l}_{\eta\eta}\right]$, we compute

$$E\left[-\ddot{l}_{\eta\eta}\right] = \frac{1}{2}\frac{\partial^2 \ln \det(V)}{\partial \eta^2} + \frac{1}{2\gamma^2} E\left[Y' \frac{\partial^2 V^{-1}}{\partial \eta^2} Y\right] \tag{C.6}$$

and evaluate both terms. For the first term in (C.6), we have from (1.27):

$$\begin{aligned} \frac{\partial^2 \ln \det(V)}{\partial \eta^2} &= \frac{1}{(1-\eta^{2+2N})^2} \left\{ \frac{2\left(1+\eta^2+\eta^{2+2N}\left(1-3\eta^2\right)\right)\left(1-\eta^{2N}\right)}{(1-\eta^2)^2} - \right. \\ &\quad \left. - 2N\eta^{2N}\left(3+\eta^{2+2N}\right) - 4N^2\eta^{2N} \right\} \qquad \text{(C.7)} \\ &= \frac{2\left(1+\eta^2\right)}{\left(1-\eta^2\right)^2} + o(1) \,. \end{aligned}$$

For the second term, we have for any non-random $N \times N$ matrix Q:

$$\begin{aligned} E\left[Y'QY\right] &= E\left[Tr\left[Y'QY\right]\right] = E\left[Tr\left[QYY'\right]\right] = Tr\left[E\left[QYY'\right]\right] \\ &= Tr\left[QE\left[YY'\right]\right] = Tr\left[Q\gamma^2 V\right] = \gamma^2 Tr\left[QV\right], \end{aligned}$$

where Tr denotes the matrix trace, which satisfies $Tr[AB] = Tr[BA]$. Therefore

$$\begin{aligned} E\left[Y'\frac{\partial^2 V^{-1}}{\partial \eta^2}Y\right] &= \gamma^2 Tr\left[\frac{\partial^2 V^{-1}}{\partial \eta^2}V\right] = \gamma^2\left(\sum_{i=1}^{N}\sum_{j=1}^{N}\frac{\partial^2 v^{ij}}{\partial \eta^2}v_{ij}\right) \\ &= \gamma^2\left(\sum_{i=1}^{N}\frac{\partial^2 v^{ii}}{\partial \eta^2}\left(1+\eta^2\right) + \sum_{i=1}^{N-1}\frac{\partial^2 v^{i,i+1}}{\partial \eta^2}\eta + \sum_{i=2}^{N}\frac{\partial^2 v^{i,i-1}}{\partial \eta^2}\eta\right) \\ &= \frac{\gamma^2}{\left(1-\eta^{2+2N}\right)^2}\left\{-\frac{4\left(1+2\eta^2+\eta^{2+2N}\left(1-4\eta^2\right)\right)\left(1-\eta^{2N}\right)}{\left(1-\eta^2\right)^2} + \right. \\ &\quad \left. + \frac{2N\left(1+\eta^{2N}\left(6-6\eta^2+2\eta^{2+2N}-3\eta^{4+2N}\right)\right)}{\left(1-\eta^2\right)} + 8N^2\eta^{2N}\right\} \\ &= \frac{2\gamma^2 N}{\left(1-\eta^2\right)} + o(N). \end{aligned} \tag{C.8}$$

Combining (C.7) and (C.8) into (C.6), it follows that

$$E\left[-\ddot{l}_{\eta\eta}\right] = \frac{1}{2}\frac{\partial^2 \ln\det(V_N)}{\partial \eta^2} + \frac{1}{2\gamma^2}E\left[Y'\frac{\partial^2 V_N^{-1}}{\partial \eta^2}Y\right] \underset{N\longrightarrow\infty}{\sim} \frac{N}{\left(1-\eta^2\right)} + o(N). \tag{C.9}$$

In light of that and (C.5), the asymptotic variance of $T^{1/2}(\hat{\eta}-\eta)$ is the same as in the γ^2 known case, that is, $(1-\eta^2)\Delta$ (which of course confirms the result of [15] for this parameter).

We can now retrieve the asymptotic covariance matrix for the original parameters (σ^2, a^2) from that of the parameters (γ^2, η). This follows from the delta method applied to the change of variable (1.9)-(1.10):

$$\begin{pmatrix}\sigma^2 \\ a^2\end{pmatrix} = f(\gamma^2, \eta) = \begin{pmatrix}\Delta^{-1}\gamma^2(1+\eta)^2 \\ -\gamma^2\eta\end{pmatrix}. \tag{C.10}$$

Hence

$$T^{1/2}\left(\begin{pmatrix}\hat{\sigma}^2 \\ \hat{a}^2\end{pmatrix} - \begin{pmatrix}\sigma^2 \\ a^2\end{pmatrix}\right) \underset{T\longrightarrow\infty}{\longrightarrow} N\left(0, \text{AVAR}(\hat{\sigma}^2, \hat{a}^2)\right),$$

where

$$\begin{aligned} \text{AVAR}(\hat{\sigma}^2, \hat{a}^2) &= \nabla f(\gamma^2, \eta).\,\text{AVAR}(\hat{\gamma}^2, \hat{\eta}).\nabla f(\gamma^2, \eta)' \\ &= \begin{pmatrix}\frac{(1+\eta)^2}{\Delta} & \frac{2\gamma^2(1+\eta)}{\Delta} \\ -\eta & -\gamma^2\end{pmatrix}\begin{pmatrix}2\gamma^4\Delta & 0 \\ 0 & (1-\eta^2)\Delta\end{pmatrix}\begin{pmatrix}\frac{(1+\eta)^2}{\Delta} & -\eta \\ \frac{2\gamma^2(1+\eta)}{\Delta} & -\gamma^2\end{pmatrix} \\ &= \begin{pmatrix}4\sqrt{\sigma^6\Delta\left(4a^2+\sigma^2\Delta\right)}+2\sigma^4\Delta & -\sigma^2\Delta h(\Delta, \sigma^2, a^2) \\ \bullet & \frac{\Delta}{2}\left(2a^2+\sigma^2\Delta\right)h(\Delta, \sigma^2, a^2)\end{pmatrix}. \end{aligned}$$

Appendix D – Proof of Theorem 2

We have that

$$\begin{aligned}
E_{\text{true}}\left[\dot{l}_\eta \dot{l}_{\gamma^2}\right] &= \text{Cov}_{\text{true}}(\dot{l}_\eta, \dot{l}_{\gamma^2}) \\
&= \text{Cov}_{\text{true}}\left(-\frac{1}{2\gamma^2}\sum_{i,j=1}^{N} Y_i Y_j \frac{\partial v^{ij}}{\partial \eta}, \frac{1}{2\gamma^4}\sum_{k,l=1}^{N} Y_k Y_l v^{kl}\right) \\
&= -\frac{1}{4\gamma^6}\sum_{i,j,k,l=1}^{N} \frac{\partial v^{ij}}{\partial \eta} v^{kl}\, \text{Cov}_{\text{true}}(Y_i Y_j, Y_k Y_l) \qquad \text{(D.1)} \\
&= -\frac{1}{4\gamma^6}\sum_{i,j,k,l=1}^{N} \frac{\partial v^{ij}}{\partial \eta} v^{kl} [\text{Cum}_{\text{true}}(Y_i, Y_j, Y_k, Y_l) + \\
&\quad +2\,\text{Cov}_{\text{true}}(Y_i, Y_j)\,\text{Cov}_{\text{true}}(Y_k, Y_l)] .
\end{aligned}$$

where "true" denotes the true distribution of the $Y's$, not the incorrectly specified normal distribution, and Cum denotes the cumulants given in Lemma 1. The last transition is because

$$\begin{aligned}
\text{Cov}_{\text{true}}(Y_i Y_j, Y_k Y_l) &= E_{\text{true}}[Y_i Y_j Y_k Y_l] - E_{\text{true}}[Y_i Y_j]\, E_{\text{true}}[Y_k Y_l] \\
&= \kappa^{ijkl} - \kappa^{ij}\kappa^{kl} \\
&= \kappa^{i,j,k,l} + \kappa^{i,j}\kappa^{k,l}[3] - \kappa^{i,j}\kappa^{k,l} \\
&= \kappa^{i,j,k,l} + \kappa^{i,k}\kappa^{j,l} + \kappa^{i,l}\kappa^{j,k} \\
&= \text{Cum}_{\text{true}}(Y_i, Y_j, Y_k, Y_l) + \text{Cov}_{\text{true}}(Y_i, Y_k)\,\text{Cov}_{\text{true}}(Y_j, Y_k) \\
&\quad + \text{Cov}_{\text{true}}(Y_i, Y_l)\,\text{Cov}_{\text{true}}(Y_j, Y_k) ,
\end{aligned}$$

since Y has mean zero (see e.g., Section 2.3 of [36]). The need for permutation goes away due to the summing over all indices (i, j, k, l), and since $V^{-1} = [v^{ij}]$ is symmetric.

When looking at (D.1), note that $\text{Cum}_{\text{normal}}(Y_i, Y_j, Y_k, Y_l) = 0$, where "normal" denotes a Normal distribution with the same first and second order moments as the true distribution. That is, if the $Y's$ were normal we would have

$$E_{\text{normal}}\left[\dot{l}_\eta \dot{l}_{\gamma^2}\right] = -\frac{1}{4\gamma^6}\sum_{i,j,k,l=1}^{N} \frac{\partial v^{ij}}{\partial \eta} v^{kl} [2\,\text{Cov}_{\text{normal}}(Y_i, Y_j)\,\text{Cov}_{\text{normal}}(Y_k, Y_l)] .$$

Also, since the covariance structure does not depend on Gaussianity, $\text{Cov}_{\text{true}}(Y_i, Y_j) = \text{Cov}_{\text{normal}}(Y_i, Y_j)$. Next, we have

$$E_{\text{normal}}\left[\dot{l}_\eta \dot{l}_{\gamma^2}\right] = -E_{\text{normal}}\left[\ddot{l}_{\eta\gamma^2}\right] = -E_{\text{true}}\left[\ddot{l}_{\eta\gamma^2}\right], \qquad \text{(D.2)}$$

with the last equality following from the fact that $\ddot{l}_{\eta\gamma^2}$ depends only on the second moments of the $Y's$. (Note that in general $E_{\text{true}}\left[\dot{l}_\eta \dot{l}_{\gamma^2}\right] \neq -E_{\text{true}}\left[\ddot{l}_{\eta\gamma^2}\right]$ because the likelihood may be misspecified.) Thus, it follows from (D.1) that

$$
\begin{aligned}
E_{\text{true}}\left[\dot{l}_{\eta}\dot{l}_{\gamma^2}\right] &= E_{\text{normal}}\left[\dot{l}_{\eta}\dot{l}_{\gamma^2}\right] - \frac{1}{4\gamma^6}\sum_{i,j,k,l=1}^{N}\frac{\partial v^{ij}}{\partial\eta}v^{kl}\,\text{Cum}_{\text{true}}(Y_i,Y_j,Y_k,Y_l) \\
&= -E_{\text{true}}\left[\ddot{l}_{\eta\gamma^2}\right] - \frac{1}{4\gamma^6}\sum_{i,j,k,l=1}^{N}\frac{\partial v^{ij}}{\partial\eta}v^{kl}\,\text{Cum}_{\text{true}}(Y_i,Y_j,Y_k,Y_l)\,.
\end{aligned}
\tag{D.3}
$$

It follows similarly that

$$
\begin{aligned}
E_{\text{true}}\left[\left(\dot{l}_{\eta}\right)^2\right] &= \text{Var}_{\text{true}}(\dot{l}_{\eta}) \\
&= -E_{\text{true}}\left[\ddot{l}_{\eta\eta}\right] + \frac{1}{4\gamma^4}\sum_{i,j,k,l=1}^{N}\frac{\partial v^{ij}}{\partial\eta}\frac{\partial v^{kl}}{\partial\eta}\,\text{Cum}_{\text{true}}(Y_i,Y_j,Y_k,Y_l)\,,
\end{aligned}
\tag{D.4}
$$

and

$$
\begin{aligned}
E_{\text{true}}\left[\left(\dot{l}_{\gamma^2}\right)^2\right] &= \text{Var}_{\text{true}}(\dot{l}_{\gamma^2}) \\
&= -E_{\text{true}}\left[\ddot{l}_{\gamma^2\gamma^2}\right] + \frac{1}{4\gamma^8}\sum_{i,j,k,l=1}^{N}v^{ij}v^{kl}\,\text{Cum}_{\text{true}}(Y_i,Y_j,Y_k,Y_l)\,.
\end{aligned}
\tag{D.5}
$$

We now need to evaluate the sums that appear on the right hand sides of (D.3), (D.4) and (D.5). Consider two generic symmetric $N \times N$ matrices $\left[\nu^{i,j}\right]$ and $\left[\omega^{i,j}\right]$. We are interested in expressions of the form

$$
\begin{aligned}
\sum_{i,j,k,l:M=m+1}(-1)^s\nu^{i,j}\omega^{k,l} &= \sum_{h=1}^{N-1}\sum_{i,j,k,l:m=h,M=h+1}(-1)^s\nu^{i,j}\omega^{k,l} \\
&= \sum_{h=1}^{N-1}\sum_{r=1}^{3}\sum_{i,j,k,l:m=h,M=h+1,s=r}(-1)^r\nu^{i,j}\omega^{k,l} \\
&= \sum_{h=1}^{N-1}\Big\{-2\nu^{h,h+1}\omega^{h+1,h+1} - 2\nu^{h+1,h+1}\omega^{h,h+1} + \\
&\quad + \nu^{h,h}\omega^{h+1,h+1} + \nu^{h+1,h+1}\omega^{h,h} + 4\nu^{h,h+1}\omega^{h,h+1} - \\
&\quad -2\nu^{h+1,h}\omega^{h,h} - 2\nu^{h,h}\omega^{h+1,h}\Big\}\,.
\end{aligned}
\tag{D.6}
$$

It follows that if we set

$$
\Upsilon(\nu,\omega) = \sum_{i,j,k,l=1}^{N}\nu^{i,j}\omega^{k,l}\,\text{Cum}_{\text{true}}(Y_i,Y_j,Y_k,Y_l)\,, \tag{D.7}
$$

then $\Upsilon(\nu,\omega) = \text{Cum}_4\left[U\right]\,\Psi(\nu,\omega)$ where

$$
\begin{aligned}
\psi(\nu,\omega) = 2\sum_{h=1}^{N}\nu^{h,h}\omega^{h,h} &+ \sum_{h=1}^{N-1}\Big\{-2\nu^{h,h+1}\omega^{h+1,h+1} - 2\nu^{h+1,h+1}\omega^{h,h+1} + \\
&+ \nu^{h,h}\omega^{h+1,h+1} + \nu^{h+1,h+1}\omega^{h,h} + 4\nu^{h,h+1}\omega^{h,h+1} - \\
&-2\nu^{h+1,h}\omega^{h,h} - 2\nu^{h,h}\omega^{h+1,h}\Big\}\,.
\end{aligned}
\tag{D.8}
$$

If the two matrices $\left[\nu^{i,j}\right]$ and $\left[\omega^{i,j}\right]$ satisfy the following reversibility property: $\nu^{N+1-i,N+1-j} = \nu^{i,j}$ and $\omega^{N+1-i,N+1-j} = \omega^{i,j}$ (so long as one is within the index set), then (D.8) simplifies to:

$$\psi(\nu,\omega) = 2\sum_{h=1}^{N} \nu^{h,h}\omega^{h,h} + \sum_{h=1}^{N-1} \left\{-4\nu^{h,h+1}\omega^{h+1,h+1} - 4\nu^{h+1,h+1}\omega^{h,h+1} + \right.$$
$$\left. +2\nu^{h,h}\omega^{h+1,h+1} + 4\nu^{h,h+1}\omega^{h,h+1}\right\}.$$

This is the case for V^{-1} and its derivative $\partial V^{-1}/\partial\eta$, as can be seen from the expression for $v^{i,j}$ given in (1.28), and consequently for $\partial v^{i,j}/\partial\eta$.

If we wish to compute the sums in equations (D.3), (D.4), and (D.5), therefore, we need, respectively, to find the three quantities $\psi(\partial v/\partial\eta, v)$, $\psi(\partial v/\partial\eta, \partial v/\partial\eta)$, and $\psi(v,v)$ respectively. All are of order $O(N)$, and only the first term is needed. Replacing the terms $v^{i,j}$ and $\partial v^{i,j}/\partial\eta$ by their expressions from (1.28), we obtain:

$$\begin{aligned}\psi(v,v) &= \frac{2}{(1+\eta^2)(1-\eta)^3(1-\eta^{2(1+N)})^2} \\ &\quad \left\{-(1+\eta)\left(1-\eta^{2N}\right)\left(1+2\eta^2+2\eta^{2(1+N)}+\eta^{2(2+N)}\right) + \right. \\ &\quad \left. + N(1-\eta)\left(1+\eta^2\right)\left(2+\eta^{2N}+\eta^{2(1+N)}+6\eta^{1+2N}+2\eta^{2+4N}\right)\right\} \\ &= \frac{4N}{(1-\eta)^2} + o(N),\end{aligned} \tag{D.9}$$

$$\begin{aligned}\psi\left(\frac{\partial v}{\partial\eta}, v\right) &= \frac{2\left(O(1) + 2N(1-\eta)\left(1+\eta^2\right)\eta\left(1+\eta^2+O(\eta^{2N})\right) + N^2 O(\eta^{2N})\right)}{\eta(1-\eta)^4(1+\eta^2)^2(1-\eta^{2(1+N)})^3} \\ &= \frac{4N}{(1-\eta)^3} + o(N),\end{aligned} \tag{D.10}$$

$$\begin{aligned}&\psi\left(\frac{\partial v}{\partial\eta}, \frac{\partial v}{\partial\eta}\right) \\ &= \frac{4\left(O(1) + 3N\left(1-\eta^4\right)\eta^2\left(\left(1+\eta^2\right)^2 + O(\eta^{2N})\right) + N^2 O(\eta^{2N}) + N^3 O(\eta^{2N})\right)}{3\eta^2(1+\eta)(1+\eta^2)^3(1-\eta)^5(1-\eta^{2(1+N)})^4} \\ &= \frac{4N}{(1-\eta)^4} + o(N).\end{aligned} \tag{D.11}$$

The asymptotic variance of the estimator $(\hat{\gamma}^2, \hat{\eta})$ obtained by maximizing the (incorrectly-specified) log-likelihood (1.25) that assumes Gaussianity of the $U's$ is given by

$$\mathrm{AVAR}_{\mathrm{true}}(\hat{\gamma}^2, \hat{\eta}) = \Delta\left(D'S^{-1}D\right)^{-1},$$

where, from (C.4), (C.5) and (C.9) we have

$$\begin{aligned}D = D' &= -\frac{1}{N}E_{\mathrm{true}}\left[\ddot{l}\right] = -\frac{1}{N}E_{\mathrm{normal}}\left[\ddot{l}\right] = \frac{1}{N}E_{\mathrm{normal}}\left[\dot{l}\dot{l}'\right] \\ &= \begin{pmatrix} \frac{1}{2\gamma^4} & -\frac{\eta}{N\gamma^2(1-\eta^2)} + o\left(\frac{1}{N}\right) \\ \bullet & \frac{1}{(1-\eta^2)} + o(1)\end{pmatrix},\end{aligned} \tag{D.12}$$

and, in light of (D.3), (D.4), and (D.5),

$$S = \frac{1}{N} E_{\text{true}} \left[\dot{l} \dot{l}' \right] = -\frac{1}{N} E_{\text{true}} \left[\ddot{l} \right] + \text{Cum}_4 \left[U \right] \ \Psi = D + \text{Cum}_4 \left[U \right] \ \Psi \,, \tag{D.13}$$

where

$$\begin{aligned} \Psi &= \frac{1}{4N} \begin{pmatrix} \frac{1}{\gamma^8} \psi \left(v, v \right) & -\frac{1}{\gamma^6} \psi \left(\frac{\partial v}{\partial \eta}, v \right) \\ \bullet & \frac{1}{\gamma^4} \psi \left(\frac{\partial v}{\partial \eta}, \frac{\partial v}{\partial \eta} \right) \end{pmatrix} \\ &= \begin{pmatrix} \frac{1}{\gamma^8 (1-\eta)^2} + o(1) & \frac{-1}{\gamma^6 (1-\eta)^3} + o(1) \\ \bullet & \frac{1}{\gamma^4 (1-\eta)^4} + o(1) \end{pmatrix} , \end{aligned} \tag{D.14}$$

from the expressions just computed. It follows that

$$\begin{aligned} \text{AVAR}_{\text{true}}(\hat{\gamma}^2, \hat{\eta}) &= \Delta \ \left(D \left(D + \text{Cum}_4 \left[U \right] \ \Psi \right)^{-1} D \right)^{-1} \\ &= \Delta \ \left(D \left(Id + \text{Cum}_4 \left[U \right] \ D^{-1} \Psi \right)^{-1} \right)^{-1} \\ &= \Delta \ \left(Id + \text{Cum}_4 \left[U \right] \ D^{-1} \Psi \right) D^{-1} \\ &= \Delta \ \left(Id + \text{Cum}_4 \left[U \right] \ D^{-1} \Psi \right) D^{-1} \\ &= \text{AVAR}_{\text{normal}}(\hat{\gamma}^2, \hat{\eta}) + \Delta \ \text{Cum}_4 \left[U \right] \ D^{-1} \Psi D^{-1} , \end{aligned}$$

where Id denotes the identity matrix and

$$\text{AVAR}_{\text{normal}}(\hat{\gamma}^2, \hat{\eta}) = \begin{pmatrix} 2\gamma^4 \Delta & 0 \\ 0 & (1-\eta^2)\Delta \end{pmatrix} , \quad D^{-1} \Psi D^{-1} = \begin{pmatrix} \frac{4}{(1-\eta)^2} & \frac{-2(1+\eta)}{\gamma^2 (1-\eta)^2} \\ \bullet & \frac{(1+\eta)^2}{\gamma^4 (1-\eta)^2} \end{pmatrix} ,$$

so that

$$\text{AVAR}_{\text{true}}(\hat{\gamma}^2, \hat{\eta}) = \Delta \begin{pmatrix} 2\gamma^4 & 0 \\ 0 & (1-\eta^2) \end{pmatrix} + \Delta \, \text{Cum}_4 \left[U \right] \begin{pmatrix} \frac{4}{(1-\eta)^2} & \frac{-2(1+\eta)}{\gamma^2 (1-\eta)^2} \\ \bullet & \frac{(1+\eta)^2}{\gamma^4 (1-\eta)^2} \end{pmatrix} .$$

By applying the delta method to change the parametrization, we now recover the asymptotic variance of the estimates of the original parameters:

$$\begin{aligned} &\text{AVAR}_{\text{true}}(\hat{\sigma}^2, \hat{a}^2) \\ &= \nabla f(\gamma^2, \eta) . \, \text{AVAR}_{\text{true}}(\hat{\gamma}^2, \hat{\eta}) . \nabla f(\gamma^2, \eta)' \\ &= \begin{pmatrix} 4\sqrt{\sigma^6 \Delta \left(4a^2 + \sigma^2 \Delta \right)} + 2\sigma^4 \Delta & -\sigma^2 \Delta h(\Delta, \sigma^2, a^2) \\ \bullet & \frac{\Delta}{2} \left(2a^2 + \sigma^2 \Delta \right) h(\Delta, \sigma^2, a^2) + \Delta \, \text{Cum}_4 \left[U \right] \end{pmatrix} . \end{aligned}$$

Appendix E –Derivations for Section 1.7

To see (1.39), let "orig" (E.7) denote parametrization in (and differentiation with respect to) the original parameters σ^2 and a^2, while "transf" denotes parametrization and differentiation in γ^2 and η, and f_{inv} denotes the inverse of the change of variable function defined in (C.10), namely

$$\begin{pmatrix} \gamma^2 \\ \eta \end{pmatrix} = f_{\text{inv}}(\sigma^2, a^2) = \begin{pmatrix} \frac{1}{2}\left\{2a^2 + \sigma^2 \Delta + \sqrt{\sigma^2 \Delta \left(4a^2 + \sigma^2 \Delta\right)}\right\} \\ \frac{1}{2a^2}\left\{-2a^2 - \sigma^2 \Delta + \sqrt{\sigma^2 \Delta \left(4a^2 + \sigma^2 \Delta\right)}\right\} \end{pmatrix}. \tag{E.1}$$

and ∇f_{inv} its Jacobian matrix. Then, from $\dot{l}_{\text{orig}} = \nabla f_{\text{inv}}(\sigma^2, a^2)'.\dot{l}_{\text{transf}}$, we have

$$\ddot{l}_{\text{orig}} = \nabla f_{\text{inv}}(\sigma^2, a^2)'.\ddot{l}_{\text{transf}}.\nabla f_{\text{inv}}(\sigma^2, a^2) + H[\dot{l}_{\text{transf}}],$$

where $H[\dot{l}_{\text{transf}}]$ is a 2×2 matrix whose terms are linear in $\dot{l}_{\text{transf}}$ and the second partial derivatives of f_{inv}. Now $E_{\text{true}}[\dot{l}_{\text{orig}}] = E_{\text{true}}[\dot{l}_{\text{transf}}] = 0$, and so $E_{\text{true}}[H[\dot{l}_{\text{transf}}]] = 0$ from which it follows that

$$\begin{aligned} D_{\text{orig}} &= N^{-1} E_{\text{true}}[-\ddot{l}_{\text{orig}}] \\ &= \nabla f_{\text{inv}}(\sigma^2, a^2)'.D_{\text{transf}}.\nabla f_{\text{inv}}(\sigma^2, a^2) \\ &= \begin{pmatrix} \frac{\Delta^{1/2}(2a^2+\sigma^2\Delta)}{2\sigma^3(4a^2+\sigma^2\Delta)^{3/2}} & \frac{\Delta^{1/2}}{\sigma(4a^2+\sigma^2\Delta)^{3/2}} \\ \bullet & \frac{1}{2a^4}\left(1 - \frac{\Delta^{1/2}\sigma(6a^2+\sigma^2\Delta)}{(4a^2+\sigma^2\Delta)^{3/2}}\right) \end{pmatrix} + o(1), \end{aligned} \tag{E.2}$$

with

$$D_{\text{transf}} = N^{-1} E_{\text{true}}[-\ddot{l}_{\text{transf}}],$$

given in (D.12). Similarly,

$$\dot{l}_{\text{orig}}\dot{l}'_{\text{orig}} = \nabla f_{\text{inv}}(\sigma^2, a^2)'.\dot{l}_{\text{transf}}\dot{l}'_{\text{transf}}.\nabla f_{\text{inv}}(\sigma^2, a^2),$$

and so

$$\begin{aligned} S_{\text{orig}} &= \nabla f_{\text{inv}}(\sigma^2, a^2)'.S_{\text{transf}}.\nabla f_{\text{inv}}(\sigma^2, a^2) \\ &= \nabla f_{\text{inv}}(\sigma^2, a^2)'.(D_{\text{transf}} + \text{Cum}_4[U]\ \Psi).\nabla f_{\text{inv}}(\sigma^2, a^2) \\ &= D_{\text{orig}} + \text{Cum}_4[U]\, \nabla f_{\text{inv}}(\sigma^2, a^2)'.\Psi.\nabla f_{\text{inv}}(\sigma^2, a^2), \end{aligned} \tag{E.3}$$

with the second equality following from the expression for S_{transf} given in (D.13).

To complete the calculation, note from (D.14) that

$$\Psi = g_{\text{transf}}.g'_{\text{transf}} + o(1),$$

where

$$g_{\text{transf}} = \begin{pmatrix} \gamma^{-4}(1-\eta)^{-1} \\ -\gamma^{-2}(1-\eta)^{-2} \end{pmatrix}.$$

Thus

$$\nabla f_{\text{inv}}(\sigma^2, a^2)'.\Psi.\nabla f_{\text{inv}}(\sigma^2, a^2) = g_{\text{orig}}.g'_{\text{orig}} + o(1), \tag{E.4}$$

where

$$g = g_{\text{orig}} = \nabla f_{\text{inv}}(\sigma^2, a^2)'.g_{\text{transf}} = \begin{pmatrix} \frac{\Delta^{1/2}}{\sigma(4a^2+\sigma^2\Delta)^{3/2}} \\ \frac{1}{2a^4}\left(1 - \frac{\Delta^{1/2}\sigma(6a^2+\sigma^2\Delta)}{(4a^2+\sigma^2\Delta)^{3/2}}\right) \end{pmatrix}, \tag{E.5}$$

which is the result (1.40). Inserting (E.4) into (E.3) yields the result (1.39).

For the profile likelihood λ, let $\hat{a}^2_{\sigma^2}$ denote the maximizer of $l(\sigma^2, a^2)$ for given σ^2. Thus by definition $\lambda(\sigma^2) = l(\sigma^2, \hat{a}^2_{\sigma^2})$. From now on, all differentiation takes place with respect to the original parameters, and we will omit the subscript "orig" in what follows. Since $0 = \dot{l}_{a^2}(\sigma^2, \hat{a}^2_{\sigma^2})$, it follows that

$$\begin{aligned} 0 &= \frac{\partial}{\partial\sigma^2}\dot{l}_{a^2}(\sigma^2, \hat{a}^2_{\sigma^2}) \\ &= \ddot{l}_{\sigma^2 a^2}(\sigma^2, \hat{a}^2_{\sigma^2}) + \ddot{l}_{a^2 a^2}(\sigma^2, \hat{a}^2_{\sigma^2})\frac{\partial \hat{a}^2_{\sigma^2}}{\partial\sigma^2}\,, \end{aligned}$$

so that

$$\frac{\partial \hat{a}^2_{\sigma^2}}{\partial\sigma^2} = -\frac{\ddot{l}_{\sigma^2 a^2}(\sigma^2, \hat{a}^2_{\sigma^2})}{\ddot{l}_{a^2 a^2}(\sigma^2, \hat{a}^2_{\sigma^2})}\,. \tag{E.6}$$

The profile score then follows

$$\dot{\lambda}(\sigma^2) = \dot{l}_{\sigma^2}(\sigma^2, \hat{a}^2_{\sigma^2}) + \dot{l}_{a^2}(\sigma^2, \hat{a}^2_{\sigma^2})\frac{\partial \hat{a}^2_{\sigma^2}}{\partial\sigma^2}\,, \tag{E.7}$$

so that at the true value of (σ^2, a^2),

$$\dot{\lambda}(\sigma^2) = \dot{l}_{\sigma^2}(\sigma^2, a^2) - \frac{E_{\text{true}}[\ddot{l}_{\sigma^2 a^2}]}{E_{\text{true}}[\ddot{l}_{a^2 a^2}]}\dot{l}_{a^2}(\sigma^2, a^2) + O_p(1)\,, \tag{E.8}$$

since $\hat{a}^2 = a^2 + O_p(N^{-1/2})$ and

$$\begin{aligned} \Delta\ddot{l}_{\sigma^2 a^2} &\equiv N^{-1}\ddot{l}_{\sigma^2 a^2}(\sigma^2, \hat{a}^2_{\sigma^2}) - N^{-1}E_{\text{true}}[\ddot{l}_{\sigma^2 a^2}] = O_p(N^{-1/2})\,, \\ \Delta\ddot{l}_{a^2 a^2} &\equiv N^{-1}\ddot{l}_{a^2 a^2}(\sigma^2, \hat{a}^2_{\sigma^2}) - N^{-1}E_{\text{true}}[\ddot{l}_{a^2 a^2}] = O_p(N^{-1/2})\,, \end{aligned}$$

as sums of random variables with expected value zero, so that

$$\begin{aligned} -\frac{\partial \hat{a}^2_{\sigma^2}}{\partial\sigma^2} &= \frac{N^{-1}\ddot{l}_{\sigma^2 a^2}(\sigma^2, \hat{a}^2_{\sigma^2})}{N^{-1}\ddot{l}_{a^2 a^2}(\sigma^2, \hat{a}^2_{\sigma^2})} \\ &= \frac{N^{-1}E_{\text{true}}[\ddot{l}_{\sigma^2 a^2}] + \Delta\ddot{l}_{\sigma^2 a^2}}{N^{-1}E_{\text{true}}[\ddot{l}_{a^2 a^2}] + \Delta\ddot{l}_{a^2 a^2}} \\ &= \frac{E_{\text{true}}[\ddot{l}_{\sigma^2 a^2}]}{E_{\text{true}}[\ddot{l}_{a^2 a^2}]} + \left(\Delta\ddot{l}_{\sigma^2 a^2} - \Delta\ddot{l}_{a^2 a^2}\right) + o_p(N^{-1/2}) \\ &= \frac{E_{\text{true}}[\ddot{l}_{\sigma^2 a^2}]}{E_{\text{true}}[\ddot{l}_{a^2 a^2}]} + O_p(N^{-1/2})\,, \end{aligned}$$

while

$$\dot{l}_{a^2}(\sigma^2, a^2) = O_p(N^{1/2})\,,$$

also as a sum of random variables with expected value zero.

Therefore

$$\begin{aligned} E_{\text{true}}[\dot{\lambda}(\sigma^2)] &= E_{\text{true}}[\dot{l}_{\sigma^2}(\sigma^2, a^2)] - \frac{E_{\text{true}}[\ddot{l}_{\sigma^2 a^2}]}{E_{\text{true}}[\ddot{l}_{a^2 a^2}]}E_{\text{true}}\left[\dot{l}_{a^2}(\sigma^2, a^2)\right] + O(1) \\ &= O(1)\,, \end{aligned}$$

since $E_{\text{true}}[\dot{l}_{\sigma^2}(\sigma^2, a^2)] = E_{\text{true}}\left[\dot{l}_{a^2}(\sigma^2, a^2)\right] = 0$. In particular, $E_{\text{true}}[\dot{\lambda}(\sigma^2)] = o(N)$ as claimed.

Further differentiating (E.7), one obtains

$$\begin{aligned}\ddot{\lambda}(\sigma^2) &= \ddot{l}_{\sigma^2\sigma^2}(\sigma^2, \hat{a}^2_{\sigma^2}) + \ddot{l}_{a^2a^2}(\sigma^2, \hat{a}^2_{\sigma^2}) \left(\frac{\partial \hat{a}^2_{\sigma^2}}{\partial \sigma^2}\right)^2 + \\ &+ 2\ddot{l}_{\sigma^2a^2}(\sigma^2, \hat{a}^2_{\sigma^2}) \frac{\partial \hat{a}^2_{\sigma^2}}{\partial \sigma^2} + \dot{l}_{a^2}(\sigma^2, \hat{a}^2_{\sigma^2}) \frac{\partial^2 \hat{a}^2_{\sigma^2}}{\partial^2 \sigma^2} \\ &= \ddot{l}_{\sigma^2\sigma^2}(\sigma^2, \hat{a}^2_{\sigma^2}) - \frac{\ddot{l}_{\sigma^2a^2}(\sigma^2, \hat{a}^2_{\sigma^2})^2}{\ddot{l}_{a^2a^2}(\sigma^2, \hat{a}^2_{\sigma^2})} + \dot{l}_{a^2}(\sigma^2, \hat{a}^2_{\sigma^2}) \frac{\partial^2 \hat{a}^2_{\sigma^2}}{\partial^2 \sigma^2}\end{aligned}$$

from (E.6). Evaluated at $\sigma^2 = \hat{\sigma}^2$, one gets $\hat{a}^2_{\sigma^2} = \hat{a}^2$ and $\dot{l}_{a^2}(\hat{\sigma}^2, \hat{a}^2) = 0$, and so

$$\begin{aligned}\ddot{\lambda}(\hat{\sigma}^2) &= \ddot{l}_{\sigma^2\sigma^2}(\hat{\sigma}^2, \hat{a}^2) - \frac{\ddot{l}_{\sigma^2a^2}(\hat{\sigma}^2, \hat{a}^2)^2}{\ddot{l}_{a^2a^2}(\hat{\sigma}^2, \hat{a}^2)} \\ &= \frac{1}{\left[\ddot{l}(\hat{\sigma}^2, \hat{a}^2)^{-1}\right]_{\sigma^2\sigma^2}}, \end{aligned} \tag{E.9}$$

where $\left[\ddot{l}(\hat{\sigma}^2, \hat{a}^2)^{-1}\right]_{\sigma^2\sigma^2}$ is the upper left element of the matrix $\ddot{l}(\hat{\sigma}^2, \hat{a}^2)^{-1}$. Thus (1.42) is valid.

Alternatively, we can see that the profile likelihood λ satisfies the Bartlett identity to first order, i.e., (1.43). Note that by (E.8),

$$\begin{aligned}N^{-1}E_{\text{true}}[\dot{\lambda}(\sigma^2)^2] &= N^{-1}E_{\text{true}}\left[\left(\dot{l}_{\sigma^2}(\sigma^2, a^2) - \frac{E_{\text{true}}[\ddot{l}_{\sigma^2a^2}]}{E_{\text{true}}[\ddot{l}_{a^2a^2}]}\dot{l}_{a^2}(\sigma^2, a^2) + O_p(1)\right)^2\right] \\ &= N^{-1}E_{\text{true}}\left[\left(\dot{l}_{\sigma^2}(\sigma^2, a^2) - \frac{E_{\text{true}}[\ddot{l}_{\sigma^2a^2}]}{E_{\text{true}}[\ddot{l}_{a^2a^2}]}\dot{l}_{a^2}(\sigma^2, a^2)\right)^2\right] + o(1) \\ &= N^{-1}E_{\text{true}}\left[\dot{l}_{\sigma^2}(\sigma^2, a^2)^2 + \left(\frac{E_{\text{true}}[\ddot{l}_{\sigma^2a^2}]}{E_{\text{true}}[\ddot{l}_{a^2a^2}]}\dot{l}_{a^2}(\sigma^2, a^2)\right)^2 - \right. \\ &\left. -2\frac{E_{\text{true}}[\ddot{l}_{\sigma^2a^2}]}{E_{\text{true}}[\ddot{l}_{a^2a^2}]}\dot{l}_{a^2}(\sigma^2, a^2)\dot{l}_{\sigma^2}(\sigma^2, a^2)\right] + o(1),\end{aligned}$$

so that

$$\begin{aligned}N^{-1}E_{\text{true}}[\dot{\lambda}(\sigma^2)^2] &= S_{\sigma^2\sigma^2} + \left(\frac{D_{\sigma^2a^2}}{D_{a^2a^2}}\right)^2 S_{a^2a^2} - 2\frac{D_{\sigma^2a^2}}{D_{a^2a^2}}S_{a^2\sigma^2} + o_p(1) \\ &= \left(D_{\sigma^2\sigma^2} + \left(\frac{D_{\sigma^2a^2}}{D_{a^2a^2}}\right)^2 D_{a^2a^2} - 2\frac{D_{\sigma^2a^2}}{D_{a^2a^2}}D_{a^2\sigma^2}\right) + \\ &+ \mathrm{Cum}_4[U]\left(g^2_{\sigma^2} + \left(\frac{D_{\sigma^2a^2}}{D_{a^2a^2}}\right)^2 g^2_{a^2} - 2\frac{D_{\sigma^2a^2}}{D_{a^2a^2}}g_{\sigma^2}g_{a^2}\right) + o_p(1),\end{aligned}$$

by invoking (1.39).

Continuing the calculation,

$$\begin{aligned}N^{-1}E_{\text{true}}[\dot{\lambda}(\sigma^2)^2] &= \left(D_{\sigma^2\sigma^2} - \frac{D^2_{\sigma^2a^2}}{D_{a^2a^2}}\right) + \mathrm{Cum}_4[U]\left(g_{\sigma^2} - \frac{D_{\sigma^2a^2}}{D_{a^2a^2}}g_{a^2}\right)^2 + o(1) \\ &= 1/\left[D^{-1}\right]_{\sigma^2\sigma^2} + o(1),\end{aligned} \tag{E.10}$$

since from the expressions for D_{orig} and g_{orig} in (E.2) and (E.5) we have

$$g_{\sigma^2} - \frac{D_{\sigma^2 a^2}}{D_{a^2 a^2}} g_{a^2} = 0\,. \tag{E.11}$$

Then by (E.9) and the law of large numbers, we have

$$N^{-1} E_{\text{true}}[\ddot{\lambda}(\sigma^2)] = -1/\left[D^{-1}\right]_{\sigma^2\sigma^2} + o(1)\,, \tag{E.12}$$

and (1.43) follows from combining (E.10) with (E.12).

Appendix F –Proof of Lemma 2

$\Sigma\Sigma^{-1} \equiv Id$ implies that

$$\frac{\partial \Sigma^{-1}}{\partial \beta_1} = -\Sigma^{-1}\frac{\partial \Sigma}{\partial \beta_1}\Sigma^{-1} \tag{F.1}$$

and, since Σ is linear in the parameters σ^2 and a^2 (see (1.45)) we have

$$\frac{\partial^2 \Sigma}{\partial \beta_2 \partial \beta_1} = 0\,, \tag{F.2}$$

so that

$$\begin{aligned}
\frac{\partial^2 \Sigma^{-1}}{\partial \beta_2 \partial \beta_1} &= \frac{\partial}{\partial \beta_2}\left(\frac{\partial \Sigma^{-1}}{\partial \beta_1}\right) \\
&= \Sigma^{-1}\frac{\partial \Sigma}{\partial \beta_2}\Sigma^{-1}\frac{\partial \Sigma}{\partial \beta_1}\Sigma^{-1} + \Sigma^{-1}\frac{\partial \Sigma}{\partial \beta_1}\Sigma^{-1}\frac{\partial \Sigma}{\partial \beta_2}\Sigma^{-1} - \Sigma^{-1}\frac{\partial^2 \Sigma}{\partial \beta_1 \partial \beta_2}\Sigma^{-1} \\
&= \Sigma^{-1}\frac{\partial \Sigma}{\partial \beta_2}\Sigma^{-1}\frac{\partial \Sigma}{\partial \beta_1}\Sigma^{-1} + \Sigma^{-1}\frac{\partial \Sigma}{\partial \beta_1}\Sigma^{-1}\frac{\partial \Sigma}{\partial \beta_2}\Sigma^{-1}\,.
\end{aligned} \tag{F.3}$$

In the rest of this lemma, let expectations be conditional on the $\Delta's$. We use the notation $E[\cdot|\,\Delta]$ as a shortcut for $E[\cdot|\,\Delta_N, ..., \Delta_1]$. At the true value of the parameter vector, we have,

$$\begin{aligned}
0 &= E[\dot{l}_{\beta_1}\big|\,\Delta] \\
&= -\frac{1}{2}\frac{\partial \ln \det \Sigma}{\partial \beta_1} - \frac{1}{2}E\left[Y'\frac{\partial \Sigma^{-1}}{\partial \beta_1}Y\bigg|\,\Delta\right].
\end{aligned} \tag{F.4}$$

with the second equality following from (1.46). Then, for any nonrandom Q, we have

$$E\left[Y'QY\right] = Tr\left[QE\left[YY'\right]\right] = Tr\left[Q\Sigma\right]. \tag{F.5}$$

This can be applied to Q that depends on the $\Delta's$, even when they are random, because the expected value is conditional on the $\Delta's$. Therefore it follows from (F.4) that

$$\frac{\partial \ln \det \Sigma}{\partial \beta_1} = -E\left[Y'\frac{\partial \Sigma^{-1}}{\partial \beta_1}Y\bigg|\,\Delta\right] = -Tr\left[\frac{\partial \Sigma^{-1}}{\partial \beta_1}\Sigma\right] = Tr\left[\Sigma^{-1}\frac{\partial \Sigma}{\partial \beta_1}\right], \tag{F.6}$$

with the last equality following from (F.1) and so

$$
\begin{aligned}
\frac{\partial^2 \ln\det\Sigma}{\partial\beta_2\partial\beta_1} &= \frac{\partial}{\partial\beta_2} Tr\left[\Sigma^{-1}\frac{\partial\Sigma}{\partial\beta_1}\right] \\
&= Tr\left[\frac{\partial}{\partial\beta_2}\left(\Sigma^{-1}\frac{\partial\Sigma}{\partial\beta_1}\right)\right] \\
&= -Tr\left[\Sigma^{-1}\frac{\partial\Sigma}{\partial\beta_2}\Sigma^{-1}\frac{\partial\Sigma}{\partial\beta_1} + \Sigma^{-1}\frac{\partial^2\Sigma}{\partial\beta_2\partial\beta_1}\right] \\
&= -Tr\left[\Sigma^{-1}\frac{\partial\Sigma}{\partial\beta_2}\Sigma^{-1}\frac{\partial\Sigma}{\partial\beta_1}\right],
\end{aligned}
\tag{F.7}
$$

again because of (F.2).

In light of (1.46), the expected information (conditional on the $\Delta's$) is given by

$$
E\left[-\ddot{l}_{\beta_2\beta_1}\middle|\Delta\right] = \frac{1}{2}\frac{\partial^2\ln\det\Sigma}{\partial\beta_2\beta_1} + \frac{1}{2}E\left[Y'\frac{\partial^2\Sigma^{-1}}{\partial\beta_2\beta_1}Y\middle|\Delta\right].
$$

Then,

$$
\begin{aligned}
E\left[Y'\frac{\partial^2\Sigma^{-1}}{\partial\beta_2\beta_1}Y\middle|\Delta\right] &= Tr\left[\frac{\partial^2\Sigma^{-1}}{\partial\beta_2\beta_1}\Sigma\right] \\
&= Tr\left[\Sigma^{-1}\frac{\partial\Sigma}{\partial\beta_2}\Sigma^{-1}\frac{\partial\Sigma}{\partial\beta_1} + \Sigma^{-1}\frac{\partial\Sigma}{\partial\beta_1}\Sigma^{-1}\frac{\partial\Sigma}{\partial\beta_2}\right] \\
&= 2Tr\left[\Sigma^{-1}\frac{\partial\Sigma}{\partial\beta_2}\Sigma^{-1}\frac{\partial\Sigma}{\partial\beta_1}\right],
\end{aligned}
$$

with the first equality following from (F.5) applied to $Q = \partial^2\Sigma^{-1}/\partial\beta_2\beta_1$, the second from (F.3) and the third from the fact that $Tr[AB] = Tr[BA]$. It follows that

$$
\begin{aligned}
E\left[-\ddot{l}_{\beta_2\beta_1}\middle|\Delta\right] &= -\frac{1}{2}Tr\left[\Sigma^{-1}\frac{\partial\Sigma}{\partial\beta_2}\Sigma^{-1}\frac{\partial\Sigma}{\partial\beta_1}\right] + Tr\left[\Sigma^{-1}\frac{\partial\Sigma}{\partial\beta_2}\Sigma^{-1}\frac{\partial\Sigma}{\partial\beta_1}\right] \\
&= \frac{1}{2}Tr\left[\Sigma^{-1}\frac{\partial\Sigma}{\partial\beta_2}\Sigma^{-1}\frac{\partial\Sigma}{\partial\beta_1}\right] \\
&= -\frac{1}{2}\frac{\partial^2\ln\det\Sigma}{\partial\beta_2\beta_1}.
\end{aligned}
$$

Appendix G – Proof of Theorem 3

In light of (1.45) and (1.52),

$$
\Sigma = \Sigma_0 + \varepsilon\sigma^2\Xi, \tag{G.1}
$$

from which it follows that

$$
\begin{aligned}
\Sigma^{-1} &= \left(\Sigma_0\left(Id + \varepsilon\sigma^2\Sigma_0^{-1}\Xi\right)\right)^{-1} \\
&= \left(Id + \varepsilon\sigma^2\Sigma_0^{-1}\Xi\right)^{-1}\Sigma_0^{-1} \\
&= \Sigma_0^{-1} - \varepsilon\sigma^2\Sigma_0^{-1}\Xi\Sigma_0^{-1} + \varepsilon^2\sigma^4\left(\Sigma_0^{-1}\Xi\right)^2\Sigma_0^{-1} + O(\varepsilon^3),
\end{aligned}
\tag{G.2}
$$

since

$$
(Id + \varepsilon A)^{-1} = Id - \varepsilon A + \varepsilon^2 A^2 + O(\varepsilon^3).
$$

Also,

$$\frac{\partial \Sigma}{\partial \beta_1} = \frac{\partial \Sigma_0}{\partial \beta_1} + \varepsilon \frac{\partial \sigma^2}{\partial \beta_1} \Xi .$$

Therefore, recalling (F.6), we have

$$\begin{aligned}
\frac{\partial \ln \det \Sigma}{\partial \beta_1} &= Tr\left[\Sigma^{-1} \frac{\partial \Sigma}{\partial \beta_1}\right] \\
&= Tr\left[\left(\Sigma_0^{-1} - \varepsilon\sigma^2 \Sigma_0^{-1} \Xi \Sigma_0^{-1} + \varepsilon^2 \sigma^4 \left(\Sigma_0^{-1}\Xi\right)^2 \Sigma_0^{-1} + O(\varepsilon^3)\right)\right. \\
&\quad \left.\left(\frac{\partial \Sigma_0}{\partial \beta_1} + \varepsilon \frac{\partial \sigma^2}{\partial \beta_1}\Xi\right)\right] \\
&= Tr\left[\Sigma_0^{-1}\frac{\partial \Sigma_0}{\partial \beta_1}\right] + \varepsilon Tr\left[-\sigma^2 \Sigma_0^{-1}\Xi\Sigma_0^{-1}\frac{\partial \Sigma_0}{\partial \beta_1} + \frac{\partial \sigma^2}{\partial \beta_1}\Sigma_0^{-1}\Xi\right] + \\
&\quad + \varepsilon^2 Tr\left[\sigma^4 \left(\Sigma_0^{-1}\Xi\right)^2 \Sigma_0^{-1}\frac{\partial \Sigma_0}{\partial \beta_1} - \sigma^2 \frac{\partial \sigma^2}{\partial \beta_1}\Sigma_0^{-1}\Xi\Sigma_0^{-1}\Xi\right] + \\
&\quad + O_p(\varepsilon^3)\,.
\end{aligned} \tag{G.3}$$

We now consider the behavior as $N \to \infty$ of the terms up to order ε^2. The remainder term is handled similarly.

Two things can be determined from the above expansion. Since the $\xi_i's$ are i.i.d. with mean 0, $E[\Xi] = 0$, and so, taking unconditional expectations with respect to the law of the $\Delta_i's$, we obtain that the coefficient of order ε is

$$\begin{aligned}
& E\left[Tr\left[-\sigma^2 \Sigma_0^{-1}\Xi\Sigma_0^{-1}\frac{\partial \Sigma_0}{\partial \beta_1} + \frac{\partial \sigma^2}{\partial \beta_1}\Sigma_0^{-1}\Xi\right]\right] \\
&= Tr\left[E\left[-\sigma^2 \Sigma_0^{-1}\Xi\Sigma_0^{-1}\frac{\partial \Sigma_0}{\partial \beta_1} + \frac{\partial \sigma^2}{\partial \beta_1}\Sigma_0^{-1}\Xi\right]\right] \\
&= Tr\left[-\sigma^2 \Sigma_0^{-1}E\left[\Xi\right]\Sigma_0^{-1}\frac{\partial \Sigma_0}{\partial \beta_1} + \frac{\partial \sigma^2}{\partial \beta_1}\Sigma_0^{-1}E\left[\Xi\right]\right] \\
&= 0\,.
\end{aligned}$$

Similarly, the coefficient of order ε^2 is

$$\begin{aligned}
& E\left[Tr\left[\sigma^4 \left(\Sigma_0^{-1}\Xi\right)^2 \Sigma_0^{-1}\frac{\partial \Sigma_0}{\partial \beta_1} - \sigma^2 \frac{\partial \sigma^2}{\partial \beta_1}\left(\Sigma_0^{-1}\Xi\right)^2\right]\right] \\
&= Tr\left[\sigma^4 E\left[\left(\Sigma_0^{-1}\Xi\right)^2\right]\Sigma_0^{-1}\frac{\partial \Sigma_0}{\partial \beta_1} - \sigma^2 \frac{\partial \sigma^2}{\partial \beta_1}E\left[\left(\Sigma_0^{-1}\Xi\right)^2\right]\right] \\
&= Tr\left[\sigma^2 E\left[\left(\Sigma_0^{-1}\Xi\right)^2\right]\left(\sigma^2\Sigma_0^{-1}\frac{\partial \Sigma_0}{\partial \beta_1} - \frac{\partial \sigma^2}{\partial \beta_1}Id\right)\right] \\
&= Tr\left[\sigma^2 \Sigma_0^{-1}E\left[\Xi\Sigma_0^{-1}\Xi\right]\left(\sigma^2\Sigma_0^{-1}\frac{\partial \Sigma_0}{\partial \beta_1} - \frac{\partial \sigma^2}{\partial \beta_1}Id\right)\right].
\end{aligned}$$

The matrix $E\left[\Xi\Sigma_0^{-1}\Xi\right]$ has the following terms

$$\left[\Xi\Sigma_0^{-1}\Xi\right]_{i,j} = \sum_{k=1}^{N}\sum_{l=1}^{N}\Xi_{ik}\left[\Sigma_0^{-1}\right]_{kl}\Xi_{lj} = \Delta_0^2\xi_i\xi_j\left[\Sigma_0^{-1}\right]_{ij},$$

and since $E[\xi_i \xi_j] = \delta_{ij} \operatorname{Var}[\xi]$ (where δ_{ij} denotes the Kronecker symbol), it follows that

$$E\left[\Xi \Sigma_0^{-1} \Xi\right] = \Delta_0^2 \operatorname{Var}[\xi]\, diag\left[\Sigma_0^{-1}\right], \tag{G.4}$$

where $diag\left[\Sigma_0^{-1}\right]$ is the diagonal matrix formed with the diagonal elements of Σ_0^{-1}. From this, we obtain that

$$\begin{aligned}
&E\left[\frac{\partial \ln \det \Sigma}{\partial \beta_1}\right] \\
&= Tr\left[\Sigma_0^{-1} \frac{\partial \Sigma_0}{\partial \beta_1}\right] + \\
&+ \varepsilon^2 Tr\left[\sigma^2 \Sigma_0^{-1} E\left[\Xi \Sigma_0^{-1} \Xi\right] \left(\sigma^2 \Sigma_0^{-1} \frac{\partial \Sigma_0}{\partial \beta_1} - \frac{\partial \sigma^2}{\partial \beta_1} Id\right)\right] + O(\varepsilon^3) \\
&= Tr\left[\Sigma_0^{-1} \frac{\partial \Sigma_0}{\partial \beta_1}\right] + \qquad\qquad (G.5) \\
&+ \varepsilon^2 \Delta_0^2 \operatorname{Var}[\xi] Tr\left[\sigma^2 \Sigma_0^{-1} diag\left[\Sigma_0^{-1}\right] \left(\sigma^2 \Sigma_0^{-1} \frac{\partial \Sigma_0}{\partial \beta_1} - \frac{\partial \sigma^2}{\partial \beta_1} Id\right)\right] + O(\varepsilon^3).
\end{aligned}$$

To calculate $E\left[\ddot{l}_{\beta_2 \beta_1}\right]$, in light of (1.51), we need to differentiate $E\left[\partial \ln \det \Sigma / \partial \beta_1\right]$ with respect to β_2. Indeed

$$E\left[-\ddot{l}_{\beta_2\beta_1}\right] = E\left[E\left[-\ddot{l}_{\beta_2\beta_1} \middle| \Delta\right]\right] = -\frac{1}{2} E\left[\frac{\partial^2 \ln \det \Sigma}{\partial \beta_2 \partial \beta_1}\right] = -\frac{1}{2} \frac{\partial}{\partial \beta_2}\left(E\left[\frac{\partial \ln \det \Sigma}{\partial \beta_1}\right]\right),$$

where we can interchange the unconditional expectation and the differentiation with respect to β_2 because the unconditional expectation is taken with respect to the law of the $\Delta_i' s$, which is independent of the β parameters (i.e., σ^2 and a^2). Therefore, differentiating (G.5) with respect to β_2 will produce the result we need. (The reader may wonder why we take the expected value before differentiating, rather than the other way around. As just discussed, the results are identical. However, it turns out that taking expectations first reduces the computational burden quite substantially.)

Combining with (G.5), we therefore have

$$\begin{aligned}
&E\left[-\ddot{l}_{\beta_2\beta_1}\right] \\
&= -\frac{1}{2} \frac{\partial}{\partial \beta_2}\left(E\left[\frac{\partial \ln \det \Sigma}{\partial \beta_1}\right]\right) \\
&= -\frac{1}{2} \frac{\partial}{\partial \beta_2} Tr\left[\Sigma_0^{-1} \frac{\partial \Sigma_0}{\partial \beta_1}\right] - \\
&- \frac{1}{2} \varepsilon^2 \Delta_0^2 \operatorname{Var}[\xi] \frac{\partial}{\partial \beta_2}\left(Tr\left[\sigma^2 \Sigma_0^{-1} diag\left[\Sigma_0^{-1}\right] \left(\sigma^2 \Sigma_0^{-1} \frac{\partial \Sigma_0}{\partial \beta_1} - \frac{\partial \sigma^2}{\partial \beta_1} Id\right)\right]\right) + \\
&+ O(\varepsilon^3) \\
&\equiv \phi^{(0)} + \varepsilon^2 \phi^{(2)} + O(\varepsilon^3). \qquad\qquad (G.6)
\end{aligned}$$

It is useful now to introduce the same transformed parameters (γ^2, η) as in previous sections and write $\Sigma_0 = \gamma^2 V$ with the parameters and V defined as in (1.9)-(1.10) and (1.26), except that Δ is replaced by Δ_0 in these expressions. To compute $\phi^{(0)}$, we start with

$$
\begin{aligned}
Tr\left[\Sigma_0^{-1}\frac{\partial\Sigma_0}{\partial\beta_1}\right] &= Tr\left[\gamma^{-2}V^{-1}\frac{\partial\left(\gamma^2V\right)}{\partial\beta_1}\right] \\
&= Tr\left[V^{-1}\frac{\partial V}{\partial\beta_1}\right] + Tr\left[\gamma^{-2}V^{-1}V\frac{\partial\gamma^2}{\partial\beta_1}\right] \\
&= Tr\left[V^{-1}\frac{\partial V}{\partial\eta}\right]\frac{\partial\eta}{\partial\beta_1} + N\gamma^{-2}\frac{\partial\gamma^2}{\partial\beta_1}\,, \qquad \text{(G.7)}
\end{aligned}
$$

with $\partial\gamma^2/\partial\beta_1$ and $\partial\eta/\partial\beta_1$ to be computed from (1.11)-(1.12). If Id denotes the identity matrix and J the matrix with 1 on the infra and supra-diagonal lines and 0 everywhere else, we have $V = \eta^2 Id + \eta J$, so that $\partial V/\partial\eta = 2\eta Id + J$. Therefore

$$
\begin{aligned}
Tr\left[V^{-1}\frac{\partial V}{\partial\eta}\right] &= 2\eta Tr\left[V^{-1}\right] + Tr\left[V^{-1}J\right] \\
&= 2\eta\sum_{i=1}^{N} v^{i,i} + \sum_{i=2}^{N-1}\left\{v^{i,i-1} + v^{i,i+1}\right\} + v^{1,2} + v^{N,N-1} \\
&= \frac{2\eta\left(1-\eta^{2N}\left(N\left(1-\eta^2\right)+1\right)\right)}{\left(1-\eta^2\right)\left(1-\eta^{2(1+N)}\right)} \\
&= \frac{2\eta}{\left(1-\eta^2\right)} + o(1)\,. \qquad \text{(G.8)}
\end{aligned}
$$

Therefore the first term in (G.7) is $O(1)$ while the second term is $O(N)$ and hence

$$
Tr\left[\Sigma_0^{-1}\frac{\partial\Sigma_0}{\partial\beta_1}\right] = N\gamma^{-2}\frac{\partial\gamma^2}{\partial\beta_1} + O(1)\,.
$$

This holds also for the partial derivative of (G.7) with respect to β_2. Indeed, given the form of (G.8), we have that

$$
\frac{\partial}{\partial\beta_2}\left(Tr\left[V^{-1}\frac{\partial V}{\partial\eta}\right]\right) = \frac{\partial}{\partial\beta_2}\left(\frac{2\eta}{\left(1-\eta^2\right)}\right) + o(1) = O(1)\,,
$$

since the remainder term in (G.8) is of the form $p(N)\eta^{q(N)}$, where p and q are polynomials in N or order greater than or equal to 0 and 1 respectively, whose differentiation with respect to η will produce terms that are of order $o(N)$. Thus it follows that

$$
\begin{aligned}
\frac{\partial}{\partial\beta_2}\left(Tr\left[\Sigma_0^{-1}\frac{\partial\Sigma_0}{\partial\beta_1}\right]\right) &= N\frac{\partial}{\partial\beta_2}\left(\gamma^{-2}\frac{\partial\gamma^2}{\partial\beta_1}\right) + o(N) \\
&= N\left\{\frac{\partial\gamma^{-2}}{\partial\beta_2}\frac{\partial\gamma^2}{\partial\beta_1} + \gamma^{-2}\frac{\partial^2\gamma^2}{\partial\beta_2\partial\beta_1}\right\} + o(N)\,. \qquad \text{(G.9)}
\end{aligned}
$$

Writing the result in matrix form, where the $(1,1)$ element corresponds to $(\beta_1,\beta_2) = (\sigma^2,\sigma^2)$, the $(1,2)$ and $(2,1)$ elements to $(\beta_1,\beta_2) = (\sigma^2,a^2)$ and the $(2,2)$ element to $(\beta_1,\beta_2) = (a^2,a^2)$, and computing the partial derivatives in (G.9), we have

$$\phi^{(0)} = -\frac{1}{2}\frac{\partial}{\partial \beta_2}\left(Tr\left[\Sigma_0^{-1}\frac{\partial \Sigma_0}{\partial \beta_1}\right]\right)$$
$$= N\begin{pmatrix} \frac{\Delta_0^{1/2}(2a^2+\sigma^2\Delta_0)}{2\sigma^3(4a^2+\sigma^2\Delta_0)^{3/2}} & \frac{\Delta_0^{1/2}}{\sigma(4a^2+\sigma^2\Delta_0)^{3/2}} \\ \bullet & \frac{1}{2a^4}\left(1-\frac{\Delta_0^{1/2}\sigma(6a^2+\sigma^2\Delta_0)}{(4a^2+\sigma^2\Delta_0)^{3/2}}\right) \end{pmatrix} + o(N). \quad (G.10)$$

As for the coefficient of order ε^2, that is $\phi^{(2)}$ in (G.6), define

$$\alpha \equiv Tr\left[\sigma^2\Sigma_0^{-1}\text{diag}\left[\Sigma_0^{-1}\right]\left(\sigma^2\Sigma_0^{-1}\frac{\partial \Sigma_0}{\partial \beta_1} - \frac{\partial \sigma^2}{\partial \beta_1}Id\right)\right], \quad (G.11)$$

so that

$$\phi^{(2)} = -\frac{1}{2}\Delta_0^2 \,\text{Var}[\xi]\frac{\partial \alpha}{\partial \beta_2}.$$

We have

$$\begin{aligned}
\alpha &= \sigma^4 Tr\left[\Sigma_0^{-1}\text{diag}\left[\Sigma_0^{-1}\right]\Sigma_0^{-1}\frac{\partial \Sigma_0}{\partial \beta_1}\right] - \sigma^2\frac{\partial \sigma^2}{\partial \beta_1}Tr\left[\Sigma_0^{-1}\text{diag}\left[\Sigma_0^{-1}\right]\right] \\
&= \sigma^4\gamma^{-6}Tr\left[V^{-1}\text{diag}\left[V^{-1}\right]V^{-1}\frac{\partial(\gamma^2 V)}{\partial \beta_1}\right] - \sigma^2\frac{\partial \sigma^2}{\partial \beta_1}\gamma^{-4}Tr\left[V^{-1}\text{diag}\left[V^{-1}\right]\right] \\
&= \sigma^4\gamma^{-4}Tr\left[V^{-1}\text{diag}\left[V^{-1}\right]V^{-1}\frac{\partial V}{\partial \eta}\frac{\partial \eta}{\partial \beta_1}\right] + \\
&\quad + \sigma^4\gamma^{-6}\frac{\partial \gamma^2}{\partial \beta_1}Tr\left[V^{-1}\text{diag}\left[V^{-1}\right]\right] - \sigma^2\frac{\partial \sigma^2}{\partial \beta_1}\gamma^{-4}Tr\left[V^{-1}\text{diag}\left[V^{-1}\right]\right] \\
&= \sigma^4\gamma^{-4}\frac{\partial \eta}{\partial \beta_1}Tr\left[V^{-1}\text{diag}\left[V^{-1}\right]V^{-1}\frac{\partial V}{\partial \eta}\right] + \\
&\quad + \left(\sigma^4\gamma^{-6}\frac{\partial \gamma^2}{\partial \beta_1} - \sigma^2\frac{\partial \sigma^2}{\partial \beta_1}\gamma^{-4}\right)Tr\left[V^{-1}\text{diag}\left[V^{-1}\right]\right].
\end{aligned}$$

Next, we compute separately

$$\begin{aligned}
Tr\left[V^{-1}\text{diag}\left[V^{-1}\right]V^{-1}\frac{\partial V}{\partial \eta}\right] &= Tr\left[\text{diag}\left[V^{-1}\right]V^{-1}\frac{\partial V}{\partial \eta}V^{-1}\right] \\
&= -Tr\left[\text{diag}\left[V^{-1}\right]\frac{\partial V^{-1}}{\partial \eta}\right] \\
&= -\sum_{i=1}^{N} v^{i,i}\frac{\partial v^{i,i}}{\partial \eta} \\
&= \frac{O(1) - 2N\eta\left(1+\eta^2-\eta^4-\eta^6+O(\eta^{2N})\right) + N^2 O(\eta^{2N})}{(1+\eta^2)^2(1-\eta^2)^4\left(1-\eta^{2(1+N)}\right)^3} \\
&= \frac{-2N\eta}{(1-\eta^2)^3} + o(N),
\end{aligned}$$

and

$$\begin{aligned} Tr\left[V^{-1}\text{diag}\left[V^{-1}\right]\right] &= \sum_{i=1}^{N}\left(v^{i,i}\right)^2 \\ &= \frac{O(1)+N\left(1-\eta^4+O(\eta^{2N})\right)}{(1+\eta^2)\left(1-\eta^2\right)^3\left(1-\eta^{2(1+N)}\right)^2} \\ &= \frac{N}{\left(1-\eta^2\right)^2}+o(N). \end{aligned}$$

Therefore

$$\alpha = \sigma^4\gamma^{-4}\frac{\partial\eta}{\partial\beta_1}\left(\frac{-2N\eta}{(1-\eta^2)^3}\right)+\left(\sigma^4\gamma^{-6}\frac{\partial\gamma^2}{\partial\beta_1}-\sigma^2\frac{\partial\sigma^2}{\partial\beta_1}\gamma^{-4}\right)\left(\frac{N}{\left(1-\eta^2\right)^2}\right)+o(N),$$

which can be differentiated with respect to β_2 to produce $\partial\alpha/\partial\beta_2$. As above, differentiation of the remainder term $o(N)$ still produces a $o(N)$ term because of the structure of the terms there (they are again of the form $p(N)\eta^{q(N)}$.)

Note that an alternative expression for α can be obtained as follows. Going back to the definition (G.11),

$$\alpha = \sigma^4 Tr\left[\Sigma_0^{-1}\text{diag}\left[\Sigma_0^{-1}\right]\Sigma_0^{-1}\frac{\partial\Sigma_0}{\partial\beta_1}\right]-\sigma^2\frac{\partial\sigma^2}{\partial\beta_1}Tr\left[\Sigma_0^{-1}\text{diag}\left[\Sigma_0^{-1}\right]\right], \qquad (G.12)$$

the first trace becomes

$$\begin{aligned} Tr\left[\Sigma_0^{-1}\text{diag}\left[\Sigma_0^{-1}\right]\Sigma_0^{-1}\frac{\partial\Sigma_0}{\partial\beta_1}\right] &= Tr\left[\text{diag}\left[\Sigma_0^{-1}\right]\Sigma_0^{-1}\frac{\partial\Sigma_0}{\partial\beta_1}\Sigma_0^{-1}\right] \\ &= -Tr\left[\text{diag}\left[\Sigma_0^{-1}\right]\frac{\partial\Sigma_0^{-1}}{\partial\beta_1}\right] \\ &= -\sum_{i=1}^{N}(\Sigma_0^{-1})_{ii}(\frac{\partial\Sigma_0^{-1}}{\partial\beta_1})_{ii} \\ &= -\frac{1}{2}\frac{\partial}{\partial\beta_1}\sum_{i=1}^{N}(\Sigma_0^{-1})_{ii}^2 \\ &= -\frac{1}{2}\frac{\partial}{\partial\beta_1}Tr\left[\Sigma_0^{-1}\text{diag}\left[\Sigma_0^{-1}\right]\right], \end{aligned}$$

so that we have

$$\begin{aligned} \alpha &= -\sigma^4\frac{1}{2}\frac{\partial}{\partial\beta_1}Tr\left[\Sigma_0^{-1}\text{diag}\left[\Sigma_0^{-1}\right]\right]-\sigma^2\frac{\partial\sigma^2}{\partial\beta_1}Tr\left[\Sigma_0^{-1}\text{diag}\left[\Sigma_0^{-1}\right]\right] \\ &= -\sigma^4\frac{1}{2}\frac{\partial}{\partial\beta_1}Tr\left[\Sigma_0^{-1}\text{diag}\left[\Sigma_0^{-1}\right]\right]-\frac{1}{2}\left(\frac{\partial\sigma^4}{\partial\beta_1}\right)Tr\left[\Sigma_0^{-1}\text{diag}\left[\Sigma_0^{-1}\right]\right] \\ &= -\frac{1}{2}\frac{\partial}{\partial\beta_1}\left(\sigma^4 Tr\left[\Sigma_0^{-1}\text{diag}\left[\Sigma_0^{-1}\right]\right]\right) \\ &= -\frac{1}{2}\frac{\partial}{\partial\beta_1}\left(\sigma^4\gamma^{-4}Tr\left[V^{-1}\text{diag}\left[V^{-1}\right]\right]\right) \\ &= -\frac{1}{2}\frac{\partial}{\partial\beta_1}\left(\sigma^4\gamma^{-4}\left(\frac{N}{\left(1-\eta^2\right)^2}+o(N)\right)\right) \\ &= -\frac{N}{2}\frac{\partial}{\partial\beta_1}\left(\frac{\sigma^4\gamma^{-4}}{\left(1-\eta^2\right)^2}\right)+o(N), \end{aligned}$$

where the calculation of $Tr\left[V^{-1}diag\left[V^{-1}\right]\right]$ is as before, and where the $o(N)$ term is a sum of terms of the form $p(N)\eta^{q(N)}$ as discussed above. From this one can interchange differentiation and the $o(N)$ term, yielding the final equality above.

Therefore

$$\begin{aligned}\frac{\partial\alpha}{\partial\beta_2} &= -\frac{1}{2}\frac{\partial^2}{\partial\beta_1\partial\beta_2}\left(\sigma^4\gamma^{-4}\left(\frac{N}{(1-\eta^2)^2}+o(N)\right)\right)\\ &= -\frac{N}{2}\frac{\partial^2}{\partial\beta_1\partial\beta_2}\left(\frac{\sigma^4\gamma^{-4}}{(1-\eta^2)^2}\right)+o(N)\,. \end{aligned} \tag{G.13}$$

Writing the result in matrix form and calculating the partial derivatives, we obtain

$$\phi^{(2)} = -\frac{1}{2}\Delta_0^2\operatorname{Var}[\xi]\frac{\partial\alpha}{\partial\beta_2} = \frac{N\ \Delta_0^2\operatorname{Var}[\xi]}{(4a^2+\sigma^2\Delta_0)^3}\begin{pmatrix}-2a^2 & -\frac{(8a^2-2\sigma^2\Delta_0)}{2\Delta_0}\\ \bullet & -\frac{8\sigma^2}{\Delta_0}\end{pmatrix}+o(N)\,. \tag{G.14}$$

Putting it all together, we have obtained

$$\begin{aligned}\frac{1}{N}E\left[-\ddot{l}_{\beta_2\beta_1}\right] &= \frac{1}{N}\left(\phi^{(0)}+\varepsilon^2\phi^{(2)}+O(\varepsilon^3)\right)\\ &\equiv F^{(0)}+\varepsilon^2F^{(2)}+O(\varepsilon^3)+o(1)\,, \end{aligned} \tag{G.15}$$

where

$$F^{(0)} = \begin{pmatrix}\frac{\Delta_0^{1/2}(2a^2+\sigma^2\Delta_0)}{2\sigma^3(4a^2+\sigma^2\Delta_0)^{3/2}} & \frac{\Delta_0^{1/2}}{\sigma(4a^2+\sigma^2\Delta_0)^{3/2}}\\ \bullet & \frac{1}{2a^4}\left(1-\frac{\Delta_0^{1/2}\sigma(6a^2+\sigma^2\Delta_0)}{(4a^2+\sigma^2\Delta_0)^{3/2}}\right)\end{pmatrix}, \tag{G.16}$$

$$F^{(2)} \equiv \frac{\Delta_0^2\operatorname{Var}[\xi]}{(4a^2+\sigma^2\Delta_0)^3}\begin{pmatrix}-2a^2 & -\frac{(8a^2-2\sigma^2\Delta_0)}{2\Delta_0}\\ \bullet & -\frac{8\sigma^2}{\Delta_0}\end{pmatrix}. \tag{G.17}$$

The asymptotic variance of the maximum-likelihood estimators $\text{AVAR}(\hat{\sigma}^2,\hat{a}^2)$ is therefore given by

$$\begin{aligned}\text{AVAR}(\hat{\sigma}^2,\hat{a}^2) &= E[\Delta]\left(F^{(0)}+\varepsilon^2F^{(2)}+O(\varepsilon^3)\right)^{-1}\\ &= \Delta_0\left(F^{(0)}\left(Id+\varepsilon^2\left[F^{(0)}\right]^{-1}F^{(2)}+O(\varepsilon^3)\right)\right)^{-1}\\ &= \Delta_0\left(Id+\varepsilon^2\left[F^{(0)}\right]^{-1}F^{(2)}+O(\varepsilon^3)\right)^{-1}\left[F^{(0)}\right]^{-1}\\ &= \Delta_0\left(Id-\varepsilon^2\left[F^{(0)}\right]^{-1}F^{(2)}+O(\varepsilon^3)\right)\left[F^{(0)}\right]^{-1}\\ &= \Delta_0\left[F^{(0)}\right]^{-1}-\varepsilon^2\Delta_0\left[F^{(0)}\right]^{-1}F^{(2)}\left[F^{(0)}\right]^{-1}+O(\varepsilon^3)\\ &\equiv A^{(0)}+\varepsilon^2A^{(2)}+O(\varepsilon^3)\,, \end{aligned}$$

where the final results for $A^{(0)}=\Delta_0\left[F^{(0)}\right]^{-1}$ and $A^{(2)}=-\Delta_0\left[F^{(0)}\right]^{-1}F^{(2)}\left[F^{(0)}\right]^{-1}$, obtained by replacing $F^{(0)}$ and $F^{(2)}$ by their expressions in (G.15), are given in the statement of the Theorem.

Appendix H – Proof of Theorem 4

It follows as in (D.3), (D.4) and (D.5) that

$$\begin{aligned}
&E_{\text{true}}\left[\dot{l}_{\beta_1}\dot{l}_{\beta_2}|\Delta\right] \\
&= \mathrm{Cov}_{\text{true}}(\dot{l}_{\beta_1}, \dot{l}_{\beta_2}|\Delta) \\
&= \mathrm{Cov}_{\text{true}}\left(-\frac{1}{2}\sum_{i,j=1}^{N} Y_iY_j\left(\frac{\partial\Sigma^{-1}}{\partial\beta_1}\right)_{ij}, -\frac{1}{2}\sum_{k,l=1}^{N} Y_kY_l\left(\frac{\partial\Sigma^{-1}}{\partial\beta_2}\right)_{kl}\right) \\
&= \frac{1}{4}\sum_{i,j,k,l=1}^{N}\left(\frac{\partial\Sigma^{-1}}{\partial\beta_1}\right)_{ij}\left(\frac{\partial\Sigma^{-1}}{\partial\beta_2}\right)_{kl}\mathrm{Cov}_{\text{true}}(Y_iY_j, Y_kY_l|\Delta) \\
&= -E_{\text{true}}\left[\ddot{l}_{\beta_1\beta_2}|\Delta\right] + \frac{1}{4}\sum_{i,j,k,l=1}^{N}\left(\frac{\partial\Sigma^{-1}}{\partial\beta_1}\right)_{ij}\left(\frac{\partial\Sigma^{-1}}{\partial\beta_2}\right)_{kl}\mathrm{Cum}_{\text{true}}(Y_i, Y_j, Y_k, Y_l|\Delta) \\
&= -E_{\text{true}}\left[\ddot{l}_{\beta_1\beta_2}|\Delta\right] + \frac{1}{4}\mathrm{Cum}_4\left[U\right]\psi\left(\frac{\partial\Sigma^{-1}}{\partial\beta_1}, \frac{\partial\Sigma^{-1}}{\partial\beta_2}\right),
\end{aligned} \tag{H.1}$$

since $\mathrm{Cum}_{\text{true}}(Y_i, Y_j, Y_k, Y_l|\Delta) = 2, \pm1$, or 0, $\times\, \mathrm{Cum}_{\text{true}}(U)$, as in (1.15), and with ψ defined in (D.8). Taking now unconditional expectations, we have

$$\begin{aligned}
E_{\text{true}}\left[\dot{l}_{\beta_1}\dot{l}_{\beta_2}\right] &= \mathrm{Cov}_{\text{true}}(\dot{l}_{\beta_1}, \dot{l}_{\beta_2}) \\
&= E\left[\mathrm{Cov}_{\text{true}}(\dot{l}_{\beta_1}, \dot{l}_{\beta_2}|\Delta)\right] + \mathrm{Cov}_{\text{true}}(E_{\text{true}}[\dot{l}_{\beta_1}|\Delta], E_{\text{true}}[\dot{l}_{\beta_2}|\Delta]) \\
&= E\left[\mathrm{Cov}_{\text{true}}(\dot{l}_{\beta_1}, \dot{l}_{\beta_2}|\Delta)\right] \\
&= -E_{\text{true}}\left[\ddot{l}_{\beta_1\beta_2}\right] + \frac{1}{4}\mathrm{Cum}_4\left[U\right]E\left[\psi\left(\frac{\partial\Sigma^{-1}}{\partial\beta_1}, \frac{\partial\Sigma^{-1}}{\partial\beta_2}\right)\right],
\end{aligned} \tag{H.2}$$

with the first and third equalities following from the fact that $E_{\text{true}}[\dot{l}_{\beta_i}|\Delta] = 0$.

Since

$$E_{\text{true}}\left[\ddot{l}_{\beta_1\beta_2}|\Delta\right] = E_{\text{normal}}\left[\ddot{l}_{\beta_1\beta_2}|\Delta\right],$$

and consequently

$$E_{\text{true}}\left[\ddot{l}_{\beta_1\beta_2}\right] = E_{\text{normal}}\left[\ddot{l}_{\beta_1\beta_2}\right]$$

have been found in the previous subsection (see (G.15)), what we need to do to obtain $E_{\text{true}}\left[\dot{l}_{\beta_1}\dot{l}_{\beta_2}\right]$ is to calculate

$$E\left[\psi\left(\frac{\partial\Sigma^{-1}}{\partial\beta_1}, \frac{\partial\Sigma^{-1}}{\partial\beta_2}\right)\right].$$

With Σ^{-1} given by (G.2), we have for $i = 1, 2$

$$\frac{\partial\Sigma^{-1}}{\partial\beta_i} = \frac{\partial\Sigma_0^{-1}}{\partial\beta_i} - \varepsilon\frac{\partial}{\partial\beta_i}\left(\sigma^2\Sigma_0^{-1}\Xi\Sigma_0^{-1}\right) + \varepsilon^2\frac{\partial}{\partial\beta_i}\left(\sigma^4\left(\Sigma_0^{-1}\Xi\right)^2\Sigma_0^{-1}\right) + O(\varepsilon^3),$$

and therefore by bilinearity of ψ we have

$$
\begin{aligned}
E\left[\psi\left(\frac{\partial\Sigma^{-1}}{\partial\beta_1},\frac{\partial\Sigma^{-1}}{\partial\beta_2}\right)\right] &= \psi\left(\frac{\partial\Sigma_0^{-1}}{\partial\beta_1},\frac{\partial\Sigma_0^{-1}}{\partial\beta_2}\right) - \\
&-\varepsilon E\left[\psi\left(\frac{\partial\Sigma_0^{-1}}{\partial\beta_1},\frac{\partial}{\partial\beta_2}\left(\sigma^2\Sigma_0^{-1}\Xi\Sigma_0^{-1}\right)\right)\right][2]+ \\
&+\varepsilon^2 E\left[\psi\left(\frac{\partial\Sigma_0^{-1}}{\partial\beta_1},\frac{\partial}{\partial\beta_2}\left(\sigma^4\left(\Sigma_0^{-1}\Xi\right)^2\Sigma_0^{-1}\right)\right)\right][2]+ \\
&+\varepsilon^2 E\left[\psi\left(\frac{\partial}{\partial\beta_1}\left(\sigma^2\Sigma_0^{-1}\Xi\Sigma_0^{-1}\right),\frac{\partial}{\partial\beta_2}\left(\sigma^2\Sigma_0^{-1}\Xi\Sigma_0^{-1}\right)\right)\right]+ \\
&+O(\varepsilon^3),
\end{aligned} \tag{H.3}
$$

where the "[2]" refers to the sum over the two terms where β_1 and β_2 are permuted.

The first (and leading) term in (H.3),

$$
\begin{aligned}
\psi\left(\frac{\partial\Sigma_0^{-1}}{\partial\beta_1},\frac{\partial\Sigma_0^{-1}}{\partial\beta_2}\right) &= \psi\left(\frac{\partial\left(\gamma^{-2}V^{-1}\right)}{\partial\beta_1},\frac{\partial\left(\gamma^{-2}V^{-1}\right)}{\partial\beta_2}\right) \\
&= \psi\left(\frac{\partial\gamma^{-2}}{\partial\beta_1}V^{-1}+\gamma^{-2}\frac{\partial V^{-1}}{\partial\beta_1},\frac{\partial\gamma^{-2}}{\partial\beta_2}V^{-1}+\gamma^{-2}\frac{\partial V^{-1}}{\partial\beta_2}\right) \\
&= \psi\left(\frac{\partial\gamma^{-2}}{\partial\beta_1}V^{-1}+\gamma^{-2}\frac{\partial V^{-1}}{\partial\eta}\frac{\partial\eta}{\partial\beta_1},\frac{\partial\gamma^{-2}}{\partial\beta_1}V^{-1}+\gamma^{-2}\frac{\partial V^{-1}}{\partial\eta}\frac{\partial\eta}{\partial\beta_1}\right),
\end{aligned}
$$

corresponds to the equally spaced, misspecified noise distribution, situation studied in Section 1.6.

The second term, linear in ε, is zero since

$$
\begin{aligned}
E\left[\psi\left(\frac{\partial\Sigma_0^{-1}}{\partial\beta_1},\frac{\partial}{\partial\beta_2}\left(\sigma^2\Sigma_0^{-1}\Xi\Sigma_0^{-1}\right)\right)\right] &= \psi\left(\frac{\partial\Sigma_0^{-1}}{\partial\beta_1},E\left[\frac{\partial}{\partial\beta_2}\left(\sigma^2\Sigma_0^{-1}\Xi\Sigma_0^{-1}\right)\right]\right) \\
&= \psi\left(\frac{\partial\Sigma_0^{-1}}{\partial\beta_1},\frac{\partial}{\partial\beta_2}\left(\sigma^2\Sigma_0^{-1}E\left[\Xi\right]\Sigma_0^{-1}\right)\right) \\
&= 0,
\end{aligned}
$$

with the first equality following from the bilinearity of ψ, the second from the fact that the unconditional expectation over the $\Delta_i's$ does not depend on the β parameters, so expectation and differentiation with respect to β_2 can be interchanged, and the third equality from the fact that $E\left[\Xi\right]=0$.

To calculate the third term in (H.3), the first of two that are quadratic in ε, note that

$$
\begin{aligned}
\alpha_1 &\equiv E\left[\psi\left(\frac{\partial\Sigma_0^{-1}}{\partial\beta_1},\frac{\partial}{\partial\beta_2}\left(\sigma^4\Sigma_0^{-1}\Xi\Sigma_0^{-1}\Xi\Sigma_0^{-1}\right)\right)\right] \\
&= \psi\left(\frac{\partial\Sigma_0^{-1}}{\partial\beta_1},\frac{\partial}{\partial\beta_2}\left(\sigma^4\Sigma_0^{-1}E\left[\Xi\Sigma_0^{-1}\Xi\right]\Sigma_0^{-1}\right)\right) \\
&= \Delta_0^2\operatorname{Var}[\xi]\psi\left(\frac{\partial\Sigma_0^{-1}}{\partial\beta_1},\frac{\partial}{\partial\beta_2}\left(\sigma^4\Sigma_0^{-1}diag(\Sigma_0^{-1})\Sigma_0^{-1}\right)\right) \\
&= \Delta_0^2\operatorname{Var}[\xi]\psi\left(\frac{\partial\left(\gamma^{-2}V^{-1}\right)}{\partial\beta_1},\frac{\partial}{\partial\beta_2}\left(\sigma^4\gamma^{-6}V^{-1}diag(V^{-1})V^{-1}\right)\right),
\end{aligned} \tag{H.4}
$$

with the second equality obtained by replacing $E\left[\Xi\Sigma_0^{-1}\Xi\right]$ with its value given in (G.4), and the third by recalling that $\Sigma_0 = \gamma^2 V$. The elements (i,j) of the two arguments of ψ in (H.4) are

$$\nu^{i,j} = \frac{\partial\left(\gamma^{-2}v^{i,j}\right)}{\partial\beta_1} = \frac{\partial\gamma^{-2}}{\partial\beta_1}v^{i,j} + \gamma^{-2}\frac{\partial v^{i,j}}{\partial\eta}\frac{\partial\eta}{\partial\beta_1},$$

and

$$\begin{aligned}\omega^{k,l} &= \frac{\partial}{\partial\beta_2}\left(\sigma^4\gamma^{-6}\sum_{m=1}^{N} v^{k,m}v^{m,m}v^{m,l}\right) \\ &= \frac{\partial\left(\sigma^4\gamma^{-6}\right)}{\partial\beta_2}\sum_{m=1}^{N} v^{k,m}v^{m,m}v^{m,l} + \sigma^4\gamma^{-6}\frac{\partial}{\partial\eta}\left(\sum_{m=1}^{N} v^{k,m}v^{m,m}v^{m,l}\right)\frac{\partial\eta}{\partial\beta_2},\end{aligned}$$

from which ψ in (H.4) can be evaluated through the sum given in (D.8).

Summing these terms, we obtain

$$\begin{aligned}&\psi\left(\frac{\partial\left(\gamma^{-2}V^{-1}\right)}{\partial\beta_1}, \frac{\partial}{\partial\beta_2}\left(\sigma^4\gamma^{-6}V^{-1}diag(V^{-1})V^{-1}\right)\right) \\ &= \frac{4N\left(C_{1V}\left(1-\eta\right)+C_{2V}C_{3V}\right)\left(C_{1W}\left(1-\eta^2\right)+2C_{2W}C_{3W}\left(1+3\eta\right)\right)}{(1-\eta)^7(1+\eta)^3} + o(N),\end{aligned}$$

where

$$\begin{aligned}C_{1V} &= \frac{\partial\gamma^{-2}}{\partial\beta_1},\; C_{2V} = \gamma^{-2},\; C_{3V} = \frac{\partial\eta}{\partial\beta_1} \\ C_{1W} &= \frac{\partial\left(\sigma^4\gamma^{-6}\right)}{\partial\beta_2},\; C_{2W} = \sigma^4\gamma^{-6},\; C_{3W} = \frac{\partial\eta}{\partial\beta_2}.\end{aligned}$$

The fourth and last term in (H.3), also quadratic in ε,

$$\alpha_2 \equiv E\left[\psi\left(\frac{\partial}{\partial\beta_1}\left(\sigma^2\Sigma_0^{-1}\Xi\Sigma_0^{-1}\right), \frac{\partial}{\partial\beta_2}\left(\sigma^2\Sigma_0^{-1}\Xi\Sigma_0^{-1}\right)\right)\right]$$

is obtained by first expressing

$$\psi\left(\frac{\partial}{\partial\beta_1}\left(\sigma^2\Sigma_0^{-1}\Xi\Sigma_0^{-1}\right), \frac{\partial}{\partial\beta_2}\left(\sigma^2\Sigma_0^{-1}\Xi\Sigma_0^{-1}\right)\right)$$

in its sum form and then taking expectations term by term. Letting now

$$\nu^{i,j} = \left(\frac{\partial}{\partial\beta_1}\left(\sigma^2\Sigma_0^{-1}\Xi\Sigma_0^{-1}\right)\right)_{ij}, \quad \omega^{k,l} = \left(\frac{\partial}{\partial\beta_2}\left(\sigma^2\Sigma_0^{-1}\Xi\Sigma_0^{-1}\right)\right)_{kl},$$

we recall our definition of $\psi(\nu,\omega)$ given in (D.8) whose unconditional expected value (over the $\Delta_i's$, i.e., over Ξ) we now need to evaluate in order to obtain α_2.

We are thus led to consider four-index tensors λ^{ijkl} and to define

$$\begin{aligned}\tilde{\psi}(\lambda) \equiv 2\sum_{h=1}^{N}\lambda^{h,h,h,h} + \sum_{h=1}^{N-1}\Big\{&-2\lambda^{h,h+1,h+1,h+1} - 2\lambda^{h+1,h+1,h,h+1} + \\ &+\lambda^{h,h,h+1,h+1} + \lambda^{h+1,h+1,h,h} + 4\lambda^{h,h+1,h,h+1} - \\ &-2\lambda^{h+1,h,h,h} - 2\lambda^{h,h,h+1,h}\Big\},\end{aligned} \tag{H.5}$$

where λ^{ijkl} is symmetric in the two first and the two last indices, respectively, i.e., $\lambda^{ijkl} = \lambda^{jikl}$ and $\lambda^{ijkl} = \lambda^{ijlk}$. In terms of our definition of ψ in (D.8), it should be noted that $\psi(\nu, \omega) = \tilde{\psi}(\lambda)$ when one takes $\lambda^{ijkl} = \nu^{i,j}\omega^{k,l}$. The expression we seek is therefore

$$\alpha_2 = E\left[\psi\left(\frac{\partial}{\partial \beta_1}\left(\sigma^2 \Sigma_0^{-1} \Xi \Sigma_0^{-1}\right), \frac{\partial}{\partial \beta_2}\left(\sigma^2 \Sigma_0^{-1} \Xi \Sigma_0^{-1}\right)\right)\right] = \tilde{\psi}(\lambda), \tag{H.6}$$

where λ^{ijkl} is taken to be the following expected value

$$\begin{aligned}
\lambda^{ijkl} &\equiv E\left[\nu^{i,j}\omega^{k,l}\right] \\
&= E\left[\left(\frac{\partial}{\partial \beta_1}\left(\sigma^2 \Sigma_0^{-1} \Xi \Sigma_0^{-1}\right)\right)_{ij}\left(\frac{\partial}{\partial \beta_2}\left(\sigma^2 \Sigma_0^{-1} \Xi \Sigma_0^{-1}\right)\right)_{kl}\right] \\
&= E\left[\sum_{r,s,t,u=1}^{N} \frac{\partial}{\partial \beta_1}\left(\sigma^2 (\Sigma_0^{-1})_{ir} \Xi_{rs} (\Sigma_0^{-1})_{sj}\right) \frac{\partial}{\partial \beta_2}\left(\sigma^2 (\Sigma_0^{-1})_{kt} \Xi_{tu} (\Sigma_0^{-1})_{ul}\right)\right] \\
&= \sum_{r,s,t,u=1}^{N} \frac{\partial}{\partial \beta_1}\left(\sigma^2 (\Sigma_0^{-1})_{ir} (\Sigma_0^{-1})_{sj}\right) \frac{\partial}{\partial \beta_2}\left(\sigma^2 (\Sigma_0^{-1})_{kt} (\Sigma_0^{-1})_{ul}\right) E\left[\Xi_{rs}\Xi_{tu}\right] \\
&= \Delta_0^2 \operatorname{Var}[\xi] \sum_{r=1}^{N} \frac{\partial}{\partial \beta_1}\left(\sigma^2 (\Sigma_0^{-1})_{ir} (\Sigma_0^{-1})_{rj}\right) \frac{\partial}{\partial \beta_2}\left(\sigma^2 (\Sigma_0^{-1})_{kr} (\Sigma_0^{-1})_{rl}\right),
\end{aligned}$$

with the third equality following from the interchangeability of unconditional expectations and differentiation with respect to β, and the fourth from the fact that $E\left[\Xi_{rs}\Xi_{tu}\right] \neq 0$ only when $r = s = t = u$, and

$$E\left[\Xi_{rr}\Xi_{rr}\right] = \Delta_0^2 \operatorname{Var}[\xi].$$

Thus we have

$$\begin{aligned}
\lambda^{ijkl} &= \Delta_0^2 \operatorname{Var}[\xi] \sum_{r=1}^{N} \frac{\partial}{\partial \beta_1}\left(\sigma^2\gamma^{-4} (V^{-1})_{ir} (V^{-1})_{rj}\right) \frac{\partial}{\partial \beta_2}\left(\sigma^2\gamma^{-4} (V^{-1})_{kr} (V^{-1})_{rl}\right) \\
&= \Delta_0^2 \operatorname{Var}[\xi] \sum_{r=1}^{N} \frac{\partial}{\partial \beta_1}\left(\sigma^2\gamma^{-4} v^{i,r} v^{r,j}\right) \frac{\partial}{\partial \beta_2}\left(\sigma^2\gamma^{-4} v^{k,r} v^{r,l}\right),
\end{aligned} \tag{H.7}$$

and

$$\begin{aligned}
&\frac{\partial}{\partial \beta_1}\left(\sigma^2\gamma^{-4} v^{i,r} v^{r,j}\right) \frac{\partial}{\partial \beta_2}\left(\sigma^2\gamma^{-4} v^{k,r} v^{r,l}\right) \\
&\qquad = \left(\frac{\partial\left(\sigma^2\gamma^{-4}\right)}{\partial \beta_1} v^{i,r} v^{r,j} + \sigma^2\gamma^{-4} \frac{\partial\left(v^{i,r} v^{r,j}\right)}{\partial \eta} \frac{\partial \eta}{\partial \beta_1}\right) \times \\
&\qquad\quad \times \left(\frac{\partial\left(\sigma^2\gamma^{-4}\right)}{\partial \beta_2} v^{k,r} v^{r,l} + \sigma^2\gamma^{-4} \frac{\partial\left(v^{k,r} v^{r,l}\right)}{\partial \eta} \frac{\partial \eta}{\partial \beta_2}\right).
\end{aligned}$$

Summing these terms, we obtain

$$\begin{aligned}\tilde{\psi}(\lambda) = & \frac{\Delta_0^2 \operatorname{Var}[\xi] \quad 2N}{(1-\eta)^7 (1+\eta)^3 (1+\eta^2)^3} \\ & \left(C_{1\lambda} \left(1-\eta^4\right) \left(2C_{5\lambda} C_{6\lambda} \left(1+\eta+\eta^2+2\eta^3\right) + C_{4\lambda} \left(1-\eta^4\right)\right) + \right. \\ & + 2C_{2\lambda} C_{3\lambda} \left(2C_{5\lambda} C_{6\lambda} \left(1+2\eta+4\eta^2+6\eta^3+5\eta^4+4\eta^5+4\eta^6\right) + \right. \\ & \left. \left. + C_{4\lambda} \left(1+\eta+\eta^2+2\eta^3-\eta^4-\eta^5-\eta^6-2\eta^7\right)\right)\right) + \\ & + o(N),\end{aligned}$$

where

$$C_{1\lambda} = \frac{\partial\left(\sigma^2\gamma^{-4}\right)}{\partial\beta_1}, \; C_{2\lambda} = \sigma^2\gamma^{-4}, \; C_{3\lambda} = \frac{\partial\eta}{\partial\beta_1},$$
$$C_{4\lambda} = \frac{\partial\left(\sigma^2\gamma^{-4}\right)}{\partial\beta_2}, \; C_{5\lambda} = \sigma^2\gamma^{-4}, \; C_{6\lambda} = \frac{\partial\eta}{\partial\beta_2}.$$

Putting it all together, we have

$$E\left[\psi\left(\frac{\partial\Sigma^{-1}}{\partial\beta_1}, \frac{\partial\Sigma^{-1}}{\partial\beta_2}\right)\right] = \psi\left(\frac{\partial\Sigma_0^{-1}}{\partial\beta_1}, \frac{\partial\Sigma_0^{-1}}{\partial\beta_2}\right) + \varepsilon^2\left(\alpha_1[2]+\alpha_2\right) + O(\varepsilon^3).$$

Finally, the asymptotic variance of the estimator $(\hat{\sigma}^2, \hat{a}^2)$ is given by

$$\mathrm{AVAR}_{\mathrm{true}}(\hat{\sigma}^2, \hat{a}^2) = E[\Delta] \; \left(D'S^{-1}D\right)^{-1}, \tag{H.8}$$

where

$$\begin{aligned}D = D' & = -\frac{1}{N} E_{\mathrm{true}}\left[\ddot{l}\right] = -\frac{1}{N} E_{\mathrm{normal}}\left[\ddot{l}\right] = \frac{1}{N} E_{\mathrm{normal}}\left[\dot{l}\dot{l}'\right] \\ & \equiv F^{(0)} + \varepsilon^2 F^{(2)} + O(\varepsilon^3)\end{aligned}$$

is given by the expression in the correctly specified case (G.15), with $F^{(0)}$ and $F^{(2)}$ given in (G.16) and (G.17) respectively. Also, in light of (H.1), we have

$$S = \frac{1}{N} E_{\mathrm{true}}\left[\dot{l}\dot{l}'\right] = -\frac{1}{N} E_{\mathrm{true}}\left[\ddot{l}\right] + \mathrm{Cum}_4[U] \; \Psi = D + \mathrm{Cum}_4[U] \; \Psi,$$

where

$$\begin{aligned}\Psi & = \frac{1}{4N}\begin{pmatrix} E\left[\psi\left(\frac{\partial\Sigma^{-1}}{\partial\sigma^2}, \frac{\partial\Sigma^{-1}}{\partial\sigma^2}\right)\right] & E\left[\psi\left(\frac{\partial\Sigma^{-1}}{\partial\sigma^2}, \frac{\partial\Sigma^{-1}}{\partial a^2}\right)\right] \\ \bullet & E\left[\psi\left(\frac{\partial\Sigma^{-1}}{\partial a^2}, \frac{\partial\Sigma^{-1}}{\partial a^2}\right)\right]\end{pmatrix} \\ & \equiv \Psi^{(0)} + \varepsilon^2\Psi^{(2)} + O(\varepsilon^3).\end{aligned}$$

Since, from (H.3), we have

$$E\left[\psi\left(\frac{\partial\Sigma^{-1}}{\partial\beta_1}, \frac{\partial\Sigma^{-1}}{\partial\beta_2}\right)\right] = \psi\left(\frac{\partial\Sigma_0^{-1}}{\partial\beta_1}, \frac{\partial\Sigma_0^{-1}}{\partial\beta_2}\right) + \varepsilon^2\alpha_1[2] + \varepsilon^2\alpha_2 + O(\varepsilon^3),$$

it follows that $\Psi^{(0)}$ is the matrix with entries $\frac{1}{4N}\psi\left(\frac{\partial\Sigma_0^{-1}}{\partial\beta_1}, \frac{\partial\Sigma_0^{-1}}{\partial\beta_2}\right)$, i.e.,

$$\Psi^{(0)} = \begin{pmatrix} \frac{\Delta_0}{\sigma^2\left(4a^2+\Delta_0\sigma^2\right)^3} & \frac{\Delta_0^{1/2}}{2a^4}\left(\frac{1}{\sigma\left(4a^2+\Delta_0\sigma^2\right)^{3/2}} - \frac{\Delta_0^{1/2}\left(6a^2+\Delta_0\sigma^2\right)}{\left(4a^2+\Delta_0\sigma^2\right)^3}\right) \\ \bullet & \frac{1}{2a^8}\left(1 - \frac{\Delta_0^{1/2}\sigma\left(6a^2+\Delta_0\sigma^2\right)}{\left(4a^2+\Delta_0\sigma^2\right)^{3/2}} - \frac{2a^4\left(16a^2+3\Delta_0\sigma^2\right)}{\left(4a^2+\Delta_0\sigma^2\right)^3}\right) \end{pmatrix} + \\ + o(1),$$

and

$$\Psi^{(2)} = \frac{1}{4N}\left(\alpha_1[2] + \alpha_2\right),$$

with

$$\frac{1}{4N}\alpha_1[2] = \mathrm{Var}[\xi]\begin{pmatrix} A & B \\ \bullet & C \end{pmatrix} + o(1),$$

with

$$A = \frac{2\Delta_0^{3/2}\left(-4a^2+\Delta_0\sigma^2\right)}{\sigma\left(4a^2+\Delta_0\sigma^2\right)^{9/2}},$$

$$B = \frac{\Delta_0\left(\left(-4a^2+\Delta_0\sigma^2\right)\left(4a^2+\Delta_0\sigma^2\right)^{3/2} - \Delta_0^{1/2}\sigma\left(-40a^4+2a^2\Delta_0\sigma^2+\Delta_0^2\sigma^4\right)\right)}{2a^4\left(4a^2+\Delta_0\sigma^2\right)^{9/2}},$$

$$C = \frac{8\Delta_0\sigma^2\left(\left(4a^2+\Delta_0\sigma^2\right)^{3/2} - \Delta_0^{1/2}\sigma\left(6a^2+\Delta_0\sigma^2\right)\right)}{a^4\left(4a^2+\Delta_0\sigma^2\right)^{9/2}},$$

$$\frac{1}{4N}\alpha_2 = \mathrm{Var}[\xi] \\ \begin{pmatrix} \frac{\Delta_0^{3/2}\left(40a^8-12a^4\Delta_0{}^2\sigma^4+\Delta_0{}^4\sigma^8\right)}{2\sigma\left(2a^2+\Delta_0\sigma^2\right)^3\left(4a^2+\Delta_0\sigma^2\right)^{9/2}} & \frac{\Delta_0^{3/2}\sigma\left(-44a^6-18a^4\Delta_0\sigma^2+7a^2\Delta_0{}^2\sigma^6+3\Delta_0{}^3\sigma^6\right)}{\left(2a^2+\Delta_0\sigma^2\right)^3\left(4a^2+\Delta_0\sigma^2\right)^{9/2}} \\ \bullet & \frac{2\Delta_0^{3/2}\sigma^3\left(50a^{22}+42a^2\Delta_0\sigma^2+9\Delta_0{}^2\sigma^4\right)}{\left(2a^2+\Delta_0\sigma^2\right)^3\left(4a^2+\Delta_0\sigma^2\right)^{9/2}} \end{pmatrix} + \\ + o(1).$$

It follows from (H.8) that

$$\begin{aligned} \mathrm{AVAR}_{\mathrm{true}}(\hat{\sigma}^2, \hat{a}^2) &= E[\Delta]\ \left(D\left(D + \mathrm{Cum}_4[U]\ \Psi\right)^{-1} D\right)^{-1} \\ &= \Delta_0\ \left(D\left(Id + \mathrm{Cum}_4[U]\ D^{-1}\Psi\right)^{-1}\right)^{-1} \\ &= \Delta_0\ \left(Id + \mathrm{Cum}_4[U]\ D^{-1}\Psi\right) D^{-1} \\ &= \Delta_0\ \left(Id + \mathrm{Cum}_4[U]\ D^{-1}\Psi\right) D^{-1} \\ &= \mathrm{AVAR}_{\mathrm{normal}}(\hat{\sigma}^2, \hat{a}^2) + \Delta_0\ \mathrm{Cum}_4[U]\ D^{-1}\Psi D^{-1}, \end{aligned}$$

where

$$\mathrm{AVAR}_{\mathrm{normal}}(\hat{\sigma}^2, \hat{a}^2) = \Delta_0 D^{-1} = A^{(0)} + \varepsilon^2 A^{(2)} + O(\varepsilon^3)$$

is the result given in Theorem 3, namely (1.53).

The correction term due to the misspecification of the error distribution is determined by $\mathrm{Cum}_4[U]$ times

$$\begin{aligned}
&\Delta_0 D^{-1} \Psi D^{-1} \\
&= \Delta_0 \left(F^{(0)} + \varepsilon^2 F^{(2)} + O(\varepsilon^3)\right)^{-1} \left(\Psi^{(0)} + \varepsilon^2 \Psi^{(2)} + O(\varepsilon^3)\right) \times \\
&\quad \times \left(F^{(0)} + \varepsilon^2 F^{(2)} + O(\varepsilon^3)\right)^{-1} \\
&= \Delta_0 \left(Id - \varepsilon^2 \left[F^{(0)}\right]^{-1} F^{(2)} + O(\varepsilon^3)\right) \left[F^{(0)}\right]^{-1} \left(\Psi^{(0)} + \varepsilon^2 \Psi^{(2)} + O(\varepsilon^3)\right) \times \\
&\quad \times \left(Id - \varepsilon^2 \left[F^{(0)}\right]^{-1} F^{(2)} + O(\varepsilon^3)\right) \left[F^{(0)}\right]^{-1} \\
&= \Delta_0 \left[F^{(0)}\right]^{-1} \Psi^{(0)} \left[F^{(0)}\right]^{-1} + \\
&\quad + \varepsilon^2 \Delta_0 \left(\left[F^{(0)}\right]^{-1} \Psi^{(2)} \left[F^{(0)}\right]^{-1} - \left[F^{(0)}\right]^{-1} F^{(2)} \left[F^{(0)}\right]^{-1} \Psi^{(0)} \left[F^{(0)}\right]^{-1} - \right. \\
&\quad \left. - \left[F^{(0)}\right]^{-1} \Psi^{(0)} \left[F^{(0)}\right]^{-1} F^{(2)} \left[F^{(0)}\right]^{-1} \right) + O(\varepsilon^3) \\
&\equiv B^{(0)} + \varepsilon^2 B^{(2)} + O(\varepsilon^3),
\end{aligned}$$

where the matrices are given in the text. The asymptotic variance is then given by

$$\mathrm{AVAR}_{\mathrm{true}}(\hat{\sigma}^2, \hat{a}^2) = \left(A^{(0)} + \mathrm{Cum}_4\left[U\right] B^{(0)}\right) + \varepsilon^2 \left(A^{(2)} + \mathrm{Cum}_4\left[U\right] B^{(2)}\right) + O(\varepsilon^3),$$

with the terms $A^{(0)}$, $A^{(2)}$, $B^{(0)}$ and $B^{(2)}$ given in the statement of the Theorem.

Appendix I – Proof of Theorem 5

From

$$E\left[Y_i^2\right] = E\left[w_i^2\right] + E\left[u_i^2\right] = \sigma^2 \Delta + \frac{c^2\left(1 - e^{-b\Delta}\right)}{b},$$

it follows that the estimator (1.5) has the following expected value

$$\begin{aligned}
E\left[\hat{\sigma}^2\right] &= \frac{1}{T} \sum_{i=1}^{N} E\left[Y_i^2\right] \\
&= \frac{N}{T} \left(\sigma^2 \Delta + \frac{c^2\left(1 - e^{-b\Delta}\right)}{b}\right) \\
&= \sigma^2 + \frac{c^2\left(1 - e^{-b\Delta}\right)}{b\Delta} \\
&= \left(\sigma^2 + c^2\right) - \frac{bc^2}{2} \Delta + O(\Delta^2).
\end{aligned} \tag{I.1}$$

The estimator's variance is

$$\begin{aligned}
\mathrm{Var}\left[\hat{\sigma}^2\right] &= \frac{1}{T^2} \mathrm{Var}\left[\sum_{i=1}^{N} Y_i^2\right] \\
&= \frac{1}{T^2} \sum_{i=1}^{N} \mathrm{Var}\left[Y_i^2\right] + \frac{2}{T^2} \sum_{i=1}^{N} \sum_{j=1}^{i-1} \mathrm{Cov}\left(Y_i^2, Y_j^2\right).
\end{aligned}$$

Since the $Y_i' s$ are normal with mean zero,

$$\mathrm{Var}\left[Y_i^2\right] = 2\,\mathrm{Var}[Y_i]^2 = 2E\left[Y_i^2\right]^2$$

and for $i > j$

$$\mathrm{Cov}\left(Y_i^2, Y_j^2\right) = 2\ \mathrm{Cov}\left(Y_i, Y_j\right)^2 = 2\ E\left[u_i u_j\right]^2,$$

since

$$\mathrm{Cov}\left(Y_i, Y_j\right) = E\left[Y_i Y_j\right] = E\left[\left(w_i + u_i\right)\left(w_j + u_j\right)\right] = E\left[u_i u_j\right].$$

Now we have

$$\begin{aligned} E\left[u_i u_j\right] &= E\left[\left(U_{\tau_i} - U_{\tau_{i-1}}\right)\left(U_{\tau_j} - U_{\tau_{j-1}}\right)\right] \\ &= E\left[U_{\tau_i} U_{\tau_j}\right] - E\left[U_{\tau_i} U_{\tau_{j-1}}\right] - E\left[U_{\tau_{i-1}} U_{\tau_j}\right] + E\left[U_{\tau_{i-1}} U_{\tau_{j-1}}\right] \\ &= -\frac{c^2\left(1 - e^{-b\Delta}\right)^2 e^{-b\Delta(i-j-1)}}{2b}, \end{aligned}$$

so that

$$\begin{aligned} \mathrm{Cov}\left(Y_i^2, Y_j^2\right) &= 2\left(-\frac{c^2\left(1 - e^{-b\Delta}\right)^2 e^{-b\Delta(i-j-1)}}{2b}\right)^2 \\ &= \frac{c^4 e^{-2b\Delta(i-j-1)}\left(1 - e^{-b\Delta}\right)^4}{2b^2}, \end{aligned}$$

and consequently

$$\begin{aligned} \mathrm{Var}\left[\hat{\sigma}^2\right] = \frac{1}{T^2}\left\{\frac{c^4\left(1 - e^{-b\Delta}\right)^2\left(Ne^{-2b\Delta} - 1 + e^{-2Nb\Delta}\right)}{b^2\left(1 + e^{-b\Delta}\right)^2} + \right. \\ \left. + 2N\left(\sigma^2\Delta + \frac{c^2\left(1 - e^{-b\Delta}\right)}{b}\right)^2\right\}, \end{aligned} \tag{I.2}$$

with $N = T/\Delta$. The RMSE expression follows from (I.1) and (I.2). As in Theorem 1, these are exact small sample expressions, valid for all (T, Δ).

Tables

Value of a	T = 1 day	T = 1 year	T = 5 years
Panel A: $\sigma = 30\%$	**Stocks**		
0.01%	1 mn	4 mn	6 mn
0.05%	5 mn	31 mn	53 mn
0.1%	12 mn	1.3 hr	2.2 hr
0.15%	22 mn	2.2 hr	3.8 hr
0.2%	32 mn	3.3 hr	5.6 hr
0.3%	57 mn	5.6 hr	1.5 day
0.4%	1.4 hr	1.3 day	2.2 days
0.5%	2 hr	1.7 day	2.9 days
0.6%	2.6 hr	2.2 days	3.7 days
0.7%	3.3 hr	2.7 days	4.6 days
0.8%	4.1 hr	3.2 days	1.1 week
0.9%	4.9 hr	3.8 days	1.3 week
1.0%	5.9 hr	4.3 days	1.5 week
Panel B: $\sigma = 10\%$	**Currencies**		
0.005%	4 mn	23 mn	39 mn
0.01%	9 mn	58 mn	1.6 hr
0.02%	23 mn	2.4 hr	4.1 hr
0.05%	1.3 hr	8.2 hr	14.0 hr
0.10%	3.5 hr	20.7 hr	1.5 day

Table 1.1. Optimal Sampling Frequency

This table reports the optimal sampling frequency Δ^* given in equation (1.20) for different values of the standard deviation of the noise term a and the length of the sample T. Throughout the table, the noise is assumed to be normally distributed (hence $\mathrm{Cum}_4[U] = 0$ in formula (1.20)). In Panel A, the standard deviation of the efficient price process is $\sigma = 30\%$ per year, and at $\sigma = 10\%$ per year in Panel B. In both panels, 1 year = 252 days, but in Panel A, 1 day = 6.5 hours (both the NYSE and NASDAQ are open for 6.5 hours from 9:30 to 16:00 EST), while in Panel B, 1 day = 24 hours as is the case for major currencies. A value of $a = 0.05\%$ means that each transaction is subject to Gaussian noise with mean 0 and standard deviation equal to 0.05% of the efficient price. If the sole source of the noise were a bid/ask spread of size s, then a should be set to $s/2$. For example, a bid/ask spread of 10 cents on a \$10 stock would correspond to $a = 0.05\%$. For the dollar/euro exchange rate, a bid/ask spread of $s = 0.04\%$ translates into $a = 0.02\%$. For the bid/ask model, which is based on binomial instead of Gaussian noise, $\mathrm{Cum}_4[U] = -s^4/8$, but this quantity is negligible given the tiny size of s.

Sampling Interval	Theoretical Mean	Sample Mean	Theoretical Stand. Dev.	Sample Stand. Dev.
5 minutes	0.185256	0.185254	0.00192	0.00191
15 minutes	0.121752	0.121749	0.00208	0.00209
30 minutes	0.10588	0.10589	0.00253	0.00254
1 hour	0.097938	0.097943	0.00330	0.00331
2 hours	0.09397	0.09401	0.00448	0.00440
1 day	0.09113	0.09115	0.00812	0.00811
1 week	0.0902	0.0907	0.0177	0.0176

Table 1.2. Monte Carlo Simulations: Bias and Variance when Market Microstructure Noise is Ignored

This table reports the results of $M = 10,000$ Monte Carlo simulations of the estimator $\hat{\sigma}^2$, with market microstructure noise present but ignored. The column "theoretical mean" reports the expected value of the estimator, as given in (1.17) and similarly for the column "theoretical standard deviation" (the variance is given in (1.19)). The "sample" columns report the corresponding moments computed over the M simulated paths. The parameter values used to generate the simulated data are $\sigma^2 = 0.3^2 = 0.09$ and $a^2 = (0.15\%)^2$ and the length of each sample is $T = 1$ year.

Figures

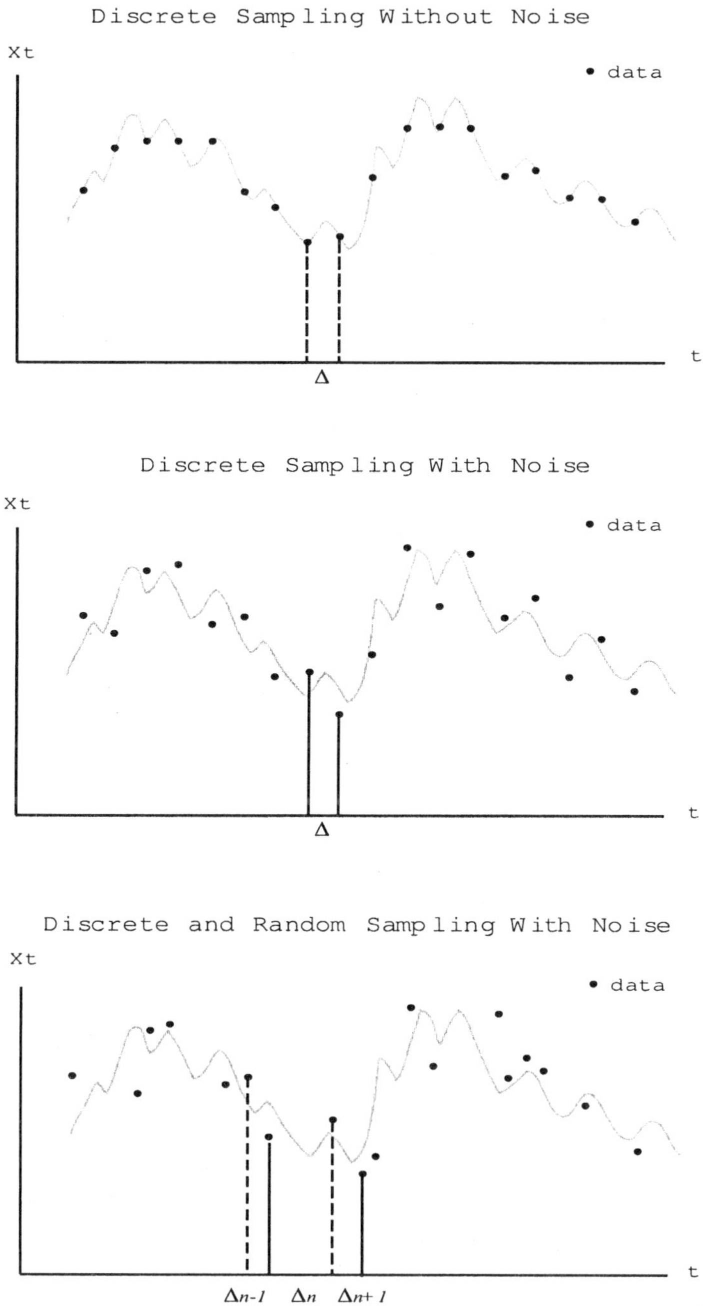

Fig. 1.1. Various discrete sampling modes – no noise (Section 1.2), with noise (Sections 1.3-1.7) and randomly spaced with noise (Section 1.8)

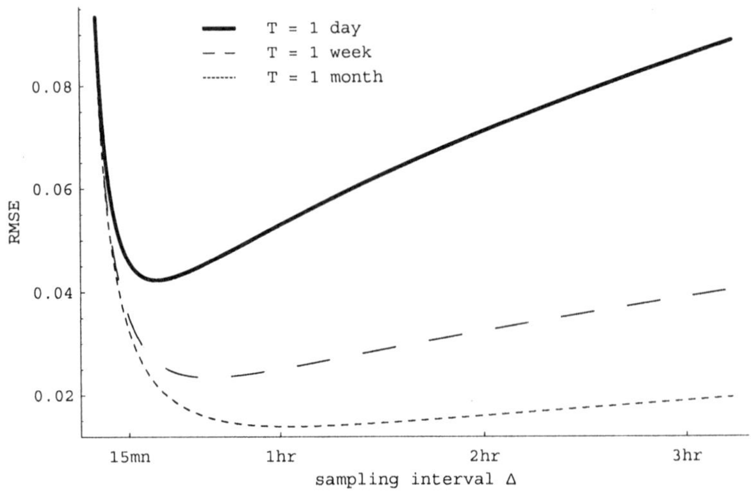

Fig. 1.2. RMSE of the estimator $\hat{\sigma}^2$ when the presence of the noise is ignored

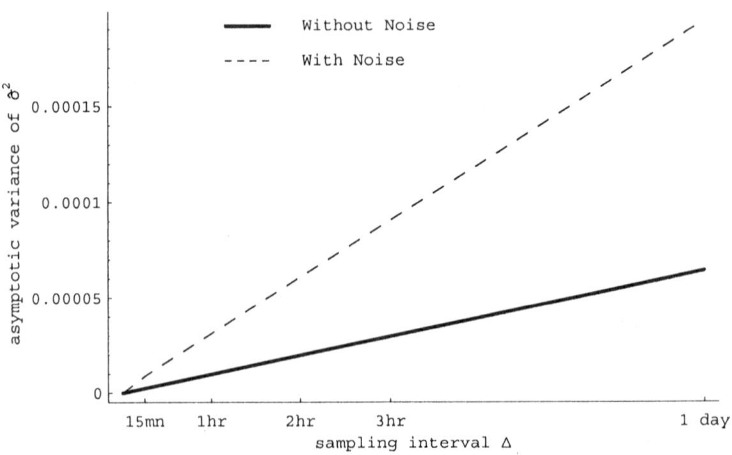

Fig. 1.3. Comparison of the asymptotic variances of the MLE $\hat{\sigma}^2$ without and with noise taken into account

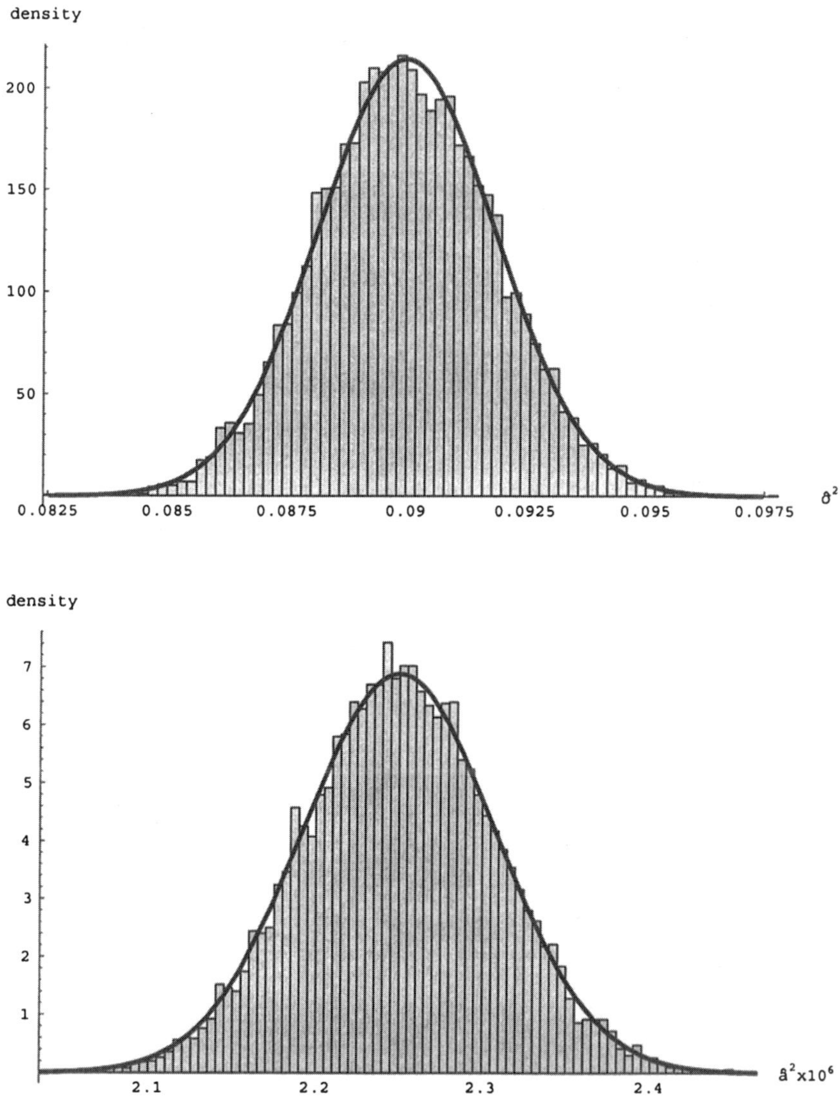

Fig. 1.4. Asymptotic and Monte Carlo distributions of the MLE $(\hat{\sigma}^2, \hat{a}^2)$ with Gaussian microstructure noise

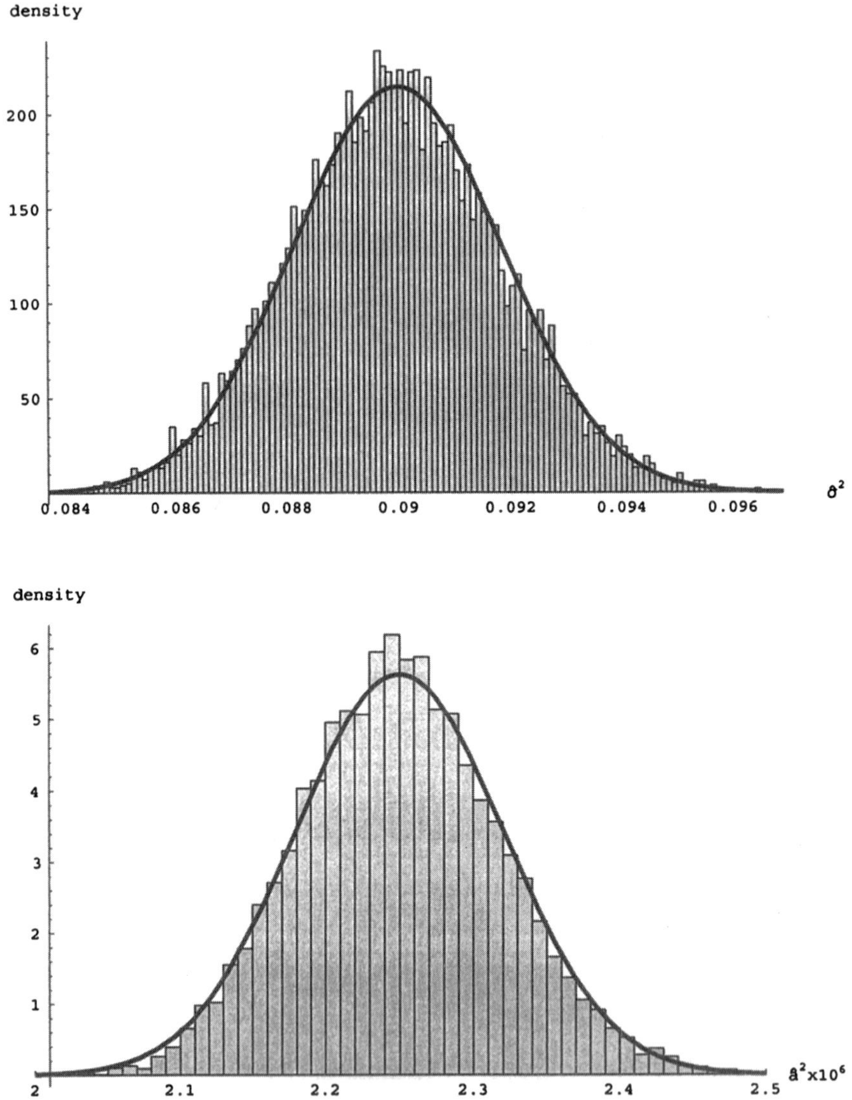

Fig. 1.5. Asymptotic and Monte Carlo distributions of the QMLE $(\hat{\sigma}^2, \hat{a}^2)$ with misspecified microstructure noise

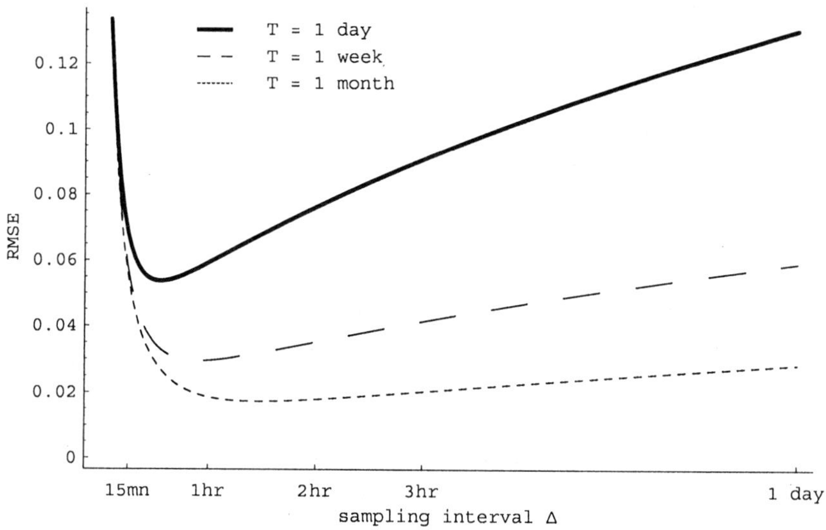

Fig. 1.6. RMSE of the estimator $\hat{\sigma}^2$ when the presence of serially correlated noise is ignored

References

[1] Abramowitz, M., and I. A. Stegun, 1972, *Handbook of Mathematical Functions.* Dover, New York, NY.

[2] Aït-Sahalia, Y., 1996, "Nonparametric Pricing of Interest Rate Derivative Securities," *Econometrica*, 64, 527–560.

[3] Aït-Sahalia, Y., 2002, "Maximum-Likelihood Estimation of Discretely-Sampled Diffusions: A Closed-Form Approximation Approach," *Econometrica*, 70, 223–262.

[4] Aït-Sahalia, Y., and P. A. Mykland, 2003, "The Effects of Random and Discrete Sampling When Estimating Continuous-Time Diffusions," *Econometrica*, 71, 483–549.

[5] Andersen, T. G., T. Bollerslev, F. X. Diebold, and P. Labys, 2001, "The Distribution of Exchange Rate Realized Volatility," *Journal of the American Statistical Association*, 96, 42–55.

[6] Andersen, T. G., T. Bollerslev, F. X. Diebold, and P. Labys, 2003, "Modeling and Forecasting Realized Volatility," *Econometrica*, 71, 579–625.

[7] Bandi, F. M., and P. C. B. Phillips, 2003, "Fully Nonparametric Estimation of Scalar Diffusion Models," *Econometrica*, 71, 241–283.

[8] Barndorff-Nielsen, O. E., and N. Shephard, 2002, "Econometric Analysis of Realized Volatility and Its Use in Estimating Stochastic Volatility Models," *Journal of the Royal Statistical Society, B*, 64, 253–280.

[9] Bessembinder, H., 1994, "Bid-Ask Spreads in the Interbank Foreign Exchange Markets," *Journal of Financial Economics*, 35, 317–348.

[10] Black, F., 1986, "Noise," *Journal of Finance*, 41, 529–543.

[11] Choi, J. Y., D. Salandro, and K. Shastri, 1988, "On the Estimation of Bid-Ask Spreads: Theory and Evidence," *The Journal of Financial and Quantitative Analysis*, 23, 219–230.

[12] Conrad, J., G. Kaul, and M. Nimalendran, 1991, "Components of Short-Horizon Individual Security Returns," *Journal of Financial Economics*, 29, 365–384.

[13] Dahlquist, G., and A. Björck, 1974, *Numerical Methods.* Prentice-Hall Series in Automatic Computation, New York.

[14] Delattre, S., and J. Jacod, 1997, "A Central Limit Theorem for Normalized Functions of the Increments of a Diffusion Process, in the Presence of Round-Off Errors," *Bernoulli*, 3, 1–28.

[15] Durbin, J., 1959, "Efficient Estimation of Parameters in Moving-Average Models," *Biometrika*, 46, 306–316.

[16] Easley, D., and M. O'Hara, 1992, "Time and the Process of Security Price Adjustment," *Journal of Finance*, 47, 577–605.

[17] Engle, R. F., and J. R. Russell, 1998, "Autoregressive Conditional Duration: A New Model for Irregularly Spaced Transaction Data," *Econometrica*, 66, 1127–1162.

[18] French, K., and R. Roll, 1986, "Stock Return Variances: The Arrival of Information and the Reaction of Traders," *Journal of Financial Economics*, 17, 5–26.

[19] Gençay, R., G. Ballocchi, M. Dacorogna, R. Olsen, and O. Pictet, 2002, "Real-Time Trading Models and the Statistical Properties of Foreign Exchange Rates," *International Economic Review*, 43, 463–491.

[20] Glosten, L. R., 1987, "Components of the Bid-Ask Spread and the Statistical Properties of Transaction Prices," *Journal of Finance*, 42, 1293–1307.

[21] Glosten, L. R., and L. E. Harris, 1988, "Estimating the Components of the Bid/Ask Spread," *Journal of Financial Economics*, 21, 123–142.

[22] Gloter, A., and J. Jacod, 2000, "Diffusions with Measurement Errors: I - Local Asymptotic Normality and II - Optimal Estimators," working paper, Université de Paris VI.

[23] Gottlieb, G., and A. Kalay, 1985, "Implications of the Discreteness of Observed Stock Prices," *Journal of Finance*, 40, 135–153.

[24] Haddad, J., 1995, "On the Closed Form of the Likelihood Function of the First Order Moving Average Model," *Biometrika*, 82, 232–234.

[25] Hamilton, J. D., 1995, *Time Series Analysis*. Princeton University Press, Princeton.

[26] Hansen, L. P., and J. A. Scheinkman, 1995, "Back to the Future: Generating Moment Implications for Continuous-Time Markov Processes," *Econometrica*, 63, 767–804.

[27] Harris, L., 1990a, "Estimation of Stock Price Variances and Serial Covariances from Discrete Observations," *Journal of Financial and Quantitative Analysis*, 25, 291–306.

[28] Harris, L., 1990b, "Statistical Properties of the Roll Serial Covariance Bid/Ask Spread Estimator," *Journal of Finance*, 45, 579–590.

[29] Hasbrouck, J., 1993, "Assessing the Quality of a Security Market: A New Approach to Transaction-Cost Measurement," *Review of Financial Studies*, 6, 191–212.

[30] Heyde, C. C., 1997, *Quasi-Likelihood and Its Application*. Springer-Verlag, New York.

[31] Jacod, J., 1996, "La Variation Quadratique du Brownien en Présence d'Erreurs d'Arrondi," *Astérisque*, 236, 155–162.

[32] Kaul, G., and M. Nimalendran, 1990, "Price Reversals: Bid-Ask Errors or Market Overreaction," *Journal of Financial Economics*, 28, 67–93.

[33] Lo, A. W., and A. C. MacKinlay, 1990, "An Econometric Analysis of Nonsynchronous Trading," *Journal of Econometrics*, 45, 181–211.

[34] Macurdy, T. E., 1982, "The Use of Time Series Processes to Model the Error Structure of Earnings in a Longitudinal Data Analysis," *Journal of Econometrics*, 18, 83–114.

[35] Madhavan, A., M. Richardson, and M. Roomans, 1997, "Why Do Security Prices Change?," *Review of Financial Studies*, 10, 1035–1064.

[36] McCullagh, P., 1987, *Tensor Methods in Statistics*. Chapman and Hall, London, U.K.

[37] McCullagh, P., and J. Nelder, 1989, *Generalized Linear Models*, second edn., Chapman and Hall, London, U.K.

[38] Merton, R. C., 1980, "On Estimating the Expected Return on the Market: An Exploratory Investigation," *Journal of Financial Economics*, 8, 323–361.

[39] Roll, R., 1984, "A Simple Model of the Implicit Bid-Ask Spread in an Efficient Market," *Journal of Finance*, 39, 1127–1139.

[40] Shaman, P., 1969, "On the Inverse of the Covariance Matrix of a First Order Moving Average," *Biometrika*, 56, 595–600.

[41] Sias, R. W., and L. T. Starks, 1997, "Return Autocorrelation and Institutional Investors," *Journal of Financial Economics*, 46, 103–131.

[42] White, H., 1982, "Maximum Likelihood Estimation of Misspecified Models," *Econometrica*, 50, 1–25.

[43] Zhang, L., P. A. Mykland, and Y. Aït-Sahalia, 2002, "A Tale of Two Time Scales: Determining Integrated Volatility with Noisy High-Frequency Data," *forthcoming in Journal of the American Statistical Association.*

2

Multipower Variation and Stochastic Volatility

Ole E. Barndorff-Nielsen and Neil Shephard

[1] Department of Mathematical Sciences, University of Aarhus, Ny Munkegade, DK-8000 Aarhus C, Denmark
oebn@imf.au.dk
[2] Nuffield College, Oxford OX1 1NF, U.K.
neil.shephard@nuf.ox.ac.uk

Summary. In this brief note we review some of our recent results on the use of high frequency financial data to estimate objects like integrated variance in stochastic volatility models. Interesting issues include multipower variation, jumps and market microstructure effects.

2.1 Introduction

This paper briefly summarises some recent and ongoing work concerning inference on stochastic volatility (see, for example, the reviews in Ghysels, Harvey, and Renault [14] and Shephard [17]), with the focus on multipower variation as a tool for such inference.

We assume that the log price process is of the form $X = Y + Z$ where Y is an Brownian semimartingale ($\mathcal{BSM}$),

$$Y_t = Y_0 + \int_0^t a_u \mathrm{d}u + \int_0^t \sigma_u \mathrm{d}W_u \tag{2.1}$$

whose quadratic variation $[Y]$, which embodies the volatile character of Y, is the main object of interest, while Z expresses effects that may be considered in some sense extraneous to the basic dynamics of the financial market. The process Z may be a jump process, representing for instance the impacts of macroeconomic announcements, or it could represent microstructure noise.

In (2.1) W is a Brownian motion, the volatility process σ is assumed to be positive and càdlàg, a is predictable and locally bounded, and we have the well known result that the quadratic variation of Y satisfies

$$[Y]_t = \int_0^t \sigma_u^2 \mathrm{d}u .$$

We shall write, for arbitrary $r > 0$,

$$\sigma_t^{r*} = \int_0^t \sigma_u^r \mathrm{d}u$$

and we wish to device inference procedures for these quantities, particularly for $\sigma_t^{2*}(= [Y]_t)$.

Although the above formulation is in terms of univariate processes, much of the theory extends rather readily to a general multivariate setting. However, we shall not consider this further here but refer to the papers Barndorff-Nielsen and Shephard [10], Barndorff-Nielsen, Graversen, Jacod, and Shephard [4] and Barndorff-Nielsen, Graversen, Jacod, Podolsky, and Shephard [3]. We shall further restrict attention to equidistant sampling schemes; the situation under more general schemes are discussed in Barndorff-Nielsen and Shephard [12] and Woerner [18]. See also Mykland and Zhang [16].

After introducing the concepts of multipower variation (MPV) and generalised multipower variation in Section 2, we discuss, in Section 3, applications of MPV to inference on volatility under $\mathcal{BSM}$ models (that is, there we suppose that $Z = 0$). Section 4 treats applications of MPV to cases where Z is a jump process, both for finite and infinite activity scenarios. The final Section 5 indicates some work in progress concerning the impact of microstructure noise.

For numerical and empirical work and illustrations of the theoretical results presented here we additionally refer to Barndorff-Nielsen, Hansen, Lunde, and Shephard [5, 6], Barndorff-Nielsen and Shephard [8, 9, 11] and Barndorff-Nielsen, Shephard, and Winkel [13].

2.2 Multipower Variation

For arbitrary continuous time processes $X = \{X_t\}_{t \geq 0}$ and equidistant subdivisions of time with lag $\delta > 0$ we define the δ-discretisation of X by

$$X_{\delta,t} = X_t - X_{\lfloor t/\delta \rfloor \delta} \,,$$

where, as usual, $\lfloor s \rfloor$ indicates the largest integer less than or equal to a real number s. Furthermore, we introduce the *realised multipower variation* (MPV) of order m for X by

$$[X_\delta]_t^{[\mathbf{r}]} = [X_\delta]_t^{[r_1,\dots,r_m]} = [X_\delta,\dots,X_\delta]_t^{[r_1,\dots,r_m]} = \sum_{j=m}^{\lfloor t/\delta \rfloor} |x_{j-m+1}|^{r_1} \cdots |x_j|^{r_m} \,,$$

where $\mathbf{r}$ is short for $r_1,\dots,r_m$, the r_l being nonnegative, and

$$x_j = X_{j\delta} - X_{(j-1)\delta} \,.$$

We shall also use the normalised version of realised MPV, defined by

$$\{X_\delta\}_t^{[\mathbf{r}]} = \{X_\delta\}_t^{[r_1,\dots,r_m]} = \{X_\delta,\dots,X_\delta\}_t^{[r_1,\dots,r_m]} = \delta^{1-r_+/2}[X_\delta]_t^{[\mathbf{r}]} \,,$$

where $r_+ = r_1 + \cdots + r_m$.

In particular we will discuss applications of the power, bipower, and tripower variations (PV, BPV and TPV):

$$[X_\delta]_t^{[r]} = \sum_{j=1}^{\lfloor t/\delta \rfloor} |x_j|^r ,$$

$$[X_\delta]_t^{[r,s]} = \sum_{j=2}^{\lfloor t/\delta \rfloor} |x_{j-1}|^r |x_j|^s ,$$

$$[X_\delta]_t^{[r,s,u]} = \sum_{j=3}^{\lfloor t/\delta \rfloor} |x_{j-2}|^r |x_{j-1}|^s |x_j|^u .$$

In the recent paper Barndorff-Nielsen, Graversen, Jacod, Podolsky, and Shephard [3] the concept of MPV is generalised to *generalised multipover variation* where one considers realised objects of the form

$$\sum_{j=m}^{\lfloor t/\delta \rfloor} g_1(\delta^{-1/2} x_{j-m+1}) \cdots g_m(\delta^{-1/2} x_j) ,$$

where $g_1, \ldots, g_m$ are real functions satisfying certain regularity conditions, powers of absolute values being a special case. While this generalisation opens up further potential for applications, the associated central limit theory for (multivariate) $\mathcal{BSM}$ models, as established in [3], is in effect not more (or less) difficult than for the MPV case. In the following Section we draw on results from [3] to establish feasible limit theory for multipower variation under the $\mathcal{BSM}$ specification.

2.3 MPV for $\mathcal{BSM}$

Let Y be a Brownian semimartingale as defined in Section 1. Important special cases are

$$Y_t = Y_0 + \int_0^t a(s, Y_s) \mathrm{d}s + \int_0^t \sigma(s, Y_s) \, \mathrm{d}W_s,$$

and

$$Y_t = Y_0 + \int_0^t a_u \mathrm{d}u + \int_0^t \sigma_u \mathrm{d}W_u,$$

with the volatility process σ satisfying a stochastic differential equation of the form

$$\sigma_t^2 = \sigma_0^2 + \int_0^t a'_s \mathrm{d}s + \int_0^t \sigma'_{s-} \mathrm{d}V_s + \int_0^t v_{s-} \mathrm{d}Z_s ,$$

where a', σ', v are adapted càdlàg processes, V is a Brownian motion, possibly correlated with W, and Z is a Lévy process. This second structure encompasses both the models of Heston type and those of non-Gaussian OU-based type introduced by Barndorff-Nielsen and Shephard [7].

Without further assumptions we then have the following convergence in probability (CiP) and central limit theorem (CLT) for MPV.

Theorem 1. *As* $\delta \to 0$

$$\{Y_\delta, ..., Y_\delta\}_t^{[r_1,...,r_m]} \xrightarrow{p} \mu_{r_1} \cdots \mu_{r_m} \sigma_t^{r_+ *} \tag{2.2}$$

and

$$\delta^{-1/2}\left(\{Y_\delta, ..., Y_\delta\}_t^{[r_1,...,r_m]} - \sigma_t^{r_+ *}\right) \xrightarrow{law} \sqrt{\nu_{\mathbf{r}}} \int_0^t \sigma_u^{r_+} \mathrm{d}B_u \,, \tag{2.3}$$

where B is a Brownian motion which is independent of Y and where

$$\nu_{\mathbf{r}} = \prod_{l=1}^m \mu_{2r_l} - (2m-1) \prod_{l=1}^m \mu_{r_l}^2 + 2 \sum_{k=1}^{m-1} \prod_{l=1}^k \mu_{r_l} \prod_{l=m-k+1}^m \mu_{r_l} \prod_{l=1}^{m-k} \mu_{r_l + r_{l+k}} \tag{2.4}$$

and $\mu_r = \mathrm{E}\{|u|^r\}$ for $u \sim N(0,1)$. The convergence in (2.3) is in fact stable as processes, which is stronger than convergence in law. □

This theorem is a special case of the results established in Barndorff-Nielsen, Graversen, Jacod, Podolsky, and Shephard [3]. The proofs given there are (unavoidably) rather long-winded and use advanced stochastic analysis. An explanatory simpler version will be given in Barndorff-Nielsen, Graversen, Jacod, and Shephard [4].

The independence between Y and B is crucial for the possibility to establish statistically feasible CLT results, such as the following :

$$\frac{\mu_{\mathbf{r}}^{-1}\{Y_\delta\}_t^{[\mathbf{r}]} - \sigma_t^{r_+ *}}{\delta^{1/2} \mu_{\mathbf{r}}^{-1} \sqrt{\nu_{\mathbf{r}} \mu_{\mathbf{s}}^{-1} \{Y_\delta\}_t^{[\mathbf{s}]}}} \xrightarrow{law} N(0,1) \,,$$

where

$$\mu_{\mathbf{r}} = \prod_{l=1}^m \mu_{r_l}$$

and $\mathbf{s} = (s_1, ..., s_m)$ with $s_+ = 2r_+$.

In particular, note that for realised PV, BPV and TPV in the case where $r_+ = 2$, $r_j = 1/r_+$, then for example,

$$\frac{\{Y_\delta\}_t^{[2]} - \sigma_t^{2*}}{\delta^{1/2} \sqrt{\nu_2 \mu_4^{-1} \{Y_\delta\}_t^{[4]}}} \xrightarrow{law} N(0,1) \,,$$

$$\frac{\mu_1^{-2}[Y_\delta]_t^{[1,1]} - \sigma_t^{2*}}{\delta^{1/2}\mu_1^{-4}\sqrt{\nu_{2,2}\{Y_\delta\}_t^{[1,1,1,1]}}} \xrightarrow{law} N(0,1)\,,$$

and

$$\frac{\mu_{2/3}^{-3}[Y_\delta]_t^{[2/3,2/3,2/3]} - \sigma_t^{2*}}{\delta^{1/2}\mu_{2/3}^{-6}\sqrt{\nu_{4/3,4/3,4/3}\{Y_\delta\}_t^{[2/3,2/3,2/3,2/3,2/3,2/3]}}} \xrightarrow{law} N(0,1)\,. \tag{2.5}$$

2.4 MPV for $\mathcal{BSM}$ + Jump Process

We now consider various extensions of the above results to one-dimensional processes of the form

$$X = Y + Z\,,$$

where $Y \in \mathcal{BSM}$ while Z is a process exhibiting jumps. The processes Y and Z are not assumed to be independent. Our discussion is based on Barndorff-Nielsen, Shephard, and Winkel [13] and is related to Barndorff-Nielsen and Shephard [11] and Woerner [18].

We assume that Y satisfies (2.2) or (2.3) for MPV and consider to which extent this limiting behaviour remains the same when Z is added to Y, i.e. whether the influence of Z is negligible (in this respect).

In other words, we ask whether:

for the CiP case,

$$\{X_\delta, ..., , X_\delta\}^{[r_1,...,r_m]} - \{Y_\delta, ..., , Y_\delta\}^{[r_1,...,r_m]} = o_p(1)\,,$$

for the CLT case,

$$\{X_\delta, ..., , X_\delta\}^{[r_1,...,r_m]} - \{Y_\delta, ..., , Y_\delta\}^{[r_1,...,r_m]} = o_p(\delta^{1/2})\,.$$

We shall use the fact that Y satisfies

$$\delta^{-1/2}|Y_{j\delta} - Y_{(j-1)\delta}| = O_p\left(|\log \delta|^{1/2}\right)$$

uniformly in j. We write $\max r$ for $\max\{r_1, ..., r_m\}$.

2.4.1 Finite Activity Case

When Z is a finite activity jump process then pathwise the number of jumps of Z is finite and, for sufficiently small δ, none of the additive terms in $[X_\delta, ..., X_\delta]^{[r_1,...,r_m]}$ involves more than one jump.

Each of the terms in $[X_\delta, ..., X_\delta]^{[r_1,...,r_m]}$ that contains no jumps is of order

$$O_p\left((|\log \delta|)^{r_+/2}\right)$$

and any of the terms that do include a jump is of order

$$O_p\left((|\log\delta|)^{(r_+ - \max r)}\right).$$

Hence

$$\begin{aligned}\delta^{1-r_+/2}\left([X_\delta]^{[r]} - [Y_\delta]^{[r]}\right) &= \delta^{1-r_+/2} O_p\left((\delta|\log\delta|)^{(r_+ - \max r)/2}\right) \\ &= O_p\left(\delta^{1-\max r/2}|\log\delta|^{(r_+ - \max r)/2}\right).\end{aligned}$$

So:

- CiP is not influenced by Z so long as $\max r < 2$, while CLT continues to hold so long as $\max r < 1$.

The bound $\max r < 2$ seems quite a tight condition for when $m = 1$ and $r = 2$

$$[X_\delta]^{[2]} \xrightarrow{p} [Y] + [Z]$$

i.e. jumps do impact the limit.

The above CiP and CLT results mean that we can use multipower variation to make inference about σ_t^{2*}, integrated variance, in the presence of finite activity jump processes so long as $\max r < 1$ and $r_+ = 2$.

An example of this is where $m = 3$ and we take $r_1 = r_2 = r_3 = 2/3$, that is using TPV – Tripower Variation, cf. relation (2.5) above.

2.4.2 Infinite Activity Case

In discussing CiP and CLT for the case where Z exhibits infinite activity, i.e. infinitely many jumps in any finite time interval, we shall for simplicity restrict consideration to the case $r_1 = \cdots = r_m = r$. Detailed calculations, using classical inequalities, show that:

- for MPVCiP it suffices that

$$\begin{gathered}\delta^{1-mr/2}[Z_\delta, ..., Z_\delta]^{[r,...,r]} = o_p(1) \\ \delta^{1-(m-1)r/2}|\log\delta|[Z_\delta, ..., Z_\delta]^{[r,...,r]} \quad \left[\binom{m}{1}\right] = o_p(1) \\ \cdots\cdots\cdot \\ \delta^{1-r/2}|\log\delta|^{m-1}[Z_\delta]^{[r]} \quad \left[\binom{m}{1}\right] = o_p(1).\end{gathered}$$

- For MPVCLT it suffices that $r \leq 1$ and

$$\begin{gathered}\delta^{(1-mr)/2}[Z_\delta, ..., Z_\delta]^{[r,...,r]} = o_p(1) \\ \delta^{(1-(m-1)r)/2}|\log\delta|[Z_\delta, ..., Z_\delta]^{[r,...,r]} \quad \left[\binom{m}{1}\right] = o_p(1) \\ \cdots\cdots\cdot \\ \delta^{(1-r)/2}|\log\delta|^{m-1}[Z_\delta]^{[r]} \quad \left[\binom{m}{1}\right] = o_p(1).\end{gathered}$$

These sufficient conditions are also close to being necessary, as the examples below will show.

2.4.3 Lévy Jumps

Now, suppose that the jump process Z is a Lévy process. Alternatively, we might consider the case of Z being an OU process with BDLP (background driving Lévy process) L. However, as shown in Barndorff-Nielsen, Shephard, and Winkel [13], the conclusions regarding CiP and CLT for $X = Y + Z$ would be the same as for $X = Y + L$.

Example 1. Let Z be the $\Gamma(\nu, \alpha)$ subordinator, i.e. Z is the Lévy process for which the law of Z_1 is the gamma distribution with pdf

$$\frac{\alpha^{\nu}}{\Gamma(\nu)} x^{\nu-1} e^{-\alpha x}.$$

This is an infinite activity process and for $t \downarrow 0$ we have

$$\mathrm{E}\left\{|Z_t|^p\right\} = \alpha^{-p} \frac{\Gamma(t\nu + p)}{\Gamma(t\nu)} \sim O(t)$$

whatever the value of $p > 0$. (Here we have used that $t\Gamma(t) \to 1$ as $t \to 0$.) Thus $[Z_\delta]^{[r]} = O_p(1)$, $[Z_\delta, Z_\delta]^{[r,r]} = O_p(\delta)$, $[Z_\delta, Z_\delta, Z_\delta]^{[r,r,r]} = O_p(\delta^2)$, etc. Consequently:

- MPVCiP is valid for all $m = 1, 2, \ldots$ and $0 < r < 2$.
- MPVCLT is valid for all $m = 1, 2, \ldots$ and $0 < r < 1$.

On the other hand we have, for example, that BPVCLT does not hold if $r = 1$ and $Y \perp\!\!\!\perp Z$.

Example 2. Let Z be the $IG(\phi, \gamma)$ subordinator, i.e. Z is the Lévy process for which the law of Z_1 is the inverse Gaussian distribution with pdf

$$\frac{\delta}{\sqrt{2\pi}} e^{\delta\gamma} x^{-3/2} e^{-\frac{1}{2}(\phi^2 x^{-1} + \gamma^2 x)}.$$

Then, as $t \downarrow 0$,

$$\mathrm{E}\left\{|Z_t|^p\right\} \sim \begin{cases} O(t) & \text{if } p > \frac{1}{2} \\ O(t|\log t|) & \text{if } p = \frac{1}{2} \\ O(t^{2p}) & \text{if } 0 < p < \frac{1}{2} \end{cases} \tag{2.6}$$

so that, for $\frac{1}{2} < r < 1$ we have $[Z_\delta]^{[r,r]} = O_p(\delta)$ and $[Z_\delta]^{[r]} = O_p(1)$. Consequently:

- MPVCiP is valid for all $m = 1, 2, \ldots$ and $0 < r < 2$.
- MPVCLT is valid for all m if $\frac{1}{2} < r < 1$.

In particular, MPVCLT holds for tripower variation with $r = \frac{2}{3}$.

Example 3. Let Z be the $NIG(\gamma, 0, 0, \phi)$ Lévy process. This is representable as the subordination of a Brownian motion B by the $IG(\phi, \gamma)$ subordinator. Hence, $E\{|Z_t|^q\}$ behaves asymptotically as in (2.6) with $p = q/2$. Consequently:

- MPVCiP is valid for all $m = 1, 2, ...$ and $0 < r < 2$.
- MPVCLT does not hold for any value of r.

What decides the possibility of MPVCiP or MPVCLT holding is essentially the degree of singularity at 0 of the Lévy measure of Z (which may be expressed in terms of the Blumenthal-Getoor index). For the three examples above the degrees are respectively x^{-1}, $x^{-3/2}$ and x^{-2}. In the latter case there are so many small jumps that the process partly resembles a diffusion, and this is what prevents separate inference on the volatility process σ.

2.5 Microstructure Noise

Zhou [21] seems to be the first paper that manifestly demonstrates the necessity to take microstructure noise into account when drawing inference on the integrated (squared) volatility of the log price process, based on high frequency data. In Andersen, Bollerslev, Diebold, and Labys [1] this was emphasised further through the introduction of the *volatility signature plot*, which made it clear that even for five minute lags the influence of the noise is generally appreciable.

However, the precise nature and influence of the noise is far from well understood and this constitutes a topic of strong current interest.

In a recent paper, Zhang, Mykland, and Aït-Sahalia [20] address the noise problem and proposes a subsampling procedure for estimating the integrated volatility of the log price process. Hansen and Lunde [15] have initiated a study of how the realised quadratic variation may be bias corrected to alleviate the noise effect. See also the work of Bandi and Russell [2]. The latter line of investigation is continued in joint ongoing work between Barndorff-Nielsen, Hansen, Lunde, and Shephard [5, 6]. That work considers a general class of kernel estimators of the quadratic variation of the log price process. It is shown, in particular, that the subsampling procedure for estimation of quadratic variation proposed by Aït-Sahalia, Mykland and Zhang is a special case of that class. However the main thrust of the Barndorff-Nielsen, Hansen, Lunde, and Shephard [5] work consists in determining, from optimality criteria, another type of kernel estimator that has turned out to yield very accurate estimates for almost all lags. The relevance of MPV for the study of microstructure noise will also be considered. In some stimulating recent work Zhang [19] has shown that subsampling can be generalised to achieve the same rate of convergence as the modified kernel suggested by Barndorff-Nielsen, Hansen, Lunde, and Shephard [5].

Acknowledgments

Ole E. Barndorff-Nielsen's work is supported by CAF (www.caf.dk), which is funded by the Danish Social Science Research Council. Neil Shephard's research is supported by the UK's ESRC through the grant "High frequency financial econometrics based upon power variation."

References

[1] Andersen, T. G., T. Bollerslev, F. X. Diebold, and P. Labys (2000). Great realizations. *Risk 13*, 105–108.

[2] Bandi, F. M. and J. R. Russell (2003). Microstructure noise, realized volatility, and optimal sampling. Unpublished paper, Graduate School of Business, University of Chicago.

[3] Barndorff-Nielsen, O. E., S. E. Graversen, J. Jacod, M. Podolskij, and N. Shephard (2005). A central limit theorem for realised power and bipower variations of continuous semimartingales. In Y. Kabanov and R. Lipster (Eds.), *From Stochastic Analysis to Mathematical Finance, Festschrift for Albert Shiryaev.* Springer. Forthcoming. Also Economics working paper 2004-W29, Nuffield College, Oxford.

[4] Barndorff-Nielsen, O. E., S. E. Graversen, J. Jacod, and N. Shephard (2005). Limit theorems for realised bipower variation in econometrics. Unpublished paper: Nuffield College, Oxford.

[5] Barndorff-Nielsen, O. E., P. R. Hansen, A. Lunde, and N. Shephard (2004). Regular and modified kernel-based estimators of integrated variance: the case with independent noise. Unpublished paper: Nuffield College, Oxford.

[6] Barndorff-Nielsen, O. E., P. R. Hansen, A. Lunde, and N. Shephard (2005). Kernel-based estimators of integrated variance: dependent noise. In preparation.

[7] Barndorff-Nielsen, O. E. and N. Shephard (2001). Non-Gaussian Ornstein–Uhlenbeck-based models and some of their uses in financial economics (with discussion). *Journal of the Royal Statistical Society, Series B 63*, 167–241.

[8] Barndorff-Nielsen, O. E. and N. Shephard (2002). Econometric analysis of realised volatility and its use in estimating stochastic volatility models. *Journal of the Royal Statistical Society, Series B 64*, 253–280.

[9] Barndorff-Nielsen, O. E. and N. Shephard (2003). Realised power variation and stochastic volatility. *Bernoulli 9*, 243–265. Correction published in pages 1109–1111.

[10] Barndorff-Nielsen, O. E. and N. Shephard (2004a). Econometric analysis of realised covariation: high frequency covariance, regression and correlation in financial economics. *Econometrica 72*, 885–925.

[11] Barndorff-Nielsen, O. E. and N. Shephard (2004b). Power and bipower variation with stochastic volatility and jumps (with discussion). *Journal of Financial Econometrics 2*, 1–48.

[12] Barndorff-Nielsen, O. E. and N. Shephard (2005). Power variation and time change. *Theory of Probability and Its Applications.* Forthcoming.

[13] Barndorff-Nielsen, O. E., N. Shephard, and M. Winkel (2004). Limit theorems for multipower variation in the presence of jumps in financial econometrics. Unpublished paper: Nuffield College, Oxford.

[14] Ghysels, E., A. C. Harvey, and E. Renault (1996). Stochastic volatility. In C. R. Rao and G. S. Maddala (Eds.), *Statistical Methods in Finance*, pp. 119–191. Amsterdam: North-Holland.

[15] Hansen, P. R. and A. Lunde (2004). An unbiased measure of realized variance. Unpublished paper: Aarhus University.

[16] Mykland, P. and L. Zhang (2005). ANOVA for diffusions. *Annals of Statistics 33*. Forthcoming.

[17] Shephard, N. (2005). *Stochastic Volatility: Selected Readings.* Oxford: Oxford University Press.

[18] Woerner, J. (2004). Power and multipower variation: inference for high frequency data. Unpublished paper.

[19] Zhang, L. (2004). Efficient estimation of stochastic volatility using noisy observations: a multi-scale approach. Unpublished paper: Department of Statistics, Carnegie Mellon University.

[20] Zhang, L., P. Mykland, and Y. Aït-Sahalia (2005). A tale of two time scales: determining integrated volatility with noisy high-frequency data. *Journal of the American Statistical Association*. Forthcoming.

[21] Zhou, B. (1996). High-frequency data and volatility in foreign-exchange rates. *Journal of Business and Economic Statistics 14*, 45–52.

3

Completeness of a General Semimartingale Market under Constrained Trading

Tomasz R. Bielecki*, Monique Jeanblanc†, and Marek Rutkowski

[1] Department of Applied Mathematics, Illinois Institute of Technology, Chicago, IL 60616, USA
bielecki@iit.edu
[2] Département de Mathématiques, Université d'Évry Val d'Essonne, 91025 Évry Cedex, France
monique.jeanblanc@univ-evry.fr
[3] School of Mathematics, University of New South Wales, Sydney, NSW 2052, Australia and
Faculty of Mathematics and Information Science, Warsaw University of Technology, 00-661 Warszawa, Poland
marekr@maths.unsw.edu.au

3.1 Introduction

In this note, we provide a rather detailed and comprehensive study of the basic properties of self-financing trading strategies in a general security market model driven by discontinuous semimartingales. Our main goal is to analyze the issue of replication of a generic contingent claim using a self-financing trading strategy that is additionally subject to an algebraic constraint, referred to as the *balance condition*. Although such portfolios may seem to be artificial at the first glance, they appear in a natural way in the analysis of hedging strategies within the reduced-form approach to credit risk.

Let us mention in this regard that in a companion paper by Bielecki et al. [1] we also include defaultable assets in our portfolio, and we show how to use constrained portfolios to derive replicating strategies for defaultable contingent claims (e.g., credit derivatives). The reader is also referred to Bielecki et al. [1], where the case of continuous semimartingale markets was studied, for some background information regarding the probabilistic and financial set-up, as well as the terminology used in this note. The main emphasis is put here on the relationship between completeness of a security market model

* The research of the first author was supported in part by NSF Grant 0202851 and by Moody's Corporation grant 5-55411.
† The research of the second author was supported in part by Moody's Corporation grant 5-55411.

with unconstrained trading and completeness of an associated model in which only trading strategies satisfying the balance condition are allowed.

3.2 Trading in Primary Assets

Let $Y_t^1, Y_t^2, \dots, Y_t^k$ represent cash values at time t of k primary assets. We postulate that the prices Y^1, $Y^2, \dots, Y^k$ follow (possibly discontinuous) semimartingales on some probability space $(\Omega, \mathcal{F}, \mathbb{P})$, endowed with a filtration $\mathbb{F}$ satisfying the usual conditions. Thus, for example, general Lévy processes, as well as jump-diffusions are covered by our analysis. Note that obviously $\mathbb{F}^Y \subseteq \mathbb{F}$, where $\mathbb{F}^Y$ is the filtration generated by the prices Y^1, $Y^2, \dots, Y^k$ of primary assets. As it is usually done, we set $X_{0-} = X_0$ for any stochastic process X, and we only consider semimartingales with càdlàg sample paths. We assume, in addition that at least one of the processes Y^1, $Y^2, \dots, Y^k$, say Y^1, is strictly positive, so that it can be chosen as a numeraire asset. We consider trading within the time interval $[0,T]$ for some finite horizon date $T > 0$. We emphasize that we do not assume the existence of a risk-free asset (a savings account).

3.2.1 Unconstrained Trading Strategies

Let $\phi = (\phi^1, \phi^2, \dots, \phi^k)$ be a trading strategy; in particular, each process ϕ^i is predictable with respect to the reference filtration $\mathbb{F}$. The component ϕ_t^i represents the number of units of the ith asset held in the portfolio at time t. Then the wealth $V_t(\phi)$ at time t of the trading strategy $\phi = (\phi^1, \phi^2, \dots, \phi^k)$ equals

$$V_t(\phi) = \sum_{i=1}^{k} \phi_t^i Y_t^i, \quad \forall\, t \in [0,T], \tag{3.1}$$

and ϕ is said to be a *self-financing strategy* if

$$V_t(\phi) = V_0(\phi) + \sum_{i=1}^{k} \int_0^t \phi_u^i \, dY_u^i, \quad \forall\, t \in [0,T]. \tag{3.2}$$

Let Φ be the class of all self-financing trading strategies. By combining the last two formulae, we obtain the following expression for the dynamics of the wealth process of a strategy $\phi \in \Phi$

$$dV_t(\phi) = \Big(V_t(\phi) - \sum_{i=2}^{k} \phi_t^i Y_t^i \Big) (Y_t^1)^{-1} \, dY_t^1 + \sum_{i=2}^{k} \phi_t^i \, dY_t^i .$$

The representation above shows that the wealth process $V(\phi)$ depends only on $k-1$ components of ϕ. Note also that, in our setting, the process $\Big(V_t(\phi) - \sum_{i=2}^{k} \phi_t^i Y_t^i \Big) (Y_t^1)^{-1}$ is predictable.

Remark 1. Let us note that Protter [4] assumes that the component of a strategy ϕ that corresponds to the savings account (which is a continuous process) is merely optional. The interested reader is referred to Protter [4] for a thorough discussion of other issues related to the regularity of sample paths of processes $\phi^1, \phi^2, \ldots, \phi^k$ and $V(\phi)$.

Choosing Y^1 as a numeraire asset, and denoting $V_t^1(\phi) = V_t(\phi)(Y_t^1)^{-1}$, $Y_t^{i,1} = Y_t^i(Y_t^1)^{-1}$, we get the following well-known result showing that the self-financing feature of a trading strategy is invariant with respect to the choice of a numeraire asset.

Lemma 1. (i) *For any $\phi \in \Phi$, we have*

$$V_t^1(\phi) = V_0^1(\phi) + \sum_{i=2}^{k} \int_0^t \phi_u^i \, dY_u^{i,1}, \quad \forall\, t \in [0,T]. \tag{3.3}$$

(ii) *Conversely, let X be an $\mathcal{F}_T$-measurable random variable, and let us assume that there exists $x \in \mathbb{R}$ and $\mathbb{F}$-predictable processes ϕ^i, $i = 2,3,\ldots,k$ such that*

$$X = Y_T^1\Big(x + \sum_{i=2}^{k} \int_0^T \phi_t^i \, dY_t^{i,1}\Big).$$

Then there exists an $\mathbb{F}$-predictable process ϕ^1 such that the strategy $\phi = (\phi^1, \phi^2, \ldots, \phi^k)$ is self-financing and replicates X. Moreover, the wealth process of ϕ satisfies $V_t(\phi) = V_t^1 Y_t^1$, where the process V^1 is given by formula (3.4) *below.*

Proof. The proof of part (i) is given, for instance, in Protter [4]. We shall thus only prove part (ii). Let us set

$$V_t^1 = x + \sum_{i=2}^{k} \int_0^t \phi_u^i \, dY_u^{i,1}, \quad \forall\, t \in [0,T], \tag{3.4}$$

and let us define the process ϕ^1 as

$$\phi_t^1 = V_t^1 - \sum_{i=2}^{k} \phi_t^i Y_t^{i,1} = (Y_t^1)^{-1}\Big(V_t - \sum_{i=2}^{k} \phi_t^i Y_t^i\Big),$$

where $V_t = V_t^1 Y_t^1$. From (3.4), we have $dV_t^1 = \sum_{i=2}^{k} \phi_t^i \, dY_t^{i,1}$, and thus

$$dV_t = d(V_t^1 Y_t^1) = V_{t-}^1 dY_t^1 + Y_{t-}^1 dV_t^1 + d[Y^1, V^1]_t \tag{3.5}$$

$$= V_{t-}^1 dY_t^1 + \sum_{i=2}^{k} \phi_t^i \big(Y_{t-}^1 \, dY_t^{i,1} + d[Y^1, Y^{i,1}]_t\big). \tag{3.6}$$

From the equality

$$dY_t^i = d(Y_t^{i,1} Y_t^1) = Y_{t-}^{i,1} dY_t^1 + Y_{t-}^1 dY_t^{i,1} + d[Y^1, Y^{i,1}]_t ,$$

it follows that

$$dV_t = V_{t-}^1 dY_t^1 + \sum_{i=2}^{k} \phi_t^i \big(dY_t^i - Y_{t-}^{i,1} dY_t^1 \big) \tag{3.7}$$

$$= \Big(V_{t-}^1 - \sum_{i=2}^{k} \phi_t^i Y_{t-}^{i,1} \Big) dY_t^1 + \sum_{i=2}^{k} \phi_t^i \, dY_t^i , \tag{3.8}$$

and our aim is to prove that

$$dV_t = \sum_{i=1}^{k} \phi_t^i \, dY_t^i .$$

The last equality holds if

$$\phi_t^1 = V_t^1 - \sum_{i=2}^{k} \phi_t^i Y_t^{i,1} = V_{t-}^1 - \sum_{i=2}^{k} \phi_t^i Y_{t-}^{i,1} , \tag{3.9}$$

i.e., if $\Delta V_t^1 = \sum_{i=2}^k \phi_t^i \Delta Y_t^{i,1}$, which is the case from the definition (3.4) of V^1. Note also that from the second equality in (3.9) it follows that the process ϕ^1 is indeed $\mathbb{F}$-predictable. Finally, the wealth process of ϕ satisfies $V_t(\phi) = V_t^1 Y_t^1$ for every $t \in [0,T]$, and thus $V_T(\phi) = X$. □

3.2.2 Constrained Trading Strategies

In this section, we make an additional assumption that the price process Y^k is strictly positive. Let $\phi = (\phi^1, \phi^2, \dots, \phi^k)$ be a self-financing trading strategy satisfying the following constraint:

$$\sum_{i=l+1}^{k} \phi_t^i Y_{t-}^i = Z_t , \quad \forall \, t \in [0,T] , \tag{3.10}$$

for some $1 \le l \le k-1$ and a predetermined, $\mathbb{F}$-predictable process Z. In the financial interpretation, equality (3.10) means that the portfolio ϕ should be rebalanced in such a way that the total wealth invested in securities $Y^{l+1}, Y^{l+2}, \dots, Y^k$ should match a predetermined stochastic process (for instance, we may assume that it is constant over time or follows a deterministic function of time). For this reason, the constraint (3.10) will be referred to as the *balance condition.*

Our first goal is to extend part (i) in Lemma 1 to the case of constrained strategies. Let $\Phi_l(Z)$ stand for the class of all self-financing trading strategies satisfying the balance condition (3.10). They will be sometimes referred to as *constrained strategies.* Since any strategy $\phi \in \Phi_l(Z)$ is self-financing, we have

$$\Delta V_t(\phi) = \sum_{i=1}^{k} \phi_t^i \Delta Y_t^i = V_t(\phi) - \sum_{i=1}^{k} \phi_t^i Y_{t-}^i ,$$

and thus we deduce from (3.10) that

$$V_{t-}(\phi) = \sum_{i=1}^{k} \phi_t^i Y_{t-}^i = \sum_{i=1}^{l} \phi_t^i Y_{t-}^i + Z_t .$$

Let us write $Y_t^{i,1} = Y_t^i (Y_t^1)^{-1}$, $Y_t^{i,k} = Y^i (Y_t^k)^{-1}$, $Z_t^1 = Z_t (Y_t^1)^{-1}$. The following result extends Lemma 1.7 in Bielecki et al. [1] from the case of continuous semimartingales to the general case. It is apparent from Proposition 1 that the wealth process $V(\phi)$ of a strategy $\phi \in \Phi_l(Z)$ depends only on $k-2$ components of ϕ.

Proposition 1. *The relative wealth* $V_t^1(\phi) = V_t(\phi)(Y_t^1)^{-1}$ *of a strategy* $\phi \in \Phi_l(Z)$ *satisfies*

$$V_t^1(\phi) = V_0^1(\phi) + \sum_{i=2}^{l} \int_0^t \phi_u^i \, dY_u^{i,1} + \sum_{i=l+1}^{k-1} \int_0^t \phi_u^i \left(dY_u^{i,1} - \frac{Y_{u-}^{i,1}}{Y_{u-}^{k,1}} \, dY_u^{k,1} \right) + \\ + \int_0^t \frac{Z_u^1}{Y_{u-}^{k,1}} \, dY_u^{k,1}. \tag{3.11}$$

Proof. Let us consider discounted values of price processes $Y^1, Y^2, \ldots, Y^k$, with Y^1 taken as a numeraire asset. By virtue of part (i) in Lemma 1, we thus have

$$V_t^1(\phi) = V_0^1(\phi) + \sum_{i=2}^{k} \int_0^t \phi_u^i \, dY_u^{i,1}. \tag{3.12}$$

The balance condition (3.10) implies that

$$\sum_{i=l+1}^{k} \phi_t^i Y_{t-}^{i,1} = Z_t^1 ,$$

and thus

$$\phi_t^k = (Y_{t-}^{k,1})^{-1} \left(Z_t^1 - \sum_{i=l+1}^{k} \phi_t^i Y_{t-}^{i,1} \right). \tag{3.13}$$

By inserting (3.13) into (3.12), we arrive at the desired formula (12.2). □

Let us take $Z = 0$, so that $\phi \in \Phi_l(0)$. Then the balance condition becomes $\sum_{i=l+1}^{k} \phi_t^i Y_{t-}^i = 0$, and (12.2) reduces to

$$dV_t^1(\phi) = \sum_{i=2}^{l} \phi_t^i \, dY_t^{i,1} + \sum_{i=l+1}^{k-1} \phi_t^i \left(dY_t^{i,1} - \frac{Y_{t-}^{i,1}}{Y_{t-}^{k,1}} \, dY_t^{k,1} \right). \tag{3.14}$$

3.2.3 Case of Continuous Semimartingales

For the sake of notational simplicity, we denote by $Y^{i,k,1}$ the process given by the formula

$$Y_t^{i,k,1} = \int_0^t \left(dY_u^{i,1} - \frac{Y_{u-}^{i,1}}{Y_{u-}^{k,1}}\, dY_u^{k,1} \right) \tag{3.15}$$

so that (12.2) becomes

$$\begin{aligned} V_t^1(\phi) = V_0^1(\phi) + \sum_{i=2}^{l} \int_0^t \phi_u^i\, dY_u^{i,1} + \sum_{i=l+1}^{k-1} \int_0^t \phi_u^i\, dY_u^{i,k,1} + \\ + \int_0^t \frac{Z_u^1}{Y_{u-}^{k,1}}\, dY_u^{k,1}. \end{aligned} \tag{3.16}$$

In Bielecki et al. [1], we postulated that the primary assets $Y^1, Y^2, \ldots, Y^k$ follow strictly positive continuous semimartingales, and we introduced the auxiliary processes $\widehat{Y}_t^{i,k,1} = Y_t^{i,k} e^{-\alpha_t^{i,k,1}}$, where

$$\alpha_t^{i,k,1} = \langle \ln Y^{i,k}, \ln Y^{1,k} \rangle_t = \int_0^t (Y_u^{i,k})^{-1} (Y_u^{1,k})^{-1}\, d\langle Y^{i,k}, Y^{1,k} \rangle_u .$$

In Lemma 1.7 in Bielecki et al. [1] (see also Vaillant [5]), we have shown that, under continuity of $Y^1, Y^2, \ldots, Y^k$, the discounted wealth of a self-financing trading strategy ϕ that satisfies the constraint $\sum_{i=l+1}^{k} \phi_t^i Y_t^i = Z_t$ can be represented as follows:

$$\begin{aligned} V_t^1(\phi) = V_0^1(\phi) + \sum_{i=2}^{l} \int_0^t \phi_u^i\, dY_u^{i,1} + \sum_{i=l+1}^{k-1} \int_0^t \widehat{\phi}_u^{i,k,1}\, d\widehat{Y}_u^{i,k,1} + \\ + \int_0^t \frac{Z_u^1}{Y_u^{k,1}}\, dY_u^{k,1}, \end{aligned} \tag{3.17}$$

where we write $\widehat{\phi}_t^{i,k,1} = \phi_t^i (Y_t^{1,k})^{-1} e^{\alpha_t^{i,k,1}}$. The following simple result reconciles expression (3.16) established in Proposition 1 with representation (3.17) derived in Bielecki et al. [1].

Lemma 2. *Assume that the prices Y^1, Y^i and Y^k follow strictly positive continuous semimartingales. Then we have*

$$Y_t^{i,k,1} = \int_0^t (Y_u^{1,k})^{-1} e^{\alpha_u^{i,k,1}}\, d\widehat{Y}_u^{i,k,1}$$

and

$$dY_t^{i,k,1} = (Y_t^{1,k})^{-1} \big(dY_t^{i,k} - Y_t^{i,k}\, d\alpha_t^{i,k,1} \big) .$$

Proof. In the case of continuous semimartingales, formula (3.15) becomes

$$Y_t^{i,k,1} = \int_0^t \Big(dY_u^{i,1} - \frac{Y_u^{i,1}}{Y_u^{k,1}}\, dY_u^{k,1} \Big) = \int_0^t \big(dY_u^{i,1} - Y_u^{i,k}\, d(Y_u^{1,k})^{-1} \big) .$$

On the other hand, an application of Itô's formula yields

$$d\widehat{Y}_t^{i,k,1} = e^{-\alpha_t^{i,k,1}} \big(dY_t^{i,k} - (Y_t^{1,k})^{-1} d\langle Y^{i,k}, Y^{1,k} \rangle_t \big)$$

and thus

$$(Y_t^{1,k})^{-1} e^{\alpha_t^{i,k,1}}\, d\widehat{Y}_u^{i,k,1} = (Y_t^{1,k})^{-1} \big(dY_t^{i,k} - (Y_t^{1,k})^{-1} d\langle Y^{i,k}, Y^{1,k} \rangle_t \big) .$$

One checks easily that for any two continuous semimartingales, say X and Y, we have

$$Y_t^{-1} \big(dX_t - Y_t^{-1} d\langle X, Y \rangle_t \big) = d(X_t Y_t^{-1}) - X_t\, dY_t^{-1} ,$$

provided that Y is strictly positive. To conclude the derivation of the first formula, it suffices to apply the last identity to processes $X = Y^{i,k}$ and $Y = Y^{1,k}$. For the second formula, note that

$$\begin{aligned} dY_t^{i,k,1} &= (Y_t^{1,k})^{-1} e^{\alpha_t^{i,k,1}}\, d\widehat{Y}_t^{i,k,1} = (Y_t^{1,k})^{-1} e^{\alpha_t^{i,k,1}}\, d(Y_t^{i,k} e^{-\alpha_t^{i,k,1}}) \\ &= (Y_t^{1,k})^{-1} \big(dY_t^{i,k} - Y_t^{i,k}\, d\alpha_t^{i,k,1} \big) , \end{aligned}$$

as required. □

It is obvious that the processes $Y^{i,k,1}$ and $\widehat{Y}^{i,k,1}$ are uniquely specified by the joint dynamics of Y^1, Y^i and Y^k. The following result shows that the converse is also true.

Corollary 1. *The price Y_t^i at time t is uniquely specified by the initial value Y_0^i and either*
(i) *the joint dynamics of processes Y^1, Y^k and $\widehat{Y}^{i,k,1}$, or*
(ii) *the joint dynamics of processes Y^1, Y^k and $Y^{i,k,1}$.*

Proof. Since $\widehat{Y}_t^{i,k,1} = Y_t^{i,k} e^{-\alpha_t^{i,k,1}}$, we have

$$\alpha_t^{i,k,1} = \langle \ln Y^{i,k}, \ln Y^{1,k} \rangle_t = \langle \ln \widehat{Y}^{i,k,1}, \ln Y^{1,k} \rangle_t ,$$

and thus

$$Y_t^i = Y_t^k \widehat{Y}_t^{i,k,1} e^{\alpha_t^{i,k,1}} = Y_t^k \widehat{Y}_t^{i,k,1} e^{\langle \ln \widehat{Y}^{i,k,1}, \ln Y^{1,k} \rangle_t} .$$

This completes the proof of part (i). For the second part, note that the process $Y^{i,1}$ satisfies

$$Y_t^{i,1} = Y_0^{i,1} + Y_t^{i,k,1} + \int_0^t \frac{Y_u^{i,1}}{Y_u^{k,1}}\, dY_u^{k,1}. \tag{3.18}$$

It is well known that the SDE

$$X_t = X_0 + H_t + \int_0^t X_u \, dY_u ,$$

where H and Y are continuous semimartingales (with $H_0 = 0$) has the unique, strong solution given by the formula

$$X_t = \mathcal{E}_t(Y)\Big(X_0 + \int_0^t \mathcal{E}_u^{-1}(Y)\, dH_u - \int_0^t \mathcal{E}_u^{-1}(Y)\, d\langle Y, H\rangle_u \Big).$$

Upon substitution, this proves (ii). □

3.3 Replication with Constrained Strategies

The next result is essentially a converse to Proposition 1. Also, it extends part (ii) of Lemma 1 to the case of constrained trading strategies. As in Section 3.2.2, we assume that $1 \le l \le k-1$, and Z is a predetermined, $\mathbb{F}$-predictable process.

Proposition 2. *Let an $\mathcal{F}_T$-measurable random variable X represent a contingent claim that settles at time T. Assume that there exist $\mathbb{F}$-predictable processes ϕ^i, $i = 2, 3, \ldots, k-1$ such that*

$$X = Y_T^1 \Big(x + \sum_{i=2}^{l} \int_0^T \phi_t^i \, dY_t^{i,1} + \sum_{i=l+1}^{k-1} \int_0^T \phi_t^i \, dY_t^{i,k,1} + \int_0^T \frac{Z_t^1}{Y_{t-}^{k,1}} \, dY_t^{k,1} \Big). \tag{3.19}$$

Then there exist the $\mathbb{F}$-predictable processes ϕ^1 and ϕ^k such that the strategy $\phi = (\phi^1, \phi^2, \ldots, \phi^k)$ belongs to $\Phi_l(Z)$ and replicates X. The wealth process of ϕ equals, for every $t \in [0, T]$,

$$V_t(\phi) = Y_t^1 \Big(x + \sum_{i=2}^{l} \int_0^t \phi_u^i \, dY_u^{i,1} + \sum_{i=l+1}^{k-1} \int_0^t \phi_u^i \, dY_u^{i,k,1} + \int_0^t \frac{Z_u^1}{Y_{u-}^{k,1}} \, dY_u^{k,1} \Big). \tag{3.20}$$

Proof. As expected, we first set (note that ϕ^k is $\mathbb{F}$-predictable)

$$\phi_t^k = \frac{1}{Y_{t-}^k} \Big(Z_t - \sum_{i=l+1}^{k-1} \phi_t^i Y_{t-}^i \Big) \tag{3.21}$$

and

$$V_t^1 = x + \sum_{i=2}^{l} \int_0^t \phi_u^i \, dY_u^{i,1} + \sum_{i=l+1}^{k-1} \int_0^t \phi_u^i \, dY_u^{i,k,1} + \int_0^t \frac{Z_u^1}{Y_{u-}^{k,1}} \, dY_u^{k,1} .$$

Arguing along the same lines as in the proof of Proposition 1, we obtain

$$V_t^1 = V_0^1 + \sum_{i=2}^{k} \int_0^t \phi_u^i \, dY_u^{i,1} \, .$$

Now, we define

$$\phi_t^1 = V_t^1 - \sum_{i=2}^{k} \phi_t^i Y_t^{i,1} = (Y_t^1)^{-1} \Big(V_t - \sum_{i=2}^{k} \phi_t^i Y_t^i \Big) \, ,$$

where $V_t = V_t^1 Y_t^1$. As in the proof of Lemma 1, we check that

$$\phi_t^1 = V_{t-}^1 - \sum_{i=2}^{k} \phi_t^i Y_{t-}^{i,1} \, ,$$

and thus the process ϕ^1 is $\mathbb{F}$-predictable. It is clear that the strategy $\phi = (\phi^1, \phi^2, \ldots, \phi^k)$ is self-financing and its wealth process satisfies $V_t(\phi) = V_t$ for every $t \in [0,T]$. In particular, $V_T(\phi) = X$, so that ϕ replicates X. Finally, equality (3.21) implies (3.10), and thus $\phi \in \Phi_l(Z)$. □

Note that equality (12.3) is a necessary (by Proposition 1) and sufficient (by Proposition 2) condition for the existence of a constrained strategy replicating a given contingent claim X.

3.3.1 Modified Balance Condition

It is tempting to replace the constraint (3.10) by a more convenient condition:

$$\sum_{i=l+1}^{k} \phi_t^i Y_t^i = Z_t \, , \quad \forall t \in [0,T] \, , \tag{3.22}$$

where Z is a predetermined, $\mathbb{F}$-predictable process. If a self-financing trading strategy ϕ satisfies the modified balance condition (3.22) then for the relative wealth process we obtain (cf. (12.2))

$$\begin{aligned} V_t^1(\phi) = V_0^1(\phi) &+ \sum_{i=2}^{l} \int_0^t \phi_u^i \, dY_u^{i,1} + \\ &+ \sum_{i=l+1}^{k-1} \int_0^t \phi_u^i \Big(dY_u^{i,1} - \frac{Y_u^{i,1}}{Y_{u-}^{k,1}} \, dY_u^{k,1} \Big) + \int_0^t \frac{Z_u^1}{Y_{u-}^{k,1}} \, dY_u^{k,1} \, . \end{aligned} \tag{3.23}$$

Note that in many cases the integrals above are meaningful, so that a counterpart of Proposition 1 with the modified balance condition can be formulated. To get a counterpart of Proposition 2, we need to replace (12.3) by the equality

$$
\begin{aligned}
X = Y_T^1 \Bigg(x + \sum_{i=2}^{l} \int_0^T \phi_t^i \, dY_t^{i,1} \, + \\
+ \sum_{i=l+1}^{k-1} \int_0^T \phi_t^i \Big(dY_t^{i,1} - \frac{Y_t^{i,1}}{Y_{t-}^{k,1}} \, dY_t^{k,1} \Big) + \int_0^T \frac{Z_t^1}{Y_{t-}^{k,1}} \, dY_t^{k,1} \Bigg),
\end{aligned} \tag{3.24}
$$

where $\phi^3, \phi^4, \dots, \phi^k$ are $\mathbb{F}$-predictable processes. We define

$$
\begin{aligned}
V_t^1 = x + \sum_{i=2}^{l} \int_0^t \phi_u^i \, dY_u^{i,1} \, + \\
+ \sum_{i=l+1}^{k-1} \int_0^t \phi_u^i \Big(dY_u^{i,1} - \frac{Y_u^{i,1}}{Y_{u-}^{k,1}} \, dY_u^{k,1} \Big) + \int_0^t \frac{Z_u^1}{Y_{u-}^{k,1}} \, dY_u^{k,1},
\end{aligned}
$$

and we set

$$
\phi_t^k = \frac{1}{Y_t^k} \Big(Z_t - \sum_{i=l+1}^{k-1} \phi_t^i Y_t^i \Big), \quad \phi_t^1 = V_t^1 - \sum_{i=2}^{k} \phi_t^i Y_t^{i,1}.
$$

Suppose, for the sake of argument, that the processes ϕ^1 and ϕ^k defined above are $\mathbb{F}$-predictable. Then the trading strategy $\phi = (\phi^1, \phi^2, \dots, \phi^k)$ is self-financing on $[0,T]$, replicates X, and satisfies the constraint (3.22). Note, however, that the predictability of ϕ^1 and ϕ^k is far from being obvious, and it is rather difficult to provide non-trivial and practically appealing sufficient conditions for this property.

3.3.2 Synthetic Assets

Let us fix i, and let us analyze the auxiliary process $Y^{i,k,1}$ given by formula (3.15). We claim that this process can be interpreted as the relative wealth of a specific self-financing trading strategy associated with $Y^1, Y^2, \dots, Y^k$. Specifically, we will show that for any $i = 2, 3, \dots, k-1$ the process $\bar{Y}^{i,k,1}$, given by the formula

$$
\bar{Y}_t^{i,k,1} = Y_t^1 Y_t^{i,k,1} = Y_t^1 \int_0^t \Big(dY_u^{i,1} - \frac{Y_{u-}^{i,1}}{Y_{u-}^{k,1}} \, dY_u^{k,1} \Big),
$$

represents the price of a *synthetic asset.* For brevity, we shall frequently write $\bar{Y}^i$ instead of $\bar{Y}^{i,k,1}$. Note that the process $\bar{Y}^i$ is not strictly positive (in fact, $\bar{Y}_0^i = 0$).

Equivalence of Primary and Synthetic Assets

Our goal is to show that trading in primary assets is formally equivalent to trading in synthetic assets. The first result shows that the process $\bar{Y}^i$ can be obtained from primary assets Y^1, Y^i and Y^k through a simple self-financing strategy. This justifies the name synthetic asset given to $\bar{Y}^i$.

Lemma 3. *For any fixed $i = 2, 3, \dots, k-1$, let an $\mathcal{F}_T$-measurable random variable $\bar{Y}^i_T$ be given as*

$$\bar{Y}^i_T = Y^1_T Y^{i,k,1}_T = Y^1_T \int_0^T \left(dY^{i,1}_t - \frac{Y^{i,1}_{t-}}{Y^{k,1}_{t-}} \, dY^{k,1}_t \right). \tag{3.25}$$

Then there exists a strategy $\phi \in \Phi_1(0)$ that replicates the claim $\bar{Y}^i_T$. Moreover, we have, for every $t \in [0,T]$,

$$V_t(\phi) = Y^1_t Y^{i,k,1}_t = Y^1_t \int_0^t \left(dY^{i,1}_u - \frac{Y^{i,1}_{u-}}{Y^{k,1}_{u-}} \, dY^{k,1}_u \right) = \bar{Y}^i_t . \tag{3.26}$$

Proof. To establish the existence of a strategy ϕ with the desired properties, it suffices to apply Proposition 2. We fix i and we start by postulating that $\phi^i = 1$ and $\phi^j = 0$ for any $2 \le j \le k-1$, $j \neq i$. Then equality (3.25) yields (12.3) with $X = \bar{Y}^i_T$, $x = 0$, $l = 1$ and $Z = 0$. Note that the balance condition becomes

$$\sum_{j=2}^{k} \phi^j_t Y^j_{t-} = Y^i_{t-} + \phi^k_t Y^k_{t-} = 0 .$$

Let us define ϕ^1 and ϕ^k by setting

$$\phi^k_t = -\frac{Y^i_{t-}}{Y^k_{t-}}, \quad \phi^1_t = V^1_t - Y^{i,1}_t - \phi^k_t Y^{k,1}_t .$$

Note that we also have

$$\phi^1_t = V^1_{t-} - Y^{i,1}_{t-} - \phi^k_t Y^{k,1}_{t-} = V^1_{t-} .$$

Hence, ϕ^1 and ϕ^k are $\mathbb{F}$-predictable processes, the strategy $\phi = (\phi^1, \phi^2, \dots, \phi^k)$ is self-financing, and it satisfies (3.10) with $l = 1$ and $Z = 0$, so that $\phi \in \Phi_1(0)$. Finally, equality (3.26) holds, and thus $V_T(\phi) = \bar{Y}^i_T$. □

Note that to replicate the claim $\bar{Y}^i_T = \bar{Y}^{i,k,1}_T$, it suffices to invest in primary assets Y^1, Y^i and Y^k. Essentially, we start with zero initial endowment, we keep at any time one unit of the ith asset, we rebalance the portfolio in such a way that the total wealth invested in the ith and kth assets is always zero, and we put the residual wealth in the first asset. Hence, we deal here with a specific strategy such that the risk of the ith asset is perfectly offset by rebalancing the investment in the kth asset, and our trades are financed by taking positions in the first asset.

Note that the process $Y^{i,1}$ satisfies the following SDE (cf. (3.18))

$$Y^{i,1}_t = Y^{i,1}_0 + \bar{Y}^{i,1}_t + \int_0^t \frac{Y^{i,1}_{u-}}{Y^{k,1}_{u-}} \, dY^{k,1}_u , \tag{3.27}$$

which is known to possess a unique strong solution. Hence, the relative price $Y_t^{i,1}$ at time t is uniquely determined by the initial value $Y_0^{i,1}$ and processes $\bar{Y}^{i,1}$ and $Y^{k,1}$. Consequently, the price Y_t^i at time t of the ith primary asset is uniquely determined by the initial value Y_0^i, the prices Y^1, Y^k of primary assets, and the price $\bar{Y}^i$ of the ith synthetic asset. We thus obtain the following result.

Lemma 4. *Filtrations generated by the primary assets $Y^1, Y^2, \ldots, Y^k$ and by the price processes $Y^1, Y^2, \ldots, Y^l, \bar{Y}^{l+1}, \ldots, \bar{Y}^{k-1}, Y^k$ coincide.*

Lemma 4 suggests that for any choice of the underlying filtration $\mathbb{F}$ (such that $\mathbb{F}^Y \subseteq \mathbb{F}$), trading in assets $Y^1, Y^2, \ldots, Y^k$ is essentially equivalent to trading in $Y^1, Y^2, \ldots, Y^l, \bar{Y}^{l+1}, \ldots, \bar{Y}^{k-1}, Y^k$. Let us first formally define the equivalence of market models.

Definition 1. *We say that the two unconstrained models, $\mathcal{M}$ and $\widetilde{\mathcal{M}}$ say, are* equivalent with respect to a filtration $\mathbb{F}$ *if both models are defined on a common probability space and every primary asset in $\mathcal{M}$ can be obtained by trading in primary assets in $\widetilde{\mathcal{M}}$ and vice versa, under the assumption that trading strategies are $\mathbb{F}$-predictable.*

Note that we do not assume that models $\mathcal{M}$ and $\widetilde{\mathcal{M}}$ have the same number of primary assets. The next result justifies our claim of equivalence of primary and synthetic assets.

Corollary 2. *Models $\mathcal{M} = (Y^1, Y^2, \ldots, Y^k; \Phi)$ and $\bar{\mathcal{M}} = (Y^1, Y^2, \ldots, Y^l, \bar{Y}^{l+1}, \ldots, \bar{Y}^{k-1}, Y^k; \Phi)$ are equivalent with respect to any filtration $\mathbb{F}$ such that $\mathbb{F}^Y \subseteq \mathbb{F}$.*

Proof. In view of Lemma 3, it suffices to show that the price process of each primary asset Y^i for $i = l, l+1, \ldots, k-1$ can be mimicked by trading in $Y^1, \bar{Y}^i$ and Y^k. To see this, note that for any fixed $i = l, l+1, \ldots, k-1$, we have (see the proof of Lemma 3)

$$\bar{Y}_t^i = V_t(\phi) = \phi_t^1 Y_t^1 + Y_t^i + \phi_t^k Y_t^k$$

with

$$d\bar{Y}_t^i = dV_t(\phi) = \phi_t^1 \, dY_t^1 + dY_t^i + \phi_t^k \, dY_t^k .$$

Consequently,

$$Y_t^i = -\phi_t^1 Y_t^1 + \bar{Y}_t^i - \phi_t^k Y_t^k$$

and

$$dY_t^i = -\phi_t^1 \, dY_t^1 + d\bar{Y}_t^i - \phi_t^k \, dY_t^k .$$

This shows that the strategy $(-\phi^1, 1, -\phi^k)$ in $Y^1, \bar{Y}^i$ and Y^k is self-financing and its wealth equals Y^i. □

Replicating Strategies with Synthetic Assets

In view of Lemma 3, the replicating trading strategy for a contingent claim X, for which (12.3) holds, can be conveniently expressed in terms of primary securities $Y^1, Y^2, \ldots, Y^l$ and Y^k, and synthetic assets $\bar{Y}^{l+1}, \bar{Y}^{l+2}, \ldots, \bar{Y}^{k-1}$. To this end, we represent (12.3)-(3.20) in the following way:

$$X = Y_T^1\Big(x + \sum_{i=2}^{l}\int_0^T \phi_t^i \, dY_t^{i,1} + \sum_{i=l+1}^{k-1}\int_0^T \phi_t^i \, d\bar{Y}_t^{i,1} + \int_0^T \frac{Z_t^1}{Y_{t-}^{k,1}} \, dY_t^{k,1}\Big) \tag{3.28}$$

where $\bar{Y}_t^{i,1} = \bar{Y}_t^i / Y_t^1 = Y_t^{i,k,1}$, and

$$V_t(\phi) = Y_t^1\Big(x + \sum_{i=2}^{l}\int_0^t \phi_u^i \, dY_u^{i,1} + \sum_{i=l+1}^{k-1}\int_0^t \phi_u^i \, d\bar{Y}_u^{i,1} + \int_0^t \frac{Z_u^1}{Y_{u-}^{k,1}} \, dY_u^{k,1}\Big). \tag{3.29}$$

Corollary 3. *Let X be an $\mathcal{F}_T$-measurable random variable such that* (3.28) *holds for some $\mathbb{F}$-predictable process Z and some $\mathbb{F}$-predictable processes $\phi^2, \phi^3, \ldots, \phi^{k-1}$. Let $\psi^i = \phi^i$ for $i = 2, 3, \ldots, k-1$,*

$$\psi_t^k = \frac{Z_t^1}{Y_{t-}^{k,1}} = \frac{Z_t}{Y_{t-}^k},$$

and

$$\begin{aligned}
\psi_t^1 &= V_t^1 - \sum_{i=2}^{l} \psi_t^i Y_t^{i,1} - \sum_{i=l+1}^{k-1} \psi_t^i \bar{Y}_t^{i,1} - \psi_t^k Y_t^{k,1} \\
&= V_{t-}^1 - \sum_{i=2}^{l} \psi_t^i Y_{t-}^{i,1} - \sum_{i=l+1}^{k-1} \psi_t^i \bar{Y}_{t-}^{i,1} - \psi_t^k Y_{t-}^{k,1}.
\end{aligned}$$

Then $\psi = (\psi^1, \psi^2, \ldots, \psi^k)$ is a self-financing trading strategy in assets $Y^1, \ldots, Y^l, \bar{Y}^{l+1}, \ldots, \bar{Y}^{k-1}, Y^k$. Moreover, ψ satisfies $\psi_t^k Y_{t-}^k = Z_t$, $t \in [0, T]$, and it replicates X.

Proof. In view of (3.28), it suffices to apply Proposition 2 with $l = k-1$. □

3.4 Model Completeness

We shall now examine the relationship between the arbitrage-free property and completeness of a market model in which trading is restricted a priori to self-financing strategies satisfying the balance condition.

3.4.1 Minimal Completeness of an Unconstrained Model

Let $\mathcal{M} = (Y^1, Y^2, \dots, Y^k; \Phi)$ be an arbitrage-free market model. Unless explicitly stated otherwise, Φ stands for the class of all $\mathbb{F}$-predictable, self-financing strategies. Note, however, that the number of traded assets and their selection may be different for each particular model. Consequently, the dimension of a strategy $\phi \in \Phi$ will depend on the number of traded assets in a given model. For the sake of brevity, this feature is not reflected in our notation.

Definition 2. *We say that a model $\mathcal{M}$ is* complete *with respect to $\mathbb{F}$ if any bounded $\mathcal{F}_T$-measurable contingent claim X is attainable in $\mathcal{M}$. Otherwise, a model $\mathcal{M}$ is said to be* incomplete *with respect to $\mathbb{F}$.*

Definition 3. *An $\mathbb{F}$-complete model $\mathcal{M} = (Y^1, Y^2, \dots, Y^k; \Phi)$ is said to be* minimally complete *with respect to $\mathbb{F}$ if for any choice of trading strategies $\phi_i \in \Phi$, $i = 1, 2, \dots, k-1$, the* reduced model *$\widehat{\mathcal{M}}^{k-1} = (\widehat{Y}^1, \widehat{Y}^2, \dots, \widehat{Y}^{k-1}; \Phi)$ where $\widehat{Y}^i = V(\phi_i)$, is incomplete with respect to $\mathbb{F}$. In this case, we say that the* degree of completeness *of $\mathcal{M}$ equals k.*

Let us stress that trading strategies in the reduced model $\widehat{\mathcal{M}}^{k-1}$ are predictable with respect to $\mathbb{F}$, rather than with respect to the filtration generated by price processes $\widehat{Y}^1, \widehat{Y}^2, \dots, \widehat{Y}^{k-1}$. Hence, by moving from $\mathcal{M}$ to $\widehat{\mathcal{M}}^{k-1}$ we reduce the number of traded asset, but we preserve the original information structure $\mathbb{F}$. Minimal completeness of a model $\mathcal{M}$ means, in particular, that all primary assets $Y^1, Y^2, \dots, Y^k$ are needed if we wish to generate the class of all (bounded) $\mathcal{F}_T$-measurable claims through $\mathbb{F}$-predictable trading strategies. The following lemma is thus an immediate consequence of Definition 3.

Lemma 5. *Assume that a model $\mathcal{M} = (Y^1, Y^2, \dots, Y^k; \Phi)$ is complete, but not minimally complete, with respect to $\mathbb{F}$. Then there exists at least one primary asset Y^i, which is redundant, in the sense that there exists a complete reduced model $\widehat{\mathcal{M}}^l$ for some $l \leq k-1$ such that $Y^i \neq \widehat{Y}^j$ for $j = 1, 2, \dots, l$.*

Complete models that are not minimally complete do not seem to describe adequately the real-life features of financial markets (in fact, it is frequently argued that the real-life markets are not even complete). Also, from the theoretical perspective, there is no advantage in keeping a redundant asset among primary securities. For this reasons, in what follows, we shall restrict our attention to market models $\mathcal{M}$ that are either incomplete or minimally complete. Lemma 6 shows that the degree of completeness is a well-defined notion, in the sense that it does not depend on the choice of traded assets, provided that the model completeness is preserved.

Lemma 6. *Let a model $\mathcal{M} = (Y^1, Y^2, \dots, Y^k; \Phi)$ be minimally complete with respect to $\mathbb{F}$. Let $\widetilde{\mathcal{M}} = (\widetilde{Y}^1, \widetilde{Y}^2, \dots, \widetilde{Y}^k; \Phi)$, where the processes $\widetilde{Y}^i =$*

$V(\phi_i)$, $i = 1, 2, \ldots, k$ represent the wealth processes of some trading strategies $\phi_1, \phi_2, \ldots, \phi_k \in \Phi$. If a model $\widetilde{\mathcal{M}}$ is complete with respect to $\mathbb{F}$ then it is also minimally complete with respect to $\mathbb{F}$, and thus its degree of completeness equals k.

Proof. The proof relies on simple algebraic considerations. By assumption, for every $i = 1, 2, \ldots, k$, we have

$$d\widetilde{Y}_t^i = \sum_{j=1}^{k} \phi_t^{ij}\, dY_t^j\,,$$

for some family ϕ^{ij}, $i, j = 1, 2, \ldots, k$ of $\mathbb{F}$-predictable stochastic processes. Assume that the model $\widetilde{\mathcal{M}}$ is complete, but not minimally complete. Then there exists $l \leq k - 1$ and trading strategies ψ^m, $m = 1, 2, \ldots, l$, such that the reduced model $\widehat{\mathcal{M}}^l = (|HATy^1, \widehat{Y}^2, \ldots, \widehat{Y}^l; \Phi)$, with asset prices satisfying

$$d\widehat{Y}_t^m = \sum_{i=1}^{l} \psi_t^{mi}\, d\widetilde{Y}_t^i\,,$$

is complete. Clearly, we have

$$d\widehat{Y}_t^m = \sum_{i=1}^{l} \psi_t^{mi} \sum_{j=1}^{k} \phi_t^{ij}\, dY_t^j = \sum_{j=1}^{k} \zeta_t^{mj} dY_t^j\,,$$

so that there exist trading strategies ζ^m, $m = 1, 2, \ldots, l$, in primary assets $Y^1, Y^2, \ldots, Y^k$ such that $\widehat{Y}^m = V(\zeta^m)$ for $m = 1, 2, \ldots, l$. This contradicts the assumption that the model $\mathcal{M}$ is minimally complete. □

By combining Lemma 6 with Corollary 2, we obtain the following result.

Corollary 4. *A model $\mathcal{M} = (Y^1, Y^2, \ldots, Y^k; \Phi)$ is minimally complete if and only if a model $\bar{\mathcal{M}} = (Y^1, Y^2, \ldots, Y^l, \bar{Y}^{l+1}, \ldots, \bar{Y}^{k-1}, Y^k; \Phi)$ has this property.*

As one might easily guess, the degree of a model completeness depends on the relationship between the number of primary assets and the number of independent sources of randomness. In the two models examined in Sections 3.5.1 and 3.5.2 below, we shall deal with $k = 4$ primary assets, but the number of independent sources of randomness will equal two and three for the first and the second model, respectively.

3.4.2 Completeness of a Constrained Model

Let $\mathcal{M} = (Y^1, Y^2, \ldots, Y^k; \Phi)$ be an arbitrage-free market model, and let us denote by $\mathcal{M}_l(Z) = (Y^1, Y^2, \ldots, Y^k; \Phi_l(Z))$ the associated model in which the class Φ is replaced by the class $\Phi_l(Z)$ of constrained strategies. We claim that if $\mathcal{M}$ is arbitrage-free and minimally complete with respect to the filtration $\mathbb{F} = \mathbb{F}^Y$, where $Y = (Y^1, Y^2, \dots, Y^k)$, then the constrained model $\mathcal{M}_l(Z)$ is arbitrage-free, but it is incomplete with respect to $\mathbb{F}$. Conversely, if the model $\mathcal{M}_l(Z)$ is arbitrage-free and complete with respect to $\mathbb{F}$, then the original model $\mathcal{M}$ is not minimally complete. To prove these claims, we need some preliminary results.

The following definition extends the notion of equivalence of security market models to the case of constrained trading.

Definition 4. *We say that the two constrained models are* equivalent with respect to a filtration $\mathbb{F}$ *if they are defined on a common probability space and the class of all wealth processes of* $\mathbb{F}$*-predictable constrained trading strategies is the same in both models.*

Corollary 5. *The constrained model*

$$\mathcal{M}_l(Z) = (Y^1, Y^2, \dots, Y^k; \Phi_l(Z))$$

is equivalent to the constrained model

$$\bar{\mathcal{M}}_{k-1}(Z) = (Y^1, Y^2, \dots, Y^l, \bar{Y}^{l+1}, \dots, \bar{Y}^{k-1}, Y^k; \Phi_{k-1}(Z)).$$

Proof. It suffices to make use of Corollaries 2 and 3. □

Note that the model $\bar{\mathcal{M}}_{k-1}(Z)$ is easier to handle than $\mathcal{M}_l(Z)$. For this reason, we shall state the next result for the model $\mathcal{M}_l(Z)$ (which is of our main interest), but we shall focus on the equivalent model $\bar{\mathcal{M}}_{k-1}(Z)$ in the proof.

Proposition 3. (i) *Assume that the model $\mathcal{M}$ is arbitrage-free and minimally complete. Then for any $\mathbb{F}$-predictable process Z and any $l = 1, 2, \dots, k-1$ the constrained model $\mathcal{M}_l(Z)$ is arbitrage-free and incomplete.*
(ii) *Assume that the constrained model $\mathcal{M}_l(Z)$ associated with $\mathcal{M}$ is arbitrage-free and complete. Then $\mathcal{M}$ is either not arbitrage-free or not minimally complete.*

Proof. The arbitrage-free property of $\mathcal{M}_l(Z)$ is an immediate consequence of Corollary 5 and the fact that $\Phi_{k-1}(Z) \subset \Phi$. In view of Corollary 4, it suffices to check that the minimal completeness of $\bar{\mathcal{M}}$ implies that $\bar{\mathcal{M}}_{k-1}(Z)$ is incomplete. By assumption, there exists a bounded, $\mathcal{F}_T$-measurable claim X that cannot be replicated in $\bar{\mathcal{M}}^k = (Y^1, Y^2, \dots, Y^l, \bar{Y}^{l+1}, \dots, \bar{Y}^{k-1}; \Phi)$ (i.e., when trading in Y^k is not allowed). Let us consider the following random variable

$$Y = X + \int_0^T \frac{Z_t}{Y_t^k} \, dY_t^k .$$

We claim that Y cannot be replicated in $\bar{\mathcal{M}}_{k-1}(Z)$. Indeed, for any trading strategy $\phi \in \Phi_{k-1}(Z)$, we have

$$V_T(\phi) = V_0(\phi) + \sum_{i=1}^{l} \int_0^T \phi_t^i \, dY_t^i + \sum_{i=l+1}^{k-1} \int_0^T \phi_t^i \, d\bar{Y}_t^i + \int_0^T \frac{Z_t}{Y_t^k} \, dY_t^k ,$$

and thus the existence of a replicating strategy for Y in $\bar{\mathcal{M}}_{k-1}(Z)$ will imply the existence of a replicating strategy for X in $\bar{\mathcal{M}}^k$, which contradicts our assumption. Part (ii) is a straightforward consequence of part (i). □

It is worth noting that the arbitrage-free property of $\mathcal{M}_l(Z)$ does not imply the same property for $\mathcal{M}$. As a trivial example, we may take $l = k - 1$ and $Z = 0$, so that trading in the asset Y^k is in fact excluded in $\mathcal{M}_l(Z)$, but it is allowed in the larger model $\mathcal{M}$.

3.5 Jump-Diffusion Case

In order to make the results of Sections 3.2-3.4 more tangible, we shall now analyze the case of jump-diffusion processes. For the sake of concreteness and simplicity, we shall take $k = 4$. Needless to say that this assumption is not essential, and the similar considerations can be done for any sufficiently large number of primary assets.

We consider a model $\mathcal{M} = (Y^1, Y^2, \ldots, Y^4; \Phi)$ with discontinuous asset prices governed by the SDE

$$dY_t^i = Y_{t-}^i \left(\mu_i \, dt + \sigma_i \, dW_t + \kappa_i \, dM_t \right) , \tag{3.30}$$

for $i = 1, \ldots, 4$, where $W_t = (W_t^1, W_t^2, \ldots, W_t^d)$, $t \in [0, T]$, is a d-dimensional standard Brownian motion and $M_t = N_t - \lambda t$, $t \in [0, T]$, is a compensated Poisson process under the actual probability $\mathbb{P}$. Let us stress that W and N are a Brownian motion and a Poisson process with respect to $\mathbb{F}$, respectively. This means, in particular, that they are independent processes. We shall assume that $\mathbb{F} = \mathbb{F}^{W,N}$ is the filtration generated by W and N.

The coefficients $\mu_i, \sigma_i = (\sigma_i^1, \sigma_i^2, \ldots, \sigma_i^d)$ and κ_i in (3.30) can be constant, deterministic or even stochastic (predictable with respect to the filtration $\mathbb{F}$). For simplicity, in what follows we shall assume that they are constant. In addition, we postulate that $\kappa_1 > -1$, so that $Y_t^1 > 0$ for every $t \in [0, T]$, provided that $Y_0^1 > 0$. Finally, let Z be a predetermined $\mathbb{F}$-predictable process. Recall that $\Phi_1(Z)$ is the class of all self-financing strategies that satisfy the balance condition

$$\sum_{i=2}^{4} \phi_t^i Y_{t-}^i = Z_t , \quad \forall \, t \in [0, T] . \tag{3.31}$$

Our goal is to present examples illustrating Proposition 3 and, more importantly, to show how to proceed if we wish to replicate a contingent claim using a trading strategy satisfying the balance condition. It should be acknowledged that in the previous sections we have not dealt at all with the issue of admissibility of trading strategies, and thus some relevant technical assumptions were not mentioned. Also, an important tool of an (equivalent) martingale measure was not yet employed.

3.5.1 Complete Constrained Model

In this subsection, it it assumed that $d = 1$, so that we have two independent sources of randomness, a one-dimensional Brownian motion W and a Poisson process N. We shall verify directly that, under natural additional conditions, the model $\mathcal{M}_1(Z)$ is arbitrage-free and complete with respect to $\mathbb{F}$, but the original model $\mathcal{M}$ is not minimally complete, so that a redundant primary asset exists in $\mathcal{M}$.

Lemma 7. *Assume that $\delta := \det A \neq 0$, where*

$$A = \begin{bmatrix} \sigma_2 - \sigma_4 & \kappa_2 - \kappa_4 \\ \sigma_3 - \sigma_4 & \kappa_3 - \kappa_4 \end{bmatrix}.$$

Then there exists a unique probability measure $\widetilde{\mathbb{P}}$, equivalent to $\mathbb{P}$ on $(\Omega, \mathcal{F}_T)$, and such that the relative prices $\bar{Y}^{2,1} = \bar{Y}^2/Y^1$ and $\bar{Y}^{3,1} = \bar{Y}^3/Y^1$ of synthetic assets $\bar{Y}^2$ and $\bar{Y}^3$ are $\widetilde{\mathbb{P}}$-martingales.

Proof. Let us write $\widehat{W}_t = W_t - \sigma_1 t$ and $\widehat{M}_t = M_t - \lambda\kappa_1 t$. By straightforward calculations, the relative value of the synthetic asset $\bar{Y}^i$ satisfies, for $i = 2, 3$,

$$\begin{aligned} d\bar{Y}_t^{i,1} = dY_t^{i,4,1} = Y_{t-}^{i,1}\Big((\mu_i - \mu_4)\, dt + \\ +(\sigma_i - \sigma_4)(dW_t - \sigma_1 dt) + \frac{\kappa_i - \kappa_4}{1+\kappa_1}\,(dM_t - \lambda\kappa_1 dt)\Big)\,, \end{aligned} \tag{3.32}$$

or equivalently,

$$d\bar{Y}_t^{i,1} = Y_{t-}^{i,1}\Big((\mu_i - \mu_4)\, dt + (\sigma_i - \sigma_4)\, d\widehat{W}_t + \frac{\kappa_i - \kappa_4}{1+\kappa_1}\, d\widehat{M}_t\Big)\,.$$

By virtue of Girsanov's theorem, there exists a unique probability measure $\widehat{\mathbb{P}}$, equivalent to $\mathbb{P}$ on $(\Omega, \mathcal{F}_T)$, and such that the processes $\widehat{W}$ and $\widehat{M}$ follow $\mathbb{F}$-martingales under $\widehat{\mathbb{P}}$. Under our assumption $\delta := \det A \neq 0$, the equations

$$\mu_4 - \mu_i = (\sigma_4 - \sigma_i)\theta + \frac{\kappa_4 - \kappa_i}{1+\kappa_1}\,\nu\lambda\,, \quad i = 2, 3\,, \tag{3.33}$$

uniquely specify θ and ν. Using once again Girsanov's theorem, we show that there exists a unique probability measure $\widetilde{\mathbb{P}}$, equivalent to $\mathbb{P}$ on $(\Omega, \mathcal{F}_T)$, and such that the processes $\widetilde{W}_t = \widehat{W}_t - \theta t = W_t - (\sigma_1 + \theta)t$ and

$$\widetilde{M}_t = \widehat{M}_t - \lambda\nu t = N_t - \lambda(1 + \kappa_1 + \nu)t$$

are $\mathbb{F}$-martingales under $\widetilde{\mathbb{P}}$. We then have, for $i = 2, 3$ and every $t \in [0, T]$,

$$d\bar{Y}_t^{i,1} = Y_{t-}^{i,1}\Big((\sigma_i - \sigma_4)\, d\widetilde{W}_t + \frac{\kappa_i - \kappa_4}{1+\kappa_1}\, d\widetilde{M}_t\Big).$$

Note that N follows under $\widetilde{\mathbb{P}}$ a Poisson process with the constant intensity $\lambda(1+\kappa_1+\nu)$, and thus $\widetilde{M}$ is the compensated Poisson process under $\widetilde{\mathbb{P}}$. Moreover, under the present assumptions, the processes $\widetilde{W}$ and $\widetilde{M}$ are independent under $\widetilde{\mathbb{P}}$. □

From now on, we postulate that $\delta = \det A \neq 0$ and $\kappa_i > -1$ for every $i = 1, 2, \ldots, 4$. Under this assumption, the filtration $\mathbb{F}$ coincides with the filtration $\mathbb{F}^Y$ generated by primary assets.

In the next result, we provide sufficient conditions for the existence of a replicating strategy satisfying the balance condition (3.31). Essentially, Proposition 4 shows that the model $\mathcal{M}_1(Z) = (Y^1, \bar{Y}^2, \bar{Y}^3, Y^4; \Phi_1(Z))$ is complete with respect to $\mathbb{F}$.

Proposition 4. *Let X be an $\mathcal{F}_T$-measurable contingent claim that settles at time T. Assume that the random variable $\widehat{X}$, given by the formula*

$$\widehat{X} = \frac{X}{Y_T^1} - \int_0^T \frac{Z_t}{Y_{t-}^4}\, dY_t^{4,1}, \tag{3.34}$$

is square-integrable under $\widetilde{\mathbb{P}}$, where $\widetilde{\mathbb{P}}$ is the unique probability measure equivalent to $\mathbb{P}$ on $(\Omega, \mathcal{F}_T)$ such that the relative prices $\bar{Y}^{2,1}$ and $\bar{Y}^{3,1}$ are $\widetilde{\mathbb{P}}$-martingales. Then X can be replicated in the model $\mathcal{M}_1(Z)$.

Proof. To prove the existence of a replicating strategy for X in the class $\Phi_1(Z)$, we may use either Proposition 2 (if we wish to work with traded assets Y^1, Y^2, Y^3, Y^4) or Corollary 3 and Lemma 7 (if we prefer to work with $Y^1, \bar{Y}^2, \bar{Y}^3, Y^4$). The second choice seems to be more convenient, and thus we shall focus on the existence a trading strategy $\psi = (\psi^1, \psi^2, \ldots, \psi^4)$ with the properties described in Corollary 3. In view of (3.28) and Corollary 3, it suffices to check that there exist a constant x, and $\mathbb{F}$-predictable processes ϕ^2 and ϕ^3 such that

$$\widehat{X} = x + \sum_{i=2}^{3} \int_0^T \phi_t^i\, d\bar{Y}_t^{i,1}. \tag{3.35}$$

To show that such processes exist, we shall use Lemma 7. It is crucial to observe that the pair $(\widetilde{W}, \widetilde{M})$, which was obtained in the proof of Lemma 8 from the original pair (W, M) by means of Girsanov's transformation, enjoys the predictable representation property (see, for example, Jacod and Shiryaev [3], Sections III.4 and III.5). Since $\widehat{X}$ is square-integrable under $\widetilde{\mathbb{P}}$, there exists a constant x and $\mathbb{F}$-predictable processes ξ and ς such that

$$\widehat{X} = x + \int_0^T \xi_t\, d\widetilde{W}_t + \int_0^T \varsigma_t\, d\widetilde{M}_t.$$

Observe that

$$d\widetilde{W}_t = \delta^{-1}\left((\kappa_3 - \kappa_4)\frac{d\bar{Y}_t^{2,1}}{Y_{t-}^{2,1}} - (\kappa_2 - \kappa_4)\frac{d\bar{Y}_t^{3,1}}{Y_{t-}^{3,1}}\right) =: \Theta_t^2\, d\bar{Y}_t^{2,1} + \Theta_t^3\, d\bar{Y}_t^{3,1}$$

and

$$d\widetilde{M}_t = (1+\kappa_1)\delta^{-1}\left((\sigma_2 - \sigma_4)\frac{d\bar{Y}_t^{3,1}}{Y_{t-}^{3,1}} - (\sigma_3 - \sigma_4)\frac{d\bar{Y}_t^{2,1}}{Y_{t-}^{2,1}}\right) =: \Psi_t^2\, d\bar{Y}_t^{2,1} + \Psi_t^3\, d\bar{Y}_t^{3,1}.$$

Hence, upon setting

$$\phi_t^2 = \xi_t \Theta_t^2 + \varsigma_t \Psi_t^2, \quad \phi_t^3 = \xi_t \Theta_t^3 + \varsigma_t \Psi_t^3,$$

we obtain the desired representation (3.35) for $\widehat{X}$. To complete the proof of the proposition, it suffices to make use of Corollary 3. □

Remark 2. If we take the class $\Phi_2(Z)$ of constrained strategies, instead of the class $\Phi_1(Z)$, then we need to show the existence of $\mathbb{F}$-predictable processes ϕ^2 and ϕ^3 such that

$$\widehat{X} = x + \int_0^T \phi_t^2\, dY_t^{2,1} + \int_0^T \phi_t^3\, d\bar{Y}_t^{3,1}. \tag{3.36}$$

To this end, it suffices to focus on an equivalent probability measure under which the relative prices $Y^{2,1}$ and $\bar{Y}^{3,1}$ are $\mathbb{F}$-martingales, and to follow the same steps as in the proof of Proposition 4.

In view of Lemma 7, the reduced model $\bar{\mathcal{M}}^4 = (Y^1, \bar{Y}^2, \bar{Y}^3; \Phi)$ admits a martingale measure $\widetilde{\mathbb{P}}$ corresponding to the choice of Y^1 as a numeraire asset, and thus it is arbitrage-free, under the usual choice of admissible trading strategies (e.g., the so-called *tame* strategies). By virtue of formula (12.2) in Proposition 1, for the arbitrage-free property of the model $\mathcal{M}_1(Z)$ to hold, it suffices, in addition, that the process

$$\int_0^t \frac{Z_u}{Y_{u-}^4}\, dY_u^{4,1}, \quad \forall\, t \in [0,T],$$

follows a martingale under $\widetilde{\mathbb{P}}$.

Note, however, that the above-mentioned property does not imply, in general, that the probability measure $\widetilde{\mathbb{P}}$ is a martingale measure for the relative price $Y^{4,1}$. Since

$$dY_t^{4,1} = Y_{t-}^{4,1}\left((\mu_4 - \mu_1)\, dt + (\sigma_4 - \sigma_1)(dW_t - \sigma_1 dt) + \frac{\kappa_4 - \kappa_1}{1+\kappa_1}\,(dM_t - \lambda\kappa_1 dt)\right), \tag{3.37}$$

a martingale measure for the relative prices $\bar{Y}^{2,1}, \bar{Y}^{3,1}$ and $Y^{4,1}$ exists if and only if for the pair (θ, ν) that solves (3.33), we also have that

$$\mu_1 - \mu_4 = (\sigma_4 - \sigma_1)\theta + \frac{\kappa_4 - \kappa_1}{1+\kappa_1}\,\nu\lambda .$$

This holds if and only if $\det \widehat{A} = 0$, where $\widehat{A}$ is the following matrix

$$\widehat{A} = \begin{bmatrix} \mu_1 - \mu_4 & \sigma_1 - \sigma_4 & \kappa_1 - \kappa_4 \\ \mu_2 - \mu_4 & \sigma_2 - \sigma_4 & \kappa_2 - \kappa_4 \\ \mu_3 - \mu_4 & \sigma_3 - \sigma_4 & \kappa_3 - \kappa_4 \end{bmatrix}.$$

Hence, the model $\bar{\mathcal{M}}$ (or, equivalently, the model $\mathcal{M}$) is not arbitrage-free, in general. In fact, $\mathcal{M}$ is arbitrage-free if and only if the primary asset Y^4 is redundant in $\mathcal{M}$. The following result summarizes our findings.

Proposition 5. *Let $\mathcal{M}$ be the model given by* (3.30). *Assume that $\kappa_i > -1$ for every $i = 1, 2, \dots, 4$ and $\delta = \det A \neq 0$. Moreover, let the process*

$$\int_0^t \frac{Z_u}{Y_{u-}^4}\, dY_u^{4,1}$$

follow a martingale under $\widetilde{\mathbb{P}}$. Then the following statements hold.
(i) *The model $\mathcal{M}_1(Z)$ is arbitrage-free and complete, in the sense of Proposition 4.*
(ii) *If the model $\mathcal{M}$ is arbitrage-free then it is complete, in the sense that any $\mathcal{F}_T$-measurable random variable X such that $X(Y_T^1)^{-1}$ is square-integrable under $\widetilde{\mathbb{P}}$ is attainable in this model, but $\mathcal{M}$ is not minimally complete.*

Example 1. Consider, for instance, a call option written on the asset Y^4, so that $X = (Y_T^4 - K)^+$, and let us assume that $Z_t = Y_{t-}^4$. Under assumptions of Proposition 5, models $\mathcal{M}$ and $\mathcal{M}_1(Z)$ are arbitrage-free and the asset Y^4 is redundant. It is thus rather clear that the option can be hedged by dynamic trading in primary assets Y^1, Y^2, Y^3 and by keeping at any time one unit of Y^4. Of course, the same conclusion applies to any European claim with Y^4 as the underlying asset.

3.5.2 Incomplete Constrained Model

We now assume that $d = 2$, so that the number of independent sources of randomness is increased to three. In view of (3.30), we have, for $i = 1, \dots, 4$,

$$dY_t^i = Y_{t-}^i \left(\mu_i\, dt + \sigma_i^1\, dW_t^1 + \sigma_i^2\, dW_t^2 + \kappa_i\, dM_t \right).$$

We are going to check that under the set of assumptions making the unconstrained model $\mathcal{M}$ arbitrage-free and minimally complete, the constrained model $\mathcal{M}_l(Z)$ is also arbitrage-free, but it is incomplete. To this end, we first examine the existence and uniqueness of a martingale measure associated with the numeraire Y^1.

Lemma 8. *Assume that* $\det \widetilde{A} \neq 0$, *where the matrix* $\widetilde{A}$ *is given as*

$$\widetilde{A} = \begin{bmatrix} \sigma_1^1 - \sigma_4^1 & \sigma_1^2 - \sigma_4^2 & \kappa_1 - \kappa_4 \\ \sigma_2^1 - \sigma_4^1 & \sigma_2^2 - \sigma_4^2 & \kappa_2 - \kappa_4 \\ \sigma_3^1 - \sigma_4^1 & \sigma_3^2 - \sigma_4^2 & \kappa_3 - \kappa_4 \end{bmatrix}.$$

Then there exists a unique probability measure $\widetilde{\mathbb{P}}$, *equivalent to* $\mathbb{P}$ *on* $(\Omega, \mathcal{F}_T)$, *and such that the relative prices* $\bar{Y}^{2,1} = \bar{Y}^2/Y^1$, $\bar{Y}^{3,1} = \bar{Y}^3/Y^1$ *of synthetic assets* $\bar{Y}^2, \bar{Y}^3$, *and the relative price* $Y^{4,1}$ *of the primary asset* Y^4 *follow martingales under* $\widetilde{\mathbb{P}}$.

Proof. Let us write

$$\widehat{W}_t = W_t - \sigma_1 t = (W_t^1, W_t^2) - (\sigma_1^1, \sigma_1^2)t$$

and $\widehat{M}_t = M_t - \lambda\kappa_1 t$. By straightforward calculations, the relative values $\bar{Y}^{i,1}$, $i = 2, 3$ and $Y^{4,1}$ satisfy

$$d\bar{Y}_t^{i,1} = Y_{t-}^{i,1}\Big((\mu_i - \mu_4)\,dt + (\sigma_i - \sigma_4)\,d\widehat{W}_t + \frac{\kappa_i - \kappa_4}{1+\kappa_1}\,d\widehat{M}_t\Big)$$

and

$$dY_t^{4,1} = Y_{t-}^{4,1}\Big((\mu_4 - \mu_1)\,dt + (\sigma_4 - \sigma_1)(dW_t - \sigma_1 dt) + \frac{\kappa_4 - \kappa_1}{1+\kappa_1}\,(dM_t - \lambda\kappa_1 dt)\Big).$$

By virtue of Girsanov's theorem, there exists a unique probability measure $\widehat{\mathbb{P}}$, equivalent to $\mathbb{P}$ on $(\Omega, \mathcal{F}_T)$, and such that the processes $\widehat{W}$ and $\widehat{M}$ follow $\mathbb{F}$-martingales under $\widehat{\mathbb{P}}$. Now, let $\theta = (\theta^1, \theta^2)$ and ν be uniquely specified by the conditions

$$\mu_4 - \mu_i = (\sigma_i - \sigma_4)\theta + \frac{\kappa_i - \kappa_4}{1+\kappa_1}\,\nu\lambda, \quad i = 2, 3, 4.$$

Another application of Girsanov's theorem yields the existence of a unique probability measure $\widetilde{\mathbb{P}}$, equivalent to $\mathbb{P}$ on $(\Omega, \mathcal{F}_T)$, such that the processes $\widetilde{W}_t = \widehat{W}_t - \theta t = W_t - (\sigma_1 + \theta)t$ and

$$\widetilde{M}_t = \widehat{M}_t - \lambda\nu t = N_t - \lambda(1 + \kappa_1 + \nu)t$$

are $\mathbb{F}$-martingales under $\widetilde{\mathbb{P}}$. We then have, for $i = 2, 3$ and every $t \in [0, T]$,

$$d\bar{Y}_t^{i,1} = Y_{t-}^{i,1}\Big((\sigma_i - \sigma_4)\,d\widetilde{W}_t + \frac{\kappa_i - \kappa_4}{1+\kappa_1}\,d\widetilde{M}_t\Big) \tag{3.38}$$

while

$$dY_t^{4,1} = Y_{t-}^{4,1}\Big((\sigma_4 - \sigma_1)d\widetilde{W}_t + \frac{\kappa_4 - \kappa_1}{1+\kappa_1}\,d\widetilde{M}_t\Big). \tag{3.39}$$

Note that N follows under $\widetilde{\mathbb{P}}$ a Poisson process with the constant intensity $\lambda(1+\kappa_1+\nu)$, and thus $\widetilde{M}$ is the compensated Poisson process under $\widetilde{\mathbb{P}}$. Moreover, under the present assumptions, the processes $\widetilde{W}$ and $\widetilde{M}$ are independent under $\widetilde{\mathbb{P}}$. □

It is clear that the inequality $\det \widetilde{A} \neq 0$ is a necessary and sufficient condition for the arbitrage-free property of the model $\mathcal{M}$. Under this assumption, we also have $\mathbb{F} = \mathbb{F}^Y$ and, as can be checked easily, the model $\mathcal{M}$ is minimally complete.

In the next result, we provide sufficient conditions for the existence of a replicating strategy satisfying the balance condition (3.31) with some predetermined process Z. In particular, it is possible to deduce from Proposition 6 that the model $\mathcal{M}_1(Z)$ is incomplete with respect to $\mathbb{F}$.

Proposition 6. *Assume that* $\det \widetilde{A} \neq 0$. *Let* X *be an* $\mathcal{F}_T$*-measurable contingent claim that settles at time* T. *Assume that the random variable* $\widehat{X}$, *given by the formula*

$$\widehat{X} := \frac{X}{Y_T^1} - \int_0^T \frac{Z_t}{Y_{t-}^4}\, dY_t^{4,1}, \tag{3.40}$$

is square-integrable under $\widetilde{\mathbb{P}}$, *where* $\widetilde{\mathbb{P}}$ *is the unique probability measure, equivalent to* $\mathbb{P}$ *on* $(\Omega, \mathcal{F}_T)$, *such that the relative prices* $\bar{Y}^{2,1}, \bar{Y}^{3,1}$ *and* $Y^{4,1}$ *follow martingales under* $\widetilde{\mathbb{P}}$. *Then* X *can be replicated in* $\mathcal{M}_1(Z)$ *if and only if the process* ϕ^4 *given by formula* (3.44) *below vanishes identically.*

Proof. We shall use similar arguments as in the proof of Proposition 4. In view of Corollary 3, we need to check that there exist a constant x, and $\mathbb{F}$-predictable processes ϕ^2 and ϕ^3 such that

$$\widehat{X} = x + \sum_{i=2}^{3} \int_0^T \phi_t^i \, d\bar{Y}_t^{i,1}. \tag{3.41}$$

Note that the pair $(\widetilde{W}, \widetilde{M})$ introduced in the proof of Lemma 8 has the predictable representation property. Since $\widehat{X}$ is square-integrable under $\widetilde{\mathbb{P}}$, there exists a constant x and $\mathbb{F}$-predictable processes ξ and ς such that

$$\widehat{X} = x + \int_0^T \xi_t \, d\widetilde{W}_t + \int_0^T \varsigma_t \, d\widetilde{M}_t. \tag{3.42}$$

In view of (3.38)-(3.39), we have

$$\begin{bmatrix} d\widetilde{W}_t^1 \\ d\widetilde{W}_t^2 \\ d\widetilde{M}_t \end{bmatrix} = \widetilde{A}^{-1} \begin{bmatrix} (Y_{t-}^{2,1})^{-1}\, d\bar{Y}_t^{2,1} \\ (Y_{t-}^{3,1})^{-1}\, d\bar{Y}_t^{3,1} \\ (Y_{t-}^{4,1})^{-1}\, dY_t^{4,1} \end{bmatrix},$$

so that there exist $\mathbb{F}$-predictable processes $\Psi^i, \Lambda^i, \Theta^i$, $i = 2, 3, 4$ such that

$$\begin{aligned} d\widetilde{W}_t^1 &= \Theta_t^2\, d\bar{Y}_t^{2,1} + \Theta_t^3\, d\bar{Y}_t^{3,1} + \Theta_t^4\, dY_t^{4,1}, \\ d\widetilde{W}_t^2 &= \Lambda_t^2\, d\bar{Y}_t^{2,1} + \Lambda_t^3\, d\bar{Y}_t^{3,1} + \Lambda_t^4\, dY_t^{4,1}, \\ d\widetilde{M}_t &= \Psi_t^2\, d\bar{Y}_t^{2,1} + \Psi_t^3\, d\bar{Y}_t^{3,1} + \Psi_t^4\, dY_t^{4,1}. \end{aligned} \tag{3.43}$$

Let us set, for $i = 2, 3, 4$,

$$\phi_t^i = \xi_t^1 \Theta_t^i + \xi_t^2 \Lambda_t^i + \varsigma_t \Psi_t^i, \quad \forall\, t \in [0, T]. \tag{3.44}$$

Suppose first that $\phi_t^4 = 0$ for every $t \in [0, T]$. Then, by combining (3.42), (3.43) and (3.44), we end up with the desired representation (3.41) for $\widehat{X}$. To show the existence of a replicating strategy for X in $\mathcal{M}_1(Z)$, it suffices to apply Corollary 3. If, on the contrary, ϕ^4 does not vanish identically, equality (3.41) cannot hold for any choice of ϕ^2 and ϕ^3. The fact that ϕ^4 is non-vanishing for some claims follows from Proposition 3. □

In general, i.e., when the component ϕ^4 does not vanish, we get the following representation

$$\frac{X}{Y_T^1} = x + \sum_{i=2}^{3} \int_0^T \phi_t^i \, d\bar{Y}_t^{i,1} + \int_0^T \widetilde{\phi}_t^4 \, dY_t^{4,1}, \tag{3.45}$$

where we set $\widetilde{\phi}_t^4 = \phi_t^4 + Z_t (Y_{t-}^4)^{-1}$. Hence, as expected any contingent claim satisfying a suitable integrability condition is attainable in the unconstrained model $\mathcal{M}$.

Example 2. To get a concrete example of a non-attainable claim in $\mathcal{M}_1(Z)$, let us take $X = (Y_T^4 - K)^+$ and $Z_t = Y_{t-}^4$. Then, for $K = Y_0^4$, we obtain $\widehat{X} = (Y_0^4 - Y_T^4)^+ (Y_T^1)^{-1}$, and thus we formally deal with the put option written on Y^4 with strike Y_0^4. We claim that $\widehat{X}$ does not admit representation (3.41). Indeed, equality (3.41) implies that the hedge ratio of a put option with respect to the underlying asset equals zero. This may happen only if the underlying asset is redundant so that hedging can be done with other primary assets, and this is not the case in our model.

References

[1] T.R. Bielecki, M. Jeanblanc and M. Rutkowski (2004a) Hedging of defaultable claims. In: *Paris-Princeton Lectures on Mathematical Finance 2003*, R.A. Carmona, E. Cinlar, I. Ekeland, E. Jouini, J.E. Scheinkman, N. Touzi, eds., Springer-Verlag, Berlin Heidelberg New York, pp. 1-132.

[2] T.R. Bielecki, M. Jeanblanc and M. Rutkowski (2004b) Completeness of a reduced-form credit risk model with discontinuous asset prices. Working paper.

[3] J. Jacod and A.N. Shiryaev (1987) *Limit Theorems for Stochastic Processes.* Springer-Verlag, Berlin Heidelberg New York.

[4] P. Protter (2001) A partial introduction to financial asset pricing theory. *Stochastic Processes and Their Applications* 91, 169-203.

[5] N. Vaillant (2001) A beginner's guide to credit derivatives. Working paper, Nomura International.

4

Extremal behavior of stochastic volatility models

Vicky Fasen, Claudia Klüppelberg, and Alexander Lindner

[1] Center for Mathematical Sciences, Munich University of Technology, D-85747 Garching, Germany
fasen@ma.tum.de
[2] *idem*, cklu@ma.tum.de
[3] *idem*, lindner@ma.tum.de

Summary. Empirical volatility changes in time and exhibits tails, which are heavier than normal. Moreover, empirical volatility has – sometimes quite substantial – upwards jumps and clusters on high levels. We investigate classical and non-classical stochastic volatility models with respect to their extreme behavior. We show that classical stochastic volatility models driven by Brownian motion can model heavy tails, but obviously they are not able to model volatility jumps. Such phenomena can be modelled by Lévy driven volatility processes as, for instance, by Lévy driven Ornstein-Uhlenbeck models. They can capture heavy tails and volatility jumps. Also volatility clusters can be found in such models, provided the driving Lévy process has regularly varying tails. This results then in a volatility model with similarly heavy tails. As the last class of stochastic volatility models, we investigate a continuous time GARCH(1,1) model. Driven by an arbitrary Lévy process it exhibits regularly varying tails, volatility upwards jumps and clusters on high levels.

Key words: COGARCH, extreme value theory, generalized Cox-Ingersoll-Ross model, Lévy process, Ornstein-Uhlenbeck process, Poisson approximation, regular variation, stochastic volatility model, subexponential distribution, tail behavior, volatility cluster.
2000 MSC Subject Classifications: primary: 60G70, 91B70, secondary: 60G10, 91B84
JEL Classifications: C23, C51.

4.1 Introduction

The classical pricing model is the Black-Scholes model given by the SDE

$$dS_t = rS_t dt + \sigma S_t dB_t\,, \quad S_0 = x \in \mathbb{R}\,, \tag{4.1}$$

where $r \in \mathbb{R}$ is the stock-appreciation rate, $\sigma > 0$ is the volatility and B is a standard Brownian motion. The Black-Scholes model is based on the assumption that the relative price changes of the asset form a Gaussian process with stationary and independent increments. The crucial parameter is the volatility σ, which is in this model assumed to be constant. However, empirical analysis of stock volatility has already shown in the 1970ies that volatility is not constant, quite the contrary, it is itself stochastic and varies in time.

This observation has led to a vast number of volatility models in discrete-time as well as in continuous-time. In this paper we concentrate on continuous-time volatility models. Moreover, we are concerned with the so-called stylized facts of volatility as e. g.

- volatility changes in time,
- volatility is random,
- volatility has heavy tails,
- volatility clusters on high levels.

Introducing a stochastic volatility extends the Black-Scholes model to

$$dS_t = rS_t dt + \sqrt{V_t} S_t dB_t \,,$$

where V can in principle be any positive stationary stochastic process.

Within the framework of SDEs quite natural models are easily defined. Common examples are the *Ornstein-Uhlenbeck (OU) process*

$$dV_t = -\lambda V_t dt + \sigma dZ_t \,, \tag{4.2}$$

where $\lambda, \sigma > 0$ and Z is the driving process, often a second Brownian motion, independent of B. As this is a Gaussian model, it is not a positive process. Alternatively, a *Cox-Ingersoll-Ross (CIR) model* has been suggested as a volatility model, defined by

$$dV_t = \lambda(a - V_t)dt + \sigma\sqrt{V_t} dZ_t \,, \tag{4.3}$$

where $\lambda, a, \sigma > 0$ and $\lambda a \geq \sigma^2/2$. The parameter a is the long-term mean of the process and λ the rate of mean reversion. Again in the classical model Z is a standard Brownian motion, independent of B.

Apart from the fact that Gaussian OU processes are not positive, another stylized fact is also violated: empirical volatility exhibits heavy tails, consequently, again the OU model as a Gaussian model seems not very appropriate. Changing the constant σ to a time dependent diffusion coefficient σV_t^γ for $\gamma \in [1/2, \infty)$ and including a linear drift yields positive stationary models with arbitrarily heavy tails. This has been shown in Borkovec and Klüppelberg [8]. Such models are called *generalized Cox-Ingersoll-Ross models*, a parameter $\gamma = 1/2$ corresponds to the classical CIR model of (4.3).

On the other hand, a constant σ is attractive, and an alternative way to generate heavy tails in the volatility is to replace the driving Gaussian process in (4.2) by a Lévy process with heavier tailed increments. Furthermore, the upward jumps often observed in empirical volatility cannot be modelled by a continuous process. So Lévy processes with jumps as driving processes seem to be quite natural. Such an OU process is positive, provided the driving Lévy process has only positive increments and no Gaussian component; i. e. it is a subordinator. This is exactly what Barndorff-Nielsen and Shephard [4, 5] have suggested, modelling the (right-continuous) volatility process as a Lévy driven *OU process*. Their stochastic volatility model is given by

$$dV_t = -\lambda V_t dt + \sigma dL_{\lambda t}\,, \tag{4.4}$$

where $a, b \in \mathbb{R}$, $\lambda > 0$ and L is a subordinator, called the *background driving Lévy process* (BDLP). The price process itself is then driven by an independent Brownian motion.

A completely different approach to obtain continuous-time volatility models starts with a GARCH model and derives from this discrete-time model a continuous-time model. A natural idea is a diffusion approximation; see e. g. Drost and Werker [16] and references therein. This approach leads to stochastic volatility models of the type

$$\begin{aligned} dS_t &= \sqrt{V_t}\, dB_t^{(1)}\,, \\ dV_t &= \lambda(a - V_t)dt + \sigma V_t dB_t^{(2)}\,, \end{aligned} \tag{4.5}$$

i. e. V is a generalized CIR model with parameter $\gamma = 1$. The two processes $B^{(1)}, B^{(2)}$ are independent Brownian motions.

A different approach has been taken by Klüppelberg, Lindner and Maller [27], who started with a discrete-time GARCH(1,1) model and replaced the noise variables by a Lévy process L with jumps $\Delta L_t = L_t - L_{t-}$, $t \geq 0$. This yields a stochastic volatility model of the type

$$\begin{aligned} dS_t &= \sqrt{V_t}\, dL_t\,, \\ dV_{t+} &= \beta dt + V_t \mathrm{e}^{X_{t-}} d(\mathrm{e}^{-X_t})\,, \end{aligned} \tag{4.6}$$

where $\beta > 0$ and V is left-continuous. The auxiliary càdlàg process X is defined by

$$X_t = \eta t - \sum_{0<s\leq t} \ln(1 + \lambda \mathrm{e}^{\eta} (\Delta L_s)^2)\,, \tag{4.7}$$

for $\eta > 0$ and $\lambda \geq 0$. This continuous-time GARCH(1,1) model is called a COGARCH(1,1) model.

Our paper focuses on the extremal behavior of stationary continuous-time stochastic volatility models. This can be described by the tail behavior of the stationary distribution and by the behavior of the process above high thresholds.

The tail behavior models the size of the fluctuations of V and determines the maximum domain of attraction (MDA) of the model. The notion of MDA is defined in Fisher-Tippett's theorem; see Theorem 8. We distinguish MDA(Φ_α), MDA(Λ) and MDA(Ψ_α), for $\alpha > 0$, respectively. Distribution functions in MDA(Φ_α) have regularly varying tails: they are heavy-tailed in the sense that not all moments are finite; see Definition 3. Distribution functions in MDA(Λ) have tails ranging from semi-heavy tails to very light tails. Distribution functions in MDA(Ψ_α) have support bounded to the right. Financial risk is usually considered as having unbounded support above, hence MDA(Ψ_α) is inappropriate in our context and will play no further role in this paper.

The description of a continuous-time process above high thresholds depends on the sample path behavior of the process. When classical volatility models driven by a Brownian motion have continuous sample paths with infinite variation, some discrete-time skeleton is introduced. A standard concept is based on so-called ϵ-upcrossings, see Definition 1, which is only valid for processes with continuous sample paths.

For Lévy driven models large jumps (for instance larger than 1) constitute a natural discrete-time skeleton, which can be utilized. Denote by $(\Gamma_k)_{k\in\mathbb{N}}$ the random time points on $[0,\infty)$, where the driving Lévy process jumps and exceeds a given threshold. The bivariate process $(\Gamma_k, V_{\Gamma_k})_{k\in\mathbb{N}}$ is interpreted as the coordinates of a point process in $[0,\infty)\times\mathbb{R}_+$. As usual we define point processes via Dirac measures. Recall that for any Borel sets $A\times B\subseteq [0,\infty)\times\mathbb{R}_+$ the measure $\sum_{k=1}^{\infty}\varepsilon\{\Gamma_k, V_{\Gamma_k}\}(A\times B)$ counts how often $\Gamma_k\in A$ and $V_{\Gamma_k}\in B$.

After appropriate normalization in time and space these point processes may converge and the limit process may allow for an interpretation, thus providing a description of the extreme behavior of the volatility process. Under weak dependence in the data we obtain as limit a Poisson random measure with mean measure ϑ (PRM(ϑ)); see Definition 9. Moreover, the two components of ϑ are independent and consist of the Poisson measure in time and the negative logarithm of an extreme value distribution in space. Under strong dependence the limit is a cluster Poisson random measure. All these considerations concern the discrete-time skeleton only and ignore the fact that we deal with continuous-time processes.

In the case of a driving Lévy process with jumps, in principle also the small jumps can influence the extreme behavior. In a very close neighborhood of a jump time Γ_k infinitely many small jumps can happen; they may contribute to the extreme behavior around Γ_k. To investigate the influence of these small jumps and the Gaussian component, we consider the process V at each point Γ_k in a surrounding interval I_k. Finally, in certain situations we investigate also the process V after it has achieved a local supremum. With each point Γ_k an excursion of V over a high threshold starts. Interesting questions concern the length of the excursion, the rate of "decrease" after Γ_k, and we shall answer such questions at least for some models considered in this paper. This is done by attaching marks to the point process $(\Gamma_k, V_{\Gamma_k})_{k\in\mathbb{N}}$. In our model marks are a vector of values of the process V after Γ_k, hence it describes the finite dimensional distributions of V after Γ_k. The limit process turns out to be different in different regimes.

Our paper is organized as follows. In Section 11 we review the extremal behavior of the generalized CIR model, which can belong to different maximum domains of attractions; i. e. such models can have arbitrary tails. Unfortunately, they are not appropriate models in case of high level volatility clusters in the data.

Section 11 deals with Lévy driven OU volatility models. Their extremal behavior is characterized by the extremal behavior of the driving Lévy process, so that we have to distinguish between different classes of BDLPes. In Section 4.3.1 this is done for subexponential Lévy processes $L = (L_t)_{t \geq 0}$. According to whether $L_1 \in \text{MDA}(\Phi_\alpha)$ for some $\alpha > 0$ or $L_1 \in \text{MDA}(\Lambda)$, the extremal behavior of the OU process is quite different. Then, in Section 4.3.2 we study OU processes with exponential tails. As a prominent example we investigate the Γ-OU process, i. e. the stationary volatility is gamma distributed. As an important larger class we study OU processes, whose BDLP belongs to $\mathcal{S}(\gamma)$ for $\gamma > 0$. This class extends subexponential Lévy processes in a natural way; see Definition 5. It turns out that for all OU processes in Section 11, high level volatility clusters are exhibited only in the case of regularly varying BDLPes.

The last class of models reviewed in this paper concerns the COGARCH process in Section 11. In contrast to the OU processes considered earlier, the COGARCH volatility has heavy tails under quite general conditions on the driving Lévy process L. Furthermore, the COGARCH exhibits high level volatility clusters.

Finally, a short conclusion is given in Section 4.5. Here we compare the models introduced in the different sections before. It turns out that there is a striking similarity concerning the extremal behaviour of models with the same stationary distribution. Here we also discuss briefly some further empirical facts of volatility not quite in the focus of our paper.

As not to disturb the flow of arguments we postpone classical definitions and concepts to an Appendix. Throughout this paper we shall use the following notation. We abbreviate distribution function by d. f. and random variable by r. v. For any d. f. F we denote its tail $\overline{F} = 1 - F$. For two r. v. s X and Y with d. f. s F and G we write $X \stackrel{d}{=} Y$ if $F = G$, and by $\stackrel{T\to\infty}{\Longrightarrow}$ we denote weak convergence for $T \to \infty$. For two functions f and g we write $f(x) \sim g(x)$ as $x \to \infty$, if $\lim_{x\to\infty} f(x)/g(x) = 1$. We also denote $\mathbb{R}_+ = (0, \infty)$. For $x \in \mathbb{R}$, we let $x^+ = \max\{x, 0\}$ and $\ln^+(x) = \ln(\max\{x, 1\})$. Integrals of the form $\int_a^b$ will be interpreted as the integral taken over the interval $(a, b]$.

4.2 Extremal behavior of generalized Cox-Ingersoll-Ross models

In this section we summarize some well-known results on classical volatility models as defined in (4.3) and (4.5) driven by a standard Brownian motion. This section is based on Borkovec and Klüppelberg [8]; for a review see [26], Section 3.

These models fall into the framework of *generalized Cox-Ingersoll-Ross* (GCIR) models. We restrict ourselves to stationary solutions of the SDE

$$dV_t = \lambda(a - V_t)dt + \sigma V_t^{\gamma} dB_t\,, \tag{4.8}$$

where $\gamma \in \left[\frac{1}{2}, \infty\right)$. For $\lambda, a, \sigma > 0$ (in the case $\gamma = 1/2$ additionally $\lambda a \geq \sigma^2/2$ is needed) these models are ergodic with state space $\mathbb{R}_+$ and have a stationary density.

Associated with the diffusion (4.8) is the scale function s and the speed measure m. The *scale function* is defined as

$$s(x) = \int_z^x \exp\left(-\frac{2\lambda}{\sigma^2}\int_z^y \frac{a-t}{t^{2\gamma}}dt\right)dy\,, \quad x \in \mathbb{R}_+\,, \tag{4.9}$$

where z is any interior point of $\mathbb{R}_+$ whose choice does not affect the extremal behavior. For the *speed measure* m we know that it is finite for the GCIR model. Moreover, m is absolutely continuous with Lebesgue density

$$m'(x) = \frac{2}{\sigma^2 x^{2\gamma} s'(x)}\,, \quad x \in \mathbb{R}_+\,,$$

where s' is the Lebesgue density of s. Then the stationary density of V is given by

$$f(x) = m'(x)/m(\mathbb{R}_+)\,, \quad x \in \mathbb{R}_+\,. \tag{4.10}$$

Proposition 1. *Let V be a* GCIR *model given by equation* (4.8) *and define $M(T) := \sup_{t\in[0,T]} V_t$ for $T > 0$. Then for any initial value $V_0 = y \in \mathbb{R}_+$ with corresponding distribution $\mathbb{P}_y$ and any $u_T \uparrow \infty$,*

$$\lim_{T\to\infty} |\mathbb{P}_y(M(T) \leq u_T) - H^T(u_T)| = 0\,,$$

where H is a d. f., defined for any $z \in \mathbb{R}_+$ by

$$H(x) = \exp\left(-\frac{1}{m(\mathbb{R}_+)s(x)}\right), \quad x > z\,. \tag{4.11}$$

The function s and the quantity $m(\mathbb{R}_+)$ depend on the choice of z.

Corollary 1 (Running maxima).
Let the assumptions of Proposition 1 hold. Assume further that $H \in \mathrm{MDA}(G)$ for $G \in \{\Phi_\alpha, \alpha > 0, \Lambda\}$ with norming constants $a_T > 0$, $b_T \in \mathbb{R}$. Then

$$\lim_{T\to\infty} \mathbb{P}(a_T^{-1}(M(T) - b_T) \leq x) = G(x)\,, \quad x \in \mathbb{R}\,.$$

It is clear that the d. f. H decides about the extremal behavior of V. We present four cases.

Example 1 (Tail behavior of GCIR *models).*
Let V be a stationary GCIR model given by equation (4.8) with stationary density f, corresponding d. f. F, and d. f. H as given in (4.11). Recall that a $\Gamma(\mu,\gamma)$ distributed r. v. has probability density

$$p(x) = \frac{\gamma^\mu}{\Gamma(\mu)} x^{\mu-1} \mathrm{e}^{-\gamma x}, \quad x > 0, \tag{4.12}$$

for $\mu > 1$ and $\gamma > 0$.

(1) $\gamma = \frac{1}{2}$: The stationary density of V is $\Gamma\left(\frac{2\lambda a}{\sigma^2}, \frac{2\lambda}{\sigma^2}\right)$. The tail of H behaves like

$$\overline{H}(x) \sim \frac{2\lambda^2 a}{\sigma^2} x \overline{F}(x), \quad x \to \infty,$$

so that the tail of H is that of a $\Gamma\left(\frac{2\lambda a}{\sigma^2} + 1, \frac{2\lambda}{\sigma^2}\right)$ distribution. Hence $H \in$ MDA(Λ) with norming constants

$$a_T = \frac{\sigma^2}{2\lambda} \quad \text{and} \quad b_T = \frac{\sigma^2}{2\lambda}\left[\ln T + \frac{2\lambda a}{\sigma^2}\ln\ln T - \ln\left(\frac{\lambda}{\Gamma(2\lambda a/\sigma^2)}\right)\right].$$

(2) $\frac{1}{2} < \gamma < 1$: The stationary density of V is given by

$$f(x) = \frac{2}{A\sigma^2} x^{-2\gamma} \exp\left(-\frac{2}{\sigma^2}\left(\frac{\lambda a}{2\gamma - 1} x^{-(2\gamma-1)} + \frac{\lambda}{2-2\gamma} x^{2-2\gamma}\right)\right), \quad x > 0,$$

with some normalizing factor $A > 0$. The d. f. H has tail

$$\overline{H}(x) \sim B x^{2(1-\gamma)} \overline{F}(x), \quad x \to \infty, \tag{4.13}$$

where $B > 0$. Hence $H \in$ MDA(Λ) with norming constants

$$\begin{aligned} a_T &= \frac{\sigma^2}{2\lambda}\left(\frac{\sigma^2(1-\gamma)}{\lambda}\ln T\right)^{\frac{2\gamma-1}{2-2\gamma}}, \\ b_T &= \left(\frac{\sigma^2(1-\gamma)}{\lambda}\ln T\right)^{\frac{1}{(2-2\gamma)}}\left(1 - \frac{2\gamma-1}{(2-2\gamma)^2}\frac{\ln\left(\frac{\sigma^2(1-\gamma)}{\lambda}\ln T\right)}{\ln T}\right) + \\ &\quad + a_T \ln\left(\frac{2\lambda}{A\sigma^2}\right). \end{aligned}$$

(3) $\gamma = 1$: The stationary density of V is inverse gamma, i. e.

$$f(x) = \left(\frac{\sigma^2}{2\lambda a}\right)^{-\frac{2\lambda}{\sigma^2}-1}\left(\Gamma\left(\frac{2\lambda}{\sigma^2}+1\right)\right)^{-1} x^{-\frac{2\lambda}{\sigma^2}-2}\exp\left(-\frac{2\lambda a}{\sigma^2}x^{-1}\right), \quad x > 0,$$

so that $V_0 \in \mathcal{R}_{-2\lambda/\sigma^2-1}$; see Definition 3. In this case $\overline{H}(x) \sim c x^{-2\lambda/\sigma^2-1}$ for $x \to \infty$ and some $c > 0$. Hence $H \in$ MDA(Φ_α) for $\alpha = 2\lambda/\sigma^2 + 1$ with norming constants $a_T = (cT)^{\sigma^2/(2\lambda+\sigma^2)}$ and $b_T = 0$.

(4) $\gamma > 1$: The stationary density f of V has the same form as in (2), but is regularly varying of index $-2\gamma + 1$. Now the tail of H becomes very extreme: $\overline{H}(x) \sim cx^{-1}$. Hence $H \in \text{MDA}(\Phi_1)$ with $a_T = cT$ and $b_T = 0$. □

Since all models (4.8) are driven by a Brownian motion, they have continuous sample paths; i. e. there is no natural discrete-time skeleton. We follow the standard approach to create a discrete-time skeleton of the process; see e. g. Leadbetter et al. [29], Chapter 12. This allows for a more profound extreme value analysis of V.

Definition 1. *Let V be a stationary version of the diffusion given by* (4.8). *V is said to have an* ϵ-upcrossing *of the level u at a point $\Gamma > 0$ if*

$$V_t < u \quad \text{for} \quad t \in (\Gamma - \epsilon, \Gamma) \quad \text{and} \quad V_\Gamma = u\,.$$

With this definition we can formulate a further result describing the extreme behavior of a stationary GCIR model.

Theorem 1 (Point process of ϵ-upcrossings).
Let V be a stationary version of the diffusion given by (4.8) *with d. f. H as in* (4.11). *Let $a_T > 0$, $b_T \in \mathbb{R}$ be the norming constants as given in Example 4.10. Let $(\Gamma_{T,k})_{k\in\mathbb{N}}$ be the time points on $\mathbb{R}_+$, where the ϵ-upcrossings of V of the level $a_T x + b_T$ occur. Let $(j_k)_{k\in\mathbb{N}}$ be the jump times of a Poisson process with intensity e^{-x} for $x \in \mathbb{R}$, if $\gamma \in [1/2, 1)$ and $x^{-\alpha}$ for $x > 0$ with $\alpha = 2\lambda/\sigma^2 + 1$ if $\gamma = 1$ and $\alpha = 1$ if $\gamma > 1$. Then*

$$\sum_{k=1}^{\infty} \varepsilon\,\{\Gamma_{T,k}/T\} \stackrel{T\to\infty}{\Longrightarrow} \sum_{k=1}^{\infty} \varepsilon\{j_k\}\,.$$

As is obvious from this result, ϵ-upcrossings of V for high levels behave like exceedances of i. i. d. data, i. e. such models do not exhibit volatility clusters. They can, however, model heavy tails as the running maxima depend on the d. f. H.

4.3 Extremal behavior of OU volatility models

We start with a precise definition of a positive OU process as a solution of (4.4). For more information on Lévy processes we refer to the excellent monographs by Sato [35], Bertoin [6] and Cont and Tankov [15]. Let L be a subordinator; i. e. L is a Lévy process with increasing sample paths, hence they are of bounded variation, and we assume that they are càdlàg. The Laplace transform is then the natural transform and has for all $t \geq 0$ the representation

$$\mathbb{E}\exp(-\lambda L_t) = \exp(t\Psi(\lambda)), \quad \lambda \geq 0.$$

The Laplace exponent Ψ has representation

$$\Psi(\lambda) = -m\lambda - \int_{(0,\infty)} (1 - \mathrm{e}^{-\lambda x})\, \nu(dx).$$

As there is no Gaussian component the characteristic triplet of arbitrary Lévy processes reduces to a pair (m, ν), where $m \geq 0$ is the drift and the Lévy measure ν has support on $\mathbb{R}_+$ and satisfies

$$\int_{(0,\infty)} (1 \wedge x)\, \nu(dx) < \infty.$$

We denote for $\lambda > 0$ by

$$V_t = \mathrm{e}^{-\lambda t} V_0 + \int_0^t \mathrm{e}^{-\lambda(t-s)}\, dL_{\lambda s}, \quad t \in \mathbb{R}, \tag{4.14}$$

the solution to the SDE (4.4). Then V becomes a càdlàg process.

A stationary solution of (4.4) exists if and only if $\int_{s>1} \ln(1+s)\, \nu(ds) < \infty$. Note that this condition is only violated for Lévy measures with very heavy tails. As all models considered in this paper have tails which are regularly varying of some negative index or lighter, all our models satisfy this stationarity condition. Stationarity is then achieved, if V_0 is taken to be independent of the driving Lévy process L and has distribution

$$V_0 \stackrel{d}{=} \int_0^\infty \mathrm{e}^{-s}\, dL_s. \tag{4.15}$$

A convenient representation for the stationary version is

$$V_t = \mathrm{e}^{-\lambda t} \int_{-\infty}^t \mathrm{e}^{\lambda s}\, dL_{\lambda s}, \quad t \geq 0. \tag{4.16}$$

In this representation, L is extended to a Lévy process on the whole real line, by letting $\widetilde{L} = (\widetilde{L}_t)_{t\geq 0}$ be an independent copy of $(L_t)_{t\geq 0}$, and defining $L_t := -\widetilde{L}_{-t-}$ for $t < 0$. The parameter λ in the process L in (4.14) ensures that the stationary marginal distribution of V is independent of λ; indeed it is given by (4.15).

The r. v. V_0 is infinitely divisible with characteristic pair (m_V, ν_V), where $m_V = m$ and

$$\nu_V\, [x, \infty) = \int_x^\infty u^{-1} \nu\, [u, \infty)\, du, \quad x > 0. \tag{4.17}$$

We are concerned with Lévy processes L, which are heavy or semi-heavy tailed; i. e. whose tails decrease not faster than exponentially. As indicated in (4.21) and (4.24) below, this induces a similar tail behavior on V, which is in accordance with empirical findings.

The structure of an OU volatility process can be best understood when considering the following example.

Example 2 (Positive shot noise process).
Let L be a positive compound Poisson process with characteristic pair $(0, \mu\mathbb{P}_F)$, where $\mu > 0$ and $\mathbb{P}_F$ is a probability measure on $\mathbb{R}_+$ with corresponding d. f. F. Then L has the representation

$$L_t = \sum_{j=1}^{N_t} \xi_j \,, \quad t > 0 \,, \tag{4.18}$$

where $(N_t)_{t\geq 0}$ is a Poisson process on $\mathbb{R}_+$ with intensity $\mu > 0$ and jump times $(\Gamma_k)_{k\in\mathbb{N}}$. The process N is independent of the i. i. d. sequence of positive r. v. s $(\xi_k)_{k\in\mathbb{N}}$ with d. f. F.

The resulting volatility process is then the positive shot noise process

$$V_t = \mathrm{e}^{-\lambda t} V_0 + \int_0^t \mathrm{e}^{-\lambda(t-s)}\, dL_{\lambda s} = \mathrm{e}^{-\lambda t} V_0 + \sum_{j=1}^{N_{\lambda t}} \mathrm{e}^{-\lambda t + \Gamma_j}\, \xi_j \,, \quad t \geq 0 \,,$$

and from (4.17) we get

$$\nu_V\, [x, \infty) = \mu \int_x^\infty u^{-1} \overline{F}(u)\, du \,, \quad x > 0 \,.$$

If $\mathbb{E} \ln(1 + \xi_1) < \infty$, a stationary solution exists in which case V can be represented as

$$V_t = \mathrm{e}^{-\lambda t} \sum_{\substack{j=-\infty \\ j\neq 0}}^{N_{\lambda t}} \mathrm{e}^{\Gamma_j} \xi_j \,, \quad t > 0 \,. \tag{4.19}$$

Here, we let $(\xi_k)_{k\in -\mathbb{N}_0}$ and $(\Gamma_k)_{k\in -\mathbb{N}_0}$ be sequences of r. v. s such that $(\xi_k)_{k\in\mathbb{Z}}$ and $(\Gamma_k)_{k\in\mathbb{Z}}$ are independent. Furthermore, $(\xi_k)_{k\in\mathbb{Z}}$ is an i. i. d. sequence and $(-\Gamma_k)_{k\in -\mathbb{N}}$ are the jump times of a Poisson process on $\mathbb{R}_+$ with intensity μ, independent of $(\Gamma_k)_{k\in\mathbb{N}}$; further, we define $\Gamma_0 := 0$.

The qualitative extreme behavior of this volatility process can be seen in Figure 4.1 in detail. The volatility jumps upwards, whenever $(N_{\lambda t})_{t\geq 0}$ jumps and decreases exponentially fast between two jumps. This means in particular that V has local suprema exactly at the jump times Γ_k/λ (and $t = 0$), i.e.

$$V_t = V_{\Gamma_k/\lambda} \mathrm{e}^{-\lambda t + \Gamma_k} \,, \quad t \in [\Gamma_k/\lambda, \Gamma_{k+1}/\lambda) \,.$$

Consequently, it is the discrete-time skeleton of V at points Γ_k/λ that determines the extreme behavior of the volatility process. □

For a general subordinator L we decompose

$$L = L^{(1)} + L^{(2)} \tag{4.20}$$

into two independent Lévy processes, where $L^{(1)}$ has characteristic pair $(0, \nu_1)$ with $\nu_1(x, \infty) = \nu(x, \infty)\, \mathbf{1}_{(1,\infty)}(x)$ and $L^{(2)}$ has characteristic pair (m, ν_2)

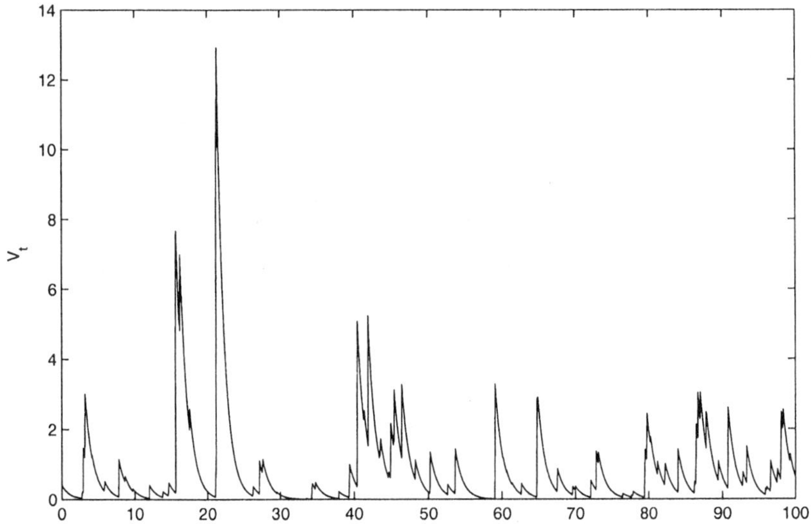

Fig. 4.1. Sample path of an OU Weibull process with L_1 as given in Example 3 with $\lambda = 1$, $\mu = 1$ and $p = 1/2$.

with $\nu_2(x,\infty) = \nu(x,1]\,\mathbf{1}_{(0,1]}(x)$. Then again $L^{(1)}$ is a compound Poisson process with intensity $\nu(1,\infty)$ and jump sizes with d. f. $\nu_1/\nu(1,\infty)$. All the small jumps and the drift are summarized in $L^{(2)}$.

What is needed, however, are the precise asymptotic links between the tails of V, L and the tail of the Lévy measure $\nu(\cdot,\infty)$. This implies immediately that we have to distinguish different regimes defined by the link between L and $\nu(\cdot,\infty)$. Any infinitely divisible distribution is asymptotically tail-equivalent to its Lévy measure, whenever it is convolution equivalent; see Theorem 7. Definitions and results concerning subexponential and convolution equivalent distributions are summarized in Appendix A.

The class $\mathcal{S}(0) = \mathcal{S}$ of subexponential d. f. s contains all d. f. s with regularly varying tails, but is much larger. Subexponential distributions can be in two different maximum domains of attractions; see Theorem 8. All d. f. s with regularly varying tails are subexponential and belong to MDA(Φ_α). Other subexponential d. f. s, as for instance the lognormal and the semi-heavy tailed Weibull d. f. s (see Example 3), belong to MDA(Λ). On the other hand, d. f. s as the gamma distribution or d. f. s in $\mathcal{S}(\gamma)$ for $\gamma > 0$ belong to MDA(Λ), but are lighter tailed than any subexponential distribution. Consequently, we also consider such exponential models below.

4.3.1 OU processes with subexponential tails

In this section we are concerned with the OU process given by (4.14), whose BDLP is subexponential. This section is based on Fasen [19, 20]; an additional reference is [18].

Proposition 2 (Tail behavior of subexponential models). *Let V be a stationary version of the OU process given by (4.14) and define $M(h) := \sup_{t\in[0,h]} V_t$ for $h > 0$.*

(a) If $L_1 \in \mathcal{S} \cap \mathrm{MDA}(\Phi_\alpha) = \mathcal{R}_{-\alpha}$, then also $V_0 \in \mathcal{R}_{-\alpha}$ and

$$\mathbb{P}(V_0 > x) \sim \alpha^{-1}\, \mathbb{P}(L_1 > x)\,, \quad x \to \infty\,. \tag{4.21}$$

Moreover,

$$\mathbb{P}(M(h) > x) \sim \left[\lambda h + \alpha^{-1}\right] \mathbb{P}(L_1 > x)\,, \quad x \to \infty\,. \tag{4.22}$$

(b) If $L_1 \in \mathcal{S} \cap \mathrm{MDA}(\Lambda)$, then also $V_0 \in \mathcal{S} \cap \mathrm{MDA}(\Lambda)$ and

$$\mathbb{P}(V_0 > x) \sim \mathbb{P}(\exp(-U)L_1 > x)\,, \quad x \to \infty\,, \tag{4.23}$$

where U is a uniform r. v. on $(0,1)$, independent of L. In particular, $\mathbb{P}(V_0 > x) = o,\ (\mathbb{P}(L_1 > x))$ as $x \to \infty$. More precisely,

$$\mathbb{P}(V_0 > x) \sim \frac{a(x)}{x}\mathbb{P}(L_1 > x), \quad x \to \infty\,, \tag{4.24}$$

where a is the function from the representation (A.1):

$$\mathbb{P}(L_1 > x) \sim c\exp\left[-\int_0^x \frac{1}{a(y)}\,dy\right]\,, \quad x \to \infty\,,$$

for some $c > 0$ and $a : \mathbb{R}_+ \to \mathbb{R}_+$ is absolutely continuous with $\lim_{x\to\infty} a'(x) = 0$ and $\lim_{x\to\infty} a(x) = \infty$. Finally,

$$\mathbb{P}(M(h) > x) \sim \lambda h \mathbb{P}(L_1 > x)\,, \quad x \to \infty\,. \tag{4.25}$$

Proof. By (4.17) we have

$$\frac{\nu_V(x,\infty)}{\nu(x,\infty)} = \frac{\int_x^\infty u^{-1}\nu(u,\infty)\,du}{\nu(x,\infty)}\,, \quad x > 0\,. \tag{4.26}$$

Assume that $L_1 \in \mathcal{R}_{-\alpha}$ for some $\alpha > 0$. Then by Theorem 7 *(i)* we have $\nu(\cdot,\infty)/\nu(1,\infty) \in \mathcal{R}_{-\alpha}$. By Karamata's theorem (e. g. Embrechts et al. [17], Theorem A 3.6) we obtain

$$\lim_{x\to\infty} \frac{\nu_V(x,\infty)}{\nu(x,\infty)} = \frac{1}{\alpha}\,.$$

This implies in particular that also $\nu_V(\cdot,\infty)/\nu_V(1,\infty) \in \mathcal{R}_{-\alpha}$ and hence, again by Theorem 7 *(i)*, $V_0 \in \mathcal{R}_{-\alpha}$ and (4.21) holds.

If $L_1 \in \mathrm{MDA}(\Lambda) \cap \mathcal{S}$ we can only conclude from (4.26) that the right hand side tends to 0. To obtain a precise result we proceed as follows. Denote by ξ_1 the jump distribution of the compound Poisson process $L^{(1)}$ as given in (4.20). Taking $\nu(\cdot,\infty)/\nu(1,\infty) \in \mathcal{R}_{-\infty}$ into account and applying l'Hospital's rule yields

$$\begin{aligned}\frac{\nu_V(x,\infty)}{\nu(1,\infty)\mathbb{P}(\exp(-U)\xi_1 > x)} &= \frac{\int_x^\infty u^{-1}\nu(u,\infty)\,du}{\int_0^1 \nu(\mathrm{e}^s x,\infty)\,ds} = \frac{\int_x^\infty u^{-1}\nu(u,\infty)\,du}{\int_x^{x\mathrm{e}} u^{-1}\nu(u,\infty)\,du} \\ &\sim \left[1 - \frac{\nu(\mathrm{e}\,x,\infty)}{\nu(x,\infty)}\right]^{-1} \longrightarrow 1\,, \quad x \to \infty\,.\end{aligned}$$

The tail-equivalence (4.23) follows then from the fact that

$$\nu(1,\infty)\,\mathbb{P}(\exp(-U)\,\xi_1 > x) \sim \mathbb{P}(\exp(-U)L_1 > x)$$

as $x \to \infty$ and Theorem 7 *(i)*.

For proving (4.24), by Theorem 7 *(i)* we may assume without loss of generality that there exists a $x_0 > 0$ such that

$$\nu(x,\infty) = c\,\exp\left[-\int_{x_0}^x \frac{1}{a(y)}\,dy\right], \quad x \geq x_0\,.$$

Then $\nu(dx) = (a(x))^{-1}\nu(x,\infty)\,dx$ and an application of l'Hospital's rule shows that

$$\begin{aligned}\frac{\nu_V(x,\infty)}{\nu(x,\infty)\,a(x)/x} &\sim \frac{-\nu(x,\infty)/x}{\nu(x,\infty)[a'(x) - a(x)/x]/x - (\nu(x,\infty)/a(x))a(x)/x} \\ &= [-a'(x) + a(x)/x + 1]^{-1} \\ &\to 1\,, \quad x \to \infty\,,\end{aligned}$$

since $a(x)/x \sim a'(x)$ and $a'(x) \to 0$ as $x \to \infty$. Theorem 7 *(i)* then gives (4.24).

The results for $M(h)$ are based on Theorem 2.1 of Rosinski and Samorodnitsky [34]. They show that for $L_{\lambda h} + V_0 \in \mathcal{S}$

$$\mathbb{P}(M(h) > x) \sim \nu_{L_{\lambda h}+V_0}(x,\infty)\,, \quad x \to \infty\,,$$

implying the result by Theorem 7 *(i)*. □

Example 3 (Semi-heavy tailed Weibull distribution).
Let L_1 have distribution tail

$$\mathbb{P}(L_1 > x) \sim K\exp(-x^p)\,, \quad x \to \infty\,,$$

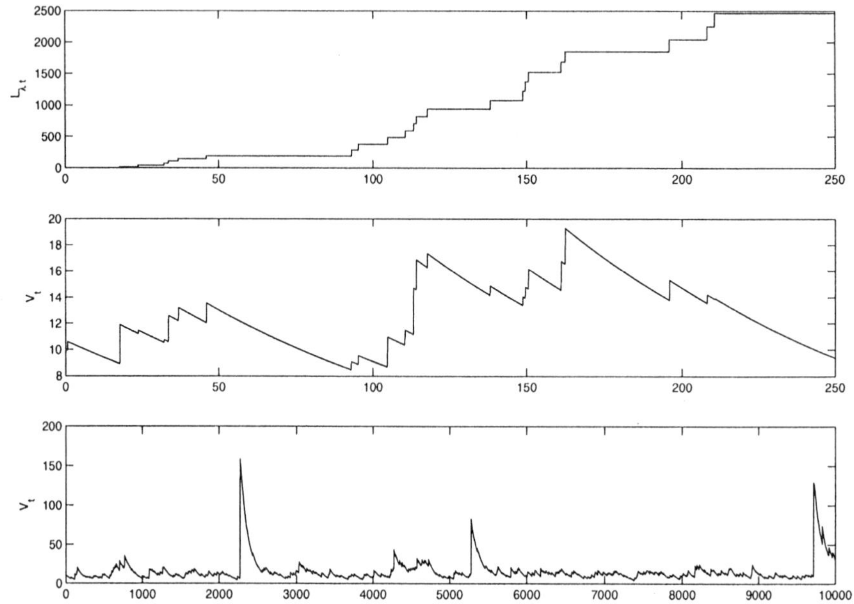

Fig. 4.2. Sample path of an OU process driven by a regularly varying compound Poisson process with $\mu = 8.5$ and $\lambda = 0.01$. The first plot shows $(L_{\lambda t})_{0 \le t \le 250}$ and the second plot shows the corresponding volatility $(V_t)_{0 \le t \le 250}$, indicating the micro-behavior of the model. The third plot gives $(V_t)_{0 \le t \le 10\,000}$ indicating the macro-behavior of the model.

for some $K > 0$ and $p \in (0,1)$. As $a(x) = x^{1-p}/p$, we obtain from (4.24) immediately

$$\mathbb{P}(V_0 > x) \sim \frac{K}{p} x^{-p} \exp(-x^p), \quad x \to \infty. \qquad \square$$

Proposition 2 shows that in the regularly varying regime the tail of V_0 is equivalent to the tail of L_1. In contrast to that, in the $\mathcal{S} \cap \mathrm{MDA}(\Lambda)$ case, the tail of V_0 becomes lighter, due to the influence of $\exp(-U)$. But in both cases V_0 is subexponential and the tail of $M(h)$ is determined by the tail of L_1, only the constants differ.

The following result gives a complete account of the extreme behavior of the volatility process V driven by a subexponential Lévy process L. There are three components considered in (4.28) and (4.29). The first one is the scaled jump time process corresponding to the jumps of $(L_{\lambda t})_{t \ge 0}$, which are larger than 1. The second component is the normalized local supremum near that jump, and the third component is a vector of normalized values of V after the jump.

Theorem 2 (Marked point process behavior of subexponential models). *Let V be a stationary version of the OU process given by (4.14). Suppose*

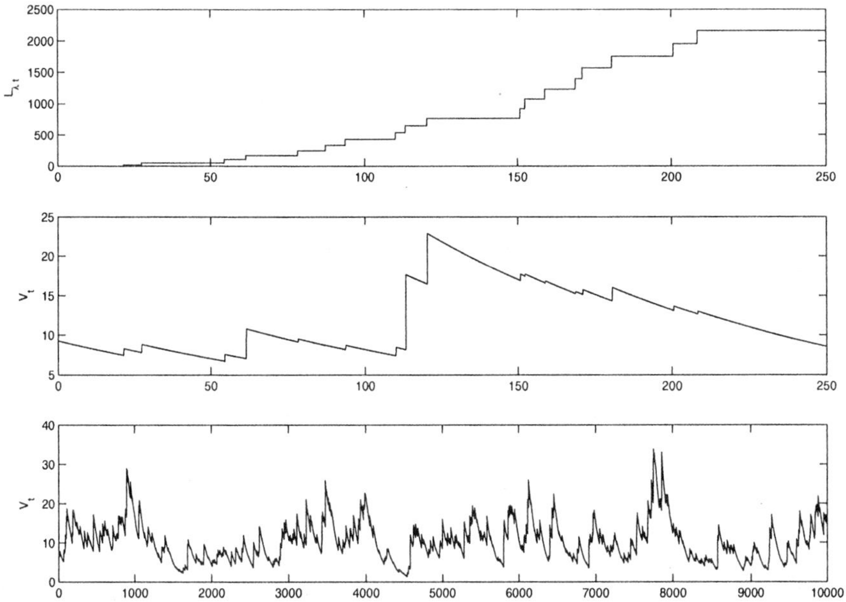

Fig. 4.3. Sample path of an OU Weibull process with $p = 0.7$, $\mu = 8.5$ and $\lambda = 0.01$. The first plot shows $(L_{\lambda t})_{0 \le t \le 250}$ and the second plot the corresponding volatility $(V_t)_{0 \le t \le 250}$, indicating the micro-behavior of the model. The third plot gives $(V_t)_{0 \le t \le 10\,000}$ indicating the macro-behavior of the model.

$\Gamma = (\Gamma_k)_{k \in \mathbb{N}}$ are the jump times of $L^{(1)}$ given by (4.20) and $I = (I_k)_{k \in \mathbb{N}}$, where $I_k = \frac{1}{2\lambda}[\Gamma_{k-1} + \Gamma_k, \Gamma_k + \Gamma_{k+1})$, $\Gamma_0 := 0$. For $m \in \mathbb{N}$ let $0 = t_0 < t_1 < \cdots < t_m$.

(a) Assume that $L_1 \in \mathcal{S} \cap \mathrm{MDA}(\Phi_\alpha)$ with norming constants $a_T > 0$ such that

$$\lim_{T \to \infty} T\mathbb{P}(L_1 > a_T x) = x^{-\alpha}, \quad x > 0. \tag{4.27}$$

Take $\Gamma^{(k)} = (\Gamma_{k,i})_{i \in \mathbb{N}}$, $k \in \mathbb{N}$, as i. i. d. copies of the sequence Γ and set $\Gamma_{k,0} = \Gamma_{k,-1} = 0$ for all $k \in \mathbb{N}$. Let $\sum_{k=1}^{\infty} \varepsilon\{s_k, P_k\}$ be a $\mathrm{PRM}(\vartheta)$ with mean measure $\vartheta(dt \times dx) = dt \times \alpha x^{-\alpha-1} \mathbf{1}_{(0,\infty)}(x)\, dx$, independent of the sequence $(\Gamma^{(k)})_{k \in \mathbb{N}}$. Then,

$$\begin{aligned} &\sum_{k=1}^{\infty} \varepsilon \left\{ \frac{\Gamma_k}{\lambda T}, a_{\lambda T}^{-1} \sup_{s \in I_k} V_s, \{a_{\lambda T}^{-1} V_{\Gamma_k/\lambda + t_i}\}_{i=0,\ldots,m} \right\} \\ &\stackrel{T\to\infty}{\Longrightarrow} \sum_{k=1}^{\infty} \sum_{j=0}^{\infty} \varepsilon \{ s_k, P_k \exp(-(\Gamma_{k,j-1} + \Gamma_{k,j})/2)\,, \\ &\qquad \{P_k \exp(-\lambda t_i - \Gamma_{k,j})\}_{i=0,\ldots,m}\}\,. \end{aligned} \tag{4.28}$$

(b) Assume that $L_1 \in \mathcal{S} \cap \mathrm{MDA}(\Lambda)$ with norming constants $a_T > 0$, $b_T \in \mathbb{R}$ such that

$$\lim_{T\to\infty} T\mathbb{P}(L_1 > a_T x + b_T) = \exp(-x), \quad x \in \mathbb{R}.$$

Let $\sum_{k=1}^{\infty} \varepsilon\{s_k, P_k\}$ be a PRM(ϑ) with $\vartheta(dt \times dx) = dt \times \mathrm{e}^{-x}\, dx$. Then

$$\begin{aligned}&\sum_{k=1}^{\infty} \varepsilon\left\{\frac{\Gamma_k}{\lambda T}, a_{\lambda T}^{-1}(\sup_{s\in I_k} V_s - b_{\lambda T}), \{a_{\lambda T}^{-1}(V_{\Gamma_k/\lambda+t_i} - b_{\lambda T})\}_{i=0,\dots,m}\right\} \\ &\stackrel{T\to\infty}{\Longrightarrow} \sum_{k=1}^{\infty} \varepsilon\{s_k, P_k, (P_k, 0, \dots, 0)\}.\end{aligned} \tag{4.29}$$

Moreover,

$$\begin{aligned}&\sum_{k=1}^{\infty} \varepsilon\left\{\frac{\Gamma_k}{\lambda T}, a_{\lambda T}^{-1}(\sup_{s\in I_k} V_s - b_{\lambda T}), \{b_{\lambda T}^{-1} V_{\Gamma_k/\lambda+t_i}\}_{i=0,\dots,m}\right\} \\ &\stackrel{T\to\infty}{\Longrightarrow} \sum_{k=1}^{\infty} \varepsilon\{s_k, P_k, \{\exp(-\lambda t_i)\}_{i=0,\dots,m}\}.\end{aligned} \tag{4.30}$$

We first give an interpretation of (4.29). The limit relations of the first two components show that the local suprema of V around the Γ_k/λ, normalized with the constants determined via L_λ, converge weakly to the same extreme value distribution as L_λ. Moreover, the third component indicates that for $t_0 = 0$ the second and third component have the same limiting behavior; i. e. the $\sup_{s\in I_k} V_s$ behaves like $V_{\Gamma_k/\lambda}$. For all $t_i > 0$ the last component is negligible, i. e. the process is considerably smaller away from $V_{\Gamma_k/\lambda}$.

In the second and third component of (4.28) all points $\Gamma_{k,j}$ and not only $\Gamma_{k,0} = 0$ like in (4.29) may influence the limit. This phenomenon has certainly its origin in the very large jumps caused by regular variation. Even though the volatility decreases between the jumps exponentially fast, huge jumps can have a long lasting influence on excursions above high thresholds. This is in contrast to the semi-heavy tailed case, where L is subexponential, but in MDA(Λ).

Both relations (4.28) and (4.30) exhibit, however, a common effect in the third component: if the Lévy process L has an exceedance over a high threshold, then the OU process decreases after this event exponentially fast.

Corollary 2 (Point process of exceedances).
Let the assumptions of Theorem 2 hold.

(a) Assume that $L_1 \in \mathcal{S} \cap \mathrm{MDA}(\Phi_\alpha)$. Let $(j_k)_{k\in\mathbb{N}}$ be the jump times of a Poisson process with intensity $x^{-\alpha}$ for $x > 0$. Let $(\zeta_k)_{k\in\mathbb{Z}}$ be i. i. d. discrete r. v. s, independent of $(j_k)_{k\in\mathbb{N}}$, with probability distribution

$$\pi_k = \mathbb{P}(\zeta_1 = k) = \mathbb{E}\exp(-\alpha(\Gamma_{k-1} + \Gamma_k)/2) - \\ -\mathbb{E}\exp(-\alpha(\Gamma_k + \Gamma_{k+1})/2), \quad k \in \mathbb{N}.$$

Then,

$$\sum_{k=1}^{\infty} \varepsilon\left\{\frac{\Gamma_k}{\lambda T}, a_{\lambda T}^{-1} \sup_{s \in I_k} V_s\right\}(\cdot \times (x, \infty)) \overset{T\to\infty}{\Longrightarrow} \sum_{k=1}^{\infty} \zeta_k \varepsilon\{j_k\}.$$

(b) Assume that $L_1 \in \mathcal{S} \cap \mathrm{MDA}(\Lambda)$. Let $(j_k)_{k\in\mathbb{N}}$ be the jump times of a Poisson process with intensity e^{-x} for $x \in \mathbb{R}$. Then,

$$\sum_{k=1}^{\infty} \varepsilon\left\{\frac{\Gamma_k}{\lambda T}, a_{\lambda T}^{-1}(\sup_{s \in I_k} V_s - b_{\lambda T})\right\}(\cdot \times (x, \infty)) \overset{T\to\infty}{\Longrightarrow} \sum_{k=1}^{\infty} \varepsilon\{j_k\}.$$

Again the qualitative difference of the two regimes is obvious. In the case of a regularly varying BDLP L the limiting process is a compound Poisson process, where at each Poisson point a cluster appears, whose size is random with distribution $(\pi_k)_{k\in\mathbb{N}}$. In contrast to this, in the MDA(Λ) case, the limit process is simply a homogeneous Poisson process; no clusters appear in the limit.

As the next result shows, the running maxima of the volatility process V have the same behavior as that of an i. i. d. sequence of copies of L_λ.

Corollary 3 (Running maxima).
Let V be a stationary version of the OU process given by (4.14), and define $M(T) := \sup_{t\in[0,T]} V_t$ for $T > 0$.

(a) Assume that $L_1 \in \mathcal{S} \cap \mathrm{MDA}(\Phi_\alpha)$ with norming constants $a_T > 0$ given by (4.27). Then

$$\lim_{T\to\infty} \mathbb{P}\left(a_{\lambda T}^{-1} M(T) \le x\right) = \exp(-x^{-\alpha}), \quad x > 0.$$

(b) Assume that $L_1 \in \mathcal{S} \cap \mathrm{MDA}(\Lambda)$ with norming constants $a_T > 0$, $b_T \in \mathbb{R}$ given by (4.29). Then

$$\lim_{T\to\infty} \mathbb{P}\left(a_{\lambda T}^{-1}(M(T) - b_{\lambda T}) \le x\right) = \exp(-\mathrm{e}^{-x}), \quad x \in \mathbb{R}.$$

Finally, we investigate the possibility of volatility clusters in the OU process. As the concept of ϵ-upcrossings is only defined for continuous-time processes, which does not fit into our framework, we shall introduce an appropriate method for describing clusters in continuous-time processes with jumps.

As our method will be motivated by the discrete-time skeleton of V, we recall that in a discrete-time process clusters are usually described by the extremal index $\theta \in (0, 1]$; see Definition 8. However, continuous-time processes are by nature dependent in small time intervals by the continuity and the structure of the process. Thus it is not adequate to adopt the extremal index concept for stochastic sequences to describe the dependence structure of the continuous-time process on a high level.

The following concept of an extremal index function has been introduced in Fasen [18].

Definition 2. *Let $(V_t)_{t\geq 0}$ be a stationary process. For $h > 0$ define the sequence $M_k(h) := \sup_{(k-1)h\leq t\leq kh} V_t$, $k \in \mathbb{N}$. Let $\theta(h)$ be the extremal index of the sequence $(M_k(h))_{k\in\mathbb{N}}$. Then we call the function $\theta : (0,\infty) \to [0,1]$* extremal index function.

The idea is to divide the positive real line into blocks of length h. By taking local suprema of the process over these blocks the natural dependence of the continuous-time process is weakened, in certain cases it even disappears. However, for fixed h the extremal index function is a measure for the expected cluster sizes among these blocks. For an extended discussion on the extremal index in the context of discrete- and continuous-time processes see Fasen [18].

Corollary 4 (Extremal index function).
(a) If $L_1 \in \mathcal{S} \cap \mathrm{MDA}(\Phi_\alpha)$, then $\theta(h) = h\alpha/(h\alpha + 1)$, $h > 0$.
(b) If $L_1 \in \mathcal{S} \cap \mathrm{MDA}(\Lambda)$, then $\theta(h) = 1$, $h > 0$.

Regularly varying OU processes exhibit clusters among blocks, since $\theta(h) < 1$. So they have the potential to model both volatility features: heavy tails and high level clusters. This is in contrast to OU processes in $\mathcal{S}\cap\mathrm{MDA}(\Lambda)$, where no clusters occur.

4.3.2 OU processes with exponential tails

In this section we investigate OU models having exponential tails, hence are lighter tailed than those considered in the previous section. More precisely, we will concentrate on two classes of models in $\mathcal{L}(\gamma)$, $\gamma > 0$; see Definition 4. The first class concerns the *class of convolution equivalent distributions* $\mathcal{S}(\gamma)$, $\gamma > 0$ (Definition 5). Here Theorem 7 provides the necessary relationship between the tails of the infinitely divisible d. f. and of its Lévy measure, which leads to a comparison between the distribution tail of the stationary r. v. V_0 and the increment L_1 of the BDLP. An important family in $\mathcal{S}(\gamma)$ are d. f. s with tail

$$\overline{F}(x) \sim x^{-\beta} l(x) \mathrm{e}^{-\gamma x - c x^p}, \quad x \to \infty,$$

where $\gamma, c \geq 0$, $p < 1$, $l(\cdot)$ is slowly varying, and if $c = 0$, then $\beta > 1$ or $\beta = 1$ and $\int_1^\infty l(x)/x \, dx < \infty$ (Klüppelberg [25], Theorem 2.1, or Pakes [32], Lemma 2.3). The *generalized inverse Gaussian distribution* (GIGD) with probability density

$$p(x) = K x^{\beta-1} \exp\left(-\left(\delta^2 x^{-1} + \gamma^2 x\right)/2\right), \quad x > 0,$$

where K is the normalizing constant, $\beta < 0$ and $\delta^2 > 0$, is a prominent example in $\mathcal{S}\left(\gamma^2/2\right)$. Further examples for distributions in $\mathcal{S}(\gamma)$ can be found e. g. in Cline [14].

The second class of processes with exponential tails, which we investigate in this section, are *Γ-OU processes.* These are defined as stationary OU processes, where V_0 is $\Gamma(\mu,\gamma)$ distributed with probability density as defined in (4.12) for $\mu > 1$ and $\gamma > 0$. The gamma distribution is infinitely divisible with absolutely continuous Lévy measure given by its density

$$\nu_V(dx) = \mu x^{-1}\mathrm{e}^{-\gamma x}dx\,, \quad x > 0\,.$$

Hence, by (4.17) the BDLP L has Lévy density

$$\nu(dx) = \mu\gamma\mathrm{e}^{-\gamma x}dx\,, \quad x > 0\,.$$

Except for the factor μ this is the probability density of an exponential d. f. Hence L is a positive compound Poisson process with Poisson rate $\mu > 0$ and exponential jumps; for details see Barndorff-Nielsen and Shephard [2]. The exponential and gamma laws with scale parameter $\gamma > 0$ belong to $\mathcal{L}(\gamma)$ but not to $\mathcal{S}(\gamma)$.

In analogy to the Γ-OU process, also for OU-$\mathcal{S}(\gamma)$ processes with $\gamma > 0$ we restrict our attention to positive compound Poisson processes as BDLPs; i. e. we work in the framework of positive shot noise processes as defined in Example 2. Note that by Proposition 4 *(b)* all d.f.s in $\mathcal{L}(\gamma)$ for $\gamma > 0$ belong to MDA(Λ).

Some of the results in this section can be found in Albin [1], who studies the extremal behavior for a larger class of OU processes by purely analytic methods.

For BDLPs in $\mathcal{S}(\gamma)$ for $\gamma > 0$ the relation of the tail of the stationary d.f. and its Lévy measure are stated in the following proposition.

Proposition 3 (Tail behavior of OU-$\mathcal{S}(\gamma)$ models for $\gamma > 0$). *Let V be a stationary version of the OU process given by (4.14).*

(a) Suppose $\nu\left(1,\cdot\right]/\nu(1,\infty) \in \mathcal{L}(\gamma)$, $\gamma > 0$. Then $\nu_V\left(1,\cdot\right]/\nu_V(1,\infty) \in \mathcal{L}(\gamma)$ with

$$\nu_V(x,\infty) \sim \frac{1}{\gamma x}\nu(x,\infty)\,, \quad x \to \infty\,.$$

(b) Suppose $L_1 \in \mathcal{S}(\gamma)$, $\gamma > 0$. Then also $V_0 \in \mathcal{S}(\gamma)$, and

$$\mathbb{P}(V_0 > x) \sim \frac{\mathbb{E}\mathrm{e}^{\gamma V_0}}{\mathbb{E}\mathrm{e}^{\gamma L_1}}\frac{1}{\gamma x}\mathbb{P}(L_1 > x)\,, \quad x \to \infty\,.$$

In particular, $\mathbb{P}(V_0 > x) = o(\mathbb{P}(L_1 > x))$ as $x \to \infty$.

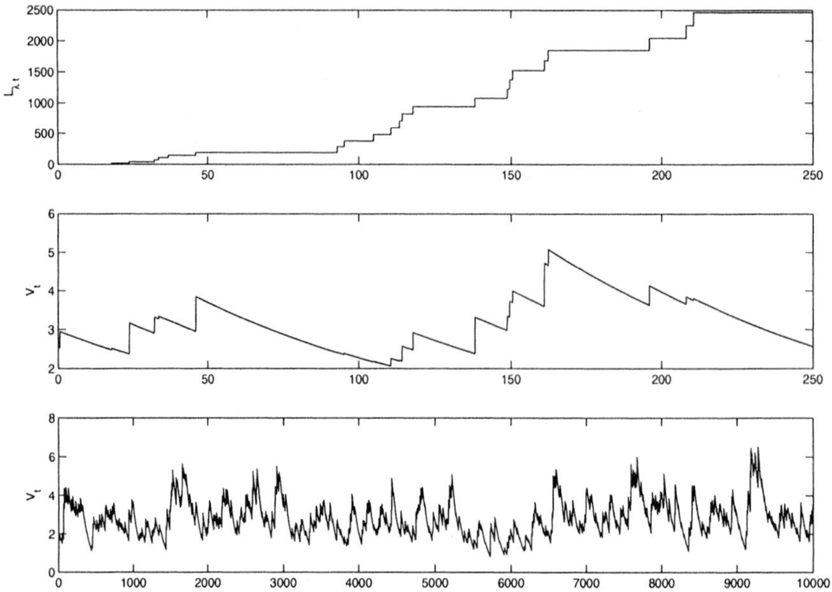

Fig. 4.4. Sample path of a Γ-OU process with $\gamma = 3$, $\mu = 8.5$ and $\lambda = 0.01$. The first plot shows $(L_{\lambda t})_{0\le t\le 250}$ and the second plot the corresponding volatility $(V_t)_{0\le t\le 250}$, indicating the micro-behavior of the model. The third plot gives $(V_t)_{0\le t\le 10\,000}$ indicating the macro-behavior of the model.

Proof. *(a)* By (A.1) the Lévy measure ν has representation

$$\nu(x,\infty) = c(x)\exp\left[-\int_1^x \frac{1}{a(y)}\,dy\right], \quad x \ge 1, \tag{4.31}$$

for functions $a, c : [1,\infty) \to \mathbb{R}_+$ with $\lim_{x\to\infty} c(x) = c > 0$ and $\lim_{x\to\infty} a(x) = 1/\gamma$, $\lim_{x\to\infty} a'(x) = 0$. Since we are only interested in the tail behavior we may assume without loss of generality that ν is absolutely continuous and $c(\cdot) \equiv c$. Recall from (4.17) that $\nu_V(dx) = x^{-1}\nu\,(x,\infty)\,dx$ and let $\nu(dx) = \nu'(x)\,dx$. Part *(a)* follows by an application of l'Hospital's rule, since

$$\frac{\nu_V(x,\infty)}{\nu(x,\infty)/(\gamma x)} \sim \frac{\nu(x,\infty)/x}{[\nu'(x)x + \nu(x,\infty)]/(\gamma x^2)} = \gamma\left[\frac{1}{a(x)} + \frac{1}{x}\right]^{-1} \to 1, \;\; x\to\infty\,.$$

(b) We first show that $V_0 \in \mathcal{S}(\gamma)$. By Theorem 7 *(i)* it suffices to show that $\nu_V\,(1,\cdot]\,/\nu(1,\infty) \in \mathcal{S}(\gamma)$. Again, we can assume without loss of generality that ν is absolutely continuous and has the representation (4.31) with constant $c(\cdot) \equiv c$. For simplicity, we further assume that $c = 1$ and $\nu(1,\infty) = 1$; the general case follows by a simple dilation. As in part *(a)* we use that $\nu_V(dx) = x^{-1}\nu\,(x,\infty)\,dx$. An application of l'Hospital's rule shows that

$$\frac{\nu_V(x-y,\infty)}{\nu_V(x,\infty)} \sim \frac{\nu(x-y,\infty)\,x}{\nu(x,\infty)\,(x-y)} \to \mathrm{e}^{\gamma y}, \quad x \to \infty\,,$$

implying $\nu_V(1,\cdot] \in \mathcal{L}(\gamma)$. Denote by ν_V^{2*} the convolution of ν_V restricted to $(1,\infty)$ with itself. Then for $1 < y' < x/2$ we use the usual decomposition of the convolution integral

$$\begin{aligned}\frac{\nu_V^{2*}(dx)}{dx} &= \int_1^x \frac{\nu(u,\infty)}{u}\frac{\nu(x-u,\infty)}{x-u}\,du \\ &= 2\int_1^{y'} \frac{\nu(u,\infty)}{u}\frac{\nu(x-u,\infty)}{x-u}\,du + \int_{y'}^{x-y'} \frac{\nu(u,\infty)}{u}\frac{\nu(x-u,\infty)}{x-u}\,du\,.\end{aligned} \tag{4.32}$$

In order to show that $\nu_V(1,\cdot] \in \mathcal{S}(\gamma)$, we calculate the limit ratio of the densities of ν_V^{2*} and ν_V. Observe that on every compact set $\nu(x-u,\infty)/\nu(x,\infty)$ converges uniformly in u to $\exp(\gamma u)$ as $x \to \infty$. For the first summand of (4.32) we thus obtain

$$\begin{aligned}\lim_{x\to\infty} 2\int_1^{y'} \frac{x}{x-u}\frac{\nu(x-u,\infty)}{\nu(x,\infty)}\frac{\nu(u,\infty)}{u}\,du &= 2\int_1^{y'} \mathrm{e}^{\gamma u}\frac{\nu(u,\infty)}{u}\,du \\ &= 2\int_1^{y'} \mathrm{e}^{\gamma u}\,\nu_V(du) < \infty\,.\end{aligned} \tag{4.33}$$

For the second summand in (4.32) we estimate

$$\begin{aligned}&\int_{y'}^{x-y'} \frac{x\nu(u,\infty)\,\nu(x-u,\infty)}{u(x-u)\,\nu(x,\infty)}\,du \\ &\le \frac{x}{y'(x-y')}\int_{y'}^{x-y'} \frac{\nu(x-u,\infty)}{\nu(x,\infty)}\,\nu(u,\infty)\,du\,.\end{aligned} \tag{4.34}$$

Furthermore, since

$$\nu(x,\infty)/\nu'(x) = a(x) \longrightarrow 1/\gamma, \quad x \to \infty\,,$$

there exist constants $K, x_0 > 0$ such that $\nu(x,\infty) \le K\nu'(x)$ for $x \ge x_0$. We obtain for $y' > x_0$

$$\begin{aligned}&\frac{x}{y'(x-y')}\int_{y'}^{x-y'} \frac{\nu(x-u,\infty)\,\nu(u,\infty)}{\nu(x,\infty)}\,du \\ &\le \frac{Kx}{y'(x-y')}\int_{y'}^{x-y'} \frac{\nu(x-u,\infty)}{\nu(x,\infty)}\,\nu(du)\,.\end{aligned} \tag{4.35}$$

Since $\nu(1,\cdot] \in \mathcal{S}(\gamma)$, the same decomposition as in (4.32) yields (for details see e. g. Pakes [32], Lemma 5.5)

$$\lim_{y'\to\infty}\lim_{x\to\infty}\int_{y'}^{x-y'}\frac{\nu(x-u,\infty)}{\nu(x,\infty)}\,\nu(du)=0\,. \tag{4.36}$$

Furthermore, $\lim_{y'\to\infty}\lim_{x\to\infty} x/[y'(x-y')] = \lim_{y'\to\infty} 1/y' = 0$. By (4.32)-(4.36) we now obtain $\nu_V^{2*}(dx) \sim (2\int_1^\infty \mathrm{e}^{\gamma u}\,\nu(du))\,\nu_V(dx)$ for $x\to\infty$, showing that $\nu_V\,(1,\cdot]$ and hence V_0 are in $\mathcal{S}(\gamma)$. The assertion on the tail behavior now follows from *(a)* and Theorem 7 *(i)*. □

The following result is an analogon to Theorem 2 and describes the extremal behavior of V completely.

Theorem 3 (Point process of exceedances of exponential models). *Let V be a stationary version of the OU process given by (4.14) with L a positive compound Poisson process as in* (4.18). *Denote by $(\Gamma_k)_{k\in\mathbb{N}}$ the jump times of the positive compound Poisson process L given by (4.18) and define $I_k := \frac{1}{\lambda}[\Gamma_k,\Gamma_{k+1})$, $k\in\mathbb{N}$. Let $\sum_{k=1}^\infty \varepsilon\{s_k,P_k\}$ be* PRM(ϑ) *with $\vartheta(dt\times dx) = dt\times \mathrm{e}^{-x}\,dx$.*

(a) Assume $L_1\in\mathcal{S}(\gamma)$, $\gamma>0$, with norming constants $a_T>0$, $b_T\in\mathbb{R}$ such that

$$\lim_{T\to\infty} T\mathbb{P}(L_1 > a_T x + b_T) = \frac{\mathbb{E}\mathrm{e}^{\gamma L_1}}{\mathbb{E}\mathrm{e}^{\gamma V_0}}\exp(-x)\,,\quad x\in\mathbb{R}\,. \tag{4.37}$$

Then

$$\sum_{k=0}^\infty \varepsilon\left\{\frac{\Gamma_k}{\lambda T}, a_{\lambda T}^{-1}(\sup_{s\in I_k} V_s - b_{\lambda T})\right\} \stackrel{T\to\infty}{\Longrightarrow} \sum_{k=1}^\infty \varepsilon\{s_k,P_k\}\,. \tag{4.38}$$

(b) Assume V is the $\Gamma(\mu,\gamma)$-OU process. Let $a_T>0, b_T\in\mathbb{R}$ be the norming constants of a $\Gamma(\mu+1,\gamma)$ distributed r. v. W, such that

$$\lim_{T\to\infty} T\mathbb{P}(W > a_T x + b_T) = \mu^{-1}\exp(-x)\,,\quad x\in\mathbb{R}\,. \tag{4.39}$$

Then

$$\sum_{k=0}^\infty \varepsilon\left\{\frac{\Gamma_k}{\lambda T}, a_{\lambda T}^{-1}(\sup_{s\in I_k} V_s - b_{\lambda T})\right\} \stackrel{T\to\infty}{\Longrightarrow} \sum_{k=1}^\infty \varepsilon\{s_k,P_k\}\,. \tag{4.40}$$

The proof is divided into several steps. We shall utilize classical results for the extreme value theory of stationary discrete-time processes. As a discrete-time skeleton $(V_{\Gamma_k/\lambda})_{k\in\mathbb{N}}$ seems to be a good candidate. However, $V_{\Gamma_k/\lambda} = \sum_{j=-\infty, j\neq 0}^{k} \mathrm{e}^{-(\Gamma_k-\Gamma_j)}\xi_j$, $k\in\mathbb{N}$, is not stationary. As we will show in Lemma 1 the process

$$\widetilde{V}_k = \sum_{j=-\infty}^{k} \mathrm{e}^{-(\Gamma_k-\Gamma_j)}\xi_j = V_{\Gamma_k/\lambda} + \mathrm{e}^{-\Gamma_k}\xi_0\,,\quad k\in\mathbb{N}\,, \tag{4.41}$$

is stationary, where $\Gamma_0 := 0$. For increasing k the process $\mathrm{e}^{-\Gamma_k}\xi_0$ tends to 0. Thus it has no influence on the extremal behavior. We shall show that the point process behavior is the same for $(V_{\Gamma_k/\lambda})_{k\in\mathbb{N}}$ and for $(\widetilde{V}_k)_{k\in\mathbb{N}}$. For the proof we also need that the D and D' conditions (Definition 9) hold for $(\widetilde{V}_k)_{k\in\mathbb{N}}$. The highly technical Lemma 2, where this is confirmed, is postponed to the Appendix.

Lemma 1. *Let V be a stationary version of the OU process given by (4.14) with L a positive compound Poisson process as in (4.18). Then $(\widetilde{V}_k)_{k\in\mathbb{Z}}$ as defined in (4.41) is stationary.*

Proof. Let $h \in \mathbb{R}$ be fixed. Note that $(\Gamma_{h+j} - \Gamma_h)_{j\in\mathbb{Z}} \stackrel{d}{=} (\Gamma_j)_{j\in\mathbb{Z}}$. Then

$$\begin{aligned}\widetilde{V}_{k+h} &= \sum_{j=-\infty}^{k+h} \mathrm{e}^{-(\Gamma_{k+h}-\Gamma_j)}\xi_j = \sum_{j=-\infty}^{k+h} \mathrm{e}^{-(\Gamma_{k+h}-\Gamma_h-(\Gamma_j-\Gamma_h))}\xi_j \stackrel{d}{=} \\ &\stackrel{d}{=} \sum_{j=-\infty}^{k+h} \mathrm{e}^{-(\Gamma_k-\Gamma_{j-h})}\xi_j = \sum_{j=-\infty}^{k} \mathrm{e}^{-(\Gamma_k-\Gamma_j)}\xi_{j+h} \stackrel{d}{=} \sum_{j=-\infty}^{k} \mathrm{e}^{-(\Gamma_k-\Gamma_j)}\xi_j = \widetilde{V}_k\,.\end{aligned}$$

Similarly, for $l \in \mathbb{N}$ we obtain

$$(\widetilde{V}_{k_1+h},\ldots,\widetilde{V}_{k_l+h}) \stackrel{d}{=} (\widetilde{V}_{k_1},\ldots,\widetilde{V}_{k_l})$$

for $k_1,\ldots,k_l \in \mathbb{N}$. □

Proof of Theorem 3. Since V is decreasing between jumps, it follows that $\sup_{s\in I_k} V_s = V_{\Gamma_k/\lambda}$. Recall that $\widetilde{V}_k = V_{\Gamma_k/\lambda} + \mathrm{e}^{-\Gamma_k}\xi_0 \stackrel{d}{=} V_0 + \xi_1$ and that $(\widetilde{V}_k)_{k\in\mathbb{N}}$ is stationary. We show first that the norming constants $a_n > 0, b_n \in \mathbb{R}$ given by (4.37) and (4.39) satisfy

$$\lim_{n\to\infty} n\mathbb{P}(\widetilde{V}_k > a_n x + b_n) = \mu^{-1}\exp(-x)\,, \quad x \in \mathbb{R}\,. \tag{4.42}$$

To show this in case *(a)*, observe that $\mathbb{P}(V_0 > x) = o(\mathbb{P}(\xi_1 > x))$ for $x \to \infty$ by Proposition 3 *(b)*, so that Theorem 7 *(i, ii)* yields

$$\begin{aligned}\mathbb{P}(\widetilde{V}_k > x) &\sim \mathbb{E}\mathrm{e}^{\gamma V_0}\mathbb{P}(\xi_1 > x) \\ &\sim \mathbb{E}\mathrm{e}^{\gamma V_0}[\mathbb{E}\mathrm{e}^{\gamma L_1}]^{-1}\mu^{-1}\mathbb{P}(L_1 > x)\,, \quad x \to \infty\,.\end{aligned} \tag{4.43}$$

From this (4.42) follows immediately, and further we see that $\widetilde{V}_k \in \mathcal{S}(\gamma)$.

In case *(b)*, $\widetilde{V}_k$ is $\Gamma(\mu+1,\gamma)$ distributed as an independent sum of a $\Gamma(\mu,\gamma)$ and an $\mathrm{Exp}(\gamma)$ r. v., and (4.42) is immediate. The norming constants of a Γ distribution can be found in Table 3.4.4 of Embrechts et al. [17].

Note that in both cases *(a)* and *(b)*, we have $\widetilde{V}_k \in \mathcal{L}(\gamma)$. Thus, by Lemma 2 and Leadbetter et al. [29], Theorem 5.5.1,

$$\sum_{k=0}^{\infty} \varepsilon\left\{k/(\mu n), a_n^{-1}(\widetilde{V}_k - b_n)\right\} \overset{n\to\infty}{\Longrightarrow} \sum_{k=1}^{\infty} \varepsilon\left\{s_k, P_k\right\}. \tag{4.44}$$

Define point processes

$$\widetilde{\kappa}_n := \sum_{k=0}^{\infty} \varepsilon\left\{k/(\mu n), a_n^{-1}(\widetilde{V}_k - b_n)\right\}$$

and

$$\kappa_n := \sum_{k=0}^{\infty} \varepsilon\left\{k/(\mu n), a_n^{-1}(V_{\Gamma_k/\lambda} - b_n)\right\}.$$

For $\epsilon > 0$ and $I = [s,t) \times (x,\infty) \subseteq \mathbb{R}_+ \times \mathbb{R}$ define $I_\epsilon := [s,t) \times (x, x+\epsilon]$. Taking into account that $V_{\Gamma_k/\lambda} \leq \widetilde{V}_k$ we have for $\delta \in (0,1)$

$$\begin{aligned}
&\mathbb{P}(\kappa_n(I) \neq \widetilde{\kappa}_n(I)) \\
&\leq \mathbb{P}(\widetilde{\kappa}_n(I_\epsilon) > 0) + \sum_{k\in[ns\mu,nt\mu)} \mathbb{P}(\widetilde{V}_k > u_n + \epsilon\, a_n, V_{\Gamma_k/\lambda} \leq u_n) \\
&\leq \mathbb{P}(\widetilde{\kappa}_n(I_\epsilon) > 0) + \sum_{k\in[0,n^\delta t\mu)} \mathbb{P}(\widetilde{V}_k > u_n + \epsilon\, a_n) + \sum_{k\in[n^\delta t\mu,nt\mu)} \mathbb{P}(\mathrm{e}^{-\Gamma_k}\xi_0 > \epsilon\, a_n).
\end{aligned}$$

We shall show below that

$$\lim_{n\to\infty} \mathbb{P}(\kappa_n(I) \neq \widetilde{\kappa}_n(I)) = 0. \tag{4.45}$$

Then by Rootzén [33], Lemma 3.3, the limit behavior of $\widetilde{\kappa}_n$ and κ_n is the same. Relation (4.40) then follows by transforming the time scale as in Hsing and Teugels [23] (for details see Fasen [18], Lemma 2.2.4).

To show (4.45), observe that by (4.44) we have

$$\lim_{n\to\infty} \mathbb{P}(\widetilde{\kappa}_n(I_\epsilon) > 0) = 1 - \exp[(t-s)(\exp(-x) - \exp(-(x+\epsilon)))] \overset{\epsilon\downarrow 0}{\longrightarrow} 0.$$

Furthermore, since $\delta < 1$,

$$\lim_{n\to\infty} \sum_{k\in[0,n^\delta t\mu)} \mathbb{P}(\widetilde{V}_k > u_n + \epsilon\, a_n) \leq \lim_{n\to\infty} n^\delta t\mu \mathbb{P}(\widetilde{V}_k > a_n(x+\epsilon) + b_n) = 0.$$

Applying (B.3) we obtain

$$\begin{aligned}
&\sum_{k\in[n^\delta t\mu,nt\mu)} \mathbb{P}(\mathrm{e}^{-\Gamma_k}\xi_0 > \epsilon\, a_n) \\
&\leq \sum_{k\in[n^\delta t\mu,nt\mu)} \left(\mathbb{P}(\mathrm{e}^{-\Gamma_k}\xi_0 > \epsilon\, a_n, \Gamma_k \geq \frac{k}{2\mu}) + \mathbb{P}(\mathrm{e}^{-\Gamma_k}\xi_0 > \epsilon\, a_n, \Gamma_k < \frac{k}{2\mu})\right) \\
&\leq \sum_{k\in[n^\delta t\mu,nt\mu)} \mathbb{P}(\mathrm{e}^{-k/(2\mu)}\xi_0 > \epsilon\, a_n) + \sum_{k\in[n^\delta t\mu,nt\mu)} \frac{\widetilde{K}}{k^3}.
\end{aligned}$$

The last summand tends to 0 as $n \to \infty$, since $\sum_{k=1}^{\infty} 1/k^3 < \infty$. Moreover, there exists an $n_0 \in \mathbb{N}$ such that $a_n \geq 1/(2\gamma)$ and $k\mathrm{e}^{-k/(2\mu)} \leq 1/2$ for $n, k \geq n_0$. Then the first exponential moment of $\gamma k \mathrm{e}^{-k/(2\mu)}\xi_0$ exists, and for $n^\delta t\mu \geq n_0$ we obtain

$$\sum_{k\in[n^\delta t\mu, nt\mu)} \mathbb{P}(\mathrm{e}^{-k/(2\mu)}\xi_0 > \epsilon\, a_n) \leq \sum_{k\in[n^\delta t\mu, nt\mu)} \mathbb{E}[\exp(\gamma k \mathrm{e}^{-k/(2\mu)})\xi_0]\mathrm{e}^{-k\gamma\epsilon\, a_n}$$

$$\leq \mathbb{E}[\exp(\gamma\xi_0/2)] \sum_{k=\lfloor n^\delta t\mu\rfloor}^{\infty} \mathrm{e}^{-k\epsilon/2} \longrightarrow 0\,, \quad n\to\infty\,,$$

since $\sum_{k=1}^{\infty} \mathrm{e}^{-k\epsilon/2} < \infty$. This proves (4.45). □

Results (4.38) and (4.40) show that local extremes of such exponential models have no cluster behavior on high levels. The following two corollaries are immediate from Theorem 3.

Corollary 5 (Point process of local maxima).
Let the assumptions of Theorem 3 hold. Denote by $(j_k)_{k\in\mathbb{N}}$ the jump times of a Poisson process with intensity e^{-x} for $x \in \mathbb{R}$. Then,

$$\sum_{k=1}^{\infty} \varepsilon\left\{\frac{\Gamma_k}{\lambda T}, a_{\lambda T}^{-1}(\sup_{s\in I_k} V_s - b_{\lambda T})\right\}(\cdot \times (x,\infty)) \overset{T\to\infty}{\Longrightarrow} \sum_{k=1}^{\infty} \varepsilon\{j_k\}\,.$$

Corollary 6 (Running maxima).
Let V be a stationary version of the OU process (4.14), where L is a positive compound Poisson process as in (4.18). Define $M(T) := \sup_{0\leq t\leq T} V_t$ for $T > 0$.

(a) Assume $L_1 \in \mathcal{S}(\gamma)$, $\gamma > 0$, with norming constants given by (4.37). Then

$$\lim_{T\to\infty} \mathbb{P}(a_{\lambda T}^{-1}(M(T) - b_{\lambda T}) \leq x) = \exp(-\mathrm{e}^{-x})\,, \quad x \in \mathbb{R}\,.$$

(b) Assume V is the $\Gamma(\mu,\gamma)$-OU process with norming constants given by (4.39). Then

$$\lim_{T\to\infty} \mathbb{P}(a_{\lambda T}^{-1}(M(T) - b_{\lambda T}) \leq x) = \exp(-\mathrm{e}^{-x})\,, \quad x \in \mathbb{R}\,. \tag{4.46}$$

For a subexponential OU process and $h > 0$ fixed the r.v. $M(h) = \sup_{0\leq t\leq h} V_t$ is tail-equivalent to the increment of the Lévy process; cf. (4.22) and (4.25). In the class $\mathcal{S}(\gamma)$, $\gamma > 0$, this is much more involved; see Braverman and Samorodnitsky [9]. Although the large jumps of the Lévy process determine the tail behavior, small jumps also have a non-negligible influence. For any $h > 0$, the tail of $M(h)$ is of the same order of magnitude as the tail of the increment of the BDLP, but in general it is only possible to give upper and lower bounds on the asymptotic ratio of the two tails. Using Corollary 6 one can calculate this constant for the OU process explicitly.

Corollary 7 (Extremal index function).
Let V be a stationary version of the OU process given by (4.14), where L is a positive compound Poisson process as in (4.18). *Define $M(h) := \sup_{0\le t\le h} V_t$ for $h > 0$.*

(a) Let $L_1 \in \mathcal{S}(\gamma)$, $\gamma > 0$. Then $M(h) \in \mathcal{L}(\gamma)$ if and only if

$$\mathbb{P}(M(h) > x) \sim \lambda h \, \mathbb{E}\mathrm{e}^{\gamma V_0} [\mathbb{E}\mathrm{e}^{\gamma L_1}]^{-1} \mathbb{P}(L_1 > x) \,, \quad x \to \infty \,. \tag{4.47}$$

In that case $M(h) \in \mathcal{S}(\gamma)$ and $\theta(\cdot) \equiv 1$.

(b) Let V be the $\Gamma(\mu,\gamma)$-OU process with norming constants given by (4.39) and let W be a $\Gamma(\mu+1,\gamma)$ r. v. Then $\theta(\cdot) \equiv 1$ and

$$\mathbb{P}(M(h) > x) \sim \lambda h \mu \mathbb{P}(W > x) \,, \quad x \to \infty \,. \tag{4.48}$$

Proof. *(a)* First we assume $M(h) \in \mathcal{L}(\gamma)$. Let $\widetilde{a}_n > 0$, $\widetilde{b}_n \in \mathbb{R}$ and $\widetilde{u}_n = \widetilde{a}_n x + \widetilde{b}_n$ be constants such that

$$\lim_{n\to\infty} n\mathbb{P}(M(h) > \widetilde{u}_n) = \exp(-x) \,.$$

Denote by $\widetilde{M_k}$ an i. i. d. sequence of copies of $M(h)$. Then we obtain from Lemma 2 *(b)* and Leadbetter et al. [29], Theorem 3.5.1, for $x \in \mathbb{R}$,

$$\begin{aligned}\lim_{n\to\infty} \mathbb{P}(\widetilde{a}_n^{-1}(M(nh) - \widetilde{b}_n) \le x) &= \lim_{n\to\infty} \mathbb{P}(\widetilde{a}_n^{-1}(\max_{k=1,\dots,n} \widetilde{M_k} - \widetilde{b}_n) \le x)\\ &= \exp(-\mathrm{e}^{-x}) \,,\end{aligned}$$

showing in particular that $\theta(h) = 1$. On the other hand, by Corollary 6,

$$\lim_{n\to\infty} \mathbb{P}(a_{\lambda nh}^{-1}(M(nh) - b_{\lambda nh}) \le x) = \exp(-\mathrm{e}^{-x}) \,, \quad x \in \mathbb{R} \,.$$

Then by the convergence to types theorem (see e. g. Theorem A 1.5 of Embrechts et al. [17]), $\widetilde{a}_n / a_{\lambda nh} \overset{n\to\infty}{\longrightarrow} 1$ and $\widetilde{b}_n - b_{\lambda nh} \overset{n\to\infty}{\longrightarrow} 0$. Applying the convergence to types theorem a second time yields

$$\lim_{n\to\infty} \mathbb{P}(a_{\lambda nh}^{-1}(\max_{k=1,\dots,n} \widetilde{M_k} - b_{\lambda nh}) \le x) = \exp(-\mathrm{e}^{-x}) \,, \quad x \in \mathbb{R} \,.$$

This implies by Leadbetter et al. [29], Theorem 1.5.1 that

$$\lim_{n\to\infty} n\mathbb{P}(M(h) > u_{\lambda nh}) = \exp(-x) \,, \quad x \in \mathbb{R} \,,$$

with $u_{\lambda nh} = a_{\lambda nh} x + b_{\lambda nh}$. By (4.42) also $\lim_{n\to\infty} n\mathbb{P}(\widetilde{V}_k > u_{\lambda nh}) = \exp(-x)/(\lambda \mu h)$. Hence $\mathbb{P}(M(h) > x) \sim h\lambda\mu \mathbb{P}(\widetilde{V}_k > x)$ for $x \to \infty$, and (4.47) follows from (4.43).

Conversely, if (4.47) holds, then it is clear that $L_1 \in \mathcal{S}(\gamma) \subseteq \mathcal{L}(\gamma)$ implies $M(h) \in \mathcal{L}(\gamma)$ by tail-equivalence. By Lemma 2 *(b)* follows $\theta(h) = 1$.

(b) We refer to Albin [1], Theorem 3, for (4.48). That $\theta(h) = 1$ follows then from (4.39), (4.46) and (4.48). □

In both cases the extremal index function is equal to one, so that for any $h > 0$ the sequence $M_k = \sup_{(k-1)h \le t \le kh} V_t$ behaves like i. i. d. data. Hence such models cannot explain volatility clusters on high levels.

4.4 Extremal behavior of the COGARCH model

The volatility of the COGARCH(1,1) process as introduced in (4.6) is the (càglàg) solution to the SDE (4.6), which is given by

$$V_t = V_0 + \beta t - \eta \int_0^t V_s \, ds + \lambda \mathrm{e}^{\eta} \sum_{0<s<t} V_s (\Delta L_s)^2 \,, \quad t \ge 0 \,, \tag{4.49}$$

see Klüppelberg, Lindner and Maller [27, 28] for details. This process is a solution of the SDE

$$dV_{t+} = (\beta - \eta V_t) dt + \lambda \mathrm{e}^{\eta} V_t \, d[L, L]_t^{(d)} \,, \tag{4.50}$$

where

$$[L, L]_t^{(d)} = \sum_{0<s\le t} (\Delta L_s)^2 \,, \quad t \ge 0 \,,$$

is the discrete part of the quadratic variation process of L. Comparing this with (4.5) we see that the COGARCH(1,1) can be interpreted as a generalized CIR model driven by the discrete part of the quadratic variation process of L. An essential feature of the COGARCH(1,1) model is that the same Lévy process drives the price process S and the volatility process V. An extension of the COGARCH(1,1) process to COGARCH(p, q) process with $1 \le p \le q$ was recently obtained by Brockwell, Chadraa and Lindner [11]. There, (4.50) is replaced by a CARMA (continuous time ARMA) type stochastic differential equation, driven by $[L, L]^{(d)}$. We shall not go into further details, and by COGARCH we shall always mean the COGARCH(1,1) process.

Denote by ν the Lévy measure of L. A stationary version of (4.49) exists if and only if

$$\int_{\mathbb{R}} \ln(1 + \lambda \mathrm{e}^{\eta} y^2) \, \nu(dy) < \eta \,. \tag{4.51}$$

With the auxiliary càdlàg process $(X_t)_{t \ge 0}$ defined in (4.7), given for $\eta > 0$, $\lambda \ge 0$ by

$$X_t = \eta t - \sum_{0<s\le t} \ln(1 + \lambda \mathrm{e}^{\eta} (\Delta L_s)^2) \,, \quad t \ge 0 \,,$$

the stationary volatility process has representation

$$V_t = \left(\beta \int_0^t \mathrm{e}^{X_s} ds + V_0 \right) \mathrm{e}^{-X_{t-}} \,, \quad t \ge 0 \,, \tag{4.52}$$

with $\beta > 0$ and $V_0 \stackrel{d}{=} \beta \int_0^\infty \mathrm{e}^{-X_t} dt$, independent of L. The auxiliary process $(X_t)_{t\geq 0}$ itself is a spectrally negative Lévy process of bounded variation with drift $\gamma_X = \eta$, no Gaussian component, and Lévy measure ν_X given by

$$\nu_X\,[0,\infty) = 0,\;\; \nu_X\,(-\infty,-x] = \nu\left(\{y \in \mathbb{R} : |y| \geq \sqrt{(\mathrm{e}^x - 1)/(\lambda \mathrm{e}^\eta)}\}\right) \;\; x > 0\,.$$

We work with the Laplace transform $\mathbb{E}\mathrm{e}^{-sX_t} = \mathrm{e}^{t\Psi(s)}$, where the Laplace exponent is

$$\Psi(s) = -\eta s + \int_{\mathbb{R}} ((1 + \lambda \mathrm{e}^\eta y^2)^s - 1)\, \nu(dy)\,, \quad s \geq 0\,. \tag{4.53}$$

For fixed $s \geq 0$, $\mathbb{E}\mathrm{e}^{-sX_t}$ exists (i. e. is finite) for one and hence all $t > 0$, if and only if the integral appearing in (4.53) is finite. This is equivalent to $\mathbb{E}|L_1|^{2s} < \infty$. Further, if there exists some $s > 0$ such that $\Psi(s) \leq 0$, then (4.51) holds, and hence a stationary version of the volatility process exists.

The qualitative extreme behavior of this volatility process can be seen in Figure 4.5, where the driving Lévy process is a compound Poisson process. As in the case of a Lévy driven OU process the volatility jumps upwards, whenever the driving Lévy process L jumps and decreases exponentially fast between two jumps.

It is instructive to observe that both, the OU process (4.14) and the right-continuous V_{t+} of the COGARCH volatility (4.52) are special cases of the generalized OU process

$$O_t = \mathrm{e}^{-\xi_t}\left(\int_0^t \mathrm{e}^{\xi_{s-}}\, d\zeta_s + O_0\right), \quad t \geq 0\,,$$

where $(\xi_t, \zeta_t)_{t\geq 0}$ is a bivariate Lévy process. The stationarity conditions for $(O_t)_{t\geq 0}$, along with other properties, have been investigated by Lindner and Maller [30].

Returning to the COGARCH volatility, the next Theorem (cf. [28], Theorem 6) shows, that under weak conditions on the moments of L, the volatility process has Pareto like tails. Since we shall apply a similar argument in the proof of Theorem 5, we sketch the idea of the proof.

Theorem 4 (Pareto tail behavior of COGARCH models).
Suppose there exists some $\alpha > 0$ such that

$$\mathbb{E}|L_1|^{2\alpha} \ln^+ |L_1| < \infty \quad \text{and} \quad \Psi(\alpha) = 0\,. \tag{4.54}$$

Let V be a stationary version of the volatility process given by (4.49). *Then for some constant $C > 0$ we have*

$$\lim_{x\to\infty} x^\alpha \mathbb{P}(V_0 > x) = C\,. \tag{4.55}$$

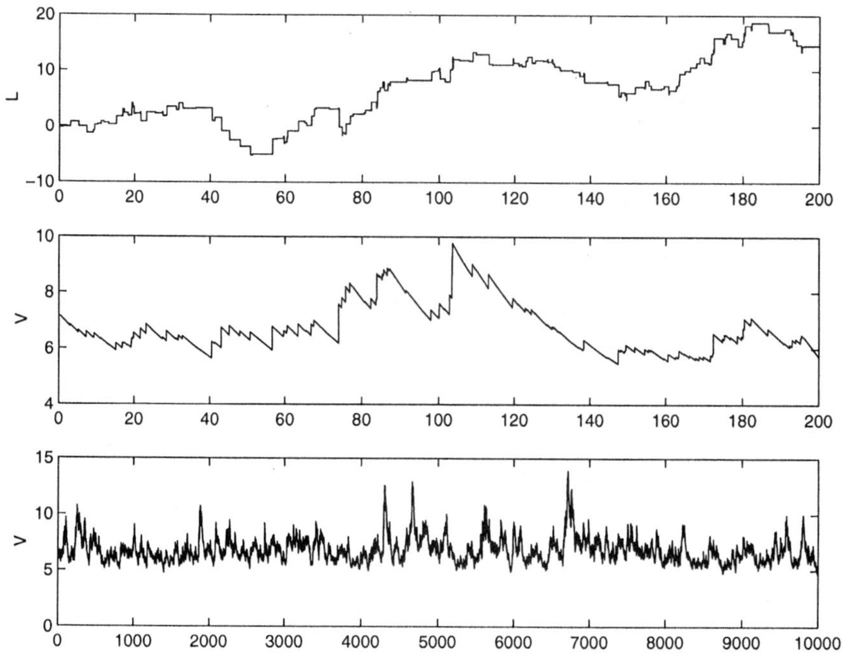

Fig. 4.5. The first plot shows the sample path of a compound Poisson driving process $(L_t)_{0 \le t \le 250}$ with rate 1 and normal jumps with mean 0 and variance 1 and the second plot the corresponding sample path of the COGARCH volatility process $(V_t)_{0 \le t \le 250}$ driven by this Lévy process. The COGARCH parameters are $\beta = 1$, $\lambda = 0.04$ and $\eta = 0.0619$. Both plots indicate the micro-behavior of the model. The third plot gives $(V_t)_{0 \le t \le 10\,000}$ indicating the macro-behavior of the model.

Proof. From (4.52) it is seen that the stationary volatility process V satisfies

$$V_t = \mathrm{e}^{-X_{t-}} V_0 + \beta \int_0^t \mathrm{e}^{X_s - X_{t-}}\, ds\,, \quad t > 0\,,$$

where V_0 is independent of $(\mathrm{e}^{-X_{t-}}, \beta \int_0^t \mathrm{e}^{X_s - X_{t-}}\, ds)$ for any $t > 0$. Thus the stationary solution V_0 satisfies for every $t > 0$ the distributional fix point equation

$$V_0 \stackrel{d}{=} A_t V_0 + B_t\,,$$

where V_0 is independent of (A_t, B_t) and

$$A_t \stackrel{d}{=} \mathrm{e}^{-X_t}, \quad B_t \stackrel{d}{=} \beta \int_0^t \mathrm{e}^{-X_s}\, ds\,.$$

The result now follows from Theorem C.1, by choosing t such that (A_t, B_t) satisfies the assumptions. This is possible because of the structure of the process and condition (4.54); for details see Klüppelberg et al. [28], Theorem 6. □

The following remark gives a simple sufficient condition for (4.54) to hold.

Remark 1. Let $D := \{d \in [0,\infty) : \mathbb{E}|L_1|^{2d} < \infty\}$ and $d_0 := \sup D \in [0,\infty]$. Suppose $d_0 \notin D$, or that there exists an $s_0 > 0$ such that $0 < \Psi(s_0) < \infty$. Further suppose that $(V_t)_{t\geq 0}$ is strictly stationary. Then (4.54) and hence (4.55) hold (cf. Klüppelberg et al. [28], Proposition 5.3).

We aim at a precise asymptotic description of the COGARCH model above a high threshold like in Section 11. It is, however, clear from the definition of V that the influence of the spectrally negative Lévy process X is hard to analyze. In particular, the influence of the small jumps of L to V needs special treatment. In this review paper we shall restrict ourselves again to the case of a compound Poisson driving process L as given in (4.18) by $L_t = \sum_{j=1}^{N_t} \xi_j$ for $t \geq 0$, where ξ has support on $\mathbb{R}$.

In this case the auxiliary process X simplifies to

$$X_t = \eta t - \sum_{k=1}^{N_t} \ln(1 + \lambda \mathrm{e}^{\eta} \xi_k^2) , \quad t \geq 0 , \tag{4.56}$$

and the Laplace exponent becomes

$$\Psi(s) = -(\eta s + \mu) + \mu \mathbb{E}(1 + \lambda \mathrm{e}^{\eta} \xi_1^2)^s . \tag{4.57}$$

In the stationary volatility model we know that $V_t \geq \beta/\eta$ a. s. and V jumps if and only if L jumps (cf. [28], Proposition 3.4 *(a)*). The jump sizes are positive and depend on the level of the process at that time. As shown in Proposition 3.4 *(b,c)* of [28],

$$V_{\Gamma_k+} - V_{\Gamma_k} = \lambda \mathrm{e}^{\eta} V_{\Gamma_k} \xi_k^2 , \quad k \in \mathbb{N} , \tag{4.58}$$

and the process decreases exponentially between jumps:

$$V_t = \frac{\beta}{\eta} + \left(V_{\Gamma_k+} - \frac{\beta}{\eta} \right) \mathrm{e}^{-(t-\Gamma_k)\eta} , \quad t \in (\Gamma_k, \Gamma_{k+1}] . \tag{4.59}$$

In analogy to the OU process driven by a compound Poisson process of Example 2, the compound Poisson driven COGARCH process V achieves local suprema only at the right limits of its jump times (and at $t = 0$). Hence it is no surprise that the discrete-time sequence $(V_{\Gamma_k+})_{k\in\mathbb{N}}$ in combination with the deterministic behavior of V between jumps suffices to describe the extremal behavior of the continuous-time COGARCH process. Consequently, we investigate the discrete-time skeleton

$$\widetilde{V}_k := V_{\Gamma_k+} , \quad k \in \mathbb{N} . \tag{4.60}$$

Using (4.58) and (4.59) we obtain

$$\widetilde{V}_{k+1} = \widetilde{V}_k \,(1+\lambda e^{\eta}\xi_{k+1}^2)e^{-(\Gamma_{k+1}-\Gamma_k)\eta} + \frac{\beta}{\eta}\,(1+\lambda e^{\eta}\xi_{k+1}^2)\left(1-e^{-(\Gamma_{k+1}-\Gamma_k)\eta}\right),$$

and we see that $(\widetilde{V}_k)_{k\in\mathbb{N}}$ satisfies the stochastic recurrence equation

$$\widetilde{V}_k = \widetilde{A}_k \widetilde{V}_{k-1} + \widetilde{B}_k \ ,k\geq 2\,, \tag{4.61}$$

with $\widetilde{V}_1$ independent of $(\widetilde{A}_k, \widetilde{B}_k)$ for any $k\geq 2$, where

$$\widetilde{A}_k = (1+\lambda e^{\eta}\,\xi_k^2)e^{-\eta\,(\Gamma_k-\Gamma_{k-1})}, \quad k\in\mathbb{N}\,, \tag{4.62}$$

$$\widetilde{B}_k = \frac{\beta}{\eta}(1+\lambda e^{\eta}\,\xi_k^2)\left(1-e^{-\eta\,(\Gamma_k-\Gamma_{k-1})}\right), \quad k\in\mathbb{N}\,, \tag{4.63}$$

and $((\widetilde{A}_k,\widetilde{B}_k))_{k\in\mathbb{N}}$ is an i.i.d. sequence. It is an interesting observation that by (4.62)

$$\ln\prod_{j=1}^k \widetilde{A}_j = \sum_{j=1}^k \ln\widetilde{A}_j = -\eta\Gamma_k + \sum_{j=1}^k \ln(1+\lambda e^{\eta}\xi_j^2) = -X_{\Gamma_k}\,.$$

On the other hand, by (4.63) and $X_s - X_{\Gamma_k} = \ln(1+\lambda e^{\eta}\xi_k^2)+\eta(s-\Gamma_k)$ for $s\in(\Gamma_k,\Gamma_{k+1})$,

$$\widetilde{B}_k = \beta\int_{\Gamma_{k-1}}^{\Gamma_k} e^{X_s - X_{\Gamma_k}}\,ds\,.$$

Denote by $(\widetilde{A},\widetilde{B})$ a copy of $(\widetilde{A}_1,\widetilde{B}_1)$, independent of L and $\widetilde{V}_1$. Then it follows that

$$\widetilde{A}\overset{d}{=}e^{-X_{\Gamma_1}} \quad\text{and}\quad \widetilde{B}\overset{d}{=}\beta e^{-X_{\Gamma_1}}\int_0^{\Gamma_1} e^{X_s}ds\,, \quad k\in\mathbb{N}\,. \tag{4.64}$$

Moreover,

$$\widetilde{V}_k = \widetilde{V}_1\prod_{j=2}^k \widetilde{A}_j + \sum_{i=2}^k \widetilde{B}_i \prod_{j=i+1}^k \widetilde{A}_j \;=\; e^{-X_{\Gamma_k}}\left[\widetilde{V}_1 + \sum_{i=2}^k \widetilde{B}_i e^{X_{\Gamma_i}}\right], \quad k\geq 2\,.$$

We are now ready to present the analogue of Theorem 4 for the sequence $(\widetilde{V}_k)_{k\in\mathbb{N}}$. As can be seen from (4.56), the process $(X_{\Gamma_k})_{k\in\mathbb{N}}$ is a random walk with increments

$$X_{\Gamma_k} - X_{\Gamma_{k-1}} = \eta\,(\Gamma_k-\Gamma_{k-1}) - \ln(1+\lambda e^{\eta}\,\xi_k^2)\,, \quad k\in\mathbb{N}.$$

Theorem 5 (Pareto tail behavior of $\widetilde{\mathrm{V}}$).
Suppose there exists some $\alpha > 0$ such that

$$\mathbb{E}|L_1|^{2\alpha} \ln^+ |L_1| < \infty \quad \text{and} \quad \Psi(\alpha) = 0\,. \tag{4.65}$$

Then a stationary solution $(\widetilde{V}_k)_{k\in\mathbb{N}}$ of (4.61) exists. Its marginal stationary distribution, denoted by $\widetilde{V}_0$, is the unique solution of the random fix point equation

$$\widetilde{V}_0 \stackrel{d}{=} \widetilde{A}\widetilde{V}_0 + \widetilde{B}\,,$$

where $(\widetilde{A}, \widetilde{B})$ is given by (4.64) and independent of L and $\widetilde{V}_0$. Furthermore,

$$\mathbb{P}(\widetilde{V}_0 > x) \sim \widetilde{C}x^{-\alpha}\,, \quad x \to \infty\,,$$

where

$$\widetilde{C} = \frac{\mathbb{E}\left[(\widetilde{A}\widetilde{V}_0 + \widetilde{B})^\alpha - (\widetilde{A}\widetilde{V}_0)^\alpha\right]}{\alpha\mathbb{E}|\widetilde{A}|^\alpha \ln^+ |\widetilde{A}|} > 0\,. \tag{4.66}$$

Proof. We shall show that conditions *(i)-(iv)* of Theorem C.1 are satisfied: by definition, $\ln \widetilde{A} \stackrel{d}{=} -\eta\Gamma_1 + \ln(1 + \lambda e^\eta \xi_1^2)$, where Γ_1 is exponentially distributed. Consequently, *(i)* follows.

To show *(ii)* note that by the independence of Γ_1 and ξ_1, for $\alpha > 0$ we have by (4.57)

$$\begin{aligned}
\mathbb{E}|\widetilde{A}|^\alpha &= \mathbb{E}e^{-\alpha\eta\Gamma_1}\mathbb{E}(1 + \lambda e^\eta\, \xi_1^2)^\alpha \\
&= \frac{\mu}{\mu + \alpha\eta}\,\frac{\mu + \alpha\eta + \Psi(\alpha)}{\mu} \\
&= 1 + \frac{1}{\mu + \alpha\eta}\Psi(\alpha) \;=\; 1\,,
\end{aligned}$$

by the second assumption in (4.65).
In order to prove *(iii)* note that

$$\mathbb{E}|\widetilde{A}|^\alpha \ln^+ |\widetilde{A}| \le \mathbb{E}|1 + \lambda e^\eta \xi_1^2|^\alpha \ln^+ \left(1 + \lambda e^\eta \xi_1^2\right) < \infty\,,$$

if and only if the first assumption in (4.65) holds, see Sato [35], Theorem 25.3. Finally, *(iv)* follows from

$$\mathbb{E}|\widetilde{B}|^\alpha \le (\beta/\eta)^\alpha \mathbb{E}|1 + \lambda e^\eta \xi_1^2|^\alpha < \infty\,.$$

That the constant $\widetilde{C}$ is indeed strictly positive follows from the fact that $\widetilde{A}$, $\widetilde{B}$ and $\widetilde{V}_0$ are strictly positive, almost surely. □

Remarks.
(i) X_{Γ_k} tends almost surely to ∞ if and only if $\mathbb{E}X_{\Gamma_1} > 0$ or, equivalently,

$\mu\mathbb{E}\ln(1+\lambda e^{\eta}\,\xi_1^2) < \eta$. Notice that the stationarity condition (4.51) is for this model equivalent to $\mathbb{E}X_{\Gamma_1} > 0$.

(ii) In a sense it is remarkable that the tail of the stationary r.v. of the continuous-time model V_0 and of the discrete-time skeleton $\widetilde{V}_0$ are so similar. As the discrete-time skeleton considers only local suprema of the process, one expects $\widetilde{V}_0$ to be stochastically larger. As the Pareto index α is the same for both models, any difference can only appear in the constants C and $\widetilde{C}$. Brockwell et al. [11] have established a precise relationship between the distributions of V_0 and $\widetilde{V}_0$, showing that

$$\left(V_0 - \frac{\beta}{\eta}\right) \stackrel{d}{=} e^{-\eta\Gamma}\left(\widetilde{V}_0 - \frac{\beta}{\eta}\right),$$

where $\Gamma \stackrel{d}{=} \Gamma_1$ is exponentially distributed with parameter μ and independent of $\widetilde{V}_0$. Using the classical result of Breiman [10], it then follows that

$$C = \mathbb{E}\left(e^{-\eta\Gamma}\right)^{\alpha}\widetilde{C} = \frac{\mu}{\mu+\alpha\eta}\widetilde{C} = \frac{1}{\mathbb{E}(1+\lambda e^{\eta}\xi_1^2)^{\alpha}}\widetilde{C}, \tag{4.67}$$

where the last equation follows from (4.57).

The extremal behavior of solutions to stochastic recurrence equations is studied in de Haan et al. [22]. Their results can be applied to the stationary discrete-time skeleton of the volatility process $(\widetilde{V}_k)_{k\in\mathbb{N}}$ as defined in (4.60).

Theorem 6 (Extremal behavior of the COGARCH model).
Let V be a stationary version of the volatility process given by (4.49) and define $M(T) := \sup_{0\le t\le T} V_t$ for $T>0$. Suppose there exists some $\alpha>0$ such that

$$\mathbb{E}|L_1|^{2\alpha}\ln^+|L_1| < \infty \quad \text{and} \quad \Psi(\alpha) = 0\,.$$

Let $\widetilde{C}$ be the constant in (4.66) and define $a_T := (\mu T)^{1/\alpha}$ for $T>0$. Then

$$\lim_{T\to\infty}\mathbb{P}(a_T^{-1}M(T)\le x) = \exp(-\widetilde{C}\theta x^{-\alpha})\,, \quad x>0\,,$$

where for $X_t^+ = \max\{0, X_t\}$

$$\theta = 1 - \mathbb{E}\left[\sup_{t\ge\Gamma_1}\{e^{-\alpha X_t^+}\}\right] \in (0,1)\,.$$

Denote by $(\Gamma_k)_{k\in\mathbb{N}}$ the jump times of the compound Poisson process L given by (4.18) and define $I_k := (\Gamma_k, \Gamma_{k+1}]$ for $k\in\mathbb{N}$. Let $(j_k)_{k\in\mathbb{N}}$ be the jump times of a Poisson process with intensity $\widetilde{C}\theta x^{-\alpha}$ for $x>0$. Then,

$$\sum_{k=1}^{\infty}\varepsilon\left\{\frac{\Gamma_k}{T}, a_T^{-1}\sup_{s\in I_k}V_s\right\}(\cdot\times(x,\infty)) \stackrel{T\to\infty}{\Longrightarrow} \sum_{k=1}^{\infty}\zeta_k\varepsilon\{j_k\}\,, \tag{4.68}$$

where $(\zeta_k)_{k\in\mathbb{N}}$ are i. i. d. discrete r. v. s, independent of $(j_k)_{k\in\mathbb{N}}$, with probability distribution

$$\pi_k = \mathbb{P}(\zeta_1 = k) = (\theta_k - \theta_{k+1})/\theta\,, \quad k \in \mathbb{N}\,.$$

Moreover,

$$\begin{aligned}\theta_k &:= \mathbb{E}\left[\exp(\alpha \min\{T_{k-1}, 0\}) - \exp(\alpha \min\{T_k, 0\})\right] \\ &= \int_0^1 \mathbb{P}\left(\operatorname{card}\left\{j \in \mathbb{N} : \mathrm{e}^{-\alpha X_{\Gamma_j}} > y\right\} = k-1\right) dy\,,\end{aligned}$$

where $\infty = T_0 \geq T_1 \geq \cdots$ are the ordered values of the sequence $(-X^+_{\Gamma_k})_{k\in\mathbb{N}}$. Finally, $\theta = \theta_1$.

Proof. Since $\sup_{s\in I_k} V_s = \widetilde{V}_k$, Theorem 5 and de Haan et al. [22], Theorem 2.1, show that

$$\sum_{k=1}^{\infty} \varepsilon\left\{\frac{k}{\mu n}, a_n^{-1}\widetilde{V}_k\right\}(\cdot \times (x,\infty)) \overset{T\to\infty}{\Longrightarrow} \sum_{k=1}^{\infty} \zeta_k \varepsilon\{j_k\}$$

and that $(\widetilde{V}_k)_{k\in\mathbb{N}}$ has extremal index $\theta \in (0,1)$, given by

$$\begin{aligned}\theta &= \alpha \int_1^\infty \mathbb{P}\left(\bigvee_{j=1}^{\infty} \prod_{k=1}^{j} \widetilde{A}_k \leq y^{-1}\right) y^{-\alpha-1}\, dy \\ &= \alpha \int_1^\infty \mathbb{P}\left(\bigvee_{j=1}^{\infty} \exp(-X_{\Gamma_j}) \leq y^{-1}\right) y^{-\alpha-1}\, dy \\ &= \int_0^1 \mathbb{P}\left(\sup_{t\geq \Gamma_1} \{\mathrm{e}^{-\alpha X_t}\} \leq z\right) dz \\ &= 1 - \mathbb{E}\left[\min\left\{1, \sup_{t\geq\Gamma_1}\{\mathrm{e}^{-\alpha X_t}\}\right\}\right].\end{aligned}$$

For the first expression for θ_k, see de Haan et al. [22], and the second expression follows by a similar calculation as above. By an application of Hsing and Teugels [23] (see also Fasen [18], Lemma 2.2.4) we transform the time scale, such that (4.68) holds. Then we obtain

$$\begin{aligned}&\lim_{T\to\infty} \mathbb{P}(a_T^{-1} M(T) \leq x) \\ &= \lim_{T\to\infty} \mathbb{P}\left(\sum_{k=1}^{\infty} \varepsilon\left\{\frac{\Gamma_k}{T}, a_T^{-1} \sup_{s\in I_k} V_s\right\}((0,1)\times(x,\infty)) = 0\right) \\ &= \mathbb{P}\left(\sum_{k=1}^{\infty} \zeta_k \varepsilon\{j_k\}((0,1)) = 0\right) = \exp(-\widetilde{C}\theta x^{-\alpha})\,.\end{aligned}$$

□

By the Poisson result (4.68) we observe clusters in local extremes of the continuous-time process. So the COGARCH is a suitable model for heavy tailed volatility models with clusters on high levels.

4.5 Conclusion

In this paper we have investigated the extremal behavior of the most popular continuous-time volatility models. We have concentrated on models with tails ranging from exponential to regularly varying; i. e. tails as they are found in empirical volatility. The quantities derived for such models include

- the tail of the stationary volatility V_0 and the relation to the tail of the distribution governing the extreme behavior,
- the asymptotic distribution of the running maxima, i. e. their MDA and the norming constants,
- the cluster behavior of the model on high levels.

We found interesting similarities in the extremal behavior of certain models, which was quite unexpected.

Recall the GCIR model of Example 1, where the tail of the stationary distribution F of V_0 is compared to the tail of H, the d. f. describing the extreme behavior. Example 1 (2) belongs to $\mathcal{S} \cap \text{MDA}(\Lambda)$, it has stationary distribution with a semi-heavy Weibull like tail. Relation (4.24) is mimicked by the fact that (4.13) can be rewritten to

$$\overline{F}(x) \sim \frac{a(x)}{x}\overline{H}(x) \quad x \to \infty .$$

Moreover, as the norming constants in the GCIR examples are calculated based on the d. f. H, analogously, by the above Corollary 3, for the OU process in $\text{MDA}(\Lambda)$, the norming constants are derived from L_1 and not from the stationary distribution of the process V.

Analogous results hold for Example 1 (3), which belongs to $\mathcal{S} \cap \text{MDA}(\Phi_\alpha)$ for some $\alpha > 0$. Here the tails of F and H are both regularly varying of the same index; this corresponds to (4.21).

Also in the case, where V_0 is gamma distributed, the behavior of the running maxima of the GCIR model in Example 1 (1) and of the Γ-OU process as given in (4.39) and (4.46), respectively, are identical.

This means also that, if the stationary distribution of a GCIR model coincides with the stationary distribution of a subexponential OU model, then also the norming constants and the behavior of the running maxima coincide. The role of $M(h)$ in $\mathcal{S}(\gamma)$, $\gamma \geq 0$, corresponds for the GCIR models to the d. f. H; the influence of the driving Brownian motion plays no role whatsoever for the extreme behavior.

Concerning volatility clusters, no model in $\text{MDA}(\Lambda)$ presented in this paper can model such clusters on high levels. And here there is a profound difference between the GCIR models with regularly varying tails and the regularly varying OU models. Whereas all GCIR models fail to model high level volatility clusters, regularly varying OU models have the potential to model them.

The COGARCH model resembles the GCIR models only in the sense that heavy tails occur, although the driving process can be very light tailed; the difference being that the COGARCH model always has heavy tails. There is no obvious relationship between the tail behavior of the stationary r. v. V_0 and L_1; the heavy tails occur by the very intrinsic dependence structure of the model.

With respect to volatility clusters, only regularly varying OU processes and COGARCH processes exhibit volatility clusters on high levels, which can be described quite precisely by the distribution of the cluster sizes; see Corollary 2 and Theorem 6.

In this paper we have refrained from discussing another important stylized fact of empirical volatility: it exhibits often long memory in the sense that the autocovariance function decreases very slowly. This phenomenon can have various reasons, as for instance discussed in Mikosch and Stărică [31]. Certainly an important issue is here a possible non-stationarity of the data. On the other hand, long range dependence is an important fact, which should not be completely ignored. All models presented in this paper have an exponentially decreasing covariance functions, which only exhibit some visual long memory, when the process is close to non-stationarity.

For diffusion models like the GCIR models, a remedy, which introduces long range dependence in such models, is to replace the driving Brownian motion by a fractional Brownian motion. This generates a new class of stationary long memory models. Such models have been suggested and analyzed in [12, 13].

For the OU process the exponentially decreasing covariance function is due to the exponential kernel function; see (4.16). The often observed long-range dependence effect in the empirical volatility cannot be modelled this way. There are two ways to introduce long memory into such models. The first one is to replace the exponential kernel function by a hyperbolic kernel function of the form $f(x) \sim |x|^{-\beta}$ for large x and for some $\beta \in (0.5, 1)$. This introduces long memory into the model, which can be modelled by regularly varying Lévy driven MA processes. The second method has been suggested by Barndorff-Nielsen and Shephard [3]: a superposition of several OU processes (supOU processes) can create long memory. Regularly varying supOU processes exhibit also volatility clusters; see Fasen [19].

Acknowledgement

The first author acknowledges gratefully the financial support by the Deutsche Forschungsgemeinschaft through the graduate program "Angewandte Algorithmische Mathematik" at the Munich University of Technology. The third author takes pleasure in thanking the Deutsche Forschungsgemeinschaft for a research grant, and the Department of Statistics at Colorado State University for its pleasant hospitality.

Appendix

In this Appendix we summarize some definitions and concepts used throughout the paper.

A Basic notation and definition

For details and further references see Embrechts et al. [17].

Definition 3. *A positive measurable function* $u : \mathbb{R} \to \mathbb{R}_+$ *is* regularly varying *with index* α, *denoted by* $u \in \mathcal{R}_\alpha$ *for* $\alpha \in \mathbb{R}$, *if*

$$\lim_{t\to\infty} \frac{u(tx)}{u(t)} = x^\alpha\,, \qquad x > 0\,.$$

The function u *is said to be* slowly varying *if* $\alpha = 0$, *and* rapidly varying, *denoted by* $u \in \mathcal{R}_{-\infty}$, *if the above limit is* 0 *for* $x > 1$ *and* ∞ *for* $0 < x < 1$.

Definition 4. *A d.f. F belongs to the class* $\mathcal{L}(\gamma)$, $\gamma \geq 0$, *if for every* $y \in \mathbb{R}$,

$$\lim_{x\to\infty} \overline{F}(x-y)/\overline{F}(x) = \mathrm{e}^{\gamma y}\,.$$

The class $\mathcal{L}(\gamma)$ is related to the class $\mathcal{R}_{-\gamma}$ by the fact that

$$F \in \mathcal{L}(\gamma) \quad \text{if and only if} \quad \overline{F} \circ \ln \in \mathcal{R}_{-\gamma}\,.$$

Thus the convergence of $\overline{F}(x-y)/\overline{F}(x)$ in Definition 4 is uniform on compact y-intervals. For an excellent monograph on regular variation we refer to Bingham, Goldie and Teugels [7].

Applying Karamata's representation for regularly varying functions to the class $\mathcal{L}(\gamma)$ we obtain for $F \in \mathcal{L}(\gamma)$, $\gamma \geq 0$, the representation

$$\overline{F}(x) = c(x) \exp\left[-\int_0^x \frac{1}{a(y)}\,dy\right], \quad x > 0\,, \tag{A.1}$$

where $a, c : \mathbb{R}_+ \to \mathbb{R}_+$ and $\lim_{x\to\infty} c(x) = c > 0$ and a is absolutely continuous with $\lim_{x\to\infty} a(x) = 1/\gamma$ and $\lim_{x\to\infty} a'(x) = 0$.

Definition 5 (Convolution equivalent distributions).
Let $\gamma \geq 0$ *and* X *have d.f.* F. *We say that* F *or* X *belongs to the class* $\mathcal{S}(\gamma)$, *if the following properties hold.*

(i) $F \in \mathcal{L}(\gamma)$,

(ii) $\lim_{x\to\infty} \frac{\overline{F^{*2}}(x)}{\overline{F}(x)} = 2\widehat{f}(\gamma) < \infty$,

where $\widehat{f}(\gamma) = \mathbb{E}e^{\gamma X}$ *is the moment generating function of* X *at* γ. *The class* $\mathcal{S} := \mathcal{S}(0)$ *is called the class of* subexponential distributions.

Theorem 7.

(i) Let F *be infinitely divisible with Lévy measure* ν *and* $\gamma \geq 0$. *Then*

$$F \in \mathcal{S}(\gamma) \quad \Leftrightarrow \quad \nu(1,\cdot]/\nu(1,\infty) \in \mathcal{S}(\gamma) \quad \Leftrightarrow \quad \lim_{x\to\infty} \overline{F}(x)/\nu(x,\infty) = \widehat{f}(\gamma).$$

(ii) Suppose $F \in \mathcal{S}(\gamma)$, $\lim_{x\to\infty} \overline{F_i}(x)/\overline{F}(x) = q_i \geq 0$ *and* $\widehat{f_i}(\gamma) < \infty$ *for* $i = 1, 2$. *Then*

$$\lim_{x\to\infty} \frac{\overline{F_1 * F_2}(x)}{\overline{F}(x)} = q_1\widehat{f_2}(\gamma) + q_2\widehat{f_1}(\gamma).$$

If $q_i > 0$ *for some* $i \in \{1, 2\}$, *then also* F_i, $F_1 * F_2 \in \mathcal{S}(\gamma)$.

(iii) Let N *be a Poisson r. v. with mean* μ *and* $(X_k)_{k\in\mathbb{N}}$ *be an i. i. d. sequence with d. f.* $F \in \mathcal{S}(\gamma)$. *The r. v.* $Y = \sum_{k=1}^{N} X_k$ *has d. f.* $G = e^{-\mu}\sum_{n=0}^{\infty} \frac{\mu^n}{n!} F^{*n}$. *Then* $G \in \mathcal{S}(\gamma)$ *and*

$$\overline{G}(x) \sim \mu\widehat{f}(\gamma)\overline{F}(x), \quad x \to \infty.$$

The following is the fundamental theorem in extreme value theory.

Theorem 8 (Fisher-Tippett Theorem).
Let $(X_n)_{n\in\mathbb{N}}$ *be an i. i. d. sequence with d. f.* F *and write* $M_n = \max_{k=1,\dots,n} X_k$. *Suppose we can find sequences of real numbers* $a_n > 0$, $b_n \in \mathbb{R}$ *such that*

$$\lim_{n\to\infty} \mathbb{P}(a_n^{-1}(M_n - b_n) \leq x) = \lim_{n\to\infty} F^n(a_n x + b_n) = G(x), \quad x \in \mathbb{R},$$

for some non-degenerate d. f. G *(we say* F *is in the* maximum domain of attraction *of* G *and write* $F \in \mathrm{MDA}(G)$*). Then there are* $a > 0$, $b \in \mathbb{R}$ *such that* $x \mapsto G(ax + b)$ *is one of the following three extreme value d. f. s:*

- *Fréchet* $\Phi_\alpha(x) = \begin{cases} 0, & x \leq 0, \\ \exp(-x^{-\alpha}), & x > 0, \end{cases}$ *for* $\alpha > 0$.
- *Gumbel* $\Lambda(x) = \exp(-e^{-x})$, $x \in \mathbb{R}$.
- *Weibull* $\Psi_\alpha(x) = \begin{cases} \exp(-(-x)^\alpha), & x \leq 0, \\ 1, & x > 0, \end{cases}$ *for* $\alpha > 0$.

We summarize some well-known facts related to domains of attraction.

Proposition 4.

(a) *The following Poisson characterization holds:* $F \in \mathrm{MDA}(G)$ *if and only if* $a_n > 0, b_n \in \mathbb{R}$ *exist such that*

$$\lim_{n\to\infty} n\overline{F}(a_n x + b_n) = -\ln G(x)\,, \quad x \in \mathbb{R}\,.$$

(b) *If* $F \in \mathcal{L}(\gamma)$ *for* $\gamma > 0$*, then* $F \in \mathrm{MDA}(\Lambda)$ *with* $a_n \to 1/\gamma$ *as* $n \to \infty$ *and* $\mathrm{e}^{b_n} \in \mathcal{R}_{1/\gamma}$.

(c) *If* $F \in \mathcal{S} \cap \mathrm{MDA}(\Lambda)$*, then* $b_n \to \infty$*,* $a_n \to \infty$ *and* $b_n/a_n \to \infty$ *as* $n \to \infty$.

(d) *If* $F \in \mathrm{MDA}(\Phi_\alpha) = \mathcal{R}_{-\alpha}$ *for* $\alpha > 0$*, then* $b_n = 0$*,* $a_n \in \mathcal{R}_{1/\alpha}$ *and* $a_n \to \infty$ *as* $n \to \infty$.

The following concept has proved useful in comparing tails.

Definition 6 (Tail-equivalence).
Two d. f. s F *and* G *(or two measures* μ *and* ν*) are called* tail-equivalent *if both have support unbounded to the right and there exists some* $c > 0$ *such that*

$$\lim_{x\to\infty} \overline{F}(x)/\overline{G}(x) = c \quad \textit{or} \quad \lim_{x\to\infty} \nu(x,\infty)/\mu(x,\infty) = c\,.$$

Important in the context of our paper is that all the following classes are closed with respect to tail-equivalence: $\mathrm{MDA}(G)$ for $G \in \{\Phi_\alpha, \alpha > 0, \Lambda)$, $\mathcal{R}_{-\alpha}$ for $\alpha \in [0,\infty)$, $\mathcal{L}(\gamma)$ for $\gamma \geq 0$, $\mathcal{S}(\gamma)$ for $\gamma \geq 0$. Moreover, for two tail-equivalent d. f. s in some $\mathrm{MDA}(G)$ one can choose the same norming constants.

Definition 7 (Poisson random measure).
Let $(A, \mathcal{A}, \vartheta)$ *be a measurable space, where* ϑ *is* σ*-finite, and* $(\Omega, \mathcal{F}, \mathbb{P})$ *be a probability space. A* Poisson random measure N with mean measure ϑ, *denoted by* $\mathrm{PRM}(\vartheta)$*, is a collection of r. v. s* $(N(A))_{A\in\mathcal{A}}$*, where* $N(A) : (\Omega, \mathcal{F}, \mathbb{P}) \to (\mathbb{N}_0, \mathcal{B}(N_0))$*, with* $N(\emptyset) = 0$*, such that:*

(a) *Given any sequence* $(A_n)_{n\in\mathbb{N}}$ *of mutually disjoint sets in* $\mathcal{A}$*:*

$$N\left(\bigcup_{n\in\mathbb{N}} A_n\right) = \sum_{n\in\mathbb{N}} N(A_n) \qquad a.s.$$

(b) $N(A)$ *is Poisson distributed with mean* $\vartheta(A)$ *for every* $A \in \mathcal{A}$.

(c) *For mutually disjoint sets* $A_1, \ldots, A_n \in \mathcal{A}$*,* $n \in \mathbb{N}$*, the r. v. s* $N(A_1), \ldots, N(A_n)$ *are independent.*

Definition 8 (Extremal index).
Let $X = (X_n)_{n\in\mathbb{Z}}$ *be a strictly stationary sequence and* $\theta \geq 0$*. If for every* $\tau > 0$ *there exists a sequence* $u_n(\tau)$ *with*

$$\lim_{n\to\infty} n\mathbb{P}(X_1 > u_n(\tau)) = \tau \quad \textit{and} \quad \lim_{n\to\infty} \mathbb{P}(\max_{k=1,\ldots,n} X_n \leq u_n(\tau)) = \mathrm{e}^{-\theta\tau}\,,$$

then θ *is called the* extremal index *of* X *and has value in* $[0,1]$.

B The conditions $D_r(u_n)$ and $D'(u_n)$

Classical results for the extremal behavior of stationary sequences are based on two conditions: the first one is a specific type of asymptotic dependence, and the second is an anti-clustering condition.

Definition 9. *Let $X = (X_n)_{n\in\mathbb{N}}$ be a strictly stationary sequence, such that for $m = 1,\dots,r$, $r \in \mathbb{N}$, the sequences of constants $(u_n^{(m)})_{n\in\mathbb{N}}$ and $(u_n)_{n\in\mathbb{N}}$ satisfy $\lim_{n\to\infty} n\overline{F}(u_n^{(m)}) = \tau^{(m)}$ and $\lim_{n\to\infty} n\overline{F}(u_n) = \tau$.*

(a) For any integers p, q and n let

$$1 \le i_1 < \cdots < i_p < j_1 < \cdots < j_q \le n$$

such that $j_1 - i_p \ge l$ and $\mathbf{v}_n = (v_n^{(1)},\dots,v_n^{(p)})$, $\mathbf{w}_n = (w_n^{(1)},\dots,w_n^{(q)})$ with $v_n^{(l)}, w_n^{(s)} \in \{u_n^{(1)},\dots,u_n^{(r)}\}$. Write $I = \{i_1,\dots,i_p\}$, $J = \{j_1,\dots,j_q\}$, $\mathbf{X}_I = (X_{i_1},\dots,X_{i_p})$ and $\mathbf{X}_J = (X_{j_1},\dots,X_{j_q})$. If for each choice of indices of I, J,

$$|\mathbb{P}(\mathbf{X}_I \le \mathbf{v}_n, \mathbf{X}_J \le \mathbf{w}_n) - \mathbb{P}(\mathbf{X}_I \le \mathbf{v}_n)\,\mathbb{P}(\mathbf{X}_J \le \mathbf{w}_n)| \le \alpha_{n,l}\,,$$

where $\alpha_{n,l_n} \to 0$ as $n \to \infty$ for some sequence $l_n = o(n)$, then X satisfies the condition $\boldsymbol{D_r(u_n)}$.

(b) X satisfies the condition $D'(u_n)$, if

$$\lim_{k\to\infty} \limsup_{n\to\infty} n \sum_{j=2}^{\lfloor n/k \rfloor} \mathbb{P}(X_1 > u_n\,,\, X_j > u_n) = 0\,.$$

We show that $(\widetilde{V}_k)_{k\in\mathbb{Z}}$ as defined in (4.41) satisfies the $\boldsymbol{D_r(u_n)}$ and $D'(u_n)$ conditions. The result is an analogon for discrete-time MA processes given in Rootzén [33], Lemma 3.2.

Lemma 2. *Let V be a stationary version of the OU process given by (4.19) with L a positive compound Poisson process as in (4.18).*

(a) Assume $\widetilde{V}_k = V_{\Gamma_k/\lambda} + \mathrm{e}^{-\Gamma_k}\xi_0 \in \mathcal{L}(\gamma)$, $\gamma > 0$, such that for $a_n > 0$, $b_n \in \mathbb{R}$ and $u_n = a_n x + b_n$,

$$\lim_{n\to\infty} n\mathbb{P}(\widetilde{V}_k > u_n) = \mathrm{e}^{-x}\,, \quad x \in \mathbb{R}\,. \tag{B.1}$$

For $r \in \mathbb{N}$ and $\mathbf{x} = (x_1,\dots,x_r)$ let $\mathbf{u}_n = (a_n x_1 + b_n,\dots,a_n x_r + b_n)$. Then $(\widetilde{V}_k)_{k\in\mathbb{N}}$ satisfies the $\boldsymbol{D_r(u_n)}$ and $D'(u_n)$ conditions.

(b) *Let L_1 be in $\mathcal{S}(\gamma)$, $\gamma > 0$. Define for $h > 0$ fixed $M_k := \sup_{(k-1)h \le t \le kh} V_t$ for $k \in \mathbb{N}$. Suppose $M_1 \in \mathcal{L}(\gamma)$ such that for $a_n > 0, b_n \in \mathbb{R}$ and $u_n = a_n x + b_n$,*

$$\lim_{n\to\infty} n\mathbb{P}(M_k > u_n) = \mathrm{e}^{-x}, \quad x \in \mathbb{R}.$$

For $r \in \mathbb{N}$ and $\mathbf{x} = (x_1, \ldots, x_r)$ let $\mathbf{u_n} = (a_n x_1 + b_n, \ldots, a_n x_r + b_n)$. Then $(M_k)_{k\in\mathbb{N}}$ satisfies the $\boldsymbol{D_r(u_n)}$ and $D'(u_n)$ conditions.

Proof. (a) To show the $\boldsymbol{D_r(u_n)}$ condition, let $u_n^{(m)} = a_n x_m + b_n$, $x_m \in \mathbb{R}$, $m = 1, \ldots, r$. Let $\mathbf{v}_n = (v_n^{(1)}, \ldots, v_n^{(p)})$, $\mathbf{w}_n = (w_n^{(1)}, \ldots, w_n^{(q)})$ with $v_n^{(l)}, w_n^{(s)} \in \{u_n^{(1)}, \ldots, u_n^{(r)}\}$ and $1 \le i_1 < \cdots < i_p < j_1 < \cdots < j_q \le n$. Define $V_k^n := \sum_{j=k-n}^{k} \mathrm{e}^{-(\Gamma_k - \Gamma_j)} \xi_j$, $\mathbf{V}_I := (\widetilde{V}_{i_1}, \ldots, \widetilde{V}_{i_p})$, $\mathbf{V}_I^n := (V_{i_1}^n, \ldots, V_{i_p}^n)$ and, similarly, $\mathbf{V}_J := (\widetilde{V}_{j_1}, \ldots, \widetilde{V}_{j_q})$, $\mathbf{V}_J^n := (V_{j_1}^n, \ldots, V_{j_q}^n)$. Then $(V_k^n)_{k\in\mathbb{Z}}$ is stationary and $\mathbf{V}_I^{\lfloor n\delta \rfloor}$ is independent of $\mathbf{V}_J^{\lfloor n\delta \rfloor}$ for $j_1 - i_p > \lfloor n\delta \rfloor$, $\delta > 0$. Since $\mathbb{P}(\xi_k < 0) = 0$, we obtain $\mathbf{V}_I^n \le \mathbf{V}_I$. Here, $\mathbf{x} = (x_1, \ldots, x_p) \le \mathbf{y} = (y_1, \ldots, y_p)$ means that $x_i \le y_i$ for all $i = 1, \ldots, p$. It now follows that for any $\epsilon > 0$,

$$\begin{aligned}
&\mathbb{P}\left(\mathbf{V}_I \le \mathbf{v}_n, \mathbf{V}_J \le \mathbf{w}_n\right) \le \mathbb{P}\left(\mathbf{V}_I^{\lfloor n\delta \rfloor} \le \mathbf{v}_n, \mathbf{V}_J^{\lfloor n\delta \rfloor} \le \mathbf{w}_n\right) \\
&= \mathbb{P}\left(\mathbf{V}_I^{\lfloor n\delta \rfloor} \le \mathbf{v}_n\right) \mathbb{P}\left(\mathbf{V}_J^{\lfloor n\delta \rfloor} \le \mathbf{w}_n\right) \\
&\le \mathbb{P}\left(\mathbf{V}_I \le \mathbf{v}_n + \epsilon(a_n, \ldots, a_n)\right) \mathbb{P}\left(\mathbf{V}_J \le \mathbf{w}_n + \epsilon(a_n, \ldots, a_n)\right) + \\
&\qquad + n\mathbb{P}\left(|\widetilde{V}_1 - V_1^{\lfloor n\delta \rfloor}| > \epsilon a_n\right) \\
&\le \mathbb{P}\left(\mathbf{V}_I \le \mathbf{v}_n\right) \mathbb{P}\left(\mathbf{V}_J \le \mathbf{w}_n\right) + n\mathbb{P}\left(|\widetilde{V}_1 - V_1^{\lfloor n\delta \rfloor}| > \epsilon a_n\right) + \\
&\qquad + \sum_{m=1}^{r} n\mathbb{P}(u_n^{(m)} \le \widetilde{V}_1 \le u_n^{(m)} + \epsilon\, a_n).
\end{aligned}$$

Similarly we can find a lower bound, such that for $j_1 - i_p > \lfloor n\delta \rfloor$,

$$\begin{aligned}
\alpha_{n,\lfloor n\delta \rfloor} &:= \left|\mathbb{P}\left(\mathbf{V}_I \le \mathbf{v}_n, \mathbf{V}_J \le \mathbf{w}_n\right) - \mathbb{P}\left(\mathbf{V}_I \le \mathbf{v}_n\right) \mathbb{P}\left(\mathbf{V}_J \le \mathbf{w}_n\right)\right| \\
&\le n\mathbb{P}(|\widetilde{V}_1 - V_1^{\lfloor n\delta \rfloor}| > \epsilon a_n) + \sum_{m=1}^{r} n\mathbb{P}(u_n^{(m)} - \epsilon a_n \le \widetilde{V}_1 \le u_n^{(m)} + \epsilon a_n) \\
&=: \widetilde{\alpha}_{n,\lfloor n\delta \rfloor,\epsilon}\,.
\end{aligned} \tag{B.2}$$

Let $X_i := \Gamma_i - \Gamma_{i-1} - 1/\mu$, $i \in \mathbb{N}$. Then $(X_i)_{i\in\mathbb{N}}$ is a centered i. i. d. sequence such that $\sum_{i=1}^{n} X_i = \Gamma_n - n/\mu$. It follows that there exists a constant $K > 0$, such that for every $n \in \mathbb{N}$,

$$\begin{aligned}
\mathbb{E}(\Gamma_n - n/\mu)^6 &= n\mathbb{E}X_1^6 + \binom{n}{3}\binom{6}{2}\binom{4}{2}(\mathbb{E}X_1^2)^3 + \\
&\quad + \binom{n}{2}\binom{6}{3}(\mathbb{E}X_1^3)^2 + \binom{n}{2}\binom{6}{2}(\mathbb{E}X_1^2)(\mathbb{E}X_1^4) \\
&\le n^3 K\,.
\end{aligned}$$

Hence, by Markov's inequality there is a constant $\widetilde{K} > 0$ such that for all $n \in \mathbb{N}$,

$$\begin{aligned}\mathbb{P}(\Gamma_n < n/(2\mu)) &\le \mathbb{P}(|\Gamma_n - n/\mu| > n/(2\mu)) \\ &\le (2\mu/n)^6 \mathbb{E}(\Gamma_n - n/\mu)^6 \le \widetilde{K}/n^3. \end{aligned} \tag{B.3}$$

Thus we obtain for $n \in \mathbb{N}$,

$$\begin{aligned} n\mathbb{P}(|\widetilde{V}_0 - V_0^{\lfloor n\delta \rfloor}| > \epsilon a_n) &= n\mathbb{P}\left(\mathrm{e}^{-\Gamma_{\lfloor n\delta \rfloor}} \sum_{j=-\infty}^{-\lfloor n\delta \rfloor - 1} \mathrm{e}^{(\Gamma_{\lfloor n\delta \rfloor} + \Gamma_j)} \xi_j > \epsilon a_n \right) \\ &\le n\mathbb{P}\left(\mathrm{e}^{-\lfloor n\delta \rfloor/(2\mu)} \widetilde{V}_1 > \epsilon a_n \right) + \frac{\widetilde{K}}{n^2}. \end{aligned} \tag{B.4}$$

Note that the first exponential moment of $\gamma\beta \mathrm{e}^{-\lfloor \delta n \rfloor/(2\mu)} \widetilde{V}_1$ exists for $\beta \mathrm{e}^{-\lfloor \delta n \rfloor/(2\mu)} < 1$. Choose $\beta_n = 2/(\epsilon(1-\epsilon)) \ln n$. There exists $n_0 = n_0(\delta, \epsilon) \in \mathbb{N}$ such that $\beta_n \mathrm{e}^{-\lfloor n\delta \rfloor/(2\mu)} < (1-\epsilon)$ and $a_n \ge (1-\epsilon)/\gamma$ for $n \ge n_0$. The first term on the right hand side of (B.4) is by Markov's inequality for $n \ge n_0$ bounded above by

$$n\mathbb{E}\exp\left[\beta_n \gamma \left(\mathrm{e}^{-\lfloor \delta n \rfloor/(2\mu)} \widetilde{V}_1 \right)\right] \mathrm{e}^{-\beta_n \epsilon \gamma a_n} \le n\mathbb{E}\exp\left[(1-\epsilon)\gamma \widetilde{V}_1\right] \mathrm{e}^{-2\ln n}, \tag{B.5}$$

which converges to 0 as $n \to \infty$. Together with (B.1), (B.2) and (B.4) this gives

$$\lim_{n\to\infty} \widetilde{\alpha}_{n,\lfloor n\delta \rfloor,\epsilon} = \sum_{m=1}^{r} [\mathrm{e}^{-(x_m - \epsilon)} - \mathrm{e}^{-(x_m + \epsilon)}],$$

so that

$$\lim_{n\to\infty} \alpha_{n,\lfloor n\delta \rfloor} \le \lim_{\epsilon \downarrow 0} \lim_{n\to\infty} \widetilde{\alpha}_{n,\lfloor n\delta \rfloor,\epsilon} = 0,$$

which implies the $\mathbf{D_r(u_n)}$ condition by Lemma 3.2.1 in Leadbetter et al. [29].

To show the $D'(u_n)$ condition, let $\epsilon > 0$. Then there exists an $x_0 > 0$ such that $\mathbb{P}(\Gamma_2 - \Gamma_1 < x_0) = \epsilon$. Since $(\Gamma_{i+1} - \Gamma_i)_{i\in\mathbb{N}}$ is a positive i. i. d. sequence, it follows that

$$\mathbb{P}(\Gamma_j - \Gamma_1 < x_0) \le \mathbb{P}(\Gamma_j - \Gamma_{j-1} < x_0, \ldots, \Gamma_2 - \Gamma_1 < x_0) = \epsilon^{j-1}, \quad j \ge 2.$$

Now choose β such that $1/2 < \beta < (1+\mathrm{e}^{-x_0})^{-1}$ and $\delta > 0$ such that $1+\delta < 2\beta$. Then, for any $k, n \in \mathbb{N}$,

$$\begin{aligned} \sum_{j=2}^{\lfloor n/k \rfloor} \mathbb{P}(\widetilde{V}_1 > u_n, \widetilde{V}_j > u_n) &= \sum_{j=2}^{\lfloor n^\delta \rfloor} \mathbb{P}(\widetilde{V}_1 > u_n, \widetilde{V}_j > u_n) + \\ &\quad + \sum_{j=\lfloor n^\delta \rfloor + 1}^{\lfloor n/k \rfloor} \mathbb{P}(\widetilde{V}_1 > u_n, \widetilde{V}_j > u_n). \end{aligned} \tag{B.6}$$

We first show that the first summand, when multiplied by n, tends to 0 as $n \to \infty$. Note that by the independence of $\widetilde{V}_1$ and $\Gamma_j - \Gamma_1$,

$$\mathbb{P}(\widetilde{V}_1 > u_n, \widetilde{V}_j > u_n) \le \epsilon^{j-1}\,\mathbb{P}(\widetilde{V}_1 > u_n) + \\ +\mathbb{P}(\widetilde{V}_1 + \widetilde{V}_j > 2u_n, \Gamma_j - \Gamma_1 \ge x_0)\,. \qquad \text{(B.7)}$$

Let $\widetilde{V}_1'$ be an independent copy of $\widetilde{V}_1$. Then

$$\begin{aligned}
&\mathbb{P}\left(\widetilde{V}_1 + \widetilde{V}_j > 2u_n, \Gamma_j - \Gamma_1 \ge x_0\right) \\
&= \mathbb{P}\left((1+\mathrm{e}^{-(\Gamma_j-\Gamma_1)})\widetilde{V}_1 + \sum_{k=2}^{j} \mathrm{e}^{-(\Gamma_j-\Gamma_k)}\xi_k > 2u_n, \Gamma_j - \Gamma_1 \ge x_0\right) \\
&\le \mathbb{P}\left((1+\mathrm{e}^{-x_0})\widetilde{V}_1 + \sum_{k=2}^{j} \mathrm{e}^{-(\Gamma_j-\Gamma_k)}\xi_k > 2u_n\right) \\
&\le \mathbb{P}\left((1+\mathrm{e}^{-x_0})\widetilde{V}_1 + \widetilde{V}_1' > 2u_n\right). \qquad \text{(B.8)}
\end{aligned}$$

As $\beta(1+\mathrm{e}^{-x_0}) < 1$, the first exponential moment of $\beta\gamma((1+\mathrm{e}^{-x_0})\widetilde{V}_1 + \widetilde{V}_1')$ exists, and by Markov's inequality the last expression can be bounded above by

$$\mathbb{E}\exp\left[\beta\gamma((1+\mathrm{e}^{-x_0})\widetilde{V}_1 + \widetilde{V}_1')\right]\mathrm{e}^{-2\beta\gamma u_n}\,. \qquad \text{(B.9)}$$

Recall that $u_n = u_n(x)$. Since $\left(n \mapsto \mathrm{e}^{-2\beta\gamma u_n}\right) \in \mathcal{R}_{-2\beta}$ for fixed x, it follows that $\left(n \mapsto n^{1+\delta}\mathrm{e}^{-2\beta\gamma u_n}\right) \in \mathcal{R}_{(1+\delta)-2\beta}$. We then obtain by (B.1), (B.7)-(B.9) and Bingham et al. [7], Proposition 1.5.1, that

$$\begin{aligned}
&n\sum_{j=2}^{\lfloor n^\delta \rfloor} \mathbb{P}(\widetilde{V}_1 > u_n, \widetilde{V}_j > u_n) \qquad \text{(B.10)} \\
&\le n\mathbb{P}(\widetilde{V}_1 > u_n)\sum_{j=2}^{\lfloor n^\delta \rfloor} \epsilon^{j-1} + \mathbb{E}[\beta\gamma((1+\mathrm{e}^{-x_0})\widetilde{V}_1 + \widetilde{V}_1')]n^{1+\delta}\mathrm{e}^{-2\beta\gamma u_n} \\
&\to \mathrm{e}^{-x}\,\frac{\epsilon}{1-\epsilon}\,, \qquad n \to \infty\,.
\end{aligned}$$

For the second term in (B.6), using the independence of $V_j^{\lfloor n^\delta \rfloor}$ and $V_1^{\lfloor n^\delta \rfloor}$ for $j > \lfloor n^\delta \rfloor$, we obtain

$$\begin{aligned}
&n\sum_{j=\lfloor n^\delta \rfloor+1}^{\lfloor n/k \rfloor} \mathbb{P}(\widetilde{V}_1 > u_n, \widetilde{V}_j > u_n) \\
&\le n\sum_{j=\lfloor n^\delta \rfloor+1}^{\lfloor n/k \rfloor} \mathbb{P}(V_1^{\lfloor n^\delta \rfloor} > u_n - \epsilon\, a_n, V_j^{\lfloor n^\delta \rfloor} > u_n - \epsilon\, a_n) + \\
&+ 2n^2\mathbb{P}(|\widetilde{V}_1 - V_1^{\lfloor n^\delta \rfloor}| > \epsilon\, a_n)
\end{aligned}$$

$$
\begin{aligned}
&= n \sum_{j=\lfloor n^\delta \rfloor+1}^{\lfloor n/k \rfloor} \mathbb{P}(V_1^{\lfloor n^\delta \rfloor} > u_n - \epsilon\, a_n)\mathbb{P}(V_j^{\lfloor n^\delta \rfloor} > u_n - \epsilon\, a_n) + \\
&\quad + 2n^2 \mathbb{P}(|\widetilde{V}_1 - V_1^{\lfloor n^\delta \rfloor}| > \epsilon\, a_n) \\
&\leq (n^2/k)\, \mathbb{P}(\widetilde{V}_1 > u_n - \epsilon\, a_n)^2 + 2n^2 \mathbb{P}(|\widetilde{V}_1 - V_1^{\lfloor n^\delta \rfloor}| > \epsilon\, a_n)\,. \qquad \text{(B.11)}
\end{aligned}
$$

Analogously to (B.4) and (B.5), with $\beta_n = 3/(\epsilon(1-\epsilon))\ln n$, we have $n^2\mathbb{P}(|\widetilde{V}_1 - V_1^{\lfloor n^\delta \rfloor}| > \epsilon\, a_n) \to 0$ as $n \to \infty$. Using (B.1), we also have

$$
\lim_{n\to\infty} (n^2/k)\, \mathbb{P}(\widetilde{V}_1 > u_n - \epsilon\, a_n)^2 = \exp(-2(x-\epsilon))/k\,, \qquad \text{(B.12)}
$$

which converges to 0 as $k \to \infty$. Then by (B.6), (B.10)-(B.12), and letting $\epsilon \downarrow 0$, the $D'(u_n)$ condition holds.

(b) To prove condition $\boldsymbol{D_r(u_n)}$, we replace V_k^n in *(a)* by

$$
M_k^n := \sup_{(k-1)h \leq t \leq kh} \int_{t-nh}^{t} \mathrm{e}^{-\lambda(t-s)}\, dL_{\lambda s}\,.
$$

We then obtain an analogue result to (B.2). Further, since

$$
\begin{aligned}
|M_k - M_k^n| &\leq \sup_{(k-1)h \leq t \leq kh} \int_{-\infty}^{t-nh} \mathrm{e}^{-\lambda(t-s)}\, dL_{\lambda s} \\
&\stackrel{d}{=} \mathrm{e}^{-\lambda nh} \sup_{(k-1)h \leq t \leq kh} \int_{-\infty}^{t} \mathrm{e}^{-\lambda(t-s)}\, dL_{\lambda s} = \mathrm{e}^{-\lambda nh} M_k\,,
\end{aligned}
$$

we obtain for any $\delta > 0$ that

$$
n\mathbb{P}(|M_k - M_k^{\lfloor n\delta \rfloor}| > \epsilon\, a_n) \leq n\mathbb{P}(\mathrm{e}^{-\lambda \lfloor n\delta \rfloor h} M_k > \epsilon\, a_n)\,.
$$

Since the first exponential moment of $\beta\gamma \mathrm{e}^{-\lambda\lfloor n\delta \rfloor h} M_k$ exists for $\beta \mathrm{e}^{-\lambda \lfloor n\delta \rfloor h} < 1$, similar reasoning as in (B.4) and (B.5) shows that $\lim_{n\to\infty} n\mathbb{P}(|M_k - M_k^{\lfloor n\delta \rfloor}| > \epsilon\, a_n) = 0$. As in the proof of *(a)* we then conclude that the $\boldsymbol{D_r(u_n)}$ condition holds.

For the proof of the $D'(u_n)$ condition we use

$$
\begin{aligned}
M_k &\leq \int_{-\infty}^{\infty} \sup_{(k-1)h \leq t \leq kh} \mathrm{e}^{-\lambda(t-s)}\, \mathbf{1}_{(-\infty,t]}(s)\, dL_{\lambda s} \\
&= L_{\lambda kh} - L_{\lambda(k-1)h} + \int_{-\infty}^{(k-1)h} \mathrm{e}^{-\lambda((k-1)h-s)}\, dL_{\lambda s} =: \overline{V}_k\,.
\end{aligned}
$$

Let $j \geq 3$. Then we have the upper bound

$$\begin{aligned}
\overline{V}_j &= L_{\lambda jh} - L_{\lambda(j-1)h} + \mathrm{e}^{-\lambda(j-1)h} \int_{-\infty}^{0} \mathrm{e}^{\lambda s}\, dL_{\lambda s} + \\
&\quad + \mathrm{e}^{-\lambda(j-2)h} \int_0^h \mathrm{e}^{-\lambda(h-s)}\, dL_{\lambda s} + \int_h^{(j-1)h} \mathrm{e}^{-\lambda((j-1)h-s)}\, dL_{\lambda s} \\
&\le L_{\lambda jh} - L_{\lambda(j-1)h} + \mathrm{e}^{-\lambda(j-2)h} \int_{-\infty}^{0} \mathrm{e}^{\lambda s}\, dL_{\lambda s} + \\
&\quad + \mathrm{e}^{-\lambda(j-2)h} L_{\lambda h} + \int_h^{(j-1)h} \mathrm{e}^{-\lambda((j-1)h-s)}\, dL_{\lambda s} \\
&\le L_{\lambda jh} - L_{\lambda(j-1)h} + \mathrm{e}^{-\lambda h}\overline{V}_1 + \int_h^{(j-1)h} \mathrm{e}^{-\lambda((j-1)h-s)}\, dL_{\lambda s}\,.
\end{aligned}$$

Let $\overline{V}_1'$ be an independent copy of $\overline{V}_1$. Then

$$\begin{aligned}
n \sum_{j=3}^{\lfloor n^\delta \rfloor} \mathbb{P}(M_1 > u_n, M_j > u_n) &\le n \sum_{j=3}^{\lfloor n^\delta \rfloor} \mathbb{P}(\overline{V}_1 > u_n, \overline{V}_j > u_n) \\
&\le n \sum_{j=3}^{\lfloor n^\delta \rfloor} \mathbb{P}(\overline{V}_1 + \overline{V}_j > 2u_n) \\
&\le n^{1+\delta}\, \mathbb{P}((1+\mathrm{e}^{-\lambda h})\overline{V}_1 + \overline{V}_1' > 2u_n)\,.
\end{aligned}$$

The tail of $\overline{V}_1$ behaves by Proposition 3 *(b)* and Theorem 7 *(ii)* like

$$\mathbb{P}(\overline{V}_1 > x) = \mathbb{P}(L_{\lambda h} + V_0 > x) \sim \mathbb{E}\mathrm{e}^{\gamma V_0}\mathbb{P}(L_{\lambda h} > x)\,, \quad x \to \infty\,,$$

so that $\overline{V}_1 \in \mathcal{S}(\gamma)$. An analogue result to (B.8) and (B.9) gives

$$\lim_{n\to\infty} n^{1+\delta}\, \mathbb{P}((1+\mathrm{e}^{-\lambda h})\overline{V}_1 + \overline{V}_1' > 2u_n) = 0\,,$$

and argueing similarly as in (B.11) and (B.12), we obtain

$$\lim_{k\to\infty} \limsup_{n\to\infty} n \sum_{j=3}^{\lfloor n/k \rfloor} \mathbb{P}(M_1 > u_n, M_j > u_n) = 0\,.$$

It remains to show that

$$\lim_{n\to\infty} n\mathbb{P}(M_1 > u_n, M_2 > u_n) = 0\,. \tag{B.13}$$

Note that

$$\begin{aligned}
\mathbb{P}(M_1 > u_n) &= \mathbb{P}(M_1 > u_n, N_{\lambda h} > 0) + \mathbb{P}(M_1 > u_n, N_{\lambda h} = 0) \\
&\ge \mathbb{P}(V_{\Gamma_1/\lambda} > u_n, N_{\lambda h} > 0) \\
&\ge \mathbb{P}(\xi_1 > u_n)\mathbb{P}(N_{\lambda h} > 0)\,,
\end{aligned}$$

so that

$$n\mathbb{P}(M_1 > u_n, M_2 > u_n) \leq \frac{n\mathbb{P}(M_1 > u_n)}{\mathbb{P}(N_{\lambda h} > 0)} \frac{\mathbb{P}(M_1 > u_n, M_2 > u_n)}{\mathbb{P}(\xi_1 > u_n)}. \quad \text{(B.14)}$$

Furthermore, we have the upper bound

$$\begin{aligned}\mathbb{P}(M_1 > u_n, M_2 > u_n) &\leq \mathbb{P}(\overline{V}_1 + \overline{V}_2 > 2u_n) \\ &\leq \mathbb{P}\left(\frac{L_{2\lambda h} - L_{\lambda h}}{2} + \frac{1 + \mathrm{e}^{-\lambda h}}{2} \int_{-\infty}^{0} \mathrm{e}^{\lambda s}\, dL_{\lambda s} + \right. \\ &\quad \left. + \frac{1}{2} \int_0^h (1 + \mathrm{e}^{-\lambda h + \lambda s})\, dL_{\lambda s} > u_n \right).\end{aligned}$$

The three summands are independent, and we shall show that for each of them the probability to be greater than u_n is of order $o(\mathbb{P}(\xi_1 > u_n))$ for $n \to \infty$, so that by Theorem 7 *(ii)*

$$\lim_{n\to\infty} \mathbb{P}(M_1 > u_n, M_2 > u_n)/\mathbb{P}(\xi_1 > u_n) = 0\,. \quad \text{(B.15)}$$

Equation (B.13) and hence condition $D'(u_n)$ then follow from (B.2), (B.14) and (B.15).

The rapidly varying tails and Theorem 7 *(i)* give

$$\lim_{x\to\infty} \frac{\mathbb{P}(L_{2\lambda h} - L_{\lambda h} > 2x)}{\mathbb{P}(\xi_1 > x)} = \lim_{x\to\infty} \mu \frac{\mathbb{P}(L_{\lambda h} > x)}{\nu(x,\infty)} \frac{\mathbb{P}(L_{\lambda h} > 2x)}{\mathbb{P}(L_{\lambda h} > x)} = 0\,,$$

which is the assertion for the first summand. Further, also by the rapidly varying tails, Proposition 3 *(b)* and Theorem 7 *(i)*,

$$\begin{aligned}&\frac{\mathbb{P}\left((1 + \mathrm{e}^{-\lambda h}) \int_{-\infty}^{0} \mathrm{e}^{\lambda s}\, dL_{\lambda s} > 2x\right)}{\mathbb{P}(\xi_1 > x)} \\ &\quad = \frac{\mathbb{P}\left((1 + \mathrm{e}^{-\lambda h}) V_0 > 2x\right)}{\mathbb{P}(V_0 > x)} \frac{\mu \mathbb{P}(V_0 > x)}{\nu(x,\infty)} \stackrel{x\to\infty}{\longrightarrow} 0\,.\end{aligned}$$

For the last summand we use that

$$X := \int_0^h (1 + \mathrm{e}^{-\lambda h}\, \mathrm{e}^{\lambda s})\, dL_{\lambda s} \stackrel{d}{=} \sum_{i=1}^{N_{\lambda h}} (1 + \mathrm{e}^{-\lambda h} \mathrm{e}^{\lambda h U_i}) \xi_i \stackrel{d}{=} \sum_{i=1}^{N_{\lambda h}} (1 + \mathrm{e}^{-\lambda h U_i}) \xi_i\,,$$

where $(U_i)_{i\in\mathbb{N}}$, U are i. i. d. uniform on $(0,1)$ and independent of L (cf. e. g. Sato [35], Proposition 3.4).

From Theorem 7 *(iii)* then follows

$$\frac{\mathbb{P}\left(\int_0^h (1 + \mathrm{e}^{-\lambda s})\, dL_{\lambda s} > 2x\right)}{\mathbb{P}(\xi_1 > x)} \sim \mu \lambda h \mathbb{E} \mathrm{e}^{\gamma X} \frac{\mathbb{P}\left(\xi_1 (1 + \mathrm{e}^{-\lambda h U})/2 > x\right)}{\mathbb{P}(\xi_1 > x)} \stackrel{x\to\infty}{\longrightarrow} 0\,.$$

□

C Stationary solution of a random recurrence equation

The following result is the central result for proving stationarity and the tail behavior of a stochastic process defined by a random recurrence equation; it goes back to seminal work by Kesten [24] and Vervaat [36].

Theorem C.1 (Goldie [21], Theorem 4.1, Lemma 2.2).
Let $(Y_k)_{k\in\mathbb{N}}$ be a stochastic process defined by $Y_k = A_k Y_{k-1} + B_k$, where $((A_k, B_k))_{k\in\mathbb{N}}$, (A, B) are i. i. d. sequences. Assume that the following conditions are satisfied for some $\alpha > 0$:

(i) *The law of $\ln|A|$, given $|A| \neq 0$, is not concentrated on a lattice $-\infty \cap r\mathbb{Z}$ for any $r > 0$.*

(ii) $\mathbb{E}|A|^\alpha = 1$.

(iii) $\mathbb{E}|A|^\alpha \ln^+ |A| < \infty$.

(iv) $\mathbb{E}|B|^\alpha < \infty$.

Then the equation $Y_\infty \stackrel{d}{=} AY_\infty + B$, where Y_∞ is independent of (A, B), has the solution unique in distribution

$$Y_\infty \stackrel{d}{=} \sum_{m=1}^{\infty} B_m \prod_{k=1}^{m} A_k\,.$$

The process $(Y_k)_{k\in\mathbb{N}}$ with $Y_0 \stackrel{d}{=} Y_\infty$ is stationary and has tails

$$\mathbb{P}(Y_\infty > x) \sim \frac{\mathbb{E}\left[((AY_\infty + B)^+)^\alpha - ((AY_\infty)^+)^\alpha\right]}{\alpha \mathbb{E}|A|^\alpha \ln^+ |A|} x^{-\alpha}\,, \quad x \to \infty\,.$$

References

[1] Albin, J. M. P., On extremes of infinitely divisible Ornstein-Uhlenbeck processes, Preprint, available at `http://www.math.chalmers.se/~palbin/`.

[2] Barndorff-Nielsen, O. E., (1998), Processes of normal inverse Gaussian type, *Finance Stoch.*, **2**, 41–68.

[3] Barndorff-Nielsen, O. E. and Shephard, N., (2001), Modelling by Lévy processes for financial econometrics. In: O. E. Bandorff-Nielsen, T. Mikosch and S. I. Resnick (Eds.), *Lévy Processes: Theory and Applications*, 283–318, Boston, Birkhäuser.

[4] Barndorff-Nielsen, O. E. and Shephard, N., (2001), Non-Gaussian Ornstein-Uhlenbeck based models and some of their uses in financial economics (with discussion), *J. Roy. Statist. Soc. Ser. B*, **63** (2), 167–241.

[5] Barndorff-Nielsen, O. E. and Shephard, N., (2002), Econometric analysis of realised volatility and its use in estimating stochastic volatility models, *J. Roy. Statist. Soc. Ser. B*, **64**, 253–280.

[6] Bertoin, J., (1996), *Lévy Processes*, Cambridge University Press, Cambridge.

[7] Bingham, N. H. and Goldie, C. M. and Teugels, J. L., (1987), *Regular Variation*, Cambridge University Press, Cambridge.

[8] Borkovec, M. and Klüppelberg, C., (1998), Extremal behavior of diffusions models in finance, *Extremes*, **1** (1), 47–80.

[9] Braverman, M. and Samorodnitsky, G., (1995), Functionals of infinitely divisible stochastic processes with exponential tails, *Stochastic Process. Appl.*, **56** (2), 207–231.

[10] Breiman, L. , (1965), On some limit theorems similar to the arc-sine law, *Theory Probab. Appl.*, **10**, 323–331.

[11] Brockwell, P. J. and Chadraa, E. and Lindner, A. M., (2005), Continuous time GARCH processes of higher order, Preprint, available at `http://www.ma.tum.de/stat/`.

[12] Buchmann, B. and Klüppelberg, C., (2004), Fractional integral equation and state space transforms, *Bernoulli*, to appear.

[13] Buchmann, B. and Klüppelberg, C., (2004), Maxima of stochastic processes driven by fractional Brownian motion, *Adv. Appl. Probab.*, to appear.

[14] Cline, D. B. H., (1987), Convolution of distributions with exponential and subexponential tails, *J. Austral. Math. Soc. Ser. A*, **43** (3), 347–365.

[15] Cont, R. and Tankov, P., (2004), *Financial Modelling with Jump Processes*, Chapman & Hall, Boca Raton.

[16] Drost, F.C. and Werker, B. J. M., (1996), Closing the GARCH gap: continuous time GARCH modelling, *J. Econometrics*, **74**, 31–57.

[17] Embrechts, P. and Klüppelberg, C. and Mikosch, T., (1997), *Modelling Extremal Events for Insurance and Finance*, Springer, Berlin.

[18] Fasen, V., (2004), *Extremes of Lévy Driven MA Processes with Applications in Finance*, Ph.D. thesis, Munich University of Technology.

[19] Fasen, V., (2005), Extremes of regularly varying mixed moving average processes, Preprint, available at `http://www.ma.tum.de/stat/`.

[20] Fasen, V., (2005), Extremes of subexponential Lévy driven moving average processes, Preprint, available at `http://www.ma.tum.de/stat/`.

[21] Goldie, C. M., (1991), Implicit renewal theory and tails of solutions of random equations, *Ann. Appl. Probab.* **1** (1), 126–166.

[22] Haan, L. de and Resnick, S. I. and Rootzén, H. and Vries, C. G., (1989), Extremal behavior of solutions to a stochastic difference equation with applications to ARCH processes, *Stochastic Process. Appl.* **32**, 213–224.

[23] Hsing, T. and Teugels, J. L., (1989), Extremal properties of shot noise processes, *Adv. Appl. Probability* **21**, 513–525.

[24] Kesten, H., (1973), Random difference equations and renewal theory for products of random matrices, *Acta Math.* **131**, 207–248.

[25] Klüppelberg, C., (1989), Subexponential distributions and characterizations of related classes, *Probab. Theory Relat. Fields* **82**, 259–269.

[26] Klüppelberg, C., (2004), Risk management with extreme value theory, In: B. Finkenstädt and H. Rootzén (Eds), *Extreme Values in Finance, Telecommunication and the Environment*, 101–168, Chapman & Hall/CRC, Boca Raton.

[27] Klüppelberg, C. and Lindner, A. and Maller, R., (2004), A continuous time GARCH process driven by a Lévy process: stationarity and second order behaviour, *J. Appl. Probab.* **41** (3), 601–622.

[28] Klüppelberg, C. and Lindner, A. and Maller, R., (2004), Continuous time volatility modelling: COGARCH versus Ornstein-Uhlenbeck models, In: *From Stochastic Calculus to Mathematical Finance.* The Shiryaev Festschrift (Eds. Yu. Kabanov, R. Liptser and J. Stoyanov). Springer, to appear.

[29] Leadbetter, M. R. and Lindgren, G. and Rootzén, H., (1983), *Extremes and Related Properties of Random Sequences and Processes*, Springer, New York.

[30] Lindner, A. and Maller, R., (2004), Lévy integrals and the stationarity of generalised Ornstein-Uhlenbeck processes, Preprint, available at `http://www.ma.tum.de/stat/`.

[31] Mikosch, T. and Stărică, C., (2000), Limit theory for the sample autocorrelations and extremes of a GARCH(1,1) process, *Ann. Statist.* **28**, 1427–1451.

[32] Pakes, A. G., (2004), Convolution equivalence and infinite divisibility, *J. Appl. Probab.* **41** (2), 407–424.

[33] Rootzén, H., (1986), Extreme value theory for moving average processes, *Ann. Probab.* **14** (2), 612–652.

[34] Rosinski, J. and Samorodnitsky, G., (1993), Distributions of subadditive functionals of sample paths of infinitely divisible processes, *Ann. Probab.* **21** (2), 996–1014.

[35] Sato, K., (1999), *Lévy Processes and Infinitely Divisible Distributions*, Cambridge University Press, Cambridge.

[36] Vervaat, W., (1979), On a stochastic difference equation and a representation of non-negative infinitely divisible random variables, *Adv. Appl. Probability* **11**, 750–783.

5

Capital Asset Pricing for Markets with Intensity Based Jumps

Eckhard Platen

University of Technology Sydney, School of Finance & Economics and Department of Mathematical Sciences, PO Box 123, Broadway, NSW, 2007, Australia
eckhard.platen@uts.edu.au

Summary. This paper proposes a unified framework for portfolio optimization, derivative pricing, modeling and risk measurement in financial markets with security price processes that exhibit intensity based jumps. It is based on the natural assumption that investors prefer more for less, in the sense that for two given portfolios with the same variance of its increments, the one with the higher expected increment is preferred. If one additionally assumes that the market together with its monetary authority acts to maximize the long term growth of the market portfolio, then this portfolio exhibits a very particular dynamics. In a market without jumps the resulting dynamics equals that of the growth optimal portfolio (GOP). Conditions are formulated under which the well-known capital asset pricing model is generalized for markets with intensity based jumps. Furthermore, the Markowitz efficient frontier and the Sharpe ratio are recovered in this continuous time setting. In this paper the numeraire for derivative pricing is chosen to be the GOP. Primary security account prices, when expressed in units of the GOP, turn out to be supermartingales. In the proposed framework an equivalent risk neutral martingale measure need not exist. Fair derivative prices are obtained as conditional expectations of future payoff structures under the real world probability measure. The concept of fair pricing is shown to generalize the classical risk neutral and the actuarial net present value pricing methodologies.

Key words: benchmark model, jump diffusions, growth optimal portfolio, market portfolio, efficient frontier, Sharpe ratio, fair pricing, actuarial pricing.
1991 *Mathematics Subject Classification:* primary 90A12; secondary 60G30, 62P20.
JEL Classification: G10, G13.

5.1 Introduction

This paper proposes an integrated approach that can be applied to portfolio optimization, credit risk and derivative and insurance pricing. It uses the *growth optimal portfolio* (GOP) as the benchmark or reference unit and establishes a class of benchmark models with intensity based jumps. In the case of diffusions without jumps, Long [23] and Bajeux-Besnainou & Portait [1] introduced the GOP, first considered in Kelly [20] as the *numeraire portfolio.* It allows for the pricing of derivatives under the real world probability measure.

Under the standard risk neutral approach a major problem arises in modeling credit and insurance risk due to the difficulty in choosing an appropriate equivalent risk neutral pricing measure. Furthermore, actuarial approaches have focused over centuries on the modeling and pricing of insurance risk under the real world probability measure, see for instance, Gerber [13] and Bühlmann [2]. On the other hand, in risk management and investment management the quantitative methods rely on the real world probability measure. It has been shown in Platen [29] and Heath & Platen [16, 17, 18] that there exist reasonable market models that cannot be treated under the classical risk neutral approach. The currently very topical subject of credit risk provides an interesting case study on the conflict inherent in pricing under a risk neutral measure, while calculating risk statistics under the real world measure. Here the question of how to reconcile real world probabilities of default with credit spreads, which are often interpreted via risk neutral default probabilities, has become a technical minefield, see Duffie & Singleton [11]. Challenges are therefore emerging from the need to have a consistent approach to the modeling of continuous and event driven risk in the combined fields of finance and insurance.

In Markowitz [24] the mean-variance portfolio theory with its well-known *efficient frontier* was introduced. This led to the *capital asset pricing model* (CAPM), see Sharpe [38], Lintner [22] and Merton [25]. The CAPM is based on the *market portfolio* as reference unit and represents an equilibrium model of exchange. In a continuous time setting Merton [25] derived the *intertemporal* CAPM from the portfolio selection behavior of investors who maximize equilibrium expected utility. The current paper aims to avoid equilibrium and utility based arguments in deriving the CAPM for a jump diffusion market. It generalizes fundamental results on the Markowitz efficient frontier as well as the CAPM and the Sharpe ratio.

In this paper we construct a class of benchmark models, see Platen [29, 30], for security prices that follow *diffusions with intensity based jumps.* One can refer to a wide range of literature on derivative pricing for jump diffusions, starting with Merton [26] and leading to a considerable variety of papers and monographs, see, for instance, Cont & Tankov [7]. We will avoid the standard assumption on the existence of an equivalent risk neutral martingale measure. In this way important freedom is gained for financial modeling, as has become clear, for instance, in Heath & Platen [16, 17, 18].

As in Long [23], where prices denominated in units of the GOP are martingales, the fair pricing concept, which we advocate in this paper defines benchmarked fair derivative price processes as martingales under the real world probability measure. Therefore, derivative prices can be obtained as conditional expectations of future benchmarked prices without any measure transformation. We will show that fair prices coincide with the corresponding risk neutral prices if an equivalent risk neutral martingale measure exists. A natural generalization of the standard risk neutral framework is therefore obtained by fair pricing under the benchmark approach. Also, the classical actuarial pricing methodology turns out to be a particular case of fair pricing when the payoff is independent of the GOP.

Section 5.2 introduces a class of benchmark models with intensity based jumps. A portfolio choice theorem is presented in Section 5.3. Capital asset pricing with jumps is considered in Section 5.4. Fair contingent claim pricing is studied in Section 5.5. Finally, the evolution of the expectation of the market portfolio is analyzed in Section 5.6.

5.2 Benchmark Model with Jumps

5.2.1 Continuous and Event Driven Uncertainty

We consider a market containing *continuously evolving uncertainty* represented by m independent standard Wiener processes $W^k = \{W_t^k,\, t \in [0,T]\}$, $k \in \{1,2,\ldots,m\}$, $m \in \{1,2,\ldots,d\}$, $d \in \{1,2,\ldots\}$. These are defined on a filtered probability space $(\Omega, \mathcal{A}_T, \underline{\mathcal{A}}, P)$ with finite time horizon $T \in (0,\infty)$. We also consider events of certain types, for instance, corporate defaults, operational failures or specified insured events that are reflected in the movements of traded securities. Events of the kth type are counted by the $\underline{\mathcal{A}}$-adapted kth *counting process* $p^k = \{p_t^k,\, t \in [0,T]\}$, whose *intensity* $h^k = \{h_t^k,\, t \in [0,T]\}$ is a given predictable, strictly positive process with

$$h_t^k > 0 \tag{5.1}$$

and

$$\int_0^T h_s^k \, ds < \infty \,, \tag{5.2}$$

almost surely for $t \in [0,T]$ and $k \in \{m+1,\ldots,d\}$. Furthermore, we introduce the kth *jump martingale* $W^k = \{W_t^k,\, t \in [0,T]\}$ with stochastic differential

$$dW_t^k = \left(dp_t^k - h_t^k \, dt\right) \left(h_t^k\right)^{-\frac{1}{2}} \,, \tag{5.3}$$

for $k \in \{m+1,\ldots,d\}$ and $t \in [0,T]$. It is assumed that the above jump martingales do not jump at the same time. They represent the compensated and normalized sources of *event driven uncertainty*.

Let us denote by $A^\top$ the transpose of a vector or matrix A. The evolution of traded uncertainty is modeled by the vector process of independent $(\underline{\mathcal{A}}, P)$-martingales $W = \{W_t = (W_t^1,\ldots,W_t^d)^\top,\, t \in [0,T]\}$. Note that $W^1, \ldots, W^m$ are Wiener processes, while $W^{m+1}, \ldots, W^d$ are compensated normalized counting processes. The filtration $\underline{\mathcal{A}} = (\mathcal{A}_t)_{t\in[0,T]}$ is the augmentation under P of the natural filtration $\mathcal{A}^W$ generated by the vector process W. It satisfies the usual conditions and $\mathcal{A}_0$ is the trivial σ-algebra, see Protter [34]. Note that the conditional variance of the kth source of uncertainty is

$$E\left(\left(W_{t+\varepsilon}^k - W_t^k\right)^2 \Big| \mathcal{A}_t\right) = \varepsilon\,, \tag{5.4}$$

for all $t \in [0,T]$, $k \in \{1,2,\ldots,d\}$ and $\varepsilon \in [0, T-t]$.

5.2.2 Primary Security Accounts

A primary security account is a particular investment account, consisting only of one kind of security, with all proceeds reinvested. For the securitization of the d sources of uncertainty, let us introduce d *risky primary security accounts*, whose values at time t are denoted by $S_t^{(j)}$, for $j \in \{1,2,\ldots,d\}$. Each of these contains shares of some kind. These security accounts represent the evolution of wealth due to the ownership of assets, with all dividends or income reinvested. The 0th primary security account $S^{(0)} = \{S_t^{(0)},\, t \in [0,T]\}$ is the riskless *savings account*, which continuously accrues the short term interest rate r_t. In this case the underlying asset is the domestic currency.

Without loss of generality we assume that the nonnegative jth primary security account value $S_t^{(j)}$ satisfies the stochastic differential equation (SDE)

$$dS_t^{(j)} = S_{t-}^{(j)} \left(a^j(t)\, dt + \sum_{k=1}^{d} b^{j,k}(t)\, dW_t^k \right), \tag{5.5}$$

for $t \in [0,T]$ with initial value $S_0^{(j)} > 0$ and $j \in \{0,1,\ldots,d\}$, see Protter [34]. Since $S_t^{(0)}$ is the savings account, we have

$$a^0(t) = r_t \tag{5.6}$$

and

$$b^{0,k}(t) = 0\,, \tag{5.7}$$

for $t \in [0,T]$ and $k \in \{1,2,\ldots,d\}$. One may interpret a roll-over treasure bill account as a suitable proxy for the savings account. We assume that the processes r, a^j, $b^{j,k}$ and h^k are finite and predictable, and that a unique strong solution for the system of SDEs (5.5) exists, see Protter [34]. To ensure nonnegativity for each primary security account we assume that

$$b^{j,k}(t) \geq -\sqrt{h_t^k}\,, \tag{5.8}$$

for all $t \in [0,T]$, $j \in \{1,2,\ldots,d\}$ and $k \in \{m+1, m+2, \ldots, d\}$.

To securitize the sources of uncertainty properly, we make the following assumption.

Assumption 1. *The* generalized volatility matrix $b(t) = [b^{j,k}(t)]_{j,k=1}^d$ *is* invertible *for Lebesgue-almost-every* $t \in [0,T]$.

Assumption 1 allows us to introduce the *market price for risk* vector

$$\theta(t) = (\theta^1(t), \ldots, \theta^d(t))^\top = b^{-1}(t)\,[a(t) - r_t\,\mathbf{1}], \tag{5.9}$$

for $t \in [0, T]$. Here $a(t) = (a^1(t), \ldots, a^d(t))^\top$ is the *appreciation rate vector* and $\mathbf{1} = (1, \ldots, 1)^\top$ the *unit vector*. Using (5.9), we can rewrite the SDE (5.5) in the form

$$dS_t^{(j)} = S_{t-}^{(j)} \left(r_t\,dt + \sum_{k=1}^{d} b^{j,k}(t)\,(\theta^k(t)\,dt + dW_t^k) \right), \tag{5.10}$$

for $t \in [0, T]$ and $j \in \{0, 1, \ldots, d\}$. For $k \in \{1, 2, \ldots, m\}$, the quantity $\theta^k(t)$ expresses the *market price for risk* with respect to the kth Wiener process W^k. If $k \in \{m+1, \ldots, d\}$, then $\theta^k(t)$ can be interpreted as the *market price for kth event risk*. We will see later that the market prices for risk play a central role in our modeling framework, and that one needs a further condition on the market prices for event risk to avoid arbitrage.

The vector process $S = \{S_t = (S_t^{(0)}, \ldots, S_t^{(d)})^\top,\ t \in [0, T]\}$ characterizes the evolution of all primary security accounts. We say that a predictable stochastic process $\delta = \{\delta(t) = (\delta^0(t), \ldots, \delta^d(t))^\top,\ t \in [0, T]\}$ is a *strategy* if it is S-integrable, see Protter [34]. The jth component of δ denotes the number of units of the jth primary security account held at time $t \in [0, T]$ in a portfolio, $j \in \{0, 1, \ldots, d\}$. For a strategy δ we denote by $S_t^{(\delta)}$ the value of the corresponding portfolio process at time t, when measured in units of the domestic currency. Thus, we set

$$S_t^{(\delta)} = \sum_{j=0}^{d} \delta^j(t)\,S_t^{(j)}, \tag{5.11}$$

for $t \in [0, T]$.

Definition 1. *A strategy δ and the corresponding portfolio process $S^{(\delta)} = \{S_t^{(\delta)},\ t \in [0, T]\}$ are called* self-financing *if*

$$dS_t^{(\delta)} = \sum_{j=0}^{d} \delta^j(t)\,dS_t^{(j)}, \tag{5.12}$$

for all $t \in [0, T]$.

All changes in value of a self-financing portfolio process are due to changes in value of underlying primary security accounts. In what follows we will only consider self-financing portfolios. Therefore, from now on we omit the phrase "self-financing".

5.2.3 Growth Optimal Portfolio

For a given strategy δ with strictly positive portfolio process $S_t^{(\delta)}$ let $\pi_\delta^j(t)$ denote the *fraction* of wealth that is invested in the jth primary security account at time t. It is defined by the relation

$$\pi_\delta^j(t) = \delta^j(t)\,\frac{S_t^{(j)}}{S_t^{(\delta)}}\,, \tag{5.13}$$

for $t \in [0,T]$ and $j \in \{0,1,\dots,d\}$. Furthermore, by (5.11) these fractions always add to one. That is

$$\sum_{j=0}^{d} \pi_\delta^j(t) = 1\,, \tag{5.14}$$

for $t \in [0,T]$. In terms of the vector of fractions $\pi_\delta(t) = (\pi_\delta^1(t),\dots,\ \pi_\delta^d(t))^\top$ we obtain for $S_t^{(\delta)}$ from (5.12), (5.10), (5.7) and (5.13) the SDE

$$dS_t^{(\delta)} = S_{t-}^{(\delta)}\left\{r_t\,dt + \pi_\delta(t-)^\top\,b(t)\,(\theta(t)\,dt + dW_t)\right\}\,, \tag{5.15}$$

for $t \in [0,T]$. Note that a portfolio process $S^{(\delta)}$ remains strictly positive if and only if

$$\sum_{j=1}^{d} \pi_\delta^j(t)\,b^{j,k}(t)\,\left(h_t^k\right)^{-\frac{1}{2}} > -1\,, \tag{5.16}$$

a.s. for all $k \in \{m+1, m+2,\dots,d\}$ and $t \in [0,T]$. For a strictly positive portfolio process $S^{(\delta)}$ we obtain by an application of Itô's formula the following SDE for its logarithm

$$\begin{aligned} d\ln(S_t^{(\delta)}) &= g_\delta(t)\,dt + \sum_{k=1}^{m}\sum_{j=1}^{d} \pi_\delta^j(t)\,b^{j,k}(t)\,dW_t^k + \\ &\quad + \sum_{k=m+1}^{d} \ln\left(1 + \sum_{j=1}^{d} \pi_\delta^j(t-)\,\frac{b^{j,k}(t)}{\sqrt{h_t^k}}\right)\sqrt{h_t^k}\,dW_t^k\,, \end{aligned} \tag{5.17}$$

for $t \in [0,T]$. The *growth rate* in this expression is given by

$$\begin{aligned} g_\delta(t) &= r_t + \sum_{k=1}^{m}\left[\sum_{j=1}^{d} \pi_\delta^j(t)\,b^{j,k}(t)\,\theta^k(t) - \frac{1}{2}\left(\sum_{j=1}^{d} \pi_\delta^j(t)\,b^{j,k}(t)\right)^2\right] + \\ &+ \sum_{k=m+1}^{d}\left[\sum_{j=1}^{d} \pi_\delta^j(t)\,b^{j,k}(t)\left(\theta^k(t) - \sqrt{h_t^k}\right) + \ln\left(1 + \sum_{j=1}^{d} \pi_\delta^j(t)\,\frac{b^{j,k}(t)}{\sqrt{h_t^k}}\right)h_t^k\right], \end{aligned} \tag{5.18}$$

for $t \in [0,T]$. Note that for the first sum on the right hand side of (5.18) a unique maximum naturally exists because it has a quadratic form with respect to the fractions. Careful inspection of the terms in the second sum reveals that, in general, a unique maximum growth rate only exists if the market prices of event risks are less than the square roots of the corresponding jump intensities. This leads to the following assumption.

Assumption 2. *Assume that*

$$\sqrt{h_t^k} > \theta^k(t)\,, \tag{5.19}$$

for $t \in [0,T]$ *and* $k \in \{m+1,\dots,d\}$.

Assumption 2 guarantees that the market is arbitrage free in the sense of Platen [30]. Furthermore, it allows us to introduce the predictable vector process $c(t) = (c^1(t),\dots, c^d(t))^\top$ with components

$$c^k(t) = \begin{cases} \theta^k(t) & \text{for} \quad k \in \{1,2,\dots,m\}\,, \\ \frac{\theta^k(t)}{1-\theta^k(t)\,(h_t^k)^{-\frac{1}{2}}} & \text{for} \quad k \in \{m+1,\dots,d\}\,, \end{cases} \tag{5.20}$$

for $t \in [0,T]$. Note that a divergent jump intensity, with $h_t^k \to \infty$ a.s. for any $t \in [0,T]$ and $k \in \{m+1,\dots,d\}$, causes the corresponding component $c^k(t)$ to approach the market price for jump risk $\theta^k(t)$ asymptotically. In this case the component is similar to the market price for risk with respect to a Wiener process.

We now define the fractions of a portfolio $S^{(\delta_*)}$ by the relation

$$\pi_{\delta_*}(t) = (\pi^1_{\delta_*}(t),\dots,\pi^d_{\delta_*}(t))^\top = \left(c(t)^\top\, b^{-1}(t)\right)^\top, \tag{5.21}$$

for $t \in [0,T]$. By (5.15) and (5.20) it follows that $S_t^{(\delta_*)}$ satisfies the SDE

$$\begin{aligned} dS_t^{(\delta_*)} &= S_{t-}^{(\delta_*)}\Big(r_t\,dt + c(t)^\top\,(\theta(t)\,dt + dW_t)\Big) \\ &= S_{t-}^{(\delta_*)}\left(r_t\,dt + \sum_{k=1}^{m} \theta^k(t)\,(\theta^k(t)\,dt + dW_t^k) + \right. \\ &\quad \left. + \sum_{k=m+1}^{d} \frac{\theta^k(t)}{1-\theta^k(t)\,(h_t^k)^{-\frac{1}{2}}}\,(\theta^k(t)\,dt + dW_t^k)\right), \end{aligned} \tag{5.22}$$

for $t \in [0,T]$, with $S_0^{(\delta_*)} > 0$. Note from (5.22) that Assumption 2 keeps the portfolio process $S^{(\delta_*)}$ strictly positive. Let us now define a *growth optimal portfolio* (GOP).

Definition 2. *A portfolio process that maximizes the growth rate* (5.18) *among all positive portfolio processes is called a* GOP.

There is an increasing literature on the GOP and other related diversified portfolios. We refer the interested reader to Korn & Schäl [21] and Platen [29, 31, 32] for recent information on this topic. The following corollary is a consequence of results in Platen [30].

Corollary 1. *Under Assumptions 1 and 2 the portfolio process* $S^{(\delta_*)} = \{S_t^{(\delta_*)},\, t \in [0,T]\}$ *satisfying* (5.22), *with wealth fractions given by* (5.21), *is a GOP. Furthermore, for any given fixed initial value* $S_0^{(\delta_*)} > 0$, *the GOP is uniquely determined.*

5.2.4 Benchmark Model with Intensity Based Jumps

We use the GOP $S^{(\delta_*)}$ as benchmark or numeraire, and call prices expressed in units of $S^{(\delta_*)}$ *benchmarked prices*. By the Itô formula, (5.15) and (5.22), a *benchmarked portfolio process* $\hat{S}^{(\delta)} = \{\hat{S}_t^{(\delta)},\, t \in [0,T]\}$, with

$$\hat{S}_t^{(\delta)} = \frac{S_t^{(\delta)}}{S_t^{(\delta_*)}}\,, \tag{5.23}$$

for $t \in [0,T]$, satisfies the SDE

$$\begin{aligned} d\hat{S}_t^{(\delta)} = \hat{S}_{t-}^{(\delta)} \Bigg(& \sum_{k=1}^{m} \left\{ \sum_{j=1}^{d} \pi_\delta^j(t)\, b^{j,k}(t) - \theta^k(t) \right\} dW_t^k \, + \\ & + \sum_{k=m+1}^{d} \left\{ \left(\sum_{j=1}^{d} \pi_\delta^j(t-)\, b^{j,k}(t) \right) \left(1 - \frac{\theta^k(t)}{\sqrt{h_t^k}} \right) - \theta^k(t) \right\} dW_t^k \Bigg), \end{aligned} \tag{5.24}$$

for $t \in [0,T]$. To obtain a simpler form of the above SDE we write $\sigma^{0,k}(t)$ instead of $\theta^k(t)$, for $t \in [0,T]$ and $k \in \{1,2,\dots,d\}$. Now, define the matrix process $\sigma = \{\sigma(t) = [\sigma^{j,k}(t)]_{j,k=0,1}^d,\, t \in [0,T]\}$ by setting

$$\sigma^{0,k}(t) = \theta^k(t) \tag{5.25}$$

and

$$\sigma^{j,k}(t) = \begin{cases} \sigma^{0,k}(t) - b^{j,k}(t) & \text{for} \quad k \in \{1,2,\dots,m\}\,, \\ \sigma^{0,k}(t) - b^{j,k}(t) \left(1 - \frac{\sigma^{0,k}(t)}{\sqrt{h_t^k}} \right) & \text{for} \quad k \in \{m+1,\dots,d\}\,, \end{cases} \tag{5.26}$$

for $t \in [0,T]$ and $j \in \{1,2,\dots,d\}$. Using (5.9) and (5.25) one can then rewrite (5.24) as

$$d\hat{S}_t^{(\delta)} = -\hat{S}_{t-}^{(\delta)} \sum_{k=1}^{d} \sum_{j=0}^{d} \pi_\delta^j(t-)\, \sigma^{j,k}(t)\, dW_t^k\,, \tag{5.27}$$

for $t \in [0,T]$. This SDE governs the dynamics of any benchmarked portfolio.

Note that the right hand side of (5.27) is driftless. Thus, a nonnegative benchmarked portfolio $\widehat{S}^{(\delta)}$ forms an $(\underline{\mathcal{A}}, P)$-local martingale. This also means that a nonnegative benchmarked portfolio process $\widehat{S}^{(\delta)}$ is always an $(\underline{\mathcal{A}}, P)$-supermartingale, see Rogers & Williams [36], that is

$$\widehat{S}_t^{(\delta)} \geq E\left(\widehat{S}_\tau^{(\delta)} \,\Big|\, \mathcal{A}_t\right), \tag{5.28}$$

for all $\tau \in [0,T]$ and $t \in [0,\tau]$. One can show that whenever a nonnegative supermartingale reaches the value zero it almost surely remains at zero. Based on this observation the above model is arbitrage free in the sense of Platen [29, 30]. We call a model of the form prescribed above, which is based on the Assumptions 1 and 2, a *benchmark model with intensity based jumps.* This notion acknowledges the fact that for this model the GOP exists and is used as the benchmark. A generalization of the benchmark model for event driven risk with respect to Poisson jump measures is given in Christensen & Platen [4].

5.3 Maximizing the Portfolio Drift

5.3.1 Optimal Portfolios

Given a strictly positive portfolio $S^{(\delta)}$, its discounted value

$$\bar{S}_t^{(\delta)} = \frac{S_t^{(\delta)}}{S_t^{(0)}} \tag{5.29}$$

satisfies the SDE

$$d\bar{S}_t^{(\delta)} = \sum_{k=1}^{d} \psi_\delta^k(t)\left(\theta^k(t)\,dt + dW_t^k\right) \tag{5.30}$$

by (5.15) and an application of the Itô formula. Here

$$\psi_\delta^k(t) = \bar{S}_{t-}^{(\delta)} \sum_{j=1}^{d} \pi_\delta^j(t-)\, b^{j,k}(t), \tag{5.31}$$

for $k \in \{1,2,\ldots,d\}$ and $t \in [0,T]$, is called the kth *generalized portfolio diffusion coefficient.* Obviously, by (5.30) and (5.31) the discounted portfolio process $\bar{S}^{(\delta)}$ has *portfolio drift*

$$\alpha_\delta(t) = \sum_{k=1}^{d} \psi_\delta^k(t)\,\theta^k(t), \tag{5.32}$$

for $t \in [0,T]$.

This drift measures the portfolio's time varying trend. The uncertainty of a discounted portfolio $\bar{S}^{(\delta)}$ can be measured by its *aggregate generalized diffusion coefficient*

$$\gamma_\delta(t) = \sqrt{\sum_{k=1}^{d} \left(\psi_\delta^k(t)\right)^2}\,, \tag{5.33}$$

at time $t \in [0,T]$. Note that by (5.4) we have normalized variances of increments of the driving martingales $W^1, W^2, \ldots, W^d$.

For a given instantaneous level of the aggregate generalized diffusion coefficient $\gamma_\delta(t)$, any rational investor who prefers more for less can be assumed to aim to maximize the portfolio drift $\alpha_\delta(t)$. Building on the seminal works by Markowitz [24] and Sharpe [38] we now aim to capture this objective mathematically for the given benchmark model. More precisely, it is our aim to identify the typical structure of the SDEs for the total portfolios of investors who prefer more for less in the following sense:

Definition 3. *A strictly positive portfolio process that maximizes the portfolio drift* (5.32) *among all strictly positive portfolio processes with the same aggregate generalized diffusion coefficient* (5.33) *is called* optimal.

For the following analysis let us introduce the *total market price for risk*,

$$|\theta(t)| = \sqrt{\sum_{k=1}^{d} \left(\theta^k(t)\right)^2}\,, \tag{5.34}$$

and the weighting factor

$$\Gamma^{(0)}(t) = \sum_{k=1}^{d}\sum_{j=1}^{d} \theta^k(t)\, b^{-1\,j,k}(t)\,, \tag{5.35}$$

for $t \in [0,T]$. As we will see later, if the total market price for risk or the weighting factor are zero, then the savings account is the optimal portfolio that an investor would naturally prefer. The following natural assumption excludes this trivial case.

Assumption 3. *Assume that*

$$0 < |\theta(t)| < \infty\,, \tag{5.36}$$

and

$$\Gamma^{(0)}(t) \neq 0 \tag{5.37}$$

almost surely for all $t \in [0,T]$.

We can now formulate a *portfolio choice theorem* in the sense of Markowitz [24], which generalizes a result in Platen [29] for continuous benchmark models. The following theorem identifies the structure of the drift and generalized diffusion coefficients in the SDE of an optimal portfolio.

Theorem C.1. *Any discounted optimal portfolio $\bar{S}^{(\delta)}$ satisfies the SDE*

$$d\bar{S}_t^{(\delta)} = \bar{S}_t^{(\delta)} \frac{\left(1 - \pi_\delta^0(t)\right)}{\Gamma^{(0)}(t)} \sum_{k=1}^{d} \theta^k(t) \left(\theta^k(t)\, dt + dW_t^k\right), \tag{5.38}$$

with optimal fractions

$$\pi_\delta^j(t) = \frac{\left(1 - \pi_\delta^0(t)\right)}{\Gamma^{(0)}(t)} \sum_{k=1}^{d} \theta^k(t)\, b^{-1\,j,k}(t), \tag{5.39}$$

for all $t \in [0,T]$ and $j \in \{1,2,\dots,d\}$.

This means that the family of discounted optimal portfolios is characterized by a single parameter, namely the fraction of wealth $\pi_\delta^0(t)$ held in the savings account. The proof of this theorem is given in the Appendix.

We obtain a particular optimal portfolio $S^{(\delta_+)}$, which we call the *mutual fund*, by choosing

$$\pi_{\delta_+}^0(t) = 1 - \Gamma^{(0)}(t), \tag{5.40}$$

for $t \in [0,T]$. By (5.38) the mutual fund satisfies the SDE

$$dS_t^{(\delta_+)} = S_t^{(\delta_+)} \left(r_t\, dt + \sum_{k=1}^{d} \theta^k(t) \left(\theta^k(t)\, dt + dW_t^k\right) \right), \tag{5.41}$$

for $t \in [0,T]$. This portfolio plays an important role in the remainder of the paper.

By Theorem C.1 it follows that any efficient portfolio $S^{(\delta)}$ can be decomposed at any time into a fraction that is invested in the mutual fund $S^{(\delta_+)}$ and a remaining fraction that is held in the savings account. Therefore, Theorem C.1 can also be interpreted as a *mutual fund theorem* or *separation theorem*, see Merton [25]. We emphasize that the assumptions of Theorem C.1 are rather weak and also realistic.

5.4 Capital Asset Pricing with Jumps

5.4.1 Markowitz Efficient Frontier and Sharpe Ratio

For a portfolio $S^{(\delta)}$ we introduce its *aggregate generalized volatility*

$$b_\delta(t) = \sqrt{\sum_{k=1}^{d} \left(\sum_{j=1}^{d} \pi_\delta^j(t)\, b^{j,k}(t) \right)^2} > 0 \tag{5.42}$$

and its *appreciation rate*

$$a_\delta(t) = r_t + \sum_{k=1}^{d} \sum_{j=1}^{d} \pi_\delta^j(t)\, b^{j,k}(t)\, \theta^k(t)\,, \tag{5.43}$$

for all $t \in [0, T]$, by inspection of (5.15). If $S^{(\delta)}$ is in fact optimal, then it follows by the Itô formula, (5.38) and (5.34) that

$$b_\delta(t) = \frac{\left(1 - \pi_\delta^0(t)\right)}{\Gamma^{(0)}(t)}\, |\theta(t)|\,, \tag{5.44}$$

and

$$a_\delta(t) = r_t + b_\delta(t)\, |\theta(t)|\,, \tag{5.45}$$

for $t \in [0, T]$. By analogy to the single-period mean-variance portfolio theory, developed in Markowitz [24], we introduce the notion of an *efficient frontier.*

Definition 4. *A portfolio $S^{(\delta)}$ is said to lie on the efficient frontier if its appreciation rate $a_\delta(t)$, as a function of squared aggregate generalized volatility $(b_\delta(t))^2$, is of the form*

$$a_\delta(t) = a_\delta(t, (b_\delta(t))^2) = r_t + \sqrt{(b_\delta(t))^2}\; |\theta(t)|\,, \tag{5.46}$$

for all times $t \in [0, T]$.

By relations (5.44), (5.45) and (5.46) the following result is directly obtained.

Corollary 2. *An optimal portfolio is always located on the efficient frontier.*

Corollary 2 can be interpreted as a "local in time" generalization of the seminal Markowitz efficient frontier to the jump diffusion setting. Note that due to Definition 3 and Theorem C.1 it is not possible to form a positive portfolio that produces an appreciation rate located above the efficient frontier. The best that an investor can do, when searching for the maximum drift while maintaining a given generalized diffusion coefficient, is to form a portfolio on the efficient frontier. The only remaining freedom is choosing the fraction of wealth that resides in the savings account. This fraction expresses the investor's *degree of risk aversion.* Note that this approach is more general than expected utility maximization, where the risk aversion at a certain time is indirectly specified via the chosen utility function and the given time horizon. In forthcoming work it will be shown how the above concept of optimal portfolios generalizes that of expected utility maximization.

5.4.2 Sharpe Ratio

For a portfolio $S^{(\delta)}$ the notation introduced in (5.42), (5.43), (5.32) and (5.33) leads to another important investment characteristic, the *Sharpe ratio* $s_\delta(t)$, which is defined by

$$s_\delta(t) = \frac{\alpha_\delta(t)}{\gamma_\delta(t)} = \frac{a_\delta(t) - r_t}{b_\delta(t)}, \tag{5.47}$$

for $t \in [0, T]$, see Sharpe [38]. By (5.44), (5.45), (5.47) and Theorem C.1 we obtain the following practically important result.

Corollary 3. *The maximum Sharpe ratio is obtained by optimal portfolios and equals the total market price for risk. For all strictly positive portfolios $S^{(\delta)}$, one has*

$$s_\delta(t) \leq |\theta(t)|, \tag{5.48}$$

for all $t \in [0, T]$, with equality attained in (5.48) *when $S^{(\delta)}$ is optimal.*

The Markowitz efficient frontier and the Sharpe ratio are fundamental tools for investment management, whose natural meaning are preserved in the benchmark model with intensity based jumps. Note that we have not specified any particular dynamics for the stochastic quantities involved. In this sense the benchmark model presented so far provides a general jump diffusion framework for modeling event driven risk.

5.4.3 Capital Asset Pricing Model

Let us define the *market portfolio* $S^{(\delta_M)}$ as the portfolio consisting of all primary security accounts weighted according to market capitalization. The seminal *capital asset pricing model* (CAPM) was developed by Sharpe [38], Lintner [22] and Merton [25] as a utility based equilibrium model of exchange with the market portfolio $S^{(\delta_M)}$ as reference unit. As we will demonstrate below we do not need to use any equilibrium or expected utility function arguments for generalizing the CAPM to the case of a continuous benchmark model with intensity based jumps. As in the classical CAPM, one can introduce the systematic risk parameter $\beta_\delta(t)$, called the *portfolio beta*, for a portfolio $S^{(\delta)}$. It is here defined as the ratio of the time derivative of the conditional covariation of the logarithms of the portfolio and the market portfolio over the time derivative of the conditional variance of the logarithm of the market portfolio, that is,

$$\beta_\delta(t) = \frac{\frac{d}{dt}\langle \ln(S^{(\delta)}), \ln(S^{(\delta_M)})\rangle_t}{\frac{d}{dt}\langle \ln(S^{(\delta_M)})\rangle_t}, \tag{5.49}$$

for all $t \in [0,T]$.

By (5.43) the instantaneous *risk premium* $p_\delta(t)$ of a portfolio $S^{(\delta)}$ is given by the expression

$$p_\delta(t) = a_\delta(t) - r_t = \sum_{k=1}^{d}\sum_{j=1}^{d} \pi_\delta^j(t)\, b^{j,k}(t)\, \theta^k(t), \tag{5.50}$$

for all $t \in [0,T]$. Note that by (5.15) and (5.41) the risk premium equals the covariance of the returns of the mutual fund with those of the portfolio.

As described in the literature, the CAPM states that the portfolio beta $\beta_\delta(t)$ equals the ratio of the portfolio risk premium over the market portfolio risk premium $\frac{p_\delta(t)}{p_{\delta_M}(t)}$, see Merton [25]. However, we note from (5.41) by using the mutual fund as reference unit that this fundamental CAPM relationship holds true when one uses the mutual fund as reference unit instead of the market portfolio, that is,

$$\frac{\frac{d}{dt}\langle \ln(S^{(\delta)}), \ln(S^{(\delta_+)})\rangle_t}{\frac{d}{dt}\langle \ln(S^{(\delta_+)})\rangle_t} = \frac{\sum_{k=1}^d \sum_{j=1}^d \pi_\delta^j(t)\, b^{j,k}(t)\, \theta^k(t)}{\sum_{k=1}^d |\theta^k(t)|^2} = \frac{p_\delta(t)}{p_{\delta_+}(t)}, \tag{5.51}$$

for all $t \in [0,T]$.

The form of the portfolio risk premium (5.50) and the portfolio beta (5.51) are exactly what the intertemporal CAPM suggests if the market portfolio equals the mutual fund. In what follows we will identify conditions which ensure that the market portfolio equals the mutual fund. This provides a basis for the derivation of the CAPM in the presence of intensity based jumps. The CAPM then arises purely out of the natural structure of a benchmark model with intensity based jumps. We make the following natural assumption.

Assumption 4. *Each market participant constructs an optimal portfolio with his or her total available wealth.*

This assumption essentially means that all investors are informed and prefer more for less. The total portfolio of the ℓth market participant, which is optimal by Assumption 4, is denoted by $S^{(\delta_\ell)}$, $\ell \in \{1,2,\ldots,n\}$. The portfolio $S_t^{(\delta_M)}$ of all market participants is the market portfolio at time t with value given by the sum

$$S_t^{(\delta_M)} = \sum_{\ell=1}^{n} S_t^{(\delta_\ell)}, \tag{5.52}$$

for all $t \in [0,T]$. It is reasonable to assume a strictly positive market portfolio. The dynamics of the discounted market portfolio $\bar{S}_t^{(\delta_M)} = \frac{S_t^{(\delta_M)}}{S_t^{(0)}}$ is characterized by the SDE

$$\begin{aligned} d\bar{S}_t^{(\delta_M)} &= \sum_{\ell=1}^{n} d\bar{S}_t^{(\delta_\ell)} \\ &= \sum_{\ell=1}^{n} \frac{\left(\bar{S}_t^{(\delta_\ell)} - \delta_\ell^0\right)}{\Gamma^{(0)}(t)} \sum_{k=1}^{d} \theta^k(t)\left(\theta^k(t)\,dt + dW_t^k\right) \\ &= \bar{S}_t^{(\delta_M)} \frac{\left(1 - \pi^0_{\delta_M}(t)\right)}{\Gamma^{(0)}(t)} \sum_{k=1}^{d} \theta^k(t)\left(\theta^k(t)\,dt + dW_t^k\right), \end{aligned} \tag{5.53}$$

for $t \in [0,T]$, by Theorem C.1 and (5.52). Thus, by Definition 3 and Theorem C.1 one can show that the market portfolio $S_t^{(\delta_M)}$ is optimal. By (5.41) the fraction $\frac{1-\pi^0_{\delta_M}(t)}{\Gamma^{(0)}(t)}$ of the market portfolio is invested in the mutual fund $S^{(\delta_+)}$ at time $t \in [0,T]$, and the fraction invested in the domestic savings account $S^{(0)}$ is given by the expression $\frac{\pi^0_{\delta_M}(t)-1+\Gamma^{(0)}(t)}{\Gamma^{(0)}(t)}$. By comparing the SDEs (5.41) and (5.53) we obtain the following result.

Corollary 4. *Given equal initial values* $S_0^{(\delta_M)} = S_0^{(\delta_+)}$, *the value of the market portfolio* $S_t^{(\delta_M)}$ *equals the value of the mutual fund* $S_t^{(\delta_+)}$ *at all times* $t \in [0,T]$ *if and only if their fractions held in the savings account are equal, that is*

$$\pi^0_{\delta_M}(t) = \pi^0_{\delta_+}(t) = 1 - \Gamma^{(0)}(t), \tag{5.54}$$

for all $t \in [0,T]$.

By (5.51), this leads to the following conclusion, which provides a generalization of the intertemporal CAPM of Merton [25] to the case of jump diffusion markets.

Corollary 5. *As long as condition* (5.54) *is satisfied, a generalized intertemporal CAPM holds. That is for any given portfolio its systematic risk parameter, given by* (5.51), *captures the ratio of the risk premium of the portfolio over that of the market portfolio.*

This means that the well-known CAPM holds under a benchmark model with intensity based jumps if the fraction invested in the savings account of the market portfolio equals the fraction invested in the savings account of the mutual fund.

There are several lines of argument for justifying condition (5.54). Realistically, the monetary authorities can be assumed to control the fraction of the market portfolio, which is held in the savings account, in such a way that the relationship (5.54) is obtained. This is typically achieved by influencing the short rate level or treasury bill supply. As we will see below, if jumps do not occur for the GOP, then condition (5.54) is equivalent to the case of having a monetary policy that maximizes the growth rate of the market portfolio and, thus, that of the economy. This seems to be a natural assumption. Because of the highly diversified nature of the market portfolio one could argue that jumps are either absent or not significant. Note however that this is an empirical issue which needs to be tested.

5.4.4 Mutual Fund and GOP

In the case where the GOP does not have jumps, it is clear from (5.22) and (5.41) that the GOP and the mutual fund coincide since this only happens when all market prices for event risks are zero. Suppose that the market prices for event risks are non-zero. From (5.20) and (5.21) we obtain the following expressions for the fractions of the GOP:

$$\pi^j_{\delta_*}(t) = \sum_{k=1}^{m} \theta^k(t)\, b^{-1\, j,k}(t) + \sum_{k=m+1}^{d} \frac{\theta^k(t)}{1 - \frac{\theta^k(t)}{\sqrt{h^k_t}}}\, b^{-1\, j,k}(t)\,, \tag{5.55}$$

for all $t \in [0,T]$ and $j \in \{1,2,\ldots,d\}$. On the other hand, according to (5.39) the mutual fund is characterized by the fractions

$$\pi^j_{\delta_+}(t) = \sum_{k=1}^{d} \theta^k(t)\, b^{-1\, j,k}(t)\,, \tag{5.56}$$

for all $t \in [0,T]$ and $j \in \{1,2,\ldots,d\}$. Consequently, one obtains

$$\pi^j_{\delta_+}(t) = \pi^j_{\delta_*}(t) + \sum_{k=m+1}^{d} \frac{(\theta^k(t))^2\, b^{-1\, j,k}}{\sqrt{h^k_t} - \theta^k(t)}\,, \tag{5.57}$$

for $t \in [0,T]$ and $j \in \{1,2,\ldots,d\}$. As already indicated, one notes from (5.57) that the mutual fund and the GOP coincide if the market prices for event risk are zero. The portfolios $S^{(\delta_+)}$ and $S^{(\delta_*)}$ are approximately similar if the intensities for the event risks are extremely high compared to their corresponding market prices for risk.

5.5 Fair Contingent Claim Pricing

5.5.1 Fair Pricing

It will now be shown that the direct observability of the GOP in the form of the market portfolio can be exploited for consistent derivative pricing. As demonstrated in Platen [29, 30], for the class of benchmark models with intensity based jumps under consideration, one does not, in general, have an equivalent risk neutral martingale measure. Therefore, the widely used risk neutral pricing methodology may break down for certain benchmark models. This is the case when the benchmarked savings account process $\widehat{S}^{(0)}$ forms a *strict local martingale* and not a martingale. For realistic benchmark models where this happens see Platen[29], Heath & Platen [16, 17, 18] and Miller & Platen [27].

Since risk neutral pricing is not available, one needs a consistent and realistic alternative concept for pricing contingent claims that generalizes the standard risk neutral approach. To value derivatives uniquely, we apply the concept of *fair pricing*, as introduced in Platen [29]. It employs the GOP as benchmark or numeraire and forms conditional expectations under the real world probability measure. In some sense it generalizes the numeraire portfolio approach of Long [23], as well as the well-known state price density, deflator, pricing kernel and discount factor approaches described in Constatinides [6], Duffie [10] and Cochrane [5], for instance.

Definition 5. *We call a price process $U = \{U_t,\ t \in [0,T]\}$* fair *if the corresponding benchmarked price process $\widehat{U} = \{\widehat{U}_t = \frac{U_t}{S_t^{(\delta_*)}},\ t \in [0,T]\}$ forms an $(\underline{\mathcal{A}}, P)$-martingale. That is, it satisfies the conditions*

$$E(|\widehat{U}_T|) < \infty$$

and

$$\widehat{U}_t = E\left(\widehat{U}_s \,\middle|\, \mathcal{A}_t\right), \tag{5.58}$$

for all $0 \leq t \leq s \leq T$.

Under the presented benchmark model with intensity based jumps we do *not* require the existence of an equivalent risk neutral martingale measure. Therefore, standard risk neutral pricing is, in general, not applicable. However, fair pricing generalizes standard risk neutral pricing, as shown in Platen [29, 31, 32]. Furthermore, in a benchmark model a *free lunch with vanishing risk*, in the sense of Delbaen & Schachermayer [8], may arise for certain model specifications, see Heath & Platen [16, 17, 18]. However, due to the supermartingale property of nonnegative benchmarked portfolios, see (5.28), one is unable to generate strictly positive terminal wealth from zero initial capital using a nonnegative portfolio. A benchmark model with intensity based jumps is arbitrage free in the sense that all nonnegative benchmarked portfolios are supermartingales, as described in Platen [29].

Definition 6. *We define a* contingent claim *H_τ, which matures at a stopping time $\tau \in [0,T]$, as a nonnegative $\mathcal{A}_\tau$-measurable random payoff with*

$$E\left(\frac{H_\tau}{S_\tau^{(\delta_*)}}\right) < \infty \tag{5.59}$$

almost surely.

With reference to Definition 5, we define the *fair price* of a contingent claim H_τ, as in Definition 6, by the process $U_{H_\tau} = \{U_{H_\tau}(t),\, t \in [0,T]\}$, determined by

$$U_{H_\tau}(t) = S_t^{(\delta_*)}\, E\left(\frac{H_\tau}{S_\tau^{(\delta_*)}} \,\Big|\, \mathcal{A}_t\right), \tag{5.60}$$

for $t \in [0, \tau]$. It will be shown below that if an equivalent risk neutral martingale measure exists, then the fair price coincides with the corresponding risk neutral price, see also Platen [29]. The benchmark approach enlarges the range of models that can be used if compared to what is possible under the risk neutral approach, see Heath & Platen [17].

5.5.2 Risk Neutral and Actuarial Pricing

Let us assume that condition (5.54) is satisfied and the last term in (5.57) is negligible. Then the market portfolio is a good proxy for the GOP. The direct observability of the market portfolio leads naturally to a practical fair pricing methodology, generalizing the well-known *arbitrage pricing theory* (APT) introduced by Ross [37], Harrison & Kreps [14] and Harrison & Pliska [15]. However, fair pricing does not require an equivalent risk neutral martingale measure to exist. Note that the Radon-Nikodym derivative process $\Lambda_Q = \{\Lambda_Q(t),\, t \in [0,T]\}$ for the presumed risk neutral measure Q can be expressed as inverse of the discounted GOP,

$$\Lambda_Q(t) = \frac{dQ}{dP}\Big|_{\mathcal{A}_t} = \frac{\bar{S}_0^{(\delta_*)}}{\bar{S}_t^{(\delta_*)}}, \tag{5.61}$$

for $t \in [0,T]$, see Karatzas & Shreve [19]. By the Itô formula and (5.22) we obtain the SDE

$$d\Lambda_Q(t) = -\Lambda_Q(t) \sum_{k=1}^{d} \theta^k(t)\, dW_t^k, \tag{5.62}$$

for $t \in [0,T]$ with $\Lambda_Q(0) = 1$. This shows that Λ_Q is an $(\underline{\mathcal{A}}, P)$-local martingale. Furthermore, by (5.27) it follows that $\bar{S}_t^{(\delta)} \Lambda_Q(t) = \frac{S_t^{(\delta)}}{S_t^{(\delta_*)}} = \widehat{S}_t^{(\delta)}$ forms an $(\underline{\mathcal{A}}, P)$-local martingale for any portfolio $S^{(\delta)}$. We emphasize that this does *not* mean that $\widehat{S}^{(\delta)}$ is automatically an $(\underline{\mathcal{A}}, P)$-martingale.

To demonstrate that the standard risk neutral approach is covered by the fair pricing concept of the benchmark approach, let us consider a fair portfolio $S^{(\delta)}$, where by Definition 5 and (5.61) we get

$$S_t^{(\delta)} = S_t^{(\delta_*)}\, E\left(\widehat{S}_s^{(\delta)} \,|\, \mathcal{A}_t\right) = E\left(\frac{\Lambda_Q(s)}{\Lambda_Q(t)} \frac{S_t^{(0)}}{S_s^{(0)}}\, S_s^{(\delta)} \,\Big|\, \mathcal{A}_t\right), \tag{5.63}$$

for $0 \leq t \leq s \leq T$. Then, by application of the Girsanov theorem, see Protter [34], one obtains the risk neutral pricing formula

$$S_t^{(\delta)} = E_Q \left(\frac{S_t^{(0)}}{S_s^{(0)}} \, S_s^{(\delta)} \,\middle|\, \mathcal{A}_t \right), \tag{5.64}$$

for all $t \in [0, T]$ and $s \in [t, T]$, if Λ_Q is in fact an $(\underline{\mathcal{A}}, P)$-martingale. Here E_Q denotes expectation under the risk neutral measure Q.

In the above sense, one recovers the risk neutral pricing methodology of the APT as a special case of fair pricing. Furthermore, this approach uses as numeraire an observable quantity in form of the market portfolio. As would be expected, this is rather important for realistic modeling and contingent claim pricing.

Let us briefly mention some empirical evidence which supports our view that we need to go beyond the APT. By (5.61) the putative Radon-Nikodym derivative Λ_Q for the candidate risk neutral measure equals the ratio of the savings account over the GOP. In the long run the market portfolio and thus the GOP is by rational investors expected to outperform the savings account. This has been also empirically confirmed by Dimson, Marsh & Staunton [9] in a detailed empirical study of all major markets over the last century. This finding demonstrates that the trajectory of the process Λ_Q *decreases systematically* over long periods. The empirical fact of a systematic decline of this process for all major currency denominations surely cannot be ignored. As a consequence, it is not likely that Λ_Q can in reality be successfully modeled as an $(\underline{\mathcal{A}}, P)$-martingale. This contradicts a core assumption of the APT. We emphasize that a decreasing graph for Λ_Q is still consistent with it being a nonnegative strict $(\underline{\mathcal{A}}, P)$-local martingale and hence a supermartingale, see Protter [34]. The proposed benchmark approach can accommodate this fully . For derivative pricing under the benchmark model with intensity based jumps, where no equivalent risk neutral martingale measure is assumed to exist, we therefore advocate the fair pricing methodology.

Fair prices are uniquely determined even in incomplete markets. Under the existence of a minimal equivalent martingale measure, see Fölmer & Schweizer [12], fair prices have been shown to correspond to local risk minimizing prices, see Platen [33]. Fair pricing is practicable since one can model and calibrate the GOP when interpreted as the market portfolio. This enables us to calculate the real world expectations in (5.60) directly.

For the practically important case where a contingent claim H_T is *independent* of the GOP $S_T^{(\delta_*)}$, one obtains directly the following *actuarial pricing formula* from the fair pricing formula (5.60).

Corollary 6. *For a contingent claim H_T that is independent of the GOP value $S_T^{(\delta_*)}$, the fair price $U_{H_T}(t)$ satisfies the* actuarial pricing formula

$$\begin{aligned} U_{H_T}(t) &= E\left(\frac{S_t^{(\delta_*)}}{S_T^{(\delta_*)}} \,\Big|\, \mathcal{A}_t\right) E\left(H_T \,|\, \mathcal{A}_t\right) \\ &= P(t,T)\, E\left(H_T \,|\, \mathcal{A}_t\right), \end{aligned} \tag{5.65}$$

where $P(t,T)$ denotes the fair price at time $t \in [0,T]$ of a zero coupon bond with maturity date T.

One may regard (5.65) as a generalized net present value pricing formula. It is still valid when interest rates are stochastic. In various ways formulas of the type (5.65) have been used in insurance and other areas of risk management, see, for instance, Bühlmann [3] and Gerber [13]. They appear here as a natural consequence of the benchmark approach.

5.6 Expected Discounted Mutual Fund

We conclude the paper by analyzing the dynamics of the mutual fund, which can be interpreted as the market portfolio under appropriate assumptions as previously discussed. The SDE (5.41) for the mutual fund reveals a close link between its drift and generalized diffusion coefficient. More precisely, the risk premium of the mutual fund equals the square of its total aggregate generalized volatility. To see this, one can rewrite the SDE (5.41) for the mutual fund in discounted form as

$$d\bar{S}_t^{(\delta_+)} = \bar{S}_t^{(\delta_+)}\, |\theta(t)|\, (|\theta(t)|\, dt + dW_t)\,, \tag{5.66}$$

for $t \in [0,T]$. Here,

$$dW_t = \frac{1}{|\theta(t)|} \sum_{k=1}^{d} \theta^k(t)\, dW_t^k \tag{5.67}$$

is the stochastic differential of an $(\underline{\mathcal{A}}, P)$-martingale W with conditional variance

$$E\left((W_{t+\varepsilon} - W_t)^2 \,|\, \mathcal{A}_t\right) = \varepsilon\,, \tag{5.68}$$

for all $t \in [0,T]$ and $\varepsilon \in (0, T-t]$, see Protter [34]. If W is continuous, then it is a standard Wiener process. Generally, W is a mixture of independent martingales that may exhibit some jumps. The SDE (5.66) reveals a useful structural relationship between the drift and the generalized diffusion coefficient of the mutual fund, which we will exploit below.

We reparameterize the mutual fund dynamics by using the average change per unit of time of the discounted mutual fund value, which is captured by the *discounted mutual fund drift*

$$\alpha(t) = \alpha_{\delta_+}(t) = \bar{S}_t^{(\delta_+)}\, |\theta(t)|^2\,, \tag{5.69}$$

for $t \in [0,T]$. Note that if there are no jumps in the GOP, then we have in (5.69) the drift of the discounted GOP. One can interpret $\alpha(t)$ as the change per unit time of the *accumulated underlying value* of the discounted market portfolio. Using the parametrization (5.69), we can express the total market price for risk in the form

$$|\theta(t)| = \sqrt{\frac{\alpha(t)}{\bar{S}_t^{(\delta_+)}}} \,. \tag{5.70}$$

By substituting (5.69) and (5.70) into (5.66), we obtain the following SDE for the discounted mutual fund

$$d\bar{S}_t^{(\delta_+)} = \alpha(t)\,dt + \sqrt{\bar{S}_t^{(\delta_+)}\,\alpha(t)}\,dW_t\,, \tag{5.71}$$

for $t \in [0,T]$. The solution of this SDE is a generalized time transformed squared Bessel process of dimension four. For the continuous version of this process, when W is a standard Wiener process, we refer the reader to Revuz & Yor [35]. In the current paper the process is driven by the normalized jump martingale W, given in (5.67).

The *transformed time* $\varphi(t)$ at time t for $\bar{S}^{(\delta_+)}$ in (5.71) is given by the expression

$$\varphi(t) = \varphi(0) + \int_0^t \alpha(s)\,ds\,, \tag{5.72}$$

with $\varphi(0) \geq 0$ as a possibly unobserved random initial value. We emphasize the fact that $\varphi(t)$ is not just one arbitrarily selected time transformation. The increment $\varphi(t) - \varphi(0)$ expresses the *change of accumulated underlying value* in the discounted mutual fund. This is an important economic quantity, which appears here naturally in the benchmark setup. In the case where the discounted mutual fund does not exhibit jumps, which seems to be a realistic assumption, it can be shown, see Platen [32], that the increase in accumulated underlying value can be directly observed via the equalities

$$\varphi(t) - \varphi(0) = 4\left\langle \sqrt{\bar{S}^{(\delta_+)}} \right\rangle_t = 4\int_0^t \frac{\alpha(s)}{4}\,ds\,, \tag{5.73}$$

for $t \in [0,T]$. This makes the transformed time or accumulated underlying value an observable quantity. It provides important information about the evolution of the average economic value of the market portfolio via some quadratic variation, which is readily observable, see Platen [32].

Let us decompose the discounted mutual fund value at time $t \in [0,T]$ as

$$\bar{S}_t^{(\delta_+)} = \bar{S}_0^{(\delta_+)} + \varphi(t) - \varphi(0) + M_t\,, \tag{5.74}$$

where $M = \{M_t,\, t \in [0,T]\}$ is the $(\underline{\mathcal{A}}, P)$-local martingale

$$M_t = \int_0^t \bar{S}_s^{(\delta_+)}\,|\theta(s)|\,dW_s = \int_0^t \sqrt{\bar{S}_s^{(\delta_+)}\,\alpha(s)}\,dW_s\,, \tag{5.75}$$

for $t \in [0,T]$. The discounted mutual fund value $\bar{S}_t^{(\delta_+)}$ in (5.74) consists of a *noise part* M_t, which models the trading uncertainty of the discounted mutual fund and a systematic part $\varphi(t)-\varphi(0)$, which expresses the increase of its accumulated underlying value. As previously mentioned, the accumulated underlying value can be interpreted as a measure of the discounted wealth that has been generated by the companies listed in the stock market. The fluctuating share prices then express the perception of the market about the value of each company. Over long periods the evolution of this perceived value has to be in line with the corresponding accumulated underlying value. Hence, this provides some type of measure of the degree to which the mutual fund is over or undervalued.

Remarkably, when the accumulated underlying value of the market portfolio is used as time scale, then by (5.71) the dynamics of the discounted mutual fund turn out to be those of a very particular stochastic process, the generalized time transformed squared Bessel process of dimension four. This is a pleasing result, not only mathematically, but also economically.

The above relationships lead directly to the following statement, which exploits equations (5.74) and (5.72) and is obtained by a realistic martingale assumption on the local martingale M.

Corollary 7. *If the local martingale M in (5.75) is a true $(\underline{\mathcal{A}}, P)$-martingale, then the expected change of the discounted mutual fund value over a given time period equals the expected change of its transformed time, that is,*

$$E\left(\bar{S}_s^{(\delta_+)} - \bar{S}_t^{(\delta_+)} \,\middle|\, \mathcal{A}_t\right) = E\left(\varphi(s) - \varphi(t) \,\middle|\, \mathcal{A}_t\right), \tag{5.76}$$

for all $t \in [0,T]$ and $s \in [t,T]$.

We emphasize that we have not made any major assumptions about the particular stochastic dynamics of the mutual fund. In principle, there is much modeling freedom that can be explored. However, as shown in Platen [30], in the case without jumps a natural dynamics emerges from the fact that the change of the transformed time or accumulated underlying value can be modeled realistically by a rather smooth increasing quantity. The dynamics of the discounted mutual fund is then in physical time that of a squared Bessel process of dimension four, see Platen [29, 30]. The resulting model with a slowly varying deterministic transformed time is called the *minimal market model* (MMM), see Platen [28]. It has major consequences for the nature of the dynamics of the supposed Radon-Nikodym derivative Λ_Q. This process is under the MMM a strict supermartingale and *not* a martingale and a core assumption of the APT is thus violated. The consequences of this fact were discussed in Section 5.5.2.

Conclusion

It has been shown that the growth optimal portfolio plays a central theoretical and practical role in finance. The paper assumes that investors always prefer more for less which leads them to hold optimal portfolios. A theorem is derived that characterizes any optimal portfolio as a mixture of some mutual fund and the savings account. Under the additional assumptions that the monetary authorities aim to maximize the long term growth of the market portfolio, assuming negligible jump risk, it has been shown that the market portfolio approximates the mutual fund and also the growth optimal portfolio. This observation provides a derivation of the capital asset pricing model for jump diffusion markets without requiring any equilibrium or expected utility maximization arguments. The Markowitz efficient frontier and Sharpe ratio follow naturally in a generalized form for the given benchmark model with intensity based jumps.

Without imposing any particular dynamics, the discounted mutual fund is identified as a generalized time transformed squared Bessel process of dimension four. The transformed time can be interpreted as the accumulated underlying value of the discounted market portfolio. Under appropriate assumptions the increase in expected discounted value of the market portfolio is shown to equal the expected increase of the transformed time.

For the pricing of contingent claims the GOP and under realistic assumptions its proxy, the market portfolio, can be used as numeraire, with expectations to be taken under the real world probability measure. The resulting fair pricing methodology applies also for market models where an equivalent risk neutral martingale measure does not exist and generalizes risk neutral and actuarial pricing.

Acknowledgement

The author likes to express his thanks to Morten Christensen, David Heath, Hardy Hulley and the participants at the Stochastic Finance 2004 Conference in Lisbon for their comments and stimulating interest in this work.

A Appendix

Proof of Theorem C.1

According to Definition 3, to identify an optimal portfolio, one maximizes the drift (5.32) locally in time, while keeping the diffusion coefficient (5.33) as given. Suppressing time dependence, our task is to find $\psi_\delta^1, \psi_\delta^2, \ldots, \psi_\delta^d$, which maximize $\sum_{k=1}^d \psi_\delta^k \, \theta^k$ subject to the constraint $\sum_{k=1}^d (\psi_\delta^k)^2 = C$, for some given value $C > 0$. For this purpose we use the Lagrange multiplier λ and consider the function

$$G(\theta^1, \ldots, \theta^k, C, \lambda, \psi_\delta^1, \ldots, \psi_\delta^d) = \sum_{k=1}^{d} \psi_\delta^k \, \theta^k + \lambda \left(C - \sum_{k=1}^{d} \left(\psi_\delta^k\right)^2 \right). \quad \text{(A.1)}$$

For $\psi_\delta^1, \psi_\delta^2, \ldots, \psi_\delta^d$ to provide a maximum for $G(\theta^1, \ldots, \theta^k, C, \lambda, \psi_\delta^1, \ldots, \psi_\delta^d)$ it is necessary that the first-order conditions

$$\frac{\partial G(\theta^1, \ldots, \theta^k, C, \lambda, \psi_\delta^1, \ldots, \psi_\delta^d)}{\partial \psi_\delta^k} = \theta^k - 2\,\lambda\,\psi_\delta^k = 0 \tag{A.2}$$

are satisfied for all $k \in \{1, 2, \ldots, d\}$. Consequently, we must have

$$\psi_\delta^k = \frac{\theta^k}{2\,\lambda}\,, \tag{A.3}$$

for all $k \in \{1, 2, \ldots, d\}$. We can now use the constraint together with (5.34) to obtain the relation

$$C = \sum_{k=1}^{d} \left(\psi_\delta^k\right)^2 = \left(\frac{|\theta|}{2\,\lambda}\right)^2 \tag{A.4}$$

from (A.3). Then from (A.3) and (A.4) one obtains

$$\psi_\delta^k = \frac{\sqrt{C}}{|\theta|}\,\theta^k\,, \tag{A.5}$$

for $k \in \{1, 2, \ldots, d\}$. Thus from (5.33), for $S^{(\delta)}$ to be an optimal portfolio, we must have

$$\psi_\delta^k(t) = \frac{|\gamma_\delta(t)|}{|\theta(t)|}\,\theta^k(t)\,, \tag{A.6}$$

for $t \in [0, T]$. Now, it follows from Assumption 1, (5.31) and (A.6) that

$$\begin{aligned} \pi_\delta^j(t) &= \frac{1}{\bar{S}_t^{(\delta)}} \sum_{k=1}^{d} \psi_\delta^k(t)\, b^{-1\,j,k}(t) \\ &= \frac{|\gamma_\delta(t)|}{\bar{S}_t^{(\delta)}\,|\theta(t)|} \sum_{k=1}^{d} \theta^k(t)\, b^{-1\,j,k}(t)\,. \end{aligned} \tag{A.7}$$

Therefore, by (5.14) we get

$$\begin{aligned} \pi_\delta^0(t) &= 1 - \sum_{j=1}^{d} \pi_\delta^j(t) \\ &= 1 - \frac{|\gamma_\delta(t)|\;\Gamma^{(0)}(t)}{\bar{S}_t^{(\delta)}\;|\theta(t)|}\,, \end{aligned} \tag{A.8}$$

with $\Gamma^{(0)}(t)$ as in (5.35), for $t \in [0, T]$. Thus, we have

$$|\gamma_\delta(t)| = \frac{\left(1 - \pi_\delta^0(t)\right)\bar{S}_t^{(\delta)}\;|\theta(t)|}{\Gamma^{(0)}(t)}\,. \tag{A.9}$$

Substitution into (A.7) yields (5.39) and, with the aid of (5.35) and (5.30), proves the theorem. □

References

[1] Bajeux-Besnainou, I. & R. Portait (1997). The numeraire portfolio: A new perspective on financial theory. *The European Journal of Finance* **3**, 291–309.

[2] Bühlmann, H. (1992). Stochastic discounting. *Insurance: Mathematics and Economics* **11**, 113–127.

[3] Bühlmann, H. (1995). Life insurance with stochastic interest rates. In G. Ottaviani (Ed.), *Financial Risk and Insurance*, pp. 1–24. Springer.

[4] Christensen, M. M. & E. Platen (2004). A general benchmark model for stochastic jump sizes. Technical report, University of Technology, Sydney. QFRC Research Paper 139, to appear in *Stochastic Analysis and Applications*.

[5] Cochrane, J. H. (2001). *Asset Pricing*. Princeton University Press.

[6] Constatinides, G. M. (1992). A theory of the nominal structure of interest rates. *Rev. Financial Studies* **5**, 531–552.

[7] Cont, R. & P. Tankov (2004). *Financial Modelling with Jump Processes*. Financial Mathematics Series. Chapman & Hall/CRC.

[8] Delbaen, F. & W. Schachermayer (1998). The fundamental theorem of asset pricing for unbounded stochastic processes. *Math. Ann.* **312**, 215–250.

[9] Dimson, E., P. Marsh, & M. Staunton (2002). *Triumph of the Optimists: 101 Years of Global Investment Returns*. Princeton University Press.

[10] Duffie, D. (2001). *Dynamic Asset Pricing Theory* (3rd ed.). Princeton, University Press.

[11] Duffie, D. & K. Singleton (2003). *Credit Risk : Pricing, Measurement, and Management*. Princeton Series in Finance. Princeton University Press.

[12] Föllmer, H. & M. Schweizer (1991). Hedging of contingent claims under incomplete information. In M. Davis and R. Elliott (Eds.), *Applied Stochastic Analysis*, Volume 5 of *Stochastics Monogr.*, pp. 389–414. Gordon and Breach, London/New York.

[13] Gerber, H. U. (1990). *Life Insurance Mathematics*. Springer, Berlin.

[14] Harrison, J. M. & D. M. Kreps (1979). Martingale and arbitrage in multiperiod securities markets. *J. Economic Theory* **20**, 381–408.

[15] Harrison, J. M. & S. R. Pliska (1981). Martingales and stochastic integrals in the theory of continuous trading. *Stochastic Process. Appl.* **11**(3), 215–260.

[16] Heath, D. & E. Platen (2002a). Consistent pricing and hedging for a modified constant elasticity of variance model. *Quant. Finance.* **2**(6), 459–467.

[17] Heath, D. & E. Platen (2002b). Perfect hedging of index derivatives under a minimal market model. *Int. J. Theor. Appl. Finance* **5**(7), 757–774.

[18] Heath, D. & E. Platen (2002c). Pricing and hedging of index derivatives under an alternative asset price model with endogenous stochastic volatility. In J. Yong (Ed.), *Recent Developments in Mathematical Finance*, pp. 117–126. World Scientific.

[19] Karatzas, I. & S. E. Shreve (1998). *Methods of Mathematical Finance*, Volume 39 of *Appl. Math.* Springer.

[20] Kelly, J. R. (1956). A new interpretation of information rate. *Bell Syst. Techn. J.* **35**, 917–926.

[21] Korn, R. & M. Schäl (1999). On value preserving and growth-optimal portfolios. *Math. Methods Oper. Res.* **50**(2), 189–218.

[22] Lintner, J. (1965). The valuation of risk assets and the selection of risky investments in stock portfolios and capital budgets. *Rev. Econom. Statist.* **47**, 13–37.

[23] Long, J. B. (1990). The numeraire portfolio. *J. Financial Economics* **26**, 29–69.
[24] Markowitz, H. (1959). *Portfolio Selection: Efficient Diversification of Investment.* Wiley, New York.
[25] Merton, R. C. (1973). An intertemporal capital asset pricing model. *Econometrica* **41**, 867–888.
[26] Merton, R. C. (1976). Option pricing when underlying stock returns are discontinuous. *J. Financial Economics* **2**, 125–144.
[27] Miller, S. & E. Platen (2004). A two-factor model for low interest rate regimes. Technical report, University of Technology, Sydney. QFRC Research Paper 130, to appear in *Asia-Pacific Financial Markets* **11**(1).
[28] Platen, E. (2001). A minimal financial market model. In *Trends in Mathematics*, pp. 293–301. Birkhäuser.
[29] Platen, E. (2002). Arbitrage in continuous complete markets. *Adv. in Appl. Probab.* **34**(3), 540–558.
[30] Platen, E. (2004a). A class of complete benchmark models with intensity based jumps. *J. Appl. Probab.* **41**, 19–34.
[31] Platen, E. (2004b). Diversified portfolios with jumps in a benchmark framework. Technical report, University of Technology, Sydney. QFRC Research Paper 129, to appear in *Asia-Pacific Financial Markets* **11**(1).
[32] Platen, E. (2004c). Modeling the volatility and expected value of a diversified world index. *Int. J. Theor. Appl. Finance* **7**(4), 511–529.
[33] Platen, E. (2004d). Pricing and hedging for incomplete jump diffusion benchmark models. In *Mathematics of Finance*, Volume 351 of *Contemporary Mathematics*, pp. 287–301. American Mathematical Society.
[34] Protter, P. (2004). *Stochastic Integration and Differential Equations* (2nd ed.). Springer.
[35] Revuz, D. & M. Yor (1999). *Continuous Martingales and Brownian Motion* (3rd ed.). Springer.
[36] Rogers, L. C. G. & D. Williams (2000). *Diffusions, Markov Processes and Martingales: Itô Calculus* (2nd ed.), Volume 2 of *Cambridge Mathematical Library*. Cambridge University Press.
[37] Ross, S. A. (1976). The arbitrage theory of capital asset pricing. *J. Economic Theory* **13**, 341–360.
[38] Sharpe, W. F. (1964). Capital asset prices: A theory of market equilibrium under conditions of risk. *J. Finance* **19**, 425–442.

6

Mortgage Valuation and Optimal Refinancing

Stanley R. Pliska

Department of Finance, University of Illinois at Chicago
601 S. Morgan Street, Chicago, IL 60607-7124 USA
srpliska@uic.edu

Summary. This paper summarizes recent research on a new approach, namely, an equilibrium approach, to the valuation of fixed-rate mortgage contracts. Working in a discrete time setting with the mortgagor's prepayment behavior described by a suitable intensity process and with exogenous mortgage rates, the value of the contract is derived in an explicit form that can be interpreted as the principal balance plus the value of a certain swap. This leads to a nonlinear equation for what the mortgage rate must be in a competitive market, and thus mortgage rates are endogenous and depend upon the mortgagor's prepayment behavior. The complementary problem, where mortgage rates are exogenous and the mortgagor seeks the optimal refinancing strategy, is then solved via a Markov decision chain. Finally, the equilibrium problem, where the mortgagor is a representative agent in the economy who seeks the optimal refinancing strategy and where the mortgage rates are endogenous, is developed, solved, and analysed. Existence and uniqueness results, as well as a numerical example, are provided.

Key words: mortgage valuation, endogenous mortgage rates, equilibrium, Markov decision chain, dynamic programming, intensity process, hazard rate

6.1 Introduction

While there is widespread agreement that the value of a mortgage contract subject to prepayment but not default risk should be given by an expectation of the present value of the cash flow, the devil is in the details. A wide variety of approaches have been considered, most of which are commonly classified into one of two categories. One kind of approach has been variously called a reduced form approach, an exogenous approach, an empirical approach, and an econometric approach. The basic idea is to build a stochastic model for interest rates and possibly other economic factors, and then add a statistical model describing how the mortgagor's prepayment behavior depends on the factors. While such an overall model can be quite complicated, it is usually straightforward to use Monte Carlo simulation to estimate the expected value of the discounted cash flow. Some of the many papers in this category are by Schwartz and Torous [16], [17], Deng [1], Deng, Quigley and Van Order [2], Kariya and Kobayashi [8], Kariya, Pliska and Ushiyama [9], and Kau, Keenan and Smurov [10].

Of significance is that in some of this research, dating back at least to Schwartz and Torous [16], [17], it was recognized that the random time when a mortgagor prepays can be described with a hazard rate model, that is, the conditional rate of prepayment given the current state of any factors and no prepayment to date. Perhaps this development was inspired by the engineering literature on reliability theory, as the time of mortgage prepayment is clearly analogous to the failure time of a system. In any event, Schwartz and Torous [16] used this hazard rate viewpoint in conjunction with a two-factor model in order to present a partial differential equation for the value of a mortgage contract. Moreover, as will be seen in this paper, recent developments involving the hazard rate as a model of a default time in the credit risk literature lead to new results, involving intensity processes, for mortgage contract valuation.

The other main kind of approach for the valuation of mortgage contracts is called an option-based or a structural approach. The basic idea is to incorporate some kind of optimal behavior with respect to the mortgagor's decision about when to refinance. Moreover, the way to do this is to appeal to some intuition based upon the theory of the optimal early exercise decision for American options, usually leading to a recursive valuation procedure that resembles the one used for the binomial option pricing model. For example, Kalotay, Yang, and Fabozzi [7] described the following procedure: first build an interest rate lattice, and then, starting with the final scheduled cash flows of the mortgage, work backwards through the lattice computing the mortgage's value, comparing the value with no refinancing and the value of a newly refinanced mortgage, with the latter assumed to be par plus the refinancing cost. If the latter is less, then the value of the existing mortgage at the node is replaced by the value of the new mortgage. Some others who took an option-based approach are Dunn and McConnell [3], [4], Stanton [18], Nakagawa and Shouda [12], Stanton and Wallace [19], Dunn and Spatt [5], and Longstaff [11].

The latter three papers are noteworthy because, in contrast to all the other option-based papers which assumed mortgage rates are exogenous, Stanton and Wallace [19], Dunn and Spatt [5], and Longstaff [11] allowed for endogenous values of fixed rate mortgages. These authors studied discrete time, finite horizon models, with time equal to the age of the mortgage contract. The mortgage rates were computed recursively, much like the "binomial option pricing" procedure by Kalotay, Yang, and Fabozzi [7] that was described above. But the model assumptions made by Stanton and Wallace [19], Dunn and Spatt [5], and Longstaff [11] are unclear, due in part to the limited use of mathematics in their expositions. Suffice it to say that their models are significantly different from the one in this paper, as evidenced by the fact that their endogenous mortgage rates seem to depend upon the age of the mortgage contract.

As indicated above, the theory of hazard rates and intensity processes for modelling default times in the credit risk literature has advanced considerably in recent years. Since the time of a default is analogous to the time when a mortgage balance is prepaid, it was natural to translate some of the credit risk developments to mortgage valuation. This was recently accomplished by Goncharov [6], who worked entirely in a continuous time setting. In particular, he showed how to unify the reduced form and the option-based approaches, he derived some explicit formulas for a mortgage's value, he derived a variety of partial differential equations useful for computing mortgage values, and he used an explicit valuation formula to provide a nonlinear equation for the endogenous mortgage rate.

This paper makes several contributions. First, some intensity based valuation results that Goncharov [6] derived for a continuous time environment are here derived for a discrete time financial market. In particular, with the mortgagor's prepayment behavior described by a suitable intensity process and with exogenous mortgage rates, in Section 2 the value of the contract is derived in an explicit form that can be interpreted as the principal balance plus the value of a certain swap. This leads in Section 3 to a nonlinear equation for what the mortgage rate must be in a competitive market, and thus mortgage rates are endogenous and depend upon the mortgagor's prepayment behavior. The complementary problem, where mortgage rates are exogenous and the mortgagor seeks the optimal refinancing strategy, is then solved in Section 4 via a Markov decision chain. Various theoretical results about computational algorithms and existence of solutions are included. The equilibrium problem, where the mortgagor is a representative agent in the economy who seeks the optimal refinancing strategy and where the mortgage rates are endogenous, is developed, solved, and analysed in Section 5. In particular, the existence of an equilibrium solution is established. Section 6 provides a simple computational example that illustrates various theoretical points, although it is probably not realistic enough to draw conclusions about actual mortgage markets. Finally, Section 7 provides some concluding remarks.

It should be noted that the derivations and proofs of this paper's main results are highly abbreviated. The interested reader can find a complete exposition of these missing technical details in the forthcoming companion paper (Pliska [14]).

6.2 Valuation of Mortgage Contracts

This paper focuses on the valuation of fixed rate mortgage contracts having N contracted coupon payments each of amount c dollars. If m is the mortgage rate at contract initiation (this interest rate is expressed on a per payment period, not necessarily on an annual, basis) and if $P(n, m)$ denotes the principal balance immediately after the nth coupon payment is made, then by simple time value of money considerations

$$P(n+1,m) = (1+m)P(n,m) - c\,. \tag{6.1}$$

Using this equation recursively one obtains

$$P(n,m) = \frac{c}{m}[1-(1+m)^{-k}] + \left(\frac{1}{1+m}\right)^k P(n+k,m)\,, \quad k = 1,...,N-n\,. \tag{6.2}$$

Throughout this paper it will be assumed that the mortgage contracts are fully amortizing, that is, $P(N,m) = 0$, and so (6.2) implies

$$P(n,m) = \frac{c}{m}[1-(1+m)^{n-N}]\,, \quad n = 0,1,...,N\,. \tag{6.3}$$

In particular, since $P(0,m)$ is the initial principal, the contracted coupon payment c is given in terms of the maturity N of the mortgage and the contracted mortgage rate m by

$$c = \frac{mP(0,m)}{1-(1+m)^{-N}}\,. \tag{6.4}$$

In accordance with common practice, just after any coupon payment the mortgagor can pay the principal balance, thereby terminating the mortgage contract. For simplicity it will be assumed that the mortgagor cannot pay any amount greater than the contracted coupon payment c unless it is the entire principal balance. Thus it will be assumed that immediately after the nth coupon payment the mortgagor must either pay the principal balance $P(n,m)$ or continue with the existing mortgage contract at least one more period. Also, since the focus of this paper is on the prepayment option, none of the mortgages considered here are subject to default.

From the perspective of the mortgage lender or of a third party considering the purchase of the mortgage contract, the value of the mortgage contract equals the expectation of the discounted cash flow up through the prepayment time or N, whichever is less. In accordance with standard financial valuation practice, the discounting is respect to the riskless, one-period short rate and the expectation is respect to a risk neutral probability measure. So to model this it will be assumed there is a probability space $(\Omega, \mathcal{F}, Q, \{\mathcal{F}_t\}_{t\geq 0})$, where Q is a *risk neutral probability measure* and where $\{\mathcal{F}_t\}_{t\geq 0}$ is a filtration describing how information is revealed to market participants. The *riskless interest rate process* $r = \{r_t; t = 1,2,...\}$ is a predictable process, where r_t represents the one-period riskless interest rate for loans from time $t-1$ to time t. And measureable on this probability space is the random variable τ representing the *mortgagor's prepayment time*; τ takes one of the values $1,2,...,N$, with $\tau = N$ meaning the mortgage is not prepaid early. Thus the value of the mortgage contract at contract initiation is given by risk neutral valuation to be

$$V = E[\frac{c}{1+r_1} + \frac{c}{(1+r_1)(1+r_2)} + \dots$$
$$+ \frac{c}{(1+r_1)\dots(1+r_{\tau\wedge N})} + \frac{P(\tau\wedge N, m)}{(1+r_1)\dots(1+r_{\tau\wedge N})}] . \tag{6.5}$$

A similar equation holds for the value of the contract at times subsequent to contract initiation.

Remark 1. To develop a practical model like this, one can imagine starting with a spot rate model of riskless interest rates that includes various securities such as the usual bank account process and zero coupon bonds of all maturities. With the mortgage contract and the issue of prepayment excluded, this model may or may not be complete. For example, see the lattice, Markov chain models in Pliska [13]. Then one would add to this model the mortgage contract together with the random prepayment time τ. If the initial riskless interest rate model is complete, then the final model could be too, in which case one has all the usual implications about uniqueness of the risk neutral probability measure and replication of contingent claims. But even if the model is not complete, the model will be free of arbitrage opportunities that can be obtain by trading the various securities.

To develop an expression for the contract value that is more useful than (6.6) it is convenient to introduce the *prepayment intensity process* $\gamma = \{\gamma_t; t = 1, 2, \dots, N\}$, where

$$\gamma_t := Q(\tau = t | \tau \geq t, \mathcal{F}_{t-1}), \quad t = 1, 2, ..., N .$$

Thus γ_t can be interpreted as the conditional, risk neutral probability that prepayment will occur next period given that it is now time $t-1$ and given the current history including the fact that prepayment has not yet occurred. Using this intensity process, expression (6.6), and basic properties of conditional expectation (see Pliska [14] for the details) one obtains the following result:

Proposition 1. *The initial value of a mortgage contract is given by*

$$V = E\Big[\sum_{i=1}^{N} \frac{c+\gamma_i P(i,m)}{1+r_1} \prod_{j=2}^{i} \frac{1-\gamma_{j-1}}{1+r_j}\Big], \quad where \prod_{j=2}^{1} := 1 . \tag{6.6}$$

This last expression for the initial value of a mortgage contract is still not so useful. A better formula is obtained by using (6.1) and (6.6) together with a lot of algebra (see Pliska [14] for the details), namely:

Theorem C.1. *The initial value of a mortgage contract is given by*

$$V = \frac{1+m}{1+r_1}P(0,m) + \frac{1}{1+r_1}E\Big[\sum_{i=1}^{N-1}(m - r_{i+1})P(i,m)\prod_{j=1}^{i}\frac{1-\gamma_j}{1+r_{j+1}}\Big] . \tag{6.7}$$

Remark 2. Note the factor in the first term will be approximately equal to one, so the first term will approximately equal the initial principal. The second term can be interpreted as the discounted value of an amortizing swap, where one party pays the fixed mortgage rate m and the other party pays the floating rate r, and where the swap can terminate in accordance with the random prepayment time τ. Thus the initial value of the mortgage contract is approximately equal to the initial principal plus the value of a swap. A similar expression can be obtained (see Pliska [14]) for the value of the mortgage contract at subsequent times.

6.3 Endogenous Mortgage Rates

While the mortgage market might be free of arbitrage opportunities that can be achieved by trading fixed income securities and the mortgage contract, it can be vulnerable to another kind of arbitrage opportunity if the mortgage contract value V is less than the initial principal $P(0,m)$. In other words, why would a lending institution offer a loan of $P(0,m)$ in exchange for a cash flow that is worth a strictly smaller amount? On the other hand, the possibility $V > P(0,m)$ implies attractive, profitable lending opportunities for financial institutions, and so it shall be argued that in a competitive market one will have $V = P(0,m)$.

With $V = P(0,m)$ it follows (see Pliska [14] for the details) from (6.7) that the mortgage rate m is endogenous and must satisfy a nonlinear equation, as summarized in the following:

Theorem C.2. *In a competitive mortgage market where $V = P(0,m)$ the mortgage rate m is endogenous and satisfies*

$$0 = (m - r_1)P(0,m) + E\Big[\sum_{i=1}^{N-1}(m - r_{i+1})P(i,m)\prod_{j=1}^{i}\frac{1-\gamma_j}{1+r_{j+1}}\Big] \tag{6.8}$$

or, equivalently,

$$0 = m - r_1 + E\Big[\sum_{i=1}^{N-1}(m - r_{i+1})\Big(\frac{(1+m)^N - (1+m)^i}{(1+m)^N - 1}\Big)\prod_{j=1}^{i}\frac{1-\gamma_j}{1+r_{j+1}}\Big]. \tag{6.9}$$

Remark 3. It is important to note that the endogenous value of the mortgage rate m depends upon the mortgagor's prepayment behavior via the intensity process γ. So if a mortgagor changes this prepayment behavior, mortgage market forces should cause the mortgage rates to change. Hence this leads to a certain equilibrium problem, the subject of a later section. Also in a later section attention will focus on a special case of the preceding model where the spot rate process r is a Markov chain, in which case it is appropriate to think of the expectations in (6.8) and (6.9) as being conditional on the values of the current state for r_1.

Remark 4. One sees that the value of m satisfying equation (6.8) must be some kind of weighted average of future possible values of the riskless short rate r. In particular, it must be smaller than the biggest possible value of the short rate. This suggests that when the short rate is historically high and when the yield curve is inverted, the endogenous mortgage rate will actually be smaller than the current short rate.

6.4 Optimal Refinancing

In this section attention is given to a complementary problem where the mortgage rate is an exogenous stochastic process and the mortgagor seeks to refinance the loan in an optimal fashion. In contrast to other approaches in the literature, the mortgagor here might choose to refinance several times before the loan is ultimately paid off.

For tractability it will be assumed that the riskless spot rate process $r = \{r_t; t = 1, 2, \dots\}$ and the mortgage rate process $M = \{M_t; t = 0, 1, \dots\}$ together comprise a time-homogeneous Markov chain having a finite state space. Here M_t represents the fixed rate for N-period mortgages that are initiated at time t. This two-component Markov chain can easily be generalized by adding additional factors such as a measure of property value, but this will not be done here for the sake of the exposition.

It is assumed the transaction cost $K(P)$ is incurred if and when the principal balance P is refinanced. Here $K(\cdot)$ is a specified, deterministic function. The decision to prepay is made immediately after the contracted coupon payment c has been paid. If the mortgagor decides to refinance at time t with a current principal balance of P, then at that time an additional $K(P)$ dollars are paid and a new N-period mortgage is initiated at the prevailing mortgage rate M_t. Alternatively, if the mortgagor decides to continue with the current mortgage contract then the prevailing coupon rate of c dollars will be paid for at least one more period. The mortgagor can refinance as many times as desired, but each time the amount of the new loan must equal the principal balance of the contract being terminated. Thus the mortgagor cannot "pull equity out of the property" by refinancing with a bigger loan. Nor can the mortgagor make any periodic payment, either bigger or smaller, than the currently contracted coupon payment c.

It is important to be mathematically precise about the mortgagor's objective. Here it will be assumed that the mortgagor seeks to minimize the expected present value of the cash flow (which includes any transaction costs associated with refinancings) until the loan is paid off. In particular, it is assumed for simplicity that the mortgagor does not prepay early for "external" reasons such as selling the property (although this is a strong assumption that would be nice to abandon). As usual, the expectation here will be with respect to a risk neutral probability measure and discounting will be with respect to the spot rate process r.

In summary, given the Markov chain (r, M) the mortgagor's problem is to find the refinancing schedule having the minimum expected discounted value. Hence it is natural to formulate this problem as a dynamic programming problem, specifically, as a Markov decision chain (see Puterman [15] for a comprehensive treatment of Markov decision chains). For the state variables one then needs to have the current values of the two interest rates as well as some variables describing the status of the current mortgage contract. There are four of the latter variables, namely, the contracted mortgage rate m, the number n of payments already made, the current principal balance P, and the current coupon payment c. However, in view of relationship (6.3), just three of these four variables need to be kept track of, say just m, n, and P. Hence the Markov decision chain must have five state variables: r, M, m, n, and P.

Now it is straightforward to proceed with the formulation of this Markov decision chain, but with five state variables the result will be in danger of having too many variables for computational tractability. So to reduce the number of state variables use will be made of Theorem 3.1 which suggests it is appropriate to assume $M_t = m(r_{t+1})$ for some deterministic function $m(\cdot)$ (note that since r is a predictable process, at time t the current state provides the value of r_{t+1}). Thus, at the cost of specifying the function $m(\cdot)$, the number of state variables have been reduced to four: r, m, n, and P.

Let $v(n, P, m, r)$ denote the minimum expected discounted value of the cash flow given n payments have been made on a contract having contracted mortgage rate m and remaining principal balance P and given that the riskless rate of interest for the next period is r. If it is optimal for the mortgagor to continue at least one more period with the current contract, then by (6.1)-(6.4)

$$\begin{aligned} v(n, P(n,m), m, r) &= (1+r)^{-1}[c + E_r v(n+1, P(n+1,m), m, R)] \\ &= (1+r)^{-1}\Big[\frac{mP(n,m)}{1-(1+m)^{n-N}} + \\ &\quad + E_r v(n+1, (1+m)P(n,m)\frac{(1+m)^{N-n-1}-1}{(1+m)^{N-n}-1}, m, R)\Big], \end{aligned}$$

where R here is the random short rate next period and the expectation is conditioned on the current value r of the short rate. On the other hand, if it is optimal to refinance, then

$$v(n, P(n,m), m, r) = K(P(n,m)) + v(0, P(n,m), m(r), r).$$

Hence the dynamic programming equation is

$$\begin{aligned} v(n, P, m, r) = \min\Big\{&K(P) + \\ &+ v(0, P, m(r), r), (1+r)^{-1}\Big[\frac{m(1+m)^{N-n}}{(1+m)^{N-n}-1}P + \\ &+ E_r v(n+1, \frac{(1+m)[(1+m)^{N-n-1}-1]}{(1+m)^{N-n}-1}P, m, R)\Big]\Big\}. \end{aligned} \tag{6.10}$$

Now this dynamic programming equation can perhaps be solved, but doing so is computationally challenging since there are four state variables, one of which is continuous. So to deal with this it is convenient to make one more assumption: the transaction cost $K(P) = kP$ for some positive constant k. As explained in Pliska [14], this assumption leads to the following result:

Proposition 2. *With $K(P) = kP$ the function $v(n,P,m,r)$ satisfies dynamic programming equation (6.11) if and only if the function $f(n,m,r) = P^{-1}v(n,P,m,r)$ satisfies the dynamic programming equation*

$$f(n,m,r) = min\Big\{k + f(0,m(r),r), \frac{1+m}{(1+r)[(1+m)^{N-n}-1]} \Big[m(1+m)^{N-n-1} + [(1+m)^{N-n-1}-1]E_r f(n+1,m,R)\Big]\Big\}. \tag{6.11}$$

With just three state variables, all of which are discrete and take just finitely many values, there is reason to be hopeful that dynamic programming equation (6.12) can be used to solve for the optimal value function f and thus v. However, there remain questions about existence and uniqueness of a solution, how to compute a solution, and how to identify an optimal refinancing strategy. Dynamic program (6.12) is in the infinite horizon category, so a simple recursive procedure starting with $f(N,m,r) = 0$ will not succeed. There are several standard infinite horizon sub-categories, but it is not apparent how to classify the one corresponding to (6.12). However, since $f(0,m(r),r) \neq k + f(0,m(r),r)$ it is straightforward to obtain the following useful result:

Proposition 3. *The dynamic programming equation (6.12) is equivalent to*

$$\begin{aligned} f(n,m,r) = min\Big\{&\frac{m(1+m)^{N-n}}{(1+r)[(1+m)^{N-n}-1]} + \frac{(1+m)[(1+m)^{N-n-1}-1]}{(1+r)[(1+m)^{N-n}-1]} \\ &E_r f(n+1,m,R), \quad k + \frac{m(r)(1+m(r))^N}{(1+r)[(1+m(r))^N-1]} + \\ &+\frac{(1+m(r))[(1+m(r))^{N-1}-1]}{(1+r)[(1+m(r))^N-1]} E_r f(1,m(r),R)\Big\}. \end{aligned} \tag{6.12}$$

As explained in Pliska [14], the recursive operator defined by (6.13) is a contraction, and so by dynamic programming theory (see, for instance, Puterman [15]) one concludes the following:

Theorem C.3. *The dynamic programming equation (6.13) has a unique solution f which can be computed by either the "successive approximations" algorithm or the "policy improvement" algorithm. The corresponding minimizing argument on the right hand side of (6.13) gives an optimal refinancing strategy.*

Remark 5. Each refinancing strategy, including the optimal one, will simply be a rule that specifies for each possible value of the state vector (n,m,r) either "continue" or "refinance."

6.5 A Mortgage Market Equilibrium Problem

This section studies an equilibrium problem that combines the ideas of the preceding two sections. The riskless short rate r is an exogenous, time homogeneous Markov chain and, for some deterministic function $m(\cdot)$, the mortgage rate evolves according to $M_t = m(r_{t+1})$. The mortgagor is a representative agent in the mortgage market who, based upon the dynamics of the riskless short rate r and the mortgage rate function $m(\cdot)$, seeks the best refinancing strategy. Meanwhile, based upon the mortgagor's prepayment behavior, the competitive forces in the mortgage market act so the mortgage rate function $m(\cdot)$ results in mortgage contracts having initial values V equal to the loan amounts $P(0, m(r_t))$. Thus the solution of this equilibrium problem will be a pair consisting of the mortgage rate function $m(\cdot)$ together with the mortgagor's refinancing strategy, with the properties that condition (6.9) of Theorem 3.1 for endogenous mortgage rates is satisfied for each possible value of the short rate r and the mortgagor's refinancing strategy is optimal as per Theorem 4.1.

To compute an equilibrium solution the following "naive" algorithm can be considered. Start with an arbitrary refinancing strategy, such as "never refinance." This defines an intensity process γ_r as well as a refinance time τ_r for each possible initial value of the spot rate process r. Then using equation (6.9) of Theorem 3.1 one computes the endogenous mortgage rate m for each possible initial value of the spot rate process r; this defines the mortgage rate function $m(\cdot)$. Next, one takes $m(\cdot)$ and solves the dynamic programming problem of Theorem 4.1 for the optimal refinancing strategy. If this strategy is the same as the one immediately before, then stop with an equilibrium solution. If not, then proceed with another iteration by using this new strategy and (6.9) to compute a new mortgage rate function $m(\cdot)$, and so forth. This algorithm will be illustrated by an example in the following section.

Of course, this or any other algorithm will not converge if there does not exist a solution to the equilibrium problem. However, as explained in Pliska [14], it is possible to formulate a new Markov decision chain, where the short rate process r is the single state variable, where one time period is one refinance cycle, and where the decision for each state is the choice of the stopping time representing the refinancing time. Moreover, the solution, if one exists, immediately provides a solution to the equilibrium problem. But as with the dynamic program of Theorem 4.1, the new dynamic program is known to always have a solution. Hence one has the following existence result:

Theorem C.4. *The equilibrium problem of this section always has a solution consisting of a mortgage rate function* $m(\cdot)$ *and an optimal refinancing strategy for the representative mortgagor.*

Remark 6. The equilibrium solution is not necessarily unique, because for some state the mortgagor might be indifferent between continuation and refinancing.

6.6 A Numerical Example

This section provides a simple numerical example that illustrates the preceding ideas but is not realistic enough to be taken seriously as a model of actual mortgage markets. The mortgage contract has a maturity of $N = 5$ periods, and the refinancing fee is $k = 0.03$, i.e., 3%. The short rate r takes one of four values: 2%, 3%, 4%, or 5%. The corresponding Markov chain has transition probabilities

$$Q(r_{t+1} = 2\% | r_t = 2\%) = Q(r_{t+1} = 3\% | r_t = 2\%) = 1/2\,,$$
$$Q(r_{t+1} = 2\% | r_t = 3\%) = Q(r_{t+1} = 3\% | r_t = 3\%) = Q(r_{t+1} = 4\% | r_t = 3\%) = 1/3\,,$$
$$Q(r_{t+1} = 3\% | r_t = 4\%) = Q(r_{t+1} = 4\% | r_t = 4\%) = Q(r_{t+1} = 5\% | r_t = 4\%) = 1/3\,,$$
$$Q(r_{t+1} = 4\% | r_t = 5\%) = Q(r_{t+1} = 5\% | r_t = 5\%) = 1/2\,.$$

Note the resulting Markov decision chain has $4^2 \times 5 = 80$ states, although some, e.g. $(1, m(5\%), 2\%)$, are not accessible.

To compute an equilibrium solution the "naive" algorithm is started with the trading strategy where the mortgagor does not refinance, regardless of the spot rate's value at contract initiation. To solve for the four endogenous mortgage rates, use is made of equation (6.9) which in this case can be written as

$$\frac{1 - (1+m)^{-5}}{m} = E_{r_1}\left[\sum_{i=1}^{5} \frac{1}{(1+r_1)\dots(1+r_i)}\right].$$

This results in the mortgage rate function $m(\cdot)$ taking the values 2.4733%, 3.0773%, 3.9061%, and 4.5201% corresponding to $r = 2\%$, 3%, 4% and 5%, respectively.

Next, the Markov decision chain is solved for the optimal refinancing strategy. It turns out that it is optimal to refinance in nine of the 80 states. But only two of these nine states are possible, namely, $(2, 4.5201\%, 2\%)$ and $(3, 3.9061\%, 2\%)$. In other words, with the mortgage rate function as above it is optimal to refinance a contract that started with $r = 5\%$ if and only if $r = 2\%$ at the end of the third period. And it is optimal to refinance a contract that started with $r = 4\%$ if and only if $r = 2\%$ at the end of the second period. But if a contract starts with either $r = 2\%$ or $r = 3\%$ then it should not be refinanced.

Since the refinancing strategy is new, the "naive" algorithm needs to proceed with at least one more iteration. For $r = 2\%$ and $r = 3\%$ the values of $m(r)$ will remain as before, but for $r = 4\%$ and $r = 5\%$ new values of endogenous mortgage rates must be computed using equation (6.9), which now has the form

$$1 = \frac{m}{1-(1+m)^{-5}} E_{r_1}\left[\sum_{i=1}^{\tau\wedge 5} \frac{1}{(1+r_1)\dots(1+r_i)}\right] + \\ + \frac{1}{1-(1+m)^{-5}} E_{r_1}\left[\frac{1-(1+m)^{-(5-\tau)}}{(1+r_1)\dots(1+r_{\tau\wedge 5})}\right].$$

Solving for m for the two cases $r = 4\%$ and $r = 5\%$ yields $m(4\%) = 3.9820\%$ and $m(5\%) = 4.5465\%$. The Markov decision chain is now solved with this new mortgage rate function, resulting in the same optimal refinancing strategy as with the preceding iteration. Hence the algorithm has converged to a solution of the equilibrium problem. Note the corresponding values of $f(0, m(r), r)$ in the solution of the dynamic programming equation (6.13) are 1.00000, 1.00000, 1.00194, and 1.00063 for $r = 2\%$, 3%, 4%, and 5%, respectively. This solution and other matters will be discussed in the following section.

6.7 Concluding Remarks

The equilibrium problem is like a two-person, nonzero sum game, where one player, the representative mortgagor, responds to given mortgage rates by choosing a refinancing strategy to minimize the expected present value of the cash flow and where the second player, the "market," responds to given mortgagor behavior by setting mortgage rates in a competitive fashion. It is interesting to note that as a result the mortgagor might "shoot himself in the foot" and force an equilibrium solution that has a higher expected present value. In other words, his myopic behavior of ignoring how his refinancing strategy affects the mortgage rates may result in an expected present value that is higher than necessary. This can be seen from the numerical example. The mortgage's expected present values 1.00194 and 1.00063 at contract initiation exceed one by precisely the expected present value of the refinancing costs. If the mortgagor is content with the strategy of never refinancing, then the expected present values of the mortgages will be less, namely 1.0, in states $r = 4\%$ and $r = 5\%$, even though smaller expected present values can be obtained, provided the mortgage rate function $m(\cdot)$ remains the same.

Although the numerical example is very simple, one thing it and the more general model suggest is that, in practice, mortgagors might be too hasty to refinance just because mortgage rates have dropped. In practice mortgagors probably focus on their monthly coupon payments while ignoring the expected present value of the new mortgage. The latter might not be small enough to justify refinancing because the new mortgage will have full maturity, whereas the existing one is much closer to maturity. Since mortgage rates have dropped the riskless interest rates have probably dropped too, and so the expected present values of the more distant new coupon payments might be higher than anticipated. It is desirable for future research to examine this more carefully be producing more realistic computational results, such as for a model having 180 monthly periods and several dozen levels of the interest rate r.

Of course actual mortgagors prepay for a variety of reasons, not just because they want to refinance the same principal. For example, some might choose to keep the principal amount unchanged while taking advantage of lower mortgage rates by paying the same monthly coupon amount as before, thereby paying off the loan more quickly than before. Others might choose to take advantage of increased property values by withdrawing equity and increasing the size of the loan. And many others prepay because, for a wide variety of circumstances, they sell their property. Given this variety of reasons and the heterogeneity of mortgagors, the equilibrium model and the concept of a representative mortgagor should probably not be taken too seriously. On the other hand, a more realistic goal for future research might be to generalize this paper's Markov decision chain model for optimal refinancing by incorporating additional mortgage prepayment reasons such as those enumerated above.

References

[1] Deng, Y., (1997), "Mortgage Termination: an Empirical Hazard Model with a Stochastic Term Structure," *J. of Real Estate and Economics* 14, 309-331.

[2] Deng, Y., J.M. Quigley, and R. Van Order, (2000), "Mortgage Termination, Heterogeneity, and the Exercise of Mortgage Options," *Econometrica* 68, 275-307.

[3] Dunn, K.B., and J.J. McConnell, (1981), "A Comparison of Alternative Models for Pricing GNMA Mortgage-Backed Securities," *J. of Finance* 36, 471-483.

[4] Dunn, K.B., and J.J. McConnell, (1981), "Valuation of Mortgage-Backed Securities," *J. of Finance* 36, 599-617.

[5] Dunn, K.B., and C.S. Spatt, (1999), "The Effect of Refinancing Costs and Market Imperfections on the Optimal Call Strategy and the Pricing of Debt Contracts," working paper, Carnegie Mellon University.

[6] Goncharov, Y., (2003), "An Intensity-Based Approach to Valuation of Mortgage Contracts Subject to Prepayment Risk," Ph.D. dissertation, University of Illinois at Chicago.

[7] Kalotay, A., D. Yang, and F.J. Fabozzi, (2004), "An Option-Theoretic Prepayment Model for Mortgages and Mortgage-Backed Securities," *Int. J. of Theoretical & Applied Finance*, to appear.

[8] Kariya, T., and M. Kobayashi, (2000), "Pricing Mortgage Backed Securities (MBS): a Model Describing the Burnout Effect," *Asian Pacific Financial Markets* 7, 182-204.

[9] Kariya, T., S.R. Pliska, and F. Ushiyama, (2002), "A 3-Factor Valuation Model for Mortgage Backed Securities (MBS)," working paper, Kyoto Institute of Economic Research.

[10] Kau, J.B., D.C. Keenan, and A.A. Smurov, (2004), "Reduced Form Mortgage Valuation," working paper, University of Georgia.

[11] Longstaff, F.A., (2002), "Optimal Refinancing and the Valuation of Mortgage-Backed Securities," working paper, University of California at Los Angeles.

[12] Nakagawa, H., and T. Shouda, (2004), "Valuation of Mortgage-Backed Securities Based on Unobservable Prepayment Cost Processes," working paper, Tokyo Institute of Technology.

[13] Pliska, S.R., (1997), *Introduction to Mathematical Finance: Discrete Time Models*, Blackwell, Oxford.

[14] Pliska, S.R., (2005), "Optimal Mortgage Refinancing with Endogenous Mortgage Rates: an Intensity Based, Equilibrium Approach," working paper, University of Illinois at Chicago.

[15] Puterman, M.L., (1994), *Markov Decision Processes: Discrete Stochastic Dynamic Programming*, J. Wiley, New York.

[16] Schwartz, E.S., and W.N. Torous, (1989), "Prepayment and the Valuation of Mortgage Backed Securities," *J. of Finance* 44, 375-392.

[17] Schwartz, E.S., and W.N. Torous, (1992), "Prepayment, Default, and the Valuation of Mortgage Pass-through Securities," *J. of Business* 65, 221-239.

[18] Stanton, R., (1995), "Rational Prepayment and the Valuation of Mortgage-Backed Securities," *Review of Financial Studies* 8, 677-708.

[19] Stanton, R., and N. Wallace, (1998), "Mortgage Choice: What's the Point?" *Real Estate Economics* 26, 173-205.

7

Computing efficient hedging strategies in discontinuous market models

Wolfgang J.Runggaldier and Sara Di Emidio

[1] Dipartimento di Matematica Pura ed Applicata, Universitá di Padova, Padova, Italy
runggal@math.unipd.it
[2] Dipartimento di Matematica Pura ed Applicata, Universitá di Padova, Padova, Italy.
sara.diemidio@gmail.com

Summary. We consider the problem of finding efficient hedging strategies in market models where prices evolve along discontinuous trajectories as a random jump process. We base ourselves on results in [3], that are briefly summarized, and discuss relevant computational issues. Numerical results are also presented.

7.1 Introduction

Our problem is of the following general form. Consider a market where agents may invest in a certain number N of (risky) assets, the prices of which we denote by the vector $S_t = (S_t^1, \cdots, S_t^N)$. We assume that S_t^i are already discounted with respect to a given non-risky asset (money market), thereby assuming implicitly that the short rate of interest is zero. We also assume that S_t^i admits a stochastic differential. We denote by $\xi_t = (\xi_t^1, \cdots, \xi_t^N)$ an investment strategy where ξ_t^i denotes the number of units of asset i, $(i = 1, \cdots, N)$ held in the portfolio at time t. Let V_t^ξ be the value, at time t, of the portfolio corresponding to a given strategy ξ that we assume to be self financing, i.e. such that

$$V_t^\xi = V_0^\xi + \int_0^t \xi_s dS_s \, , \quad V_0^\xi \quad \text{given} \, . \tag{7.1}$$

For simplicity we do not consider transaction costs. Given a maturity T, the problem in its most general form consists in determining ξ such that

$$E\left\{\ell\left(F(S_T), V_T^\xi\right)\right\} \quad \to \quad \min \tag{7.2}$$

for a given function $F(\cdot)$ of the asset price vector at maturity and a given loss function $\ell(\cdot,\cdot)$. In particular, we are interested in the hedging of a given claim $F(S_T)$, for which we consider more specifically

$$\ell\left(F(S_T), V_T^{\xi}\right) = \ell\left(F(S_T) - V_T^{\xi}\right), \tag{7.3}$$

namely a loss function of the hedging error. We call *efficient* a strategy that achieves the min in (12.3).

Standard price evolution models are diffusion-type models. However, especially on small time scales, the price evolution exhibits a jumping behavior. This is also the case in other situations, where one does not necessarily consider small time scales, like e.g. in the case of default sensitive assets (see [2], [5]).

A possible model for such a jumping behavior is

$$S_t^i = S_0^i \exp\left[a^i N_t^+ - b^i N_t^-\right] \ , \quad i = 1, \cdots, N \, , \tag{7.4}$$

where $a^i, b^i > 0$ and N_t^+, N_t^- are independent Poisson processes with intensities λ^+, λ^- respectively. A jump of N_t^+ causes an up-movement of the various S_t^i by a factor e^{a^i} and a jump of N_t^- a down-movement by the factor e^{-b^i}. The model thus generalizes the classical binomial market model by allowing the up- and down-movements to occur at random points in time. While the binomial model is complete, this one is incomplete. Corresponding to the multinomial generalization of the binomial model, here we could more generally consider

$$S_t^i = S_0^i \exp\left[\sum_{h=1}^{H} a^{i,h} N_t^{h,+} - \sum_{k=1}^{K} b^{i,k} N_t^{k,-}\right] \tag{7.5}$$

with $N_t^{h,+}, N_t^{k,-}$ independent Poisson jump processes.

We assume that the intensities λ^+, λ^- of N_t^+, N_t^- in (12.4) are constant over time. However, we allow them to be unknown and, taking the Bayesian point of view, we consider them as random variables, the distribution of which is continuously updated on the basis of the observed price movements. Therefore, while the intensities themselves are taken to be constant over time, their Bayesian updating gives them a dynamic aspect.

Standard approaches to solve the optimization problem (12.3) with (12.1) and (12.4) are based either on the method of Dynamic Programming (DP) or on the so-called *martingale method* (see e.g. a survey in [4]). Of the two, DP is inherently a dynamic approach. With uncertainty in the jump intensities and their dynamic Bayesian updating, DP thus turns out to be the more appropriate approach in our setting and this the more so if the purpose is to obtain quantitative results. For the standard diffusion-type price evolution models, DP leads to the solution of HJB-equations with the emphasis on finding explicit analytic solutions. In our context, DP leads to relations of the form as they appear in piecewise deterministic control problems (see e.g. [1]) and our purpose is to present a computationally feasible solution approach.

7.2 The Specific Problem

We consider the case of a single risky asset so that (12.4) becomes

$$S_t = S_0 \exp\left[a\, N_t^+ - b\, N_t^-\right] \tag{7.6}$$

with $a, b > 0$ and N_t^+, N_t^- independent Poisson processes with intensities λ^+, λ^- respectively. We allow λ^+, λ^- to be unknown and, taking the Bayesian point of view, we consider them as random variables. Since, for given t, $N_t^i, i = +, -$ are Poisson with parameters $\lambda^i t$, a convenient distribution for λ^i as random variables is a Gamma distribution (*conjugate family*), i.e.

$$f(\lambda^i; \alpha^i, \beta) = \frac{\beta^{\alpha^i}}{\Gamma(\alpha^i)} (\lambda^i)^{\alpha^i - 1} e^{-\beta\lambda^i}\,. \tag{7.7}$$

In fact, if the prior distribution for λ^i is Gamma, all updated distributions of λ^i are again Gamma : if λ^i has a prior with parameters (α_0^i, β_0) then, if at t one has observed N_t^+, the updated distribution is Gamma with parameters

$$\alpha_t^i = \alpha_0^i + N_t^i \;(i = +, -)\,, \qquad \beta_t = \beta_0 + t\,. \tag{7.8}$$

Itô's formula implies that, according to (12.6), one has

$$dS_t = S_{t-}\left[(e^a - 1)dN_t^+ + (e^{-b} - 1)dN_t^-\right] \tag{7.9}$$

and the self financing property of the strategy ξ, expressed by (12.1), becomes

$$dV_t^\xi = \xi_t dS_t = \xi_t S_{t-}\left[(e^a - 1)dN_t^+ + (e^{-b} - 1)dN_t^-\right]\,. \tag{7.10}$$

We assume V_0 to be given and, in what follows, we shall write V_t^ξ whenever we want to stress the dependence of the portfolio value on ξ. For observed portfolio values we shall simply write V_t.

We shall consider a strategy ξ to be admissible if it is predictable with respect to the filtration generated by S_t and such that $V_t^\xi \eta - c$ a.s. for a given $c > 0$. Given $T > 0$, the objective is to minimize

$$E\left\{\ell\left(F(S_T) - V_T^\xi\right)\right\} \quad \to \quad \min \tag{7.11}$$

where we suppose $F(\cdot)$ to be continuous and $\ell(\cdot)$ is considered to be a loss function, increasing, convex with $\ell(z) = 0$ for $z < 0$ (e.g. $\ell(z) = z^p$, $p\eta 1$, for $z > 0$). We shall assume that

$$E\left\{\ell\left(F(S_T) + c\right)\right\} < +\infty\,. \tag{7.12}$$

In view of the above, in particular the Bayesian updating, a sufficient statistic at the generic time t is the tuple

$$\left(V_t = v, N_t^+ = u, N_t^- = d, t\right) \tag{7.13}$$

where u, d are positive integers and $t \in [0, T]$. Concerning the range of values for v, let $C(u, d)$ denote the super-hedging capital that depends on the values of $N_t^+ = u, N_t^- = d$, but is independent of t (see [3]). It is the smallest initial capital, beyond which a given claim can always be perfectly hedged with a self financing portfolio. Our hedging problem thus looses its meaning for a value $V_t^\xi = v$ larger than $C(u, d)$. Consequently we shall consider $v \in [-c, C(u, d)]$.

We denote by $\mathcal{A}_{v,u,d,t}$ the class of admissible strategies over $[t, T]$, given the time $t-$statistic (v, u, d). Putting

$$\begin{cases} \tau_n := \inf\{t\eta 0 \mid N_t^+ + N_t^- = n\}, \\ \widehat{\tau}_n := \tau_n \wedge T, \end{cases} \tag{7.14}$$

the admissibility condition $V_t^\xi \eta - c$ then implies

$$\begin{aligned} &\xi_t \in I_{v,u,d} \\ &:= \left[-\frac{c+v}{S_0 e^{au-bd}(e^a - 1)}, \frac{c+v}{S_0 e^{au-bd}(1 - e^{-b})}\right], \; t \in \left(\widehat{\tau}_{u+d}, \widehat{\tau}_{u+d+1}\right]. \end{aligned} \tag{7.15}$$

The optimal value function (*minimal expected risk*) in (v, u, d, t) is then

$$\begin{aligned} &J^*(v, u, d, t) \\ &= \min_{\xi \in \mathcal{A}_{v,u,d,t}} E\left\{\ell\left(F(S_T) - v - \int_t^T \xi_s dS_s\right) \mid N_t^+ = u, N_t^- = d\right\}. \end{aligned} \tag{7.16}$$

7.3 Solution Approach

The solution approach is based on Dynamic Programming (DP). In the case of known intensities, putting $\lambda := \lambda^+ + \lambda^-$, it leads (see [3]) to the following relation for $J^*(v, u, d, t)$,

$$\begin{aligned} &J^*(v, u, d, t) = \left(T J^*\right)(v, u, d, t) := \\ &\int_0^{T-t} e^{-\lambda s} \min_{\zeta \in I_{v,u,d}} \left\{ \begin{array}{l} \lambda^+ J^*(v + \zeta S_0 e^{au-bd}(e^a - 1), u+1, d, t+s) + \\ +\lambda^- J^*(v + \zeta S_0 e^{au-bd}(e^{-b} - 1), u, d+1, t+s) \end{array} \right\} ds \\ &+ e^{-\lambda(T-t)} \ell\left(F(S_0 e^{au-bd}) - v\right). \end{aligned} \tag{7.17}$$

Intuitively, in (12.13) $J^*(v, u, d, t)$ appears as the minimum over the investment decision of the "expectation" of the value of J^* at the next jump whereby one takes into account that, over the remaining time to maturity, there may be a next jump either upwards or downwards or no jump at all. The case of no further jump does not affect the minimization; on the other hand, even if the price S remains constant between two successive jumps, the horizon shrinks and so the strategy changes to take this into account. Finally, notice that

$$e^{\lambda(T-t)} = 1 - \int_0^{T-t} (\lambda^+ + \lambda^-) e^{-\lambda s} ds \,.$$

The optimal investment decision at time $t \in (\tau_{u+d}, \tau_{u+d+1}]$ and with $v = V_{\tau_{u+d}}$ is then

$$\xi_t^* = arg \min_{\zeta \in I_{v,u,d}} \left\{ \begin{array}{l} \lambda^+ J^*(v + \zeta S_0 e^{au-bd}(e^a - 1), u+1, d, t) \\ +\lambda^- J^*(v + \zeta S_0 e^{au-bd}(e^{-b} - 1), u, d+1, t) \end{array} \right\} . \tag{7.18}$$

When the intensities are unknown, the relation (12.13) becomes

$$\begin{aligned} & J^*(v,u,d,t) = (\mathcal{T} J^*)(v,u,d,t) := \\ & \int_0^{T-t} \min_{\zeta \in I_{v,u,d}} \left\{ \begin{array}{l} p^+(u,d,t,s)\, J^*(v + \zeta S_0 e^{au-bd}(e^a - 1), u+1, d, t+s) \\ +p^-(u,d,t,s)\, J^*(v + \zeta S_0 e^{au-bd}(e^{-b} - 1), u, d+1, t+s) \end{array} \right\} ds \\ & +p^0(u+d,t)\, \ell\left(F(S_0 e^{au-bd}) - v\right), \end{aligned} \tag{7.19}$$

where $p^+(u,d,t,s), p^-(u,d,t,s)$ are the updated probabilities for an up- respectively down-jump at $t+s$, given that $N_t^+ = u, N_t^- = d$. Notice that, even if S remains constant from t to $t+s$, this still reveals additional information regarding the jump intensities. Furthermore, $p^0(u+d,t)$ is the updated probability that no more jumps occur in $[t,T]$, given $N_t^+ = u, N_t^- = d$.

In terms of the updated Gamma distributions for λ^+, λ^- we have (see [3])

$$\begin{cases} p^+(u,d,t,s) = \left(\frac{\beta_0+t}{\beta_0+t+s}\right)^{\alpha_0+u+d} \frac{\alpha_0^+ + u}{\beta_0+t+s}, \\ p^-(u,d,t,s) = \left(\frac{\beta_0+t}{\beta_0+t+s}\right)^{\alpha_0+u+d} \frac{\alpha_0^- + d}{\beta_0+t+s}, \end{cases} \tag{7.20}$$

where $\alpha_0 := \alpha_0^+ + \alpha_0^-$. Notice that both expressions have a common factor except for $\alpha_0^+ + u$ and $\alpha_0^- + d$ respectively. Furthermore,

$$\begin{aligned} p^0(u+d,t) &= 1 - \int_0^{T-t} \left(p^+(u,d,t,s) + p^-(u,d,t,s)\right) ds \\ &= \left(\frac{\beta_0 + t}{\beta_0 + T}\right)^{\alpha_0+u+d} . \end{aligned} \tag{7.21}$$

For what concerns the optimal investment decision, its value at time t has an expression that is analogous to the case of known intensities (for explicitly computable expressions see S.i)-S.iii), respectively S.i')-S.iii') below).

7.4 Computational Aspects

From the previous section it follows that, if one is able to compute the solution $J^*(v,u,d,t)$ of (12.13) respectively (12.15) for all tuples (v,u,d,t), then one can compute also the optimal strategy and the given problem is completely solved. A direct solution of (12.13) resp. (12.15) is difficult, if not impossible to obtain and so in this section we present, extending some of the results in [3], a computationally feasible approximation approach, structured along two levels:

i) *successive iterations*
ii) *quantization coupled with interpolation*

7.4.1 Successive Iterations

The operator $\mathcal{T}$, defined in (12.13) for the case of known intensities, is a contraction operator with contraction constant $1 - e^{-\lambda T} < 1$ so that the solution J^* of (12.13) can be obtained in the limit of successive iterations of this same operator $\mathcal{T}$. However, in the case of unknown intensities, the operator $\mathcal{T}$ in (12.15) contracts with factor $1 - p^0(u + d, t)$ that, see (7.21), tends to 1 (no contraction) in the limit when the total number $u+d$ of observed jumps tends to ∞.

This situation can be circumvented as follows. Let J^n be the $n-$th iterate of $\mathcal{T}$, both for known and unknown intensities, according to

$$J^0 \equiv 0 \quad \text{and, for } h \leq n, \quad J^h = \mathcal{T} J^{h-1}. \tag{7.22}$$

It can be shown that, see [3],

$$J^n(v, u, d, t) = \min_{\xi \in \mathcal{A}_{v,u,d,t}} E\left\{\ell\left(F(S_T) - V_T^\xi\right), \tau_{u+d+n} > T \mid N_t^+ = u, N_t^- = d\right\}, \tag{7.23}$$

i.e. the $n-$th iterate can be interpreted as minimal risk in (v, u, d, t) if at most n jumps occur in the remaining interval $[t, T]$. It follows that if one fixes a priori a maximum number n of jumps then, both for known and unknown intensities, the $n-$th iterate of $\mathcal{T}$ suffices to obtain the optimal value under this restriction on the number of jumps.

One can then easily see that, for all (v, u, d, t),

$$\begin{aligned} J^n(\cdot) \leq J^*(\cdot) \leq J^n(\cdot) + \\ + E\left\{\ell\left(F(S_T) + c\right), \tau_{u+d+n} \leq T \mid N_t^+ = u, N_t^- = d\right\}. \end{aligned} \tag{7.24}$$

Having made the assumption that $E\left\{\ell\left(F(S_T) + c\right)\right\} < +\infty$, it then follows that

$$J^n(v, u, d, t) \overset{n \to \infty}{\longrightarrow} J^*(v, u, d, t), \tag{7.25}$$

uniformly in (v, t) for all (u, d).

Consider now the optimal strategy under the restriction of at most n jumps that we denote by ξ_t^n. It is computed as follows, where we consider now only the case of unknown intensities and where we take into account the fact that, by (7.20), $p^+(\cdot)$ and $p^-(\cdot)$ have a common factor and distinguish themselves only by the factors $\alpha_0^+ + u$ and $\alpha_0^- + d$ respectively:

S.i) for $t \in [0, \widehat{\tau}_1]$ and $v = V_0$ put

$$\xi_t^n = arg \min_{\zeta \in I_{v,0,0}} \left\{ \begin{array}{l} \alpha_0^+ J^{n-1}(v + \zeta S_0 e^{au-bd}(e^a - 1), 1, 0, t) \\ +\alpha_0^- J^{n-1}(v + \zeta S_0 e^{au-bd}(e^{-b} - 1), 0, 1, t) \end{array} \right\},$$

S.ii) for $t \in (\widehat{\tau}_h, \widehat{\tau}_{h+1}]$, $(1 \le h < n-2)$, having observed $N^+_{\widehat{\tau}_h} = u, N^-_{\widehat{\tau}_h} = d$, $(u + d = h)$, $V_{\widehat{\tau}_h} = v$, put:

$$\xi_t^n = arg \min_{\zeta \in I_{v,u,d}} \left\{ \begin{array}{l} (\alpha_0^+ + u)\, J^{n-h-1}(v + \zeta S_0 e^{au-bd}(e^a - 1), u+1, d, t) \\ +(\alpha_0^- + d)\, J^{n-h-1}(v + \zeta S_0 e^{au-bd}(e^{-b} - 1), u, d+1, t) \end{array} \right\},$$

S.iii) if $\tau_{n-1} < T$ then, for $t \in (\widehat{\tau}_{n-1}, T]$, put:

$$\xi_t^n \equiv 0 \quad \textit{(i.e. transfer all funds to the money account)}.$$

Let V_t^n be the wealth process associated with ξ_t^n, i.e.

$$V_t^n = V_0 + \int_0^t \xi_s^n dS_s \,. \tag{7.26}$$

As with (7.24), one can easily see that, in particular at the initial time i.e. for $(v, u, d, t) = (V_0, 0, 0, 0)$, one has

$$\left\{ \begin{array}{lllll} J^n(\cdot) & \le E\{\ell(F(S_T) - V_T^n)\} \le & J^n(\cdot) & + E\{\ell(F(S_T)+c), \tau_n \le T\} \\ \downarrow n\to\infty & & \downarrow n\to\infty & \downarrow n\to\infty \\ J^*(\cdot) & & J^*(\cdot) & 0 \end{array} \right. \tag{7.27}$$

which is a relation that specifies the sense in which the performance of the strategy ξ^n is suboptimal. In particular, (7.27) shows that, for $n \to \infty$, the performance of the strategy ξ^n tends to that of the optimal one ξ^*.

Concluding this subsection we have found that, by iterating the operator $\mathcal{T}$ (both for known as well as unknown intensities) a sufficiently large number n of times, one can approximate the optimal value and the performance of the optimal strategy as closely as possible. It remains to actually compute the iterations with the operator $\mathcal{T}$ corresponding to the various possible tuples (v, u, d, t). This is the subject of the next subsection.

7.4.2 Computation by Quantization

The iterations with the operator $\mathcal{T}$ in (12.13) respectively (12.15) have to be computed for all possible tuples (v, u, d, t). Given a maximum number n of jumps, one has $u + d \le n$ and so the pair (u, d) takes only a finite number of possible values. The pair (v, t) however takes a continuum of possible values with

$$v \in [-c, C(n)]\,, \qquad t \in [0, T]\,, \tag{7.28}$$

where, given n,

$$C(n) := \max\{C(u,d) \mid u+d \leq n\} \tag{7.29}$$

and where $C(u,d)$ is the super-hedging capital introduced after (12.10). To make the iteration in (7.22) computable, we have thus to discretize the possible values of (v,t) and we do this by *quantization* (computation over a grid of values) followed by an interpolation, in the same variables, of the computed values.

More precisely, given n, consider a finite grid G

$$G \subset D := [-c, C(n)] \times [0,T] \tag{7.30}$$

containing the extremal points of D. Define

$$(\mathcal{T}_G J)(v,u,d,t) := \begin{cases} (\mathcal{T}J)(v,u,d,t) & , \text{ if } (v,t) \in G \\ \text{cadlag interpolation} & , \text{ else} \end{cases} \tag{7.31}$$

where by *cadlag interpolation* we mean a right-continuous, piecewise constant interpolation.

Let $J^{n,G}$ denote the n−th iterate of $\mathcal{T}_G$ according to

$$J^{0,G} \equiv 0 \quad \text{and, for } h \leq n, \quad J^{h,G} = \mathcal{T}_G J^{h-1,G}. \tag{7.32}$$

More specifically, denote by v_j, $(j = 0, 1, \cdots, J)$ and t_i, $(i = 0, 1, \cdots, I)$ the points in $[-c, C(n)]$ and $[0,T]$ respectively that define the grid $G \subset D$ and let $V_j := [v_j, v_{j+1})$, $(j = 0, \cdots, J-1)$. Taking into account the definition of the operator $\mathcal{T}$ in (12.15) and the expressions (7.20), (7.21), the computation of the recursions in (7.32) can then be performed according to the formula (see also (55) in [3])

$$J^{h,G}(v_j,u,d,t_i) = J^{1,G}(v_j,u,d,t_i) + \sum_{l=0}^{I-1} 1_{\{t_i \leq t_l\}} \tag{7.33}$$

$$\min_{\zeta \in I_{v_j,u,d}} \left\{ \begin{array}{l} \gamma^u_{i,l} \sum_{m=0}^{J-1} 1_{\{V_m\}} \left(v_j + \zeta S_0 e^{au-bd}(e^a - 1)\right) J^{h-1,G}(v_m, u+1, d, t_l) \\ + \gamma^d_{i,l} \sum_{m=0}^{J-1} 1_{\{V_m\}} \left(v_j + \zeta S_0 e^{au-bd}(e^{-b} - 1)\right) J^{h-1,G}(v_m, u, d+1, t_l) \end{array} \right\}$$

where

$$\begin{cases} \gamma^u_{i,l} := \dfrac{(\beta_0 + t_i)^{\alpha_0+u+d}(\alpha_0^+ + u)}{\alpha_0 + u + d} \\ \qquad \left[(\beta_0 + t_l)^{-(\alpha_0+u+d)} - (\beta_0 + t_{l+1})^{-(\alpha_0+u+d)}\right] \\ \gamma^d_{i,l} := \gamma^u_{i,l} \dfrac{(\alpha_0^- + d)}{(\alpha_0^+ + u)} \end{cases} \tag{7.34}$$

and where (see again (12.15))

$$
\begin{aligned}
J^{1,G}(v_j,u,d,t_i) &= p^0(u+d,t_i)\,\ell\left(F(S_0e^{au-bd})-v_j\right)\\
&= \left(\frac{\beta_0+t_i}{\beta_0+T}\right)^{\alpha_0+u+d}\ell\left(F(S_0e^{au-bd})-v_j\right). \qquad (7.35)
\end{aligned}
$$

Denote by $\xi^{n,G}$ the strategy, defined by analogy to ξ^n, but corresponding to the iterations of $\mathcal{T}_G$ and let $V_t^{n,G}$ be the associated wealth process. More precisely, recalling from (7.20) that $p^+(\cdot)$ and $p^-(\cdot)$ have a common factor and distinguish themselves only by the terms $(\alpha_0^+ + u)$ and $(\alpha_0^- + d)$ respectively, we have (compare with (S.i–S.iii):

S.i') for $t \in [0,\widehat{\tau}_1]$ and $v = V_0$ put

$$
\xi_t^{n,G} = arg \min_{\zeta\in I_{v,0,0}} \left\{ \begin{array}{l} \alpha_0^+ \sum_{m=0}^{J-1} 1_{\{V_m\}}(v+\zeta S_0 e^{au-bd}(e^a-1))\,J^{n-1,G} \\ \hfill (v_m,1,0,t) \\ +\alpha_0^- \sum_{m=0}^{J-1} 1_{\{V_m\}}(v+\zeta S_0 e^{au-bd}(e^{-b}-1))\,J^{n-1,G} \\ \hfill (v_m,0,1,t) \end{array} \right\},
$$

S.ii') for $t \in (\widehat{\tau}_h,\widehat{\tau}_{h+1}]$, $(1 \le h < n-2)$, having observed $N^+_{\widehat{\tau}_h} = u, N^-_{\widehat{\tau}_h} = d$, $(u+d=h)$, $V_{\widehat{\tau}_h} = v$, put:

$$
\xi_t^{n,G} = arg \min_{\zeta\in I_{v,u,d}} \left\{ \begin{array}{l} (\alpha_0^+ + u) \sum_{m=0}^{J-1} 1_{\{V_m\}} \\ (v+\zeta S_0 e^{au-bd}(e^a-1))\,J^{n-h-1,G}(v_m,u+1,d,t)+ \\ +(\alpha_0^- + d) \sum_{m=0}^{J-1} 1_{\{V_m\}} \\ (v+\zeta S_0 e^{au-bd}(e^{-b}-1))\,J^{n-h-1,G}(v_m,u,d+1,t) \end{array} \right\},
$$

S.iii') if $\tau_{n-1} < T$ then, for $t \in (\widehat{\tau}_{n-1},T]$, put:

$$
\xi_t^{n,G} \equiv 0 \quad \textit{(i.e. transfer all funds to the money account).}
$$

With respect to S.i)-S.iii), here we have emphasized the fact that, as a function of v, $J^{n,G}(v,u,d,t)$ remains constant for all values of v between two grid points. Notice furthermore that, also as a function of t, $J^{n,G}(v,u,d,t)$ remains constant between two grid points and so $\xi_t^{n,G}$ changes only at a jump time as indicated in S.i')-S.iii') or, within a same interval $(\widehat{\tau}_h,\widehat{\tau}_{h+1}]$, when t crosses a grid point (see the numerical results below).

Since, with $(v,t) \in G$, the tuples (v,u,d,t) are now finite in number, the values $J^{n,G}$ and the strategies $\xi^{n,G}$ can actually be computed.

We next discuss the goodness of the approximation introduced above.

7.4.3 Bounds and Convergence

Given n, let

$$E^n := \{(v,u,d,t) \mid v\eta - c,\ u+d \leq n,\ t \in [0,T]\} \tag{7.36}$$

and denote by $\mathcal{D}(E^n)$ the space of cadlag functions on E^n endowed with the sup-norm $||\cdot||_{E^n}$.

We have now two facts. The first one follows straightforwardly from the continuity of $J^*(v,u,d,t)$ (recall that we had assumed $F(\cdot)$ continuous and $\ell(\cdot)$ convex), namely:

$$\begin{cases} \varepsilon(G) := ||\, J^* - \mathcal{T}_G J^*||_{E^n} \ \rightarrow\ 0\,, \\ \text{for } \delta^G := \sup\limits_{(v,t)\in D,\,(v',t')\in G} [|v-v'| + |t-t'|] \ \rightarrow\ 0\,. \end{cases} \tag{7.37}$$

The second follows from results in [3], in particular Corollary 4.3 and Section 5 (notice that, in the notation of [3], we have $H_n^* = H_n^n$ in Section 5 there), namely:

$$||\, J^* - J^{n,G}||_{E^n} \leq \frac{\varepsilon(G)}{p^0(n,0)}\,. \tag{7.38}$$

Combining (7.37) with (7.38) we have that, for given n, the upper bound in (7.38) tends to zero for $\delta^G \to 0$, i.e. for the grid G becoming finer and finer. Notice however that, since $\lim_{n\to\infty} p^0(n,0) = 0$, the convergence to zero of this upper bound becomes slower as n increases.

To evaluate the goodness of the approximation introduced by the computable quantities $J^{n,G}$ and $\xi^{n,G}$ that depend on the choice of n and G, notice that by analogy to (7.27) we have, at the initial time $t=0$,

$$\begin{aligned} J^{n,G}(\cdot) &\leq E\left\{\ell\left(F(S_T) - V_T^{n,G}\right)\right\} \\ &\leq J^{n,G}(\cdot) + E\left\{\ell\left(F(S_T)+c\right),\ \tau_n \leq T\right\} \end{aligned} \tag{7.39}$$

which, combined with (7.38) and the fact that the rightmost term tends to zero for $n \to \infty$, specifies the sub-optimality of the performance of the computable strategy $\xi^{n,G}$: having chosen n sufficiently large so that the rightmost term is small enough, choose the grid G sufficiently fine so that $J^{n,G}(\cdot)$ is close enough to $J^*(\cdot)$ in the sense of (7.38). By our approach one can thus approximate the optimal value and the performance of the optimal strategy as closely as possible.

7.5 Example and Numerical Results

7.5.1 Description of the Example

We consider here an example corresponding to example 5.1 in [3]. More precisely, we consider the geometric Poisson price model (12.6) with $S_0 = 1$ and

a, b such that $e^a = 2, e^{-b} = 1/2$. Assume the intensities λ^+, λ^- unknown and having as prior distribution a Gamma with parameters $(\alpha_0^+ = 1, \beta_0 = 1)$ and $(\alpha_0^- = 1, \beta_0 = 1)$ respectively. The claim is supposed to be a European call, namely $F(S_T) = (S_T - 1)^+$ and as loss function take a quadratic, namely $\ell(z) = [\max(z, 0)]^2$. Finally, take a horizon of $T = 2$ and let the lower bound for the portfolio value correspond to $c = 0.5$. In the given situation the value of $C(n)$ in (7.29) is bounded from above by $C(n) \leq 2^{n-2}$. As domain D for the pair (v, t) we therefore take the rectangle $D(n) = [-0.5, 2^{n-2}] \times [0, 2]$.

7.5.2 Numerical Results

For the given example we report here numerical results for portfolio values and strategies when the maximum number of jumps is supposed to be either $n = 5$ or $n = 8$ so that some comparison can be made. The portfolio values are reported also for $n = 6$ and $n = 7$. We recall from section 7.4.1 that, if the actual number of jumps turns out to be larger than the given n, then we put $\xi_t^{n,G} \equiv 0$ for $t \in (\widehat{\tau}_{n-1}, T]$, i.e. we transfer all funds to the money account.

Case of $n = 5$ and quantization given by $(v_0, \cdots, v_3) = (-\frac{1}{2}, 0, 1, 2, 4, 8)$, $(t_0, \cdots, t_3) = (0, \frac{1}{2}, 1, 2)$

The strategy $\xi_t^{5,G}$ is described in the following table, where an interval in the first column means that the strategy can be assigned any value within that interval and where the values for $V_{\widehat{\tau}_k}$ $(k = 1, 2, 3)$ are the values on the grid that correspond to the left end point of the interval that contains the actual value of $V_{\widehat{\tau}_k}$.

$$\xi_s^{5,G} = \begin{cases} 1 & \text{for } s \in [0, \widehat{\tau}_1] \cap [0, 2],\ V_0 = 0 \\ [1,2] & \text{for } s \in [0, \widehat{\tau}_1] \cap [0, 2],\ V_0 = 1 \\ [0,2] & \text{for } s \in [0, \widehat{\tau}_1] \cap [0, 2],\ V_0 = 2 \\ [-2,4] & \text{for } s \in [0, \widehat{\tau}_1] \cap [0, 2], V_0 = 4 \\ 0 & \text{for } s \in [0, \widehat{\tau}_1] \cap [0, 2],\ V_0 = 8 \\ \frac{3}{2} & \text{for } s \in (\widehat{\tau}_1, \widehat{\tau}_2] \cap [0, 2] \text{ with } u = 1, d = 0,\ V_{\widehat{\tau}_1} = 1 \\ 1 & \text{for } s \in (\widehat{\tau}_1, \widehat{\tau}_2] \cap [0, 2] \text{ with } u = 1, d = 0,\ V_{\widehat{\tau}_1} = 2 \\ 0 & \text{for } s \in (\widehat{\tau}_1, \widehat{\tau}_2] \cap [0, 2] \text{ with } u = 1, d = 0,\ V_{\widehat{\tau}_1} = 8 \\ 0 & \text{for } s \in (\widehat{\tau}_1, \widehat{\tau}_2] \cap [0, 2] \text{ with } u = 0, d = 1,\ V_{\widehat{\tau}_1} \in \{-\frac{1}{2}, 0, 8\} \\ [-2,8] & \text{for } s \in (\widehat{\tau}_1, \widehat{\tau}_2] \cap [0, 2] \text{ with } u = 0, d = 1,\ V_{\widehat{\tau}_1} = 2 \\ [-6,8] & \text{for } s \in (\widehat{\tau}_1, \widehat{\tau}_2] \cap [0, 2] \text{ with } u = 0, d = 1,\ V_{\widehat{\tau}_1} = 4 \\ 0 & \text{for } s \in (\widehat{\tau}_2, \widehat{\tau}_3] \cap [0, 2] \text{ with } u = 1, d = 1,\ V_{\widehat{\tau}_2} \in \{-\frac{1}{2}, 8\} \\ 1 & \text{for } s \in (\widehat{\tau}_2, \widehat{\tau}_3] \cap [0, 2] \text{ with } u = 1, d = 1,\ V_{\widehat{\tau}_2} = 0 \\ [1,2] & \text{for } s \in (\widehat{\tau}_2, \widehat{\tau}_3] \cap [0, 2] \text{ with } u = 1, d = 1,\ V_{\widehat{\tau}_2} = 1 \\ [0,4] & \text{for } s \in (\widehat{\tau}_2, \widehat{\tau}_3] \cap [0, 2] \text{ with } u = 1, d = 1,\ V_{\widehat{\tau}_2} = 2 \\ [-2,4] & \text{for } s \in (\widehat{\tau}_2, \widehat{\tau}_3] \cap [0, 2] \text{ with } u = 1, d = 1,\ V_{\widehat{\tau}_2} = 4 \\ 0 & \text{for } s \in (\widehat{\tau}_2, \widehat{\tau}_3] \cap [1, 2] \text{ with } u = 2, d = 0,\ V_{\widehat{\tau}_2} = 8 \\ 1 & \text{for } s \in (\widehat{\tau}_2, \widehat{\tau}_3] \cap [1, 2] \text{ with } u = 2, d = 0,\ V_{\widehat{\tau}_2} = 4 \\ 0 & \text{for } s \in (\widehat{\tau}_2, \widehat{\tau}_3] \cap [0, 2] \text{ with } u = 0, d = 2,\ V_{\widehat{\tau}_2} \in \{-\frac{1}{2}, 0, 8\} \end{cases}$$

$\cdots/\cdots$

$$\xi_s^{5,G} = \begin{cases} [-8,16] & \text{for } s \in (\widehat{\tau}_2, \widehat{\tau}_3] \cap [0,2] \text{ with } u=0, d=2,\ V_{\widehat{\tau}_2}=2 \\ [-16,16] & \text{for } s \in (\widehat{\tau}_2, \widehat{\tau}_3] \cap [0,2] \text{ with } u=0, d=2,\ V_{\widehat{\tau}_2}=4 \\ 0 & \text{for } s \in (\widehat{\tau}_3, \widehat{\tau}_4] \cap [0,2] \text{ with } u=2, d=1,\ V_{\widehat{\tau}_3} \in \{-\frac{1}{2}, 8\} \\ \frac{3}{2} & \text{for } s \in (\widehat{\tau}_3, \widehat{\tau}_4] \cap [0,2] \text{ with } u=2, d=1,\ V_{\widehat{\tau}_3}=1 \\ 1 & \text{for } s \in (\widehat{\tau}_3, \widehat{\tau}_4] \cap [0,2] \text{ with } u=2, d=1,\ V_{\widehat{\tau}_3}=2 \\ 0 & \text{for } s \in (\widehat{\tau}_3, \widehat{\tau}_4] \cap [0,2] \text{ with } u=1, d=2,\ V_{\widehat{\tau}_3} \in \{-\frac{1}{2}, 0, 8\} \\ [-4,8] & \text{for } s \in (\widehat{\tau}_3, \widehat{\tau}_4] \cap [0,2] \text{ with } u=1, d=2,\ V_{\widehat{\tau}_3}=2 \\ [-8,8] & \text{for } s \in (\widehat{\tau}_3, \widehat{\tau}_4] \cap [0,2] \text{ with } u=1, d=2,\ V_{\widehat{\tau}_3}=4 \\ 0 & \text{for } s \in (\widehat{\tau}_3, \widehat{\tau}_4] \cap [0,2] \text{ with } u=3, d=0,\ V_{\widehat{\tau}_3}=8 \\ 0 & \text{for } s \in (\widehat{\tau}_3, \widehat{\tau}_4] \cap [0,2] \text{ with } u=0, d=3,\ V_{\widehat{\tau}_3} \in \{-\frac{1}{2}, 0, 8\} \\ [-16,32] & \text{for } s \in (\widehat{\tau}_3, \widehat{\tau}_4] \cap [0,2] \text{ with } u=0, d=3,\ V_{\widehat{\tau}_3}=2 \\ [-32,32] & \text{for } s \in (\widehat{\tau}_3, \widehat{\tau}_4] \cap [0,2] \text{ with } u=0, d=3,\ V_{\widehat{\tau}_3}=4 \end{cases}$$

The values of $J^{5,G}$ for the various initial conditions $(v,0,0,0)$ corresponding to the 5 non-negative grid-values of v are shown on the next table. As expected, they are decreasing in v.

$$\begin{cases} J^{5,G}(0,0,0,0) = 0.8366 \\ J^{5,G}(1,0,0,0) = 0.7023 \\ J^{5,G}(2,0,0,0) = 0.6994 \\ J^{5,G}(4,0,0,0) = 0.6994 \\ J^{5,G}(8,0,0,0) = 0.6994 \end{cases}$$

Case of $n = 8$ and quantization given by $(v_0, \cdots, v_9) = (-\frac{1}{2}, 0, 1, 2, 4, 8, 16, 32, 64)$, $(t_0, \cdots, t_3) = (0, \frac{1}{2}, 1, 2)$

Corresponding to the previous subsection, in the next table we describe the strategy $\xi_t^{8,G}$ that now is naturally more complex.

$$\xi_s^{8,G} = \begin{cases} 1 & \text{for } s \in [0, \widehat{\tau}_1] \cap [0,2],\ V_0 = 0 \\ [1,2] & \text{for } s \in [0, \widehat{\tau}_1] \cap [0,2],\ V_0 = 1 \\ \{0,2\} & \text{for } s \in [0, \widehat{\tau}_1] \cap [0,2],\ V_0 = 2 \\ \{0,4\} & \text{for } s \in [0, \widehat{\tau}_1] \cap [0,2],\ V_0 = 4 \\ [0,8] & \text{for } s \in [0, \widehat{\tau}_1] \cap [0,2],\ V_0 = 8 \\ [-8,24] & \text{for } s \in [0, \widehat{\tau}_1] \cap [0,2],\ V_0 = 16 \\ [-24,32] & \text{for } s \in [0, \widehat{\tau}_1] \cap [0,2],\ V_0 = 32 \\ 0 & \text{for } s \in [0, \widehat{\tau}_1] \cap [0,2],\ V_0 = 64 \\ 0 & \text{for } s \in (\widehat{\tau}_1, \widehat{\tau}_2] \cap [0,2] \text{ with } u=1, d=0,\ V_{\widehat{\tau}_1}=64 \\ \frac{3}{2} & \text{for } s \in (\widehat{\tau}_1, \widehat{\tau}_2] \cap [0,2] \text{ with } u=1, d=0,\ V_{\widehat{\tau}_1}=1 \\ 1 & \text{for } s \in (\widehat{\tau}_1, \widehat{\tau}_2] \cap [0,2] \text{ with } u=1, d=0,\ V_{\widehat{\tau}_1}=2 \\ \{0,2\} & \text{for } s \in (\widehat{\tau}_1, \widehat{\tau}_2] \cap [0,2] \text{ with } u=1, d=0,\ V_{\widehat{\tau}_1}=4 \\ [0,4] & \text{for } s \in (\widehat{\tau}_1, \widehat{\tau}_2] \cap [0,2] \text{ with } u=1, d=0,\ V_{\widehat{\tau}_1}=8 \\ 0 & \text{for } s \in (\widehat{\tau}_1, \widehat{\tau}_2] \cap [0,2] \text{ with } u=0, d=1,\ V_{\widehat{\tau}_1} \in \{-\frac{1}{2}, 64\} \\ 2 & \text{for } s \in (\widehat{\tau}_1, \widehat{\tau}_2] \cap [0,1] \text{ with } u=0, d=1, V_{\widehat{\tau}_1}=0 \\ 0 & \text{for } s \in (\widehat{\tau}_1, \widehat{\tau}_2] \cap [1,2] \text{ with } u=0, d=1, V_{\widehat{\tau}_1}=0 \end{cases}$$

…/…

$$
\xi_s^{8,G} = \begin{cases}
\frac{1}{2} & \text{for } s \in (\widehat{\tau}_1, \widehat{\tau}_2] \cap [0,2] \quad \text{with } u = 0, d = 1,\ V_{\widehat{\tau}_1} = 2 \\
1 & \text{for } s \in (\widehat{\tau}_1, \widehat{\tau}_2] \cap [0,2] \quad \text{with } u = 0, d = 1,\ V_{\widehat{\tau}_1} = 4 \\
[0,2] & \text{for } s \in (\widehat{\tau}_1, \widehat{\tau}_2] \cap [0,2] \quad \text{with } u = 0, d = 1,\ V_{\widehat{\tau}_1} = 8 \\
[-2,6] & \text{for } s \in (\widehat{\tau}_1, \widehat{\tau}_2] \cap [0,2] \quad \text{with } u = 0, d = 1,\ V_{\widehat{\tau}_1} = 16 \\
[-6,8] & \text{for } s \in (\widehat{\tau}_1, \widehat{\tau}_2] \cap [0,2] \quad \text{with } u = 0, d = 1,\ V_{\widehat{\tau}_1} = 32 \\
0 & \text{for } s \in (\widehat{\tau}_2, \widehat{\tau}_3] \cap [0,2] \quad \text{with } u = 1, d = 1,\ V_{\widehat{\tau}_2} \in \{-\frac{1}{2}, 64\} \\
1 & \text{for } s \in (\widehat{\tau}_2, \widehat{\tau}_3] \cap [1,2] \quad \text{with } u = 1, d = 1,\ V_{\widehat{\tau}_2} = 0 \\
[1,2] & \text{for } s \in (\widehat{\tau}_2, \widehat{\tau}_3] \cap [0,2] \quad \text{with } u = 1, d = 1,\ V_{\widehat{\tau}_2} = 1 \\
\{0,2\} & \text{for } s \in (\widehat{\tau}_2, \widehat{\tau}_3] \cap [0,2] \quad \text{with } u = 1, d = 1,\ V_{\widehat{\tau}_2} = 2 \\
[0,4] & \text{for } s \in (\widehat{\tau}_2, \widehat{\tau}_3] \cap [0,2] \quad \text{with } u = 1, d = 1,\ V_{\widehat{\tau}_2} = 4 \\
[-4,12] & \text{for } s \in (\widehat{\tau}_2, \widehat{\tau}_3] \cap [0,2] \quad \text{with } u = 1, d = 1,\ V_{\widehat{\tau}_2} = 8 \\
[-12,24] & \text{for } s \in (\widehat{\tau}_2, \widehat{\tau}_3] \cap [0,2] \quad \text{with } u = 1, d = 1,\ V_{\widehat{\tau}_2} = 16 \\
0 & \text{for } s \in (\widehat{\tau}_2, \widehat{\tau}_3] \cap [1,2] \quad \text{with } u = 2, d = 0,\ V_{\widehat{\tau}_2} = 64 \\
1 & \text{for } s \in (\widehat{\tau}_2, \widehat{\tau}_3] \cap [1,2] \quad \text{with } u = 2, d = 0,\ V_{\widehat{\tau}_2} = 4 \\
[0,2] & \text{for } s \in (\widehat{\tau}_2, \widehat{\tau}_3] \cap [1,2] \quad \text{with } u = 2, d = 0,\ V_{\widehat{\tau}_2} = 8 \\
0 & \text{for } s \in (\widehat{\tau}_2, \widehat{\tau}_3] \cap [0,2] \quad \text{with } u = 0, d = 2,\ V_{\widehat{\tau}_2} \in \{-\frac{1}{2}, 0, 64\} \\
[4,8] & \text{for } s \in (\widehat{\tau}_2, \widehat{\tau}_3] \cap [0,2] \quad \text{with } u = 0, d = 2,\ V_{\widehat{\tau}_2} = 1 \\
[0,8] & \text{for } s \in (\widehat{\tau}_2, \widehat{\tau}_3] \cap [0,2] \quad \text{with } u = 0, d = 2,\ V_{\widehat{\tau}_2} = 2 \\
[-24,56] & \text{for } s \in (\widehat{\tau}_2, \widehat{\tau}_3] \cap [0,2] \quad \text{with } u = 0, d = 2,\ V_{\widehat{\tau}_2} = 8 \\
[-56,120] & \text{for } s \in (\widehat{\tau}_2, \widehat{\tau}_3] \cap [0,2] \quad \text{with } u = 0, d = 2,\ V_{\widehat{\tau}_2} = 16 \\
[-120,128] & \text{for } s \in (\widehat{\tau}_2, \widehat{\tau}_3] \cap [0,2] \quad \text{with } u = 0, d = 2,\ V_{\widehat{\tau}_2} = 32 \\
0 & \text{for } s \in (\widehat{\tau}_3, \widehat{\tau}_4] \cap [0,2] \quad \text{with } u = 2, d = 1,\ V_{\widehat{\tau}_3} = \{-\frac{1}{2}, 64\} \\
\frac{3}{2} & \text{for } s \in (\widehat{\tau}_3, \widehat{\tau}_4] \cap [0,2] \quad \text{with } u = 2, d = 1,\ V_{\widehat{\tau}_3} = 1 \\
1 & \text{for } s \in (\widehat{\tau}_3, \widehat{\tau}_4] \cap [0,2] \quad \text{with } u = 2, d = 1,\ V_{\widehat{\tau}_3} = 2 \\
[0,2] & \text{for } s \in (\widehat{\tau}_3, \widehat{\tau}_4] \cap [0,2] \quad \text{with } u = 2, d = 1,\ V_{\widehat{\tau}_3} = 4 \\
[-2,6] & \text{for } s \in (\widehat{\tau}_3, \widehat{\tau}_4] \cap [0,2] \quad \text{with } u = 2, d = 1,\ V_{\widehat{\tau}_3} = 8 \\
0 & \text{for } s \in (\widehat{\tau}_3, \widehat{\tau}_4] \cap [0,2] \quad \text{with } u = 1, d = 2,\ V_{\widehat{\tau}_3} \in \{-\frac{1}{2}, 0, 64\} \\
[0,4] & \text{for } s \in (\widehat{\tau}_3, \widehat{\tau}_4] \cap [0,2] \quad \text{with } u = 1, d = 2,\ V_{\widehat{\tau}_3} = 2 \\
[-4,12] & \text{for } s \in (\widehat{\tau}_3, \widehat{\tau}_4] \cap [0,2] \quad \text{with } u = 1, d = 2,\ V_{\widehat{\tau}_3} = 4 \\
[-12,28] & \text{for } s \in (\widehat{\tau}_3, \widehat{\tau}_4] \cap [0,2] \quad \text{with } u = 1, d = 2,\ V_{\widehat{\tau}_3} = 8 \\
[-28,60] & \text{for } s \in (\widehat{\tau}_3, \widehat{\tau}_4] \cap [0,2] \quad \text{with } u = 1, d = 2,\ V_{\widehat{\tau}_3} = 16 \\
0 & \text{for } s \in (\widehat{\tau}_3, \widehat{\tau}_4] \cap [0,2] \quad \text{with } u = 3, d = 0,\ V_{\widehat{\tau}_3} = 64 \\
1 & \text{for } s \in (\widehat{\tau}_3, \widehat{\tau}_4] \cap [0,2] \quad \text{with } u = 3, d = 0,\ V_{\widehat{\tau}_3} = 8 \\
0 & \text{for } s \in (\widehat{\tau}_3, \widehat{\tau}_4] \cap [0,2] \quad \text{with } u = 0, d = 3,\ V_{\widehat{\tau}_3} \in \{-\frac{1}{2}, 0, 64\} \\
[-8,32] & \text{for } s \in (\widehat{\tau}_3, \widehat{\tau}_4] \cap [0,2] \quad \text{with } u = 0, d = 3,\ V_{\widehat{\tau}_3} = 2 \\
[-56,128] & \text{for } s \in (\widehat{\tau}_3, \widehat{\tau}_4] \cap [0,2] \quad \text{with } u = 0, d = 3,\ V_{\widehat{\tau}_3} = 8 \\
[-120,256] & \text{for } s \in (\widehat{\tau}_3, \widehat{\tau}_4] \cap [0,2] \quad \text{with } u = 0, d = 3,\ V_{\widehat{\tau}_3} = 16 \\
[-248,256] & \text{for } s \in (\widehat{\tau}_3, \widehat{\tau}_4] \cap [0,2] \quad \text{with } u = 0, d = 3,\ V_{\widehat{\tau}_3} = 32 \\
0 & \text{for } s \in (\widehat{\tau}_4, \widehat{\tau}_5] \cap [0,2] \quad \text{with } u = 4, d = 0,\ V_{\widehat{\tau}_4} = 64 \\
1 & \text{for } s \in (\widehat{\tau}_4, \widehat{\tau}_5] \cap [0,2] \quad \text{with } u = 4, d = 0,\ V_{\widehat{\tau}_4} = 16 \\
0 & \text{for } s \in (\widehat{\tau}_4, \widehat{\tau}_5] \cap [0,2] \quad \text{with } u = 0, d = 4,\ V_{\widehat{\tau}_4} \in \{-\frac{1}{2}, 0, 64\} \\
[-32,64] & \text{for } s \in (\widehat{\tau}_4, \widehat{\tau}_5] \cap [0,2] \quad \text{with } u = 0, d = 4,\ V_{\widehat{\tau}_4} = 2 \\
[-128,256] & \text{for } s \in (\widehat{\tau}_4, \widehat{\tau}_5] \cap [0,2] \quad \text{with } u = 0, d = 4,\ V_{\widehat{\tau}_4} = 8 \\
[-256,512] & \text{for } s \in (\widehat{\tau}_4, \widehat{\tau}_5] \cap [0,2] \quad \text{with } u = 0, d = 4,\ V_{\widehat{\tau}_4} = 16 \\
[-512,512] & \text{for } s \in (\widehat{\tau}_4, \widehat{\tau}_5] \cap [0,2] \quad \text{with } u = 0, d = 4,\ V_{\widehat{\tau}_4} = 32 \\
0 & \text{for } s \in (\widehat{\tau}_4, \widehat{\tau}_5] \cap [0,2] \quad \text{with } u = 2, d = 2,\ V_{\widehat{\tau}_4} \in \{-\frac{1}{2}, 64\} \\
[1,2] & \text{for } s \in (\widehat{\tau}_4, \widehat{\tau}_5] \cap [0,2] \quad \text{with } u = 2, d = 2,\ V_{\widehat{\tau}_4} = 1
\end{cases}
$$

$\cdots/\cdots$

$$
\xi_s^{8,G} = \begin{cases}
1 & \text{for } s \in (\widehat{\tau}_4, \widehat{\tau}_5] \cap [0,2] \quad \text{with } u=2, d=2,\ V_{\widehat{\tau}_4} = 0 \\
[0,2] & \text{for } s \in (\widehat{\tau}_4, \widehat{\tau}_5] \cap [0,2] \quad \text{with } u=2, d=2,\ V_{\widehat{\tau}_4} = 2 \\
[-2,6] & \text{for } s \in (\widehat{\tau}_4, \widehat{\tau}_5] \cap [0,2] \quad \text{with } u=2, d=2,\ V_{\widehat{\tau}_4} = 4 \\
0 & \text{for } s \in (\widehat{\tau}_4, \widehat{\tau}_5] \cap [0,2] \quad \text{with } u=3, d=1,\ V_{\widehat{\tau}_4} \in \{-\frac{1}{2}, 64\} \\
1 & \text{for } s \in (\widehat{\tau}_4, \widehat{\tau}_5] \cap [0,2] \quad \text{with } u=3, d=1,\ V_{\widehat{\tau}_4} = 4 \\
0 & \text{for } s \in (\widehat{\tau}_4, \widehat{\tau}_5] \cap [0,2] \quad \text{with } u=1, d=3,\ V_{\widehat{\tau}_4} \in \{-\frac{1}{2}, 0, 64\} \\
[0,8] & \text{for } s \in (\widehat{\tau}_4, \widehat{\tau}_5] \cap [0,2] \quad \text{with } u=1, d=3,\ V_{\widehat{\tau}_4} = 1 \\
[-4,16] & \text{for } s \in (\widehat{\tau}_4, \widehat{\tau}_5] \cap [0,2] \quad \text{with } u=1, d=3,\ V_{\widehat{\tau}_4} = 2 \\
[-12,32] & \text{for } s \in (\widehat{\tau}_4, \widehat{\tau}_5] \cap [0,2] \quad \text{with } u=1, d=3,\ V_{\widehat{\tau}_4} = 4 \\
[-28,64] & \text{for } s \in (\widehat{\tau}_4, \widehat{\tau}_5] \cap [0,2] \quad \text{with } u=1, d=3,\ V_{\widehat{\tau}_4} = 8 \\
[-60,128] & \text{for } s \in (\widehat{\tau}_4, \widehat{\tau}_5] \cap [0,2] \quad \text{with } u=1, d=3,\ V_{\widehat{\tau}_4} = 16 \\
0 & \text{for } s \in (\widehat{\tau}_5, \widehat{\tau}_6] \cap [0,2] \quad \text{with } u=5, d=0,\ V_{\widehat{\tau}_5} = 64 \\
1 & \text{for } s \in (\widehat{\tau}_5, \widehat{\tau}_6] \cap [0,2] \quad \text{with } u=5, d=0,\ V_{\widehat{\tau}_5} = 32 \\
0 & \text{for } s \in (\widehat{\tau}_5, \widehat{\tau}_6] \cap [0,2] \quad \text{with } u=0, d=5,\ V_{\widehat{\tau}_5} \in \{-\frac{1}{2}, 0, 64\} \\
[-64,128] & \text{for } s \in (\widehat{\tau}_5, \widehat{\tau}_6] \cap [0,2] \quad \text{with } u=0, d=5,\ V_{\widehat{\tau}_5} = 2 \\
[-256,512] & \text{for } s \in (\widehat{\tau}_5, \widehat{\tau}_6] \cap [0,2] \quad \text{with } u=0, d=5,\ V_{\widehat{\tau}_5} = 8 \\
[-512,1024] & \text{for } s \in (\widehat{\tau}_5, \widehat{\tau}_6] \cap [0,2] \quad \text{with } u=0, d=5,\ V_{\widehat{\tau}_5} = 16 \\
[-1024,1024] & \text{for } s \in (\widehat{\tau}_5, \widehat{\tau}_6] \cap [0,2] \quad \text{with } u=0, d=5,\ V_{\widehat{\tau}_5} = 32 \\
0 & \text{for } s \in (\widehat{\tau}_5, \widehat{\tau}_6] \cap [0,2] \quad \text{with } u=4, d=1,\ V_{\widehat{\tau}_5} \in \{-\frac{1}{2}, 64\} \\
1 & \text{for } s \in (\widehat{\tau}_5, \widehat{\tau}_6] \cap [0,2] \quad \text{with } u=4, d=1,\ V_{\widehat{\tau}_5} = 8 \\
0 & \text{for } s \in (\widehat{\tau}_5, \widehat{\tau}_6] \cap [0,2] \quad \text{with } u=1, d=4,\ V_{\widehat{\tau}_5} \in \{-\frac{1}{2}, 0, 64\} \\
[-8,16] & \text{for } s \in (\widehat{\tau}_5, \widehat{\tau}_6] \cap [0,2] \quad \text{with } u=1, d=4,\ V_{\widehat{\tau}_5} = 1 \\
[-16,32] & \text{for } s \in (\widehat{\tau}_5, \widehat{\tau}_6] \cap [0,2] \quad \text{with } u=1, d=4,\ V_{\widehat{\tau}_5} = 2 \\
[-32,64] & \text{for } s \in (\widehat{\tau}_5, \widehat{\tau}_6] \cap [0,2] \quad \text{with } u=1, d=4,\ V_{\widehat{\tau}_5} = 4 \\
[-64,128] & \text{for } s \in (\widehat{\tau}_5, \widehat{\tau}_6] \cap [0,2] \quad \text{with } u=1, d=4,\ V_{\widehat{\tau}_5} = 8 \\
[-128,256] & \text{for } s \in (\widehat{\tau}_5, \widehat{\tau}_6] \cap [0,2] \quad \text{with } u=1, d=4,\ V_{\widehat{\tau}_5} = 16 \\
0 & \text{for } s \in (\widehat{\tau}_5, \widehat{\tau}_6] \cap [0,2] \quad \text{with } u=3, d=2,\ V_{\widehat{\tau}_5} \in \{-\frac{1}{2}, 64\} \\
\frac{3}{2} & \text{for } s \in (\widehat{\tau}_5, \widehat{\tau}_6] \cap [0,2] \quad \text{with } u=3, d=2,\ V_{\widehat{\tau}_5} = 1 \\
1 & \text{for } s \in (\widehat{\tau}_5, \widehat{\tau}_6] \cap [0,2] \quad \text{with } u=3, d=2,\ V_{\widehat{\tau}_5} = 2 \\
0 & \text{for } s \in (\widehat{\tau}_5, \widehat{\tau}_6] \cap [0,2] \quad \text{with } u=2, d=3,\ V_{\widehat{\tau}_5} \in \{-\frac{1}{2}, 0, 64\} \\
[0,4] & \text{for } s \in (\widehat{\tau}_5, \widehat{\tau}_6] \cap [0,2] \quad \text{with } u=2, d=3,\ V_{\widehat{\tau}_5} = 1 \\
[-2,8] & \text{for } s \in (\widehat{\tau}_5, \widehat{\tau}_6] \cap [0,2] \quad \text{with } u=2, d=3,\ V_{\widehat{\tau}_5} = 2 \\
[-6,16] & \text{for } s \in (\widehat{\tau}_5, \widehat{\tau}_6] \cap [0,2] \quad \text{with } u=2, d=3,\ V_{\widehat{\tau}_5} = 4 \\
0 & \text{for } s \in (\widehat{\tau}_6, \widehat{\tau}_7] \cap [0,2] \quad \text{with } u=6, d=0,\ V_{\widehat{\tau}_6} = 64 \\
\frac{1}{2} & \text{for } s \in (\widehat{\tau}_6, \widehat{\tau}_7] \cap [0,2] \quad \text{with } u=6, d=0,\ V_{\widehat{\tau}_6} = 32 \\
0 & \text{for } s \in (\widehat{\tau}_6, \widehat{\tau}_7] \cap [0,2] \quad \text{with } u=0, d=6,\ V_{\widehat{\tau}_6} \in \{-\frac{1}{2}, 0, 64\} \\
[-256,512] & \text{for } s \in (\widehat{\tau}_6, \widehat{\tau}_7] \cap [0,2] \quad \text{with } u=0, d=6,\ V_{\widehat{\tau}_6} = 4 \\
[-1024,2048] & \text{for } s \in (\widehat{\tau}_6, \widehat{\tau}_7] \cap [0,2] \quad \text{with } u=0, d=6,\ V_{\widehat{\tau}_6} = 16 \\
[-2048,2048] & \text{for } s \in (\widehat{\tau}_6, \widehat{\tau}_7] \cap [0,2] \quad \text{with } u=0, d=6,\ V_{\widehat{\tau}_6} = 32 \\
0 & \text{for } s \in (\widehat{\tau}_6, \widehat{\tau}_7] \cap [0,2] \quad \text{with } u=1, d=5,\ V_{\widehat{\tau}_6} \in \{-\frac{1}{2}, 0, 64\} \\
[-16,32] & \text{for } s \in (\widehat{\tau}_6, \widehat{\tau}_7] \cap [0,2] \quad \text{with } u=1, d=5,\ V_{\widehat{\tau}_6} = 1 \\
[-32,64] & \text{for } s \in (\widehat{\tau}_6, \widehat{\tau}_7] \cap [0,2] \quad \text{with } u=1, d=5,\ V_{\widehat{\tau}_6} = 2 \\
[-64,128] & \text{for } s \in (\widehat{\tau}_6, \widehat{\tau}_7] \cap [0,2] \quad \text{with } u=1, d=5,\ V_{\widehat{\tau}_6} = 4 \\
[-128,256] & \text{for } s \in (\widehat{\tau}_6, \widehat{\tau}_7] \cap [0,2] \quad \text{with } u=1, d=5,\ V_{\widehat{\tau}_6} = 8 \\
[-256,512] & \text{for } s \in (\widehat{\tau}_6, \widehat{\tau}_7] \cap [0,2] \quad \text{with } u=1, d=5,\ V_{\widehat{\tau}_6} = 16 \\
0 & \text{for } s \in (\widehat{\tau}_6, \widehat{\tau}_7] \cap [0,2] \quad \text{with } u=5, d=1,\ V_{\widehat{\tau}_6} = 64 \\
1 & \text{for } s \in (\widehat{\tau}_6, \widehat{\tau}_7] \cap [0,2] \quad \text{with } u=5, d=1,\ V_{\widehat{\tau}_6} = 16
\end{cases}
$$

$\cdots/\cdots$

$$\xi_s^{8,G} = \begin{cases} 0 & \text{for } s \in (\widehat{\tau}_6, \widehat{\tau}_7] \cap [0,2] \text{ with } u = 3, d = 3,\ V_{\widehat{\tau}_6} \in \{-\frac{1}{2}, 64\} \\ 1 & \text{for } s \in (\widehat{\tau}_6, \widehat{\tau}_7] \cap [0,2] \text{ with } u = 3, d = 3,\ V_{\widehat{\tau}_6} = 0 \\ [0,2] & \text{for } s \in (\widehat{\tau}_6, \widehat{\tau}_7] \cap [0,2] \text{ with } u = 3, d = 3,\ V_{\widehat{\tau}_6} = 1 \\ 0 & \text{for } s \in (\widehat{\tau}_6, \widehat{\tau}_7] \cap [0,2] \text{ with } u = 2, d = 4,\ V_{\widehat{\tau}_6} \in \{-\frac{1}{2}, 0, 64\} \\ [-4,8] & \text{for } s \in (\widehat{\tau}_6, \widehat{\tau}_7] \cap [0,2] \text{ with } u = 2, d = 4,\ V_{\widehat{\tau}_6} = 1 \\ [-8,16] & \text{for } s \in (\widehat{\tau}_6, \widehat{\tau}_7] \cap [0,2] \text{ with } u = 2, d = 4,\ V_{\widehat{\tau}_6} = 2 \\ [-16,32] & \text{for } s \in (\widehat{\tau}_6, \widehat{\tau}_7] \cap [0,2] \text{ with } u = 2, d = 4,\ V_{\widehat{\tau}_6} = 4 \\ 0 & \text{for } s \in (\widehat{\tau}_6, \widehat{\tau}_7] \cap [0,2] \text{ with } u = 4, d = 2,\ V_{\widehat{\tau}_6} = 64 \\ [1, \frac{3}{2}] & \text{for } s \in (\widehat{\tau}_6, \widehat{\tau}_7] \cap [0,2] \text{ with } u = 4, d = 2,\ V_{\widehat{\tau}_6} = 4 \end{cases}$$

In the next table we also show the values of $J^{8,G}$ for the various initial conditions $(v, 0, 0, 0)$ corresponding to the 8 non-negative grid-values of v. Since more jumps imply a riskier context, the values of $J^{8,G}$ are naturally larger than the corresponding values for $n = 5$.

$$\begin{cases} J^{8,G}(0,0,0,0) &= 16.8681 \\ J^{8,G}(1,0,0,0) &= 15.388 \\ J^{8,G}(2,0,0,0) &= 15.3612 \\ J^{8,G}(4,0,0,0) &= 15.3604 \\ J^{8,G}(8,0,0,0) &= 15.3603 \\ J^{8,G}(16,0,0,0) &= 15.3603 \\ J^{8,G}(32,0,0,0) &= 15.3603 \\ J^{8,G}(64,0,0,0) &= 15.3603 \end{cases}$$

Finally, without reporting also the strategies for the intermediate cases of $n = 6$ and $n = 7$, in the next two tables we also show the values of $J^{n,G}$ for these cases and for the various initial conditions $(v, 0, 0, 0)$ corresponding to the n non-negative grid-values of v. From these values and those shown above for $n = 5$ and $n = 8$ one can get a feeling for the increase of the minimal expected risk corresponding to an increase in the number of jumps, i.e. to an increase of the riskiness of the situation (for one more jump the minimal expected risk increases roughly by a factor of 3).

$$\begin{cases} J^{6,G}(0,0,0,0) &= 2.3432 \\ J^{6,G}(1,0,0,0) &= 2.050 \\ J^{6,G}(2,0,0,0) &= 2.0437 \\ J^{6,G}(4,0,0,0) &= 2.0435 \\ J^{6,G}(8,0,0,0) &= 2.0435 \\ J^{6,G}(16,0,0,0) &= 2.0435 \end{cases}$$

$$\begin{cases} J^{7,G}(0,0,0,0) &= 6.3283 \\ J^{7,G}(1,0,0,0) &= 5.677 \\ J^{7,G}(2,0,0,0) &= 5.663 \\ J^{7,G}(4,0,0,0) &= 5.6632 \\ J^{7,G}(8,0,0,0) &= 5.6632 \\ J^{7,G}(16,0,0,0) &= 5.6632 \\ J^{7,G}(32,0,0,0) &= 5.6632 \end{cases}$$

7.5.3 Conclusions

As can be guessed from the numerical results reported above, the calculations become increasingly heavier with increasing values of n. The grid on the other hand influences less markedly the computational complexity. One may thus conclude that the algorithm described in the paper performs sufficiently well in situations where one does not expect too many jumps to happen as in the case of default sensitive assets. If there are many jumps such as in situations of high frequency data and small time scales, then it may be advisable to model the price evolution by means of continuous trajectories (approximating the de facto discontinuous trajectories by continuous ones) and use an algorithm tailored to this latter situation.

References

[1] A.Almudevar (2001). A dynamic programming algorithm for the optimal control of piecewise deterministic Markov processes. *SIAM J. on Control and Optimization.* 40, 525-539.

[2] T.R.Bielecki (2002). M.Rutkowski, *Credit Risk: Modeling, Valuation and Hedging*, Springer Verlag.

[3] M.Kirch, W.J. Runggaldier (2004). Efficient hedging when asset prices follow a geometric Poisson process with unknown intensities. *SIAM J. on Control and Optimization.* 43, 1174-1195.

[4] R.Korn (1997). *Optimal Portfolios. Stochastic models for optimal investment and risk management in continuous time.* World Scientific Publishing Co. Pte. Ltd.

[5] P.Schönbucher (2003). *Credit Derivatives Pricing Models: Model, Pricing and Implementation*, John Wiley & Sons.

8

A Downside Risk Analysis based on Financial Index Tracking Models

Lian Yu, Shuzhong Zhang*, and Xun Yu Zhou†

[1] Department of Industrial Engineering and Operations Research, University of California, Berkeley, CA 94720
lliany@ieor.berkeley.edu

[2] Department of Systems Engineering and Engineering Management, The Chinese University of Hong Kong Shatin, Hong Kong.
zhang@se.cuhk.edu.hk

[3] Department of Systems Engineering and Engineering Management, The Chinese University of Hong Kong Shatin, Hong Kong
xyzhou@se.cuhk.edu.hk

Summary. This paper is mainly concerned with a single-stage financial index tracking problem under the downside risk constraint where short-selling is allowed. First, we formulate the portfolio selection model with the downside probability constraint to track the financial index. Due to the convexity of this problem, the optimal portfolio is derived analytically by applying the Karush-Kuhn-Tucker optimality conditions. Moreover, we extend the risk measure to higher order moment of the downside and study the corresponding portfolio optimization problem.

8.1 Introduction

In 1950's, Harry M. Markowitz [13] proposed the mean-variance portfolio selection model, which was the beginning of the modern portfolio theory. In his model, the variance of return is regarded as a risk measure and the investor should make a trade-off between the risk and the return.

Since then many financial models for portfolio selection have been proposed. However, the concept of risk is rather subjective, and different people may use different criteria to measure the risk. Intuitively, it is more sensible for an investor to be concerned with the risk of loss rather than the risk of gain. The concept of downside risk apparently has considerable impact on the investor's view point. Therefore, in Markowitz's seminal book [14], he suggested to use the semivariance as an alternative measure of risk. In 1972,

* Supported by the RGC Earmarked Grant CUHK4233/01E.

† Supported by the RGC Earmarked Grants CUHK 4175/00E and CUHK 4234/01E.

Hogan and Warren [9] used the below-target semivariance as the risk measure and formulated a portfolio selection model. Furthermore, the associated capital asset pricing model (ES-CAPM) that describes the equilibrium relationship between the market portfolio and individual assets was obtained by Hogan and Warren in 1974 [10].

In addition, as the semivariance restricts the investors to only one utility function, described by a quadratic equation, another downside risk measure, named the lower partial moment (LPM), has been proposed by Bawa [1] and Fishburn [5] respectively. The lower partial moment is defined as a general family of below-target risk measures, which encompasses a significant number of utility functions from risk-seeking to risk-neutral to risk aversion. Moreover, when a riskless asset is included in the portfolio, the capital market equilibrium was established by Bawa and Lindenberg [3].

Besides the lower partial moment risk measure, stochastic dominance (SD) is another very powerful downside risk analysis tool. It converts the probability distribution of an investment into a cumulative probability curve. In 1969, Hadar and Russell [6] showed that the SD rules provide admissible sets of alternatives under restrictions on all possible risk-averse utility assumptions. Furthermore, the consistence between the mean-semivariance model and second degree stochastic dominance was presented by Ogryczak and Ruszczynski in [18].

The purpose of this paper is to study the index tracking models based on the downside risk measure. Specifically, we first formulate the portfolio selection model under the downside probability risk measure to beat the underlying financial index, where short-selling is allowed and the assets' rates of return are assumed to be jointly normally distributed. Then, by applying the Karush-Kuhn-Tucker (KKT) optimality conditions, we obtain the analytical form of the optimal portfolio. Finally, we introduce higher order moments of the downside distribution to measure the risk, and study the index tracking problem under the m-th moment downside constraint.

8.2 Statement of the Problem

In this section, we will define the downside risk measure, and construct the financial index tracking model under this risk measure. Our goal will be to beat a certain financial index by forming a portfolio consisting of risky and riskless assets, in a single period. The financial index can be the stock index itself or a given target.

Suppose there are n risky assets S_j, $j = 1\dots,n$, one riskless asset S_0 and a financial index I in the market. For the n risky assets, let ξ_j be the return rate of the j-th asset, assumed to be random variables. For the riskless asset, let the return rate be R, which we will assume to be deterministic. For the financial index, let the return rate be ξ_0, also assumed to be a random variable. Let x_j be the proportion of the initial wealth invested into asset S_j.

A vector $\bar{x} = (x_0, x_1, \ldots, x_n)$ where $\sum_{j=0}^{n} x_j = 1$ is called a *feasible portfolio.* Therefore, when short-selling is allowed, the feasible region for the portfolio optimization problem is the set of all feasible portfolios, namely

$$F := \left\{ \bar{x} = (x_0, x_1, \ldots, x_n) \in \mathbb{R}^{n+1} : \sum_{j=0}^{n} x_j = 1 \right\}.$$

Denoting $\widehat{\xi} := (\xi_1, \ldots, \xi_n)$ and $x := (x_1, \ldots, x_n)$, the return rate of the portfolio $\bar{x}$ is $Rx_0 + \widehat{\xi}^T x$. Let $\mathrm{E}\xi$ be the mathematical expectation of a random variable ξ, and define

$$\mu_0 = \mathrm{E}\xi_0 \quad \text{and} \quad \widehat{\mu} = \mathrm{E}\widehat{\xi}\ ,$$

the expected rates of return of the financial index and the risky assets, respectively. The expected rate of return of a portfolio $\bar{x} = (x_0, \ldots, x_n)$ is

$$\mathrm{E}\left(Rx_0 + \widehat{\xi}^T x \right) = Rx_0 + \widehat{\mu}^T x\,.$$

Let the covariance matrix between ξ_0 and $\widehat{\xi}$ be

$$H := \mathrm{cov}(\xi_0, \widehat{\xi}) = \begin{bmatrix} h_0 & h^T \\ h & \widehat{H} \end{bmatrix},$$

where h_0 is the variance of the financial index, h is the covariance vector between the risky assets and the financial index, and $\widehat{H}$ is the covariance matrix of the risky assets. Without loss of generality, we assume $H \succ 0$, which implies that there is no redundant asset in the market, so any risky asset or the financial index can not be expressed as a linear combination of the others. The slack random variable is defined as follows.

Definition 1. *The slack random variable of the rate of return on a portfolio $\bar{x}$ with respect to the financial index I is defined as*

$$\eta(x) := \xi_0 - Rx_0 - \widehat{\xi}^T x. \tag{8.1}$$

To measure the risk we may use either the probability of the downside of the portfolio with respect to the financial index, $\mathrm{P}\{\eta(x) \geq 0\}$, or any higher order moment of the downside chance, $\mathrm{E}(\eta(x)_+)^m$, $m = 1, 2, \ldots$, where

$$x_+ = \begin{cases} x \text{ if } x > 0 \\ 0 \text{ if } x \leq 0\,. \end{cases}$$

By varying the value of m, we may construct different portfolio selection models that respond to different risk preferences of the investors.

8.3 Mean-Downside Chance Index Tracking Model

In this section, we will discuss the case where the probability of the downside is regarded as the risk measure. This risk measure is referred to as the downside chance, because only the right tail of the distribution of the slack random variable is used in the calculation.

As usual, we assume that an investor tries to maximize the expected rate of return of his investment, while keeping the risk below a certain level. Accordingly, the portfolio selection model (P) under the downside chance constraint is formulated as follows

$$\begin{aligned}(P)\ \text{maximize}\ \ & Rx_0 + \widehat{\mu}^T x \\ \text{subject to}\ \ & \mathrm{P}\{\eta(x) \geq 0\} \leq \delta\,, \\ & \bar{x}^T \imath = 1\,,\end{aligned} \tag{8.2}$$

where $\delta \in [0,1]$ is the expected downside chance of a potential investor, and $\imath = (1,\ldots,1)$ with appropriate dimension. A feasible portfolio is said to be *efficient* if it is optimal for the Problem (P) given a certain δ. For such a portfolio we denote by $\mu(\delta)$ the associated objective value, and the point $(\delta, \mu(\delta))$ is called an *efficient point.* Further, we call *efficient frontier* the set of all efficient points, for which Problem (P) admits an attainable optimal solution.

8.3.1 Efficient Frontier

In this section, we will derive an algorithm for approaching the efficient frontier for Problem (P). We now introduce an assumption that will be assumed throughout the article.

Assumption 1. *The return rates of the financial index I and the j-th risky assets S_j, $j = 1,\ldots,n$, are jointly normally distributed.*

The above assumption stipulates that ξ_0 and $\widehat{\xi}$ are jointly distributed as normal random vectors, and that the slack random variable $\eta(x)$ follows a normal distribution.

By substituting $x_0 = 1 - x^T \imath$, we can rewrite the slack $\eta(x)$ as

$$\eta(x) = \xi_0 - R - (\widehat{\xi} - R\imath)^T x\,.$$

Obviously, the slack variable has expectation

$$\mathrm{E}\eta(x) = \mu_0 - R - (\widehat{\mu} - R\imath)^T x\,,$$

and variance

$$\sigma^2(\eta(x)) = h_0 - 2h^T x + x^T \widehat{H} x\,.$$

Hence,

$$\eta(x) \sim N(\mu_0 - R - (\widehat{\mu} - R\imath)^T x, h_0 - 2h^T x + x^T \widehat{H} x).$$

Then, the downside chance of the rate of return on the portfolio $\bar{x}$ with respect to the financial index I can be explicitly expressed as follows

$$\mathrm{P}\{\eta(x) \geq 0\} = 1 - \Phi\left(-\frac{\mu_0 - R - (\widehat{\mu} - R\imath)^T x}{\sqrt{h_0 - 2h^T x + x^T \widehat{H} x}}\right),$$

where

$$\Phi(x) := \int_{-\infty}^{x} \frac{1}{\sqrt{2\pi}} e^{-\frac{y^2}{2}}\, dy,$$

is the distribution function of the standard normal random variable $N(0,1)$. Thus, Problem (P) is equivalent to the following,

$$\begin{array}{ll} \text{maximize} & (\widehat{\mu} - R\imath)^T x \\ \text{subject to} & 1 - \Phi(f(x)) \leq \delta \end{array} \tag{8.3}$$

where

$$f(x) := -\frac{\mu_0 - R - (\widehat{\mu} - R\imath)^T x}{\sqrt{h_0 - 2h^T x + x^T \widehat{H} x}}.$$

We assume, without loss of generality, that $\widehat{\mu} - R\imath$ is not the zero vector. In fact, if this were the case, the solution to our model would be to invest all the initial wealth on the riskless asset.

We may rewrite

$$f(x) = \frac{c_0 + c^T y}{\sqrt{y^T y + a_0}},$$

with

$$\begin{aligned} y &:= \widehat{H}^{\frac{1}{2}}(x - \widehat{H}^{-1} h),\; c := \widehat{H}^{-\frac{1}{2}}(\widehat{\mu} - R\imath), \\ a_0 &:= h_0 - h^T \widehat{H}^{-1} h, \quad c_0 := (\widehat{\mu} - R\imath)^T \widehat{H}^{-1} h - \mu_0 + R. \end{aligned}$$

Assuming $c_0 \geq 0$ we have the following result.

Lemma 1. *Suppose that* $c_0 := (\widehat{\mu} - R\imath)^T \widehat{H}^{-1} h - \mu_0 + R \geq 0$. *Then* $f(x)$ *is uniformly bounded and satisfies*

$$-\|c\| \leq f(x) \leq \frac{\sqrt{a_0^2 \|c\|^2 + a_0 c_0^2}}{a_0}.$$

The proof is long with tedious calculations. It is included in the Appendix.

On account of the notation just introduced, the optimization problem (8.3) becomes

$$\begin{array}{ll} \text{maximize} & c^T y \\ \text{subject to} & 1 - \Phi(\frac{c_0 + c^T y}{\sqrt{y^T y + a_0}}) \leq \delta. \end{array} \tag{8.4}$$

As $\Phi(\cdot)$ is increasing, this may be rewritten as

$$\begin{array}{ll} \text{maximize} & c^T y \\ \text{subject to} & \frac{c_0 + c^T y}{\sqrt{y^T y + a_0}} \geq \Phi^{-1}(1-\delta) \,. \end{array} \tag{8.5}$$

Lemma 2. *When* $\delta \in \left[1 - \Phi\left(\frac{\sqrt{a_0^2 \|c\|^2 + a_0 c_0^2}}{a_0}\right), 1 - \Phi(\|c\|)\right]$, *the portfolio selection problem* (P) *is a convex programming problem.*

Proof: As $\delta \in \left[1 - \Phi\left(\frac{\sqrt{a_0^2 \|c\|^2 + a_0 c_0^2}}{a_0}\right), 1 - \Phi(\|c\|)\right]$, we have

$$\|c\| \leq \Phi^{-1}(1-\delta) \leq \frac{\sqrt{a_0^2 \|c\|^2 + a_0 c_0^2}}{a_0} \,.$$

Therefore, Problem (8.5) becomes,

$$\begin{array}{ll} \text{maximize} & c^T y \\ \text{subject to} & \sqrt{y^T y + a_0} \leq b^T y + b_0 \,, \end{array} \tag{8.6}$$

where

$$b := \frac{c}{\Phi^{-1}(1-\delta)} \,, \qquad \text{and} \qquad b_0 := \frac{c_0}{\Phi^{-1}(1-\delta)} \,.$$

Suppose the feasible set for this problem is not empty, and let y be a feasible solution for it. As $a_0 \geq 0$ and $\|b\| := \frac{\|c\|}{\Phi^{-1}(1-\delta)}$, it follows that $\sqrt{y^T y + a_0} \geq \|y\|$ and $b^T y + b_0 \leq \|b\| \cdot \|y\| + b_0 < \|y\| + b_0$. Now y is a solution of (8.6) so it satisfies the constraint of the problem, that is

$$\|y\| \leq \sqrt{y^T y + a_0} \leq b^T y + b_0 < \|y\| + b_0 \,.$$

So, for the feasible set to be nonempty we must assume that $b_0 \geq 0$. But this follows from the assumption that $c_0 \geq 0$.

As the objective function of Problem (8.6) is linear, we only need to show that the constraint of the problem defines a convex set. Put

$$g(y) = \left\| \begin{bmatrix} \sqrt{a_0} \\ y \end{bmatrix} \right\| - b^T y - b_0 \,,$$

which is a convex function due to the convexity of the Euclidean norm. Therefore, $g(y) \leq 0$ is a convex set, which is in fact the intersection of a second order cone with an affine space. Consequently, the Problem (P) is a convex programming problem. □

The following theorem describes the optimal solution for the portfolio selection model (P).

Theorem C.1. *Depending on the expected downside chance δ, we have the following description of the optimal portfolio of the financial index tracking model under the downside chance risk:*

(a) if $\delta \in [1 - \Phi(\|c\|), 1]$, the expected rate of return of the optimal portfolio is $+\infty$;

(b) if $\delta \in \left[1 - \Phi\left(\frac{\sqrt{a_0^2\|c\|^2 + a_0 c_0^2}}{a_0}\right), 1 - \Phi(\|c\|)\right)$, the optimal portfolio is

$$x^* = \frac{\sqrt{b_0^2 - a_0 + a_0 \|b\|^2} + b_0 \|b\|}{\|b\| (1 - \|b\|^2)} \widehat{H}^{-\frac{1}{2}} b + \widehat{H}^{-1} h \,,$$

and the corresponding return rate is

$$\mu_0 + b_0 \Phi^{-1}(1 - \delta) + \frac{\sqrt{b_0^2 - a_0 + a_0 \|b\|^2} + b_0 \|b\|}{1 - \|b\|^2} \|b\| \, \Phi^{-1}(1 - \delta) \,;$$

(c) if $\delta \in \left[0, 1 - \Phi\left(\frac{\sqrt{a_0^2\|c\|^2 + a_0 c_0^2}}{a_0}\right)\right)$, there is no feasible portfolio for this portfolio selection problem.

Proof: If $\delta \in [1 - \Phi(-\|c\|), 1]$, then $\Phi^{-1}(1 - \delta) \leq -\|c\|$. According to Lemma 1, it follows that the constraint of Problem (8.5) is always satisfied. Therefore, the feasible set is the whole space $\mathbb{R}^n$ and the optimal value for the objective function is $+\infty$.

Suppose now that $\delta \in [0.5, 1 - \Phi(-\|c\|))$. Then $-\|c\| < \Phi^{-1}(1-\delta) \leq 0$, and this is a trivial case for the optimization problem (8.5). In fact, this problem becomes

$$\begin{aligned} &\text{maximize } c^T y \\ &\text{subject to } c^T y \geq \Phi^{-1}(1 - \delta) \sqrt{y^T y + a_0} - c_0 \,. \end{aligned}$$

If $\|y\| \longrightarrow +\infty$ the objective function $c^T y \longrightarrow +\infty$. Now, as $\Phi^{-1}(1 - \delta) \leq 0$ and $\sqrt{y^T y + a_0} \longrightarrow +\infty$ as $\|y\| \longrightarrow +\infty$, it follows that $\lim_{\|y\| \to +\infty} \Phi^{-1}(1 - \delta) \sqrt{y^T y + a_0} = -\infty$. Hence, when $\|y\| \longrightarrow +\infty$, the constraint of the above optimization problem is always satisfied, so the optimal value for the objective function is $+\infty$.

If $\delta \in [1 - \Phi(\|c\|), 0.5)$, then $0 < \Phi^{-1}(1 - \delta) \leq \|c\|$. Choose $y = kc$, for some fixed value of k. Then

$$\lim_{k \to +\infty} f(x) = \lim_{k \to +\infty} \frac{c_0 + k c^T c}{\sqrt{k^2 c^T c + a_0}} = \|c\| \,,$$

so that,

$$\lim_{k \to +\infty} f(x) \geq \Phi^{-1}(1 - \delta) \,.$$

This means that, as $k \longrightarrow +\infty$, there exist feasible solutions for Problem (8.5), while the optimal value for the objective function satisfies

$$\lim_{k\to+\infty} c^T y = \lim_{k\to+\infty} k \|c\|^2 = +\infty \,.$$

We proved earlier that when $\delta \in \left[1 - \Phi\left(\frac{\sqrt{a_0^2\|c\|^2 + a_0 c_0^2}}{a_0}\right), 1 - \Phi(\|c\|)\right]$, Problem (8.5) is a convex programming problem. We now derive an analytical solution for this problem by applying the KKT optimality conditions to Problem (8.6). Let

$$L = c^T y + \lambda(\sqrt{y^T y + a_0} - b^T y - b_0)$$

be the Lagrange multiplier of this problem. Then the optimal solution must satisfy

$$\begin{aligned}
&\frac{\partial L}{\partial y} = c + \lambda\left(\frac{y}{\sqrt{y^T y + a_0}} - b\right) = 0\,,\\
&\sqrt{y^T y + a_0} - b^T y - b_0 \le 0\,,\\
&\lambda \ge 0\,,\\
&\lambda(\sqrt{y^T y + a_0} - b^T y - b_0) = 0\,.
\end{aligned}$$

As $c \neq 0$, it follows that $\lambda \neq 0$. The optimal value of the objective function is attained on the boundary of the feasible set, so the KKT optimality conditions may be rewritten as

$$\frac{y}{\sqrt{y^T y + a_0}} - b = -\frac{1}{\lambda} c\,, \tag{8.7}$$

$$y^T y + a_0 = (b^T y + b_0)^2\,, \tag{8.8}$$

$$b^T y + b_0 \ge 0\,, \tag{8.9}$$

$$\lambda > 0\,. \tag{8.10}$$

From (8.7) and (8.8), we have

$$y = b(b^T y + b_0) - \frac{c(b^T y + b_0)}{\lambda}\,.$$

Define $\nu_1 = b^T y + b_0$ and $\nu_2 = \frac{c(b^T y + b_0)}{\lambda}$. Then $y = \nu_1 b - \nu_2 c$ and $\nu_1 \ge 0$, $\nu_2 \ge 0$ taking (8.8) and (8.9) into account. From (8.8) we obtain

$$\left(b^T b - 1\right)\nu_1^2 - 2b^T c \nu_1 \nu_2 + c^T c \nu_2^2 + a_0 = 0\,. \tag{8.11}$$

As $\nu_2 = \frac{b^T b - 1}{b^T c}\nu_1 + \frac{b_0}{b^T c}$, this last equation is equivalent to the quadratic equation

$$\left(\frac{1}{b^T b} - 1\right)\nu_1^2 - 2\frac{b_0}{b^T b}\nu_1 + \frac{b_0^2}{b^T b} + a_0 = 0\,, \tag{8.12}$$

whose discriminant is

$$\Delta_1 = \frac{4}{b^T b}\left(b_0^2 - a_0 + a_0 b^T b\right).$$

Any solution of Problem (8.6) must satisfy its constraints thus, in particular,

$$y^T y + a_0 \leq \left(b^T y + b_0\right)^2 \leq \|b\|^2 \cdot \|y\|^2 + 2b_0 \|b\| \cdot \|y\| + b_0^2 .$$

Hence, it holds

$$\left(1 - b^T b\right) \|y\|^2 - 2b_0 \|b\| \cdot \|y\| + a_0 - b_0^2 \leq 0 . \tag{8.13}$$

As $\|b\| < 1$, it follows that $1 - b^T b \geq 0$. Now, if a feasible solution for Problem (8.6) exists, there also exists a feasible solution for the above inequality. Therefore, the discriminant of the quadratic expression in $\|y\|$ on (8.13) should be nonnegative, that is,

$$\Delta_2 = 4\left(-a_0 + a_0 \|b\|^2 + b_0^2\right) \geq 0 ,$$

so, also, $\Delta_1 \geq 0$. Therefore the quadratic equation (8.12) has two distinct roots

$$\nu_1^* = \frac{b_0 \pm \|b\| \sqrt{b_0^2 - a_0 + a_0 \|b\|^2}}{1 - \|b\|^2} .$$

The corresponding y^* is

$$y^* = \frac{b}{b^T b}\left(\nu_1^* - b_0\right),$$

and the optimal value for the objective function of Problem (8.6) is

$$c^T y^* = \Phi^{-1}(1-\delta)(\nu_1^* - b_0) ,$$

which increases with ν_1^*. Hence we select the largest root of (8.12), that is, we choose

$$\nu_1^* = \frac{b_0 + \|b\| \sqrt{b_0^2 - a_0 + a_0 \|b\|^2}}{1 - \|b\|^2} ,$$

to which corresponds the optimal solution of Problem (8.6)

$$y^* = \frac{\sqrt{b_0^2 - a_0 + a_0 \|b\|^2} + b_0 \|b\|}{\|b\| \left(1 - \|b\|^2\right)} b .$$

Accordingly, the optimal allocation of the risky assets is

$$x^* = \frac{\sqrt{b_0^2 - a_0 + a_0 \|b\|^2} + b_0 \|b\|}{\|b\| \left(1 - \|b\|^2\right)} \widehat{H}^{-\frac{1}{2}} b + \widehat{H}^{-1} h ,$$

and the maximal expected return rate is

$$\mu_0 + b_0\Phi^{-1}(1-\delta) + \frac{\sqrt{b_0^2 - a_0 + a_0\left\|b\right\|^2} + b_0\left\|b\right\|}{1-\left\|b\right\|^2}\left\|b\right\|\Phi^{-1}(1-\delta)\,.$$

Finally, when $\delta \in \left[0, 1-\Phi\left(\frac{\sqrt{a_0^2\|c\|^2+a_0c_0^2}}{a_0}\right)\right)$, we have

$$\Phi^{-1}(1-\delta) > \frac{\sqrt{a_0^2\left\|c\right\|^2 + a_0c_0^2}}{a_0}\,.$$

Remembering the bounds on $f(x)$ from Lemma 1, it follows that the feasible set is empty. Therefore, there are no optimal solutions for problem (P). The proof is completed. □

We have characterized efficient portfolios for a given value of δ. Allowing δ to vary we may find the efficient frontier of this portfolio selection model by solving a series of convex programs (see Figure 8.1).

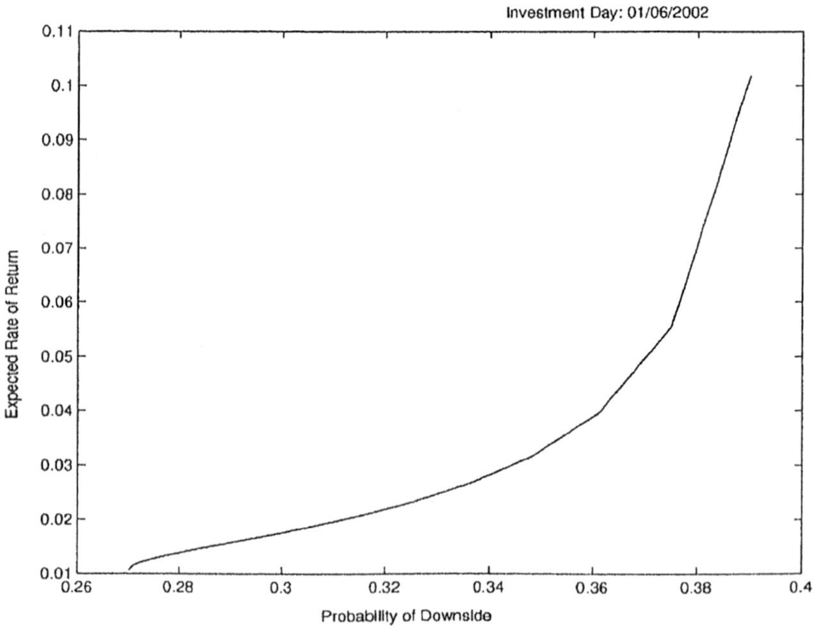

Fig. 8.1. Efficient Frontier of the Mean-Downside Chance Model

We may compare the efficient frontiers of the Mean-Downside Chance Model and the Mean-Variance Model when the financial index to be tracked is fixed as a given deterministic target.

Theorem C.2. *When we track a given target, the optimal portfolio for the Mean-Downside Chance Model is also optimal for the Mean-Variance Model.*

Proof: When the financial index is a deterministic value we have $h_0 = 0$ and $h = (0, \dots, 0)$. Thus

$$a_0 = h_0 - h^T \widehat{H}^{-1} h = 0, \quad b = \frac{\widehat{H}^{-\frac{1}{2}}(\widehat{\mu} - R\imath)}{\Phi^{-1}(1-\delta)}, \quad \text{and} \quad b_0 = \frac{R - \mu_0}{\Phi^{-1}(1-\delta)}.$$

The optimal solution for Problem (P) becomes

$$x^* = \frac{b_0 + \sqrt{b_0^2 \cdot \frac{1}{b^T b}}}{1 - b^T b} \cdot \widehat{H}^{-\frac{1}{2}} b,$$

and the corresponding value for the objective function is

$$\mu^* = R + (\widehat{\mu} - R\imath)^T \widehat{H}^{-\frac{1}{2}} b \cdot \frac{b_0 + \sqrt{b_0^2 \cdot \frac{1}{b^T b}}}{1 - b^T b}.$$

The Mean-Variance Model may be described as

$$\begin{array}{ll} \text{minimize} & \frac{1}{2} x^T \widehat{H} x \\ \text{subject to} & R x_0 + x^T \widehat{\mu} = \mu^* \\ & x_0 + x^T \imath = 1. \end{array} \tag{8.14}$$

According to the KKT optimality conditions, the optimal solution of Problem (8.14) is

$$\begin{aligned} x &= \frac{(\mu^* - R)\widehat{H}^{-1}(\widehat{\mu} - R\imath)}{(\widehat{\mu} - R\imath)\widehat{H}^{-1}(\widehat{\mu} - R\imath)} \\ &= \frac{(\widehat{\mu} - R\imath)^T \widehat{H}^{-\frac{1}{2}} b \cdot \frac{b_0 + \sqrt{b_0^2 \cdot \frac{1}{b^T b}}}{1 - b^T b} (\widehat{\mu} - R\imath)}{(\widehat{\mu} - R\imath)\widehat{H}^{-1}(\widehat{\mu} - R\imath)} \\ &= \frac{b_0 + \sqrt{b_0^2 \cdot \frac{1}{b^T b}}}{1 - b^T b} \widehat{H}^{-\frac{1}{2}} b. \end{aligned}$$

Therefore, the optimal solutions of Problems (P) and (8.14) are identical.

Hence, the efficient frontier for Problem (P) is included in the efficient frontier for Mean-Variance Model. □

8.3.2 Application to the Multi-stage Case

In this section, we will apply the algorithm for the single-stage financial index tracking model with the downside chance constraint to search for a feasible strategy for the corresponding multi-stage problem.

Consider an investor in an-stage investment period $[0, T]$, where there are m identical sub-investment periods during this time interval. Assume that the trading of assets is self-financed, and the transaction costs and consumption are ignored. We use t_0 to denote the original decision point (the starting time of the whole investment period), t_i to denote the i-th decision point (the starting time of the i-th sub-investment period), $i = 1, \ldots, m-1$, and $t_m = T$ to denote the end of the whole investment period. Suppose that the return rate of the financial index I is ξ_0 at the original decision point t_0, ξ_1 at time t_1, $\ldots$, and ξ_m at time t_m. Let $\widehat{\xi}_i := (\xi_{1i}, \ldots, \xi_{ni})$ be the return rates of the risky assets, and R_i be the riskless rate of return at the instants t_i, $i = 1, \ldots, m$. We define the expected return rates of the financial index I and the risky assets at time t_i as

$$\mu_{0i} := \mathrm{E}\xi_{0i} \qquad \text{and} \qquad \widehat{\mu}_i := \mathrm{E}\widehat{\xi}_i\,, \quad i = 1, \ldots, m\,,$$

respectively. The covariance matrix between the financial index and the risky assets at time t_i, $i = 0, 1, \ldots, m$, is

$$H_i := \mathrm{cov}(\xi_i, \widehat{\xi}_i) = \begin{bmatrix} h_{0i} & h_i^T \\ h_i & \widehat{H}_i \end{bmatrix},$$

where h_{0i} is the variance of the financial index, h_i the covariance vector between the risky assets and the financial index, and $\widehat{H}_i$ the covariance matrix of the risky assets, all at i-th decision point t_i. Without loss of generality, we assume that the covariance matrix H_i, $i = 0, 1, \ldots, m$, is positive definite.

At the original decision instant t_0, the investor constructs a portfolio consisting of riskless and risky assets, and at each decision instant t_i, $i = 1, 2, \ldots, m-1$, the investor will revise his portfolio. Depending on the observed information up to time t_i, the investor can change his existing proportions of the wealth invested in the riskless and risky assets.

Let x_{0i} be the proportion of the riskless asset, and $x_i := (x_{1i}, \ldots, x_{ni})$ be the proportions of the risky assets at time t_i, $i = 1, 2, \ldots, m-1$. The portfolio at each decision point is a vector of riskless and risky assets $\bar{x}_i := (x_{0i}, x_{1i}, \ldots, x_{ni})$ such that $\sum_{j=0}^{n} x_{ji} = 1$. When short-selling of assets is allowed, the feasible region for the portfolio selection problem at time t_i, $i = 0, 1, \ldots, m$, is the set of all feasible portfolios, that is

$$F_i := \left\{ \bar{x}_i := (x_{0i}, x_{1i}, \ldots, x_{ni}) \in \mathbb{R}^{n+1} : \sum_{j=0}^{n} x_{ji} = 1 \right\}.$$

We assume that the assets' return rates in different investment periods are independent. As $x_{0i} = 1 - x_i^T \imath$, $i = 0, 1, \ldots, m$, the portfolio optimization problem (P_T) in the m-stage investment horizon $[0, T]$, under a given risk level δ, is formulated as follows,

$$\begin{aligned} (P_T)\ \text{maximize}\ & \textstyle\prod_{i=0}^{m-1} \left[\widehat{\mu}_i^T x_i + R_i(1 - x_i^T \imath) + 1\right] \\ \text{subject to}\ & \mathrm{P}\left(\textstyle\prod_{i=0}^{m-1} (\widehat{\xi}_i^T x_i + R_i(1 - x_i^T \imath) + 1) \geq \prod_{i=0}^{m-1} (\xi_{0i} + 1)\right) \geq e^{-\delta}, \end{aligned}$$

where $\widehat{\mu}_i^T x_i + R_i(1 - x_i^T \imath) + 1$ is the expected return rate of the portfolio $\bar{x}_i$, $\widehat{\xi}_i^T x_i + R_i(1 - x_i^T \imath) + 1$ is the return rate of the portfolio $\bar{x}_i$, and $\xi_{0i} + 1$ is the return rate of the financial index, all during the i-th sub-investment horizon. For this model, our objective is to maximize the expected portfolio rate of return at time t_m, subject to the probability that the portfolio rate of return is greater than the financial index rate of return at time $t_m = T$ being greater than a certain risk level.

Consider the i-th single-stage portfolio selection problem (P_i) in the investment period $[t_i, t_{i+1}]$, $i = 0, 1, \ldots, m-1$,

$$\begin{aligned} (P_i) \text{ maximize } \quad & (\widehat{\mu}_i + R_i \imath)^T x_i \\ \text{subject to } \quad & \mathrm{P}\left(\eta(x_i) \le 0\right) \ge 1 - \tfrac{\delta}{m} \,, \end{aligned}$$

where $\eta(x_i)$ is the slack random variable at the decision point t_i,

$$\eta(x_i) := \xi_{0i} - R_i - (\widehat{\xi}_i - R_i \imath)^T x_i \,.$$

Applying the algorithm for solving the single-stage portfolio optimization model, we can find a feasible solution for the optimization problem (P_T).

Lemma 3. *The optimal strategy for the single-stage financial index tracking model (P) in each sub-investment period* $[t_i, t_{i+1}]$, $i = 0, 1, \ldots, m-1$, *provides a feasible strategy for Problem* (P_T).

Proof: Taking into account the constraint of Problem (P_T), we have

$$\begin{aligned} & \mathrm{P}\left(\prod_{i=0}^{m-1} (\widehat{\xi}_i^T x_i + R_i(1 - x_i^T \imath) + 1) \ge \prod_{i=0}^{m-1} (\xi_{0i} + 1)\right) \\ & \ge \prod_{i=0}^{m-1} \mathrm{P}\left(\widehat{\xi}_i^T x_i + R_i(1 - x_i^T \imath) \ge \xi_{0i}\right) \ge \left(1 - \frac{\delta}{m}\right)^m \approx e^{-\delta}. \end{aligned}$$

Therefore, if we optimize Problem (P_i) in each sub-investment period $[t_i, t_{i+1}]$, the corresponding optimal portfolio at each decision point t_i, $i = 0, 1, \ldots, m-1$, also satisfies the constraint of Problem (P_T). So, these portfolios provide a feasible solution for this problem. □

Although this investment strategy is not optimal, it provides a simple and sensible strategy for the investor to invest in a multi-stage investment horizon.

8.4 Mean-High Order Moment Downside Index Tracking Model

In this section, we will use the m-th moment of the downside chance as the risk measure, and study the financial index tracking model under this risk measure.

The probability of the downside can be regarded as the zero-th moment of the downside of the slack $\eta(x)$. Obviously, we do not have to be restricted to the probability of the downside chance as our risk measure. Any higher order moment of the downside can be used if the procedures are appropriately modified. In particular, we consider the m-th moment of the downside as an investor's risk measure, where the integer m is greater or equal than 1.

Let $\mathrm{E}(\eta(x)_+)^m$ be the m-th moment of the downside of the slack $\eta(x)$. Then for a given risk level δ, the portfolio selection model (P_m) for the m-th moment downside risk measure is formulated as follows

$$\begin{array}{lll}(P_m) & \text{maximize} & (\widehat{\mu} - R\imath)^T x \\ & \text{subject to} & \mathrm{E}(\eta(x)_+)^m \le \delta\,. \end{array} \tag{8.15}$$

Clearly, the m-th moment of the downside $\mathrm{E}(\eta(x)_+)^m$ is convex on the feasible region of x, so we are maximizing a linear function on a convex set. Thus, this problem is a convex optimization problem, and standard convex programming algorithms can be used to solve it.

Let $x(\delta) = (x_1(\delta), \ldots, x_n(\delta))$ be an optimal solution for Problem (P_m), which, naturally, depends on the parameter δ. Define the optimal objective function value for Problem (P_m) as

$$\nu(\delta) := (\widehat{\mu} - R\imath)^T x(\delta)\,.$$

Theorem C.3. *$\nu(\delta)$ ia a non-decreasing concave function.*

Proof: Suppose that $\delta_1 < \delta_2$, then the feasible region for Problem (8.15) corresponding to δ_1 is not larger than the one corresponding to δ_2. So, the optimal solution when $\delta = \delta_1$ is not greater than the one when $\delta = \delta_2$. That is, $\nu(\delta)$ is non-decreasing.

We now prove the convexity of $\nu(\delta)$. For a given δ_1, let $x^{(1)} = (x_1^{(1)}, \ldots, x_n^{(1)})$ be the optimal solution, and $\nu(\delta_1)$ the optimal objective function value. Let $x^{(2)} = (x_1^{(2)}, \ldots, x_n^{(2)})$, and $\nu(\delta_2)$ be the same for a given δ_2. Using the constraint of Problem (8.15), we have

$$\mathrm{E}(\eta(x^{(1)})_+)^m \le \delta_1\,, \qquad \text{and} \qquad \mathrm{E}(\eta(x^{(2)})_+)^m \le \delta_2\,.$$

Let $\delta = \lambda\delta_1 + (1-\lambda)\delta_2$, for some $\lambda \in [0,1]$, and construct the portfolio

$$x = \left(\lambda x_1^{(1)} + (1-\lambda)x_1^{(2)}, \ldots, \lambda x_n^{(1)} + (1-\lambda)x_n^{(2)}\right).$$

Then

$$\begin{aligned}\mathrm{E}\left[\eta(\lambda x^{(1)} + (1-\lambda)x^{(2)})\right] &\le \lambda \mathrm{E}\left(\eta(x^{(1)})_+\right)^m + (1-\lambda)\mathrm{E}\left(\eta(x^{(2)})_+\right)^m \\ &\le \lambda\delta_1 + (1-\lambda)\delta_2\,.\end{aligned}$$

That is $\lambda x^{(1)} + (1-\lambda)x^{(2)}$ is a feasible solution corresponding to $\delta = \lambda\delta_1 + (1-\lambda)\delta_2$ with corresponding objective function value $\lambda\nu(\delta_1) + (1-\lambda)\nu(\delta_2)$. As $\nu(\delta)$ is optimal, we have

$$\nu(\delta) \geq \lambda \nu(\delta_1) + (1 - \lambda)\nu(\delta_2) .$$

Thus, $\nu(\delta)$ is a concave function, which concludes the proof. □

The graph of the function $\nu(\delta)$ is called the efficient frontier for this portfolio selection problem (see Figure 8.2).

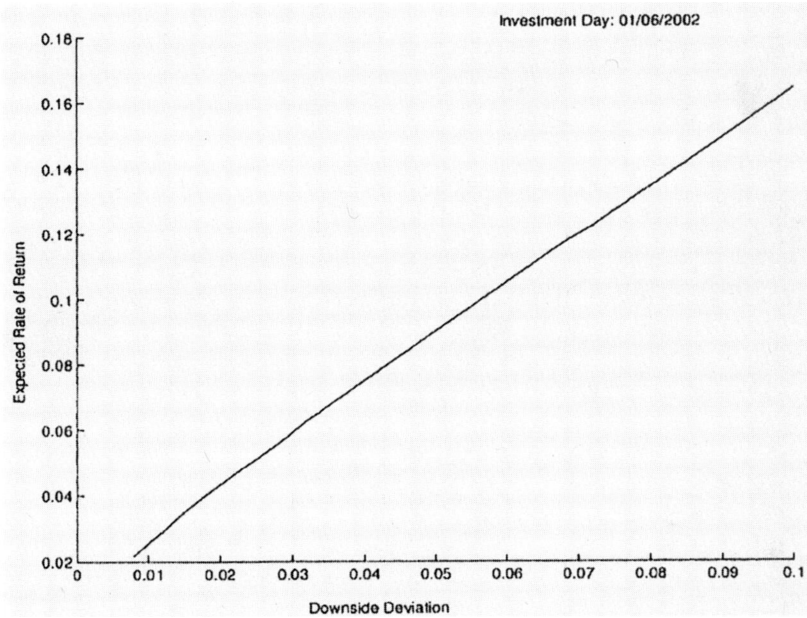

Fig. 8.2. Efficient Frontier of the Mean-Downside High Order Moment Model

As seen earlier

$$\eta(x) \sim N(\mu_0 - R - (\widehat{\mu} - R\imath)^T x, h_0 - 2h^T x + x^T \widehat{H} x) ,$$

so the m-th moment of the downside is

$$\mathrm{E}(\eta(x)_+)^m = \int_0^{+\infty} t^m \frac{1}{\sqrt{2\pi(h_0 - 2h^T x + x^T \widehat{H} x)}} \exp\left(-\frac{(t - (\mu_0 - R - (\widehat{\mu} - R\imath)^T x))^2}{2(h_0 - 2h^T x + x^T \widehat{H} x)} \right) dt .$$

In order to simplify the notation, denote

$$\mu(x) := \mu_0 - R - (\widehat{\mu} - R\imath)^T x , \qquad \text{and} \qquad \sigma(x) := \sqrt{h_0 - 2h^T x + x^T \widehat{H} x} .$$

Then, we may write

$$\mathrm{E}(\eta(x)_+)^m = \sum_{j=0}^{m} \binom{m}{j} \sigma(x)^j \mu(x)^{m-j} F_j^+ \left(-\frac{\mu(x)}{\sigma(x)} \right),$$

where

$$F_j^+(x) := \int_x^{+\infty} s^j \phi(s)\, ds, \quad j = 0, 1, \ldots, m,$$

and $\phi(x)$ is the density function of a standard normal distribution.

It can be difficult to deal with such a risk measure function for $m \geq 2$. However, it may still be possible to work with $m = 1$. In this case, we have

$$\mathrm{E}(\eta(x)_+) = \mu(x)\Phi\left(\frac{\mu(x)}{\sigma(x)}\right) + \sigma(x)\phi\left(\frac{\mu(x)}{\sigma(x)}\right).$$

For a given risk level δ, we construct the portfolio selection model (P_1) under the downside deviation risk measure as follows,

$$\begin{array}{lll} (P_1) & \text{maximize} & (\widehat{\mu} - R\imath)^T x \\ & \text{subject to} & \mu(x)\Phi\left(\frac{\mu(x)}{\sigma(x)}\right) + \sigma(x)\phi\left(\frac{\mu(x)}{\sigma(x)}\right) \leq \delta. \end{array} \tag{8.16}$$

Using the fact that $\sigma(x) > 0$, this may be restated as

$$\begin{array}{ll} \text{maximize} & (\widehat{\mu} - R\imath)^T x \\ \text{subject to} & \theta(x)\Phi(\theta(x)) + \phi(\theta(x)) \leq \frac{\delta}{\sigma(x)}, \end{array} \tag{8.17}$$

where we denote $\theta(x) = \frac{\mu(x)}{\sigma(x)}$.

To solve Problem (P_1), we only need to find the optimal solution for a two-step optimization problem (P_1') stated as follows.

Step 1. Optimize the following function

$$\max_{\widehat{\sigma}} f(\widehat{\sigma}), \tag{8.18}$$

where $f(\widehat{\sigma})$ is defined in Step 2.

Step 2. For a fixed parameter $\widehat{\sigma}$, solve the optimization problem

$$\begin{array}{ll} \text{maximize} & (\widehat{\mu} - R\imath)^T x \\ \text{subject to} & \sigma(x) = \widehat{\sigma} \\ & \theta(x)\Phi(\theta(x)) + \phi(\theta(x)) < \frac{\delta}{\widehat{\sigma}}. \end{array} \tag{8.19}$$

The optimal objective value of this problem is denoted by $f(\widehat{\sigma})$. Then, we have the following result.

Theorem C.4. *The two-step optimization problem (P_1') is equivalent to problem (P_1).*

Proof: Let x^* be the optimal solution for Problem (P_1) and $v(P_1) = v^*$ the corresponding optimal objective function value. Define x'^* and $v(P_1') = v'^*$ analogously with respect to Problem (P_1'). Obviously, x'^* is a feasible solution for Problem (P_1'), and satisfies the constraints of Problem (8.19). Then x'^* also satisfies the constraint of Problem (8.17). Therefore, x'^* is a feasible solution for Problem (P_1), so we have $v'^* \leq v'$.

Assume now that $v'^* < v'$. For Problem (8.19), let $\widehat{\sigma}^* = \sigma(x^*)$, so that x^* is a feasible solution for Problem (8.19) with respect to $\widehat{\sigma}^*$. Since v^* is the corresponding objective function value of x^*, it follows $v'^* \geq v^*$, which is a contradiction. So we have $v'^* = v'$, and Problem (P_1) is equivalent to Problem (P_1'). □

As what regards Problem (8.19), denote $G(x) = x\Phi(x) + \phi(x)$. Clearly, $G(.)$ is an increasing function, so its inverse $G^{-1}(x)$ is also an increasing function. Therefore, the last constraint of Problem (8.19) may be re-written as

$$\begin{aligned} G(\theta(x)) &\leq \frac{\delta}{\widehat{\sigma}}, \\ \Leftrightarrow (\widehat{\mu} - R\imath)^T x &\geq \mu_0 - R - \widehat{\sigma} G^{-1}(\delta/\widehat{\sigma}) . \end{aligned} \tag{8.20}$$

Consider now one more optimization problem

$$\begin{array}{ll} \text{maximize} & (\widehat{\mu} - R\imath)^T x \\ \text{subject to} & \sigma(x) = \widehat{\sigma} . \end{array} \tag{8.21}$$

For a fixed $\widehat{\sigma}$, if the optimal objective function value of Problem (8.21) is greater than the right hand side of inequality (8.20), then the optimal solution for Problem (8.21) is also the optimal solution for Problem (8.19). Otherwise, for this fixed $\widehat{\sigma}$, there is no feasible solution for Problem (8.19) and we define $f(\widehat{\sigma}) = -\infty$.

Notice that Problem (8.21) is a quadratic optimization problem, so by virtue of the KKT optimality conditions, the optimal solution is

$$x^*(\widehat{\sigma}) = \sqrt{\frac{\widehat{\sigma}^2 - h_0 + h^T \widehat{H}^{-1} h}{(\widehat{\mu} - R\imath)^T \widehat{H}^{-1} (\widehat{\mu} - R\imath)}}\, \widehat{H}^{-1} (\widehat{\mu} - R\imath) + \widehat{H}^{-1} h ,$$

and the corresponding optimal objective function value is

$$\sqrt{(\widehat{\sigma}^2 - h_0 + h^T \widehat{H}^{-1} h)(\widehat{\mu} - R\imath)^T \widehat{H}^{-1} (\widehat{\mu} - R\imath)} + (\widehat{\mu} - R\imath)^T \widehat{H}^{-1} h .$$

Consider the function

$$\begin{aligned} g(\widehat{\sigma}, \delta) = {} & \mu_0 - R - \widehat{\sigma} G^{-1}(\delta/\widehat{\sigma}) - (\widehat{\mu} - R\imath)^T \widehat{H}^{-1} h \\ & - \sqrt{(\widehat{\sigma}^2 - h_0 + h^T \widehat{H}^{-1} h)(\widehat{\mu} - R\imath)^T \widehat{H}^{-1} (\widehat{\mu} - R\imath)} . \end{aligned} \tag{8.22}$$

Proposition 1. *For any fixed $\delta > 0$, the function $g(\widehat{\sigma}, \delta)$ is a convex and coercive function in $\widehat{\sigma}$, that is*

$$\frac{\partial^2 g(\widehat{\sigma}, \delta)}{\partial \widehat{\sigma}^2} \geq 0 \qquad \text{and} \qquad \lim_{\widehat{\sigma} \to +\infty} g(\widehat{\sigma}, \delta) = +\infty .$$

Accordingly, when we fix δ, for any $\widehat{\sigma}$, consider a portfolio x. If $\widehat{\sigma} > 0$, this portfolio is infeasible for Problem (8.17). For a given risk levei δ, we have the following result.

Lemma 4. *For Problem (P_1), there exists a δ_0 such that:*

(a) if $\delta > \delta_0$, then there exists $\widehat{\sigma}$, such that $g(\widehat{\sigma}, \delta) \leq 0$, so the optimal solution for Problem (P_1) exists;
(b) if $\delta \leq \delta_0$, then for every $\widehat{\sigma}$ we have $g(\widehat{\sigma}, \delta) > 0$, so the feasible set for Problem (P_1) is empty.

Proof: As $g(\widehat{\sigma}, \delta)$ is convex in $\widehat{\sigma}$, proving that there exists $\widehat{\sigma}$, such that $g(\widehat{\sigma}, \delta) \leq 0$, is equivalent to proving that $\inf_{\widehat{\sigma}} g(\widehat{\sigma}, \delta) \leq 0$. Notice that $g(\widehat{\sigma}, \delta)$ and $\inf_{\widehat{\sigma}} g(\widehat{\sigma}, \delta)$ both decrease in δ. Finally, as $\lim_{x \to -\infty} G(x) = 0$ and $\lim_{x \to +\infty} G(x) = +\infty$, it follows

$$\lim_{\delta \to +\infty} \inf_{\widehat{\sigma}} g(\widehat{\sigma}, \delta) = - \lim_{\delta \to +\infty} \inf_{\widehat{\sigma}} G^{-1}(\delta / \widehat{\sigma}) = -\infty ,$$

and

$$\lim_{\delta \to 0+} \inf_{\widehat{\sigma}} g(\widehat{\sigma}, \delta) = - \lim_{\delta \to 0+} \inf_{\widehat{\sigma}} G^{-1}(\delta / \widehat{\sigma}) = +\infty .$$

Hence, the result follows. □

For finding such a δ_0, we consider the first moment of the downside $\mathrm{E}(\eta(x)_+)$.

Lemma 5. *There is a lower bound for the first moment of the downside of the slack $\mathrm{E}(\eta(x)_+)$.*

Proof: For every $x \in F$, we have

$$\begin{aligned} \mathrm{E}(\eta(x)_+) &= \mathrm{E}\{\eta(x) \cdot \mathbf{I}_{\eta(x)>1}\} + \mathrm{E}\{\eta(x)_+ \cdot \mathbf{I}_{\eta(x) \leq 1}\} \\ &\geq \mathrm{E}\{\eta(x) \cdot \mathbf{I}_{\eta(x)>1}\} \geq \mathrm{P}(\eta(x) > 1). \end{aligned}$$

So,

$$\mathrm{E}(\eta(x)_+) \geq \mathrm{P}\left(\frac{\eta(x) - \mu(x)}{\sigma(x)} > \frac{1 - \mu(x)}{\sigma(x)} \right) = 1 - \Phi\left(\frac{1 - \mu(x)}{\sigma(x)} \right).$$

Using Lemma 1, it is easy to prove that $\frac{1-\mu(x)}{\sigma(x)}$ is bounded. Denote the upper bound for this quotient by $\overline{u}$. Then we have

$$\mathrm{E}(\eta(x)_+) \geq 1 - \Phi(\overline{u}) ,$$

so the result follows. □

Therefore, we define

$$\delta_0 := \inf_x \mathrm{E}\left(\eta(x)_+\right). \tag{8.23}$$

As we proved that $\mathrm{E}(\eta(x)_+)$ is a smooth and convex function of x, by using Newton's method we can find the minimum value of this function. Hence, for a given $\delta \geq \delta_0$, $f(\widehat{\sigma})$ is, obviously, an increasing function of $\widehat{\sigma}$. We can search from the minimum $\widehat{\sigma}$ to $+\infty$ to find the optimal $\widehat{\sigma}^*$, and the corresponding x^* is the optimal solution for Problem (P_1).

8.5 Numerical Example

In this section, we will compare the monthly expected rates of return of the optimal portfolios generated by the mean-downside chance model and the mean-downside deviation model as well as the monthly expected rate of return of the relevant financial index. In the numerical experiments, we choose five stocks in the market in every case. First, we pick up the following five stocks: CLP Holdings, Henderson Ld. Dev., Hong Kong Electric, Swire Pacific, Whard Holdings, whose market values are around the average of the stocks. Second, we randomly choose five different stocks from the constituent of the Hang Seng Index as follows: Cathay Pacific, Cheung Kong Holdings, China Mobile (HK) Ltd., CLP Holdings, Johnson Electric Holdings. In the comparison, the investment horizon is from June 2000 to May 2001, and the performance of the Hang Seng Index is also included.

From the numerical results, we conclude that the trend of the monthly expected rate of return of the optimal portfolio derived from the mean-downside chance model fluctuates less than that of the mean-downside deviation model, and the former is closer to the underlying financial index. In addition, we observe the monthly expected rate of return graphs, and count the points derived from the mean-downside chance model which are below the points generated by the Hang Seng Index. Then, we calculate the actual probability of the downside being under the mean-downside chance framework. This downside probability is quite close to the investor's expected risk level δ. Our experience with the portfolio selection model using the downside chance risk measure is encouraging. Indeed it is a good and intuitive measure of risk for the investors to sense and feel the downside chance, so that they can effectively control their risk tolerance level if needed.

8.6 Conclusion

In this paper, we have studied the financial index tracking models under the downside risk measure. In some sense, the index tracking model with the downside risk measure is consistent with the traditional mean-variance model. If the financial index to be beaten is fixed as a constant, maximization of the

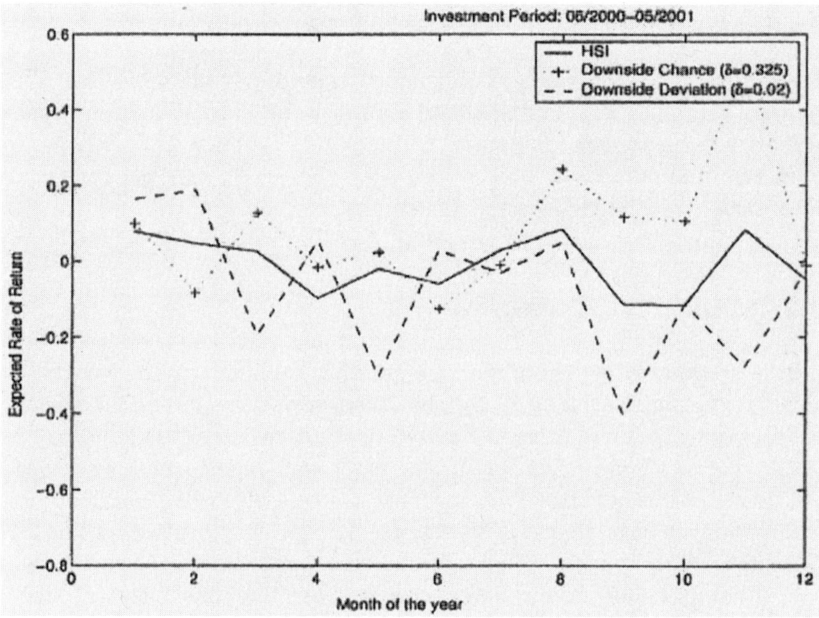

Fig. 8.3. Monthly Rate of Return Using Middle Value Stocks

index tracking model yields solutions which are efficient for the mean-variance model. However, if we randomly pick up an efficient portfolio of the mean-variance model, it may not be efficient in terms of the downside index tracking rule. In addition, there are many potential research topics to be investigated in the future. Such problems include: (1) the downside index tracking problem for multistage investment horizon; (2) the downside index tracking problem under continuous time investment framework; (3) what happens if the joint-normal distribution assumption is dropped? Can one still solve the problem at least approximately?

Appendix

Proof of Lemma 1.

First we prove that $f(x)$ is a bounded function. We may write

$$|f(x)| = \left| \frac{[\mu_0 - R \quad \widehat{\mu} - R\imath] \begin{bmatrix} 1 \\ -x \end{bmatrix}}{\left([1 \quad x^T] \begin{bmatrix} h_0 & h^T \\ h & \widehat{H} \end{bmatrix} \begin{bmatrix} 1 \\ -x \end{bmatrix} \right)^{\frac{1}{2}}} \right|.$$

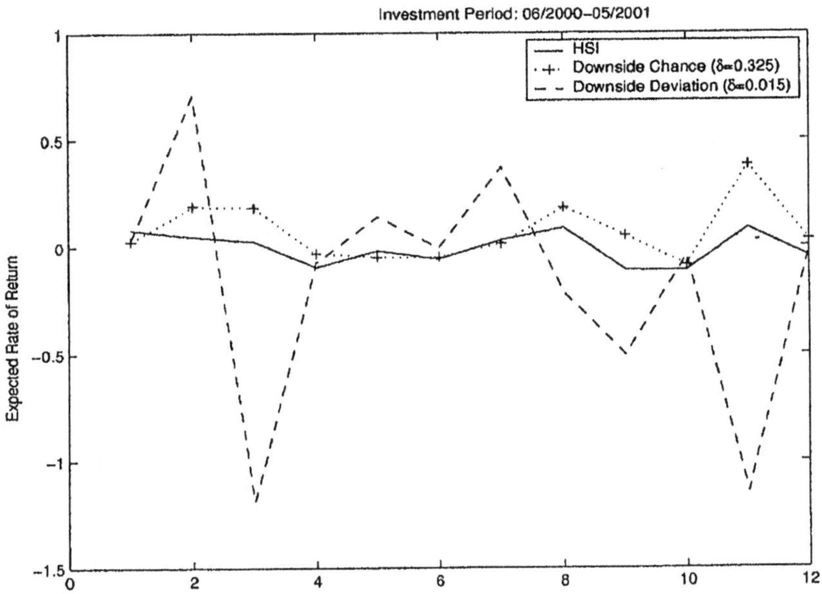

Fig. 8.4. Monthly Rate of Return Using Randomly Selected Stocks

As the matrix $\begin{bmatrix} h_0 & h^T \\ h & \widehat{H} \end{bmatrix}$ is positive definite, its smallest eigenvalue λ_{min} is strictly positive, so that we have

$$|f(x)| \le \frac{\left\| \begin{bmatrix} \mu_0 - R \\ \widehat{\mu} - R\imath \end{bmatrix} \right\| \cdot \left\| \begin{bmatrix} 1 \\ -x \end{bmatrix} \right\|}{\left\| \begin{bmatrix} 1 \\ -x \end{bmatrix} \right\| \sqrt{\lambda_{min}}} = \frac{\left\| \begin{bmatrix} \mu_0 - R \\ \widehat{\mu} - R\imath \end{bmatrix} \right\|}{\sqrt{\lambda_{min}}} .$$

Hence, $f(x)$ is bounded from above. Let us now compute sharper lower and upper bounds for the function $f(x)$. As $f(x) = \frac{c_0 + c^T y}{\sqrt{y^T y + a_0}}$, it follows that

$$\frac{c_0 - \|c\| \cdot \|y\|}{\sqrt{y^T y + a_0}} \le f(x) \le \frac{c_0 + \|c\| \cdot \|y\|}{\sqrt{y^T y + a_0}} =: \overline{f}(x) ,$$

with equality holding if and only if y is parallel to the vector c. The maximum of $\overline{f}(x)$ corresponds to a local maximum of $f(x)$. For simplicity, let

$$z := \frac{c_0}{\|c\|} + \|y\| .$$

As $\|y\| = z - \frac{c_0}{\|c\|}$, we have

$$\overline{f}(x) = \|c\| \frac{z}{\sqrt{z^2 - 2\frac{c_0}{\|c\|}z + \frac{c_0^2}{\|c\|^2} + a_0}}.$$

Recall that

$$a_0 = \begin{bmatrix} -1 & h^T \widehat{H}^{-1} \end{bmatrix} \begin{bmatrix} h_0 & h^T \\ h & \widehat{H} \end{bmatrix} \begin{bmatrix} -1 \\ h^T \widehat{H}^{-1} \end{bmatrix},$$

so $a_0 \geq 0$, from the positive definiteness of the matrix, and that $c_0 \geq 0$. Therefore, the maximum of $\overline{f}(x)$ is attained when

$$z^* = \frac{a_0 \|c\|^2 + c_0^2}{c_0 \|c\|},$$

to which corresponds y^* such that $\|y^*\| = z^* - \frac{c_0}{\|c\|} = \frac{a_0\|c\|}{c_0}$. At the optimal point y is parallel to c, so

$$y^* = \frac{a_0}{c_0} c.$$

The corresponding maximum value of the function $f(x)$ is

$$\max\{f(x)\} = \frac{\sqrt{a_0^2 \|c\|^2 + a_0 c_0^2}}{a_0}.$$

Taking into account that $a_0 \geq 0$ and $c_0 \geq 0$, we have

$$f(x) \geq \frac{c_0 - \|c\| \cdot \|y\|}{\sqrt{\|y\|^2 + a_0}} \geq -\frac{\|c\| \cdot \|y\|}{\sqrt{\|y\|^2 + a_0}} \geq -\|c\|,$$

which concludes the proof. □

Proof of Proposition 1.

We start by proving the convexity in $\widehat{\sigma}$. The function being differentiable, we compute the second derivative of g with respect to $\widehat{\sigma}$.

$$\frac{\partial g(\widehat{\sigma}, \delta)}{\partial \widehat{\sigma}} = -G^{-1}(\delta/\widehat{\sigma}) - \widehat{\sigma} \cdot \frac{\partial G^{-1}(\delta/\widehat{\sigma})}{\partial \widehat{\sigma}} - \widehat{\sigma}\sqrt{\frac{(\widehat{\mu} - R\imath)^T \widehat{H}^{-1}(\widehat{\mu} - R\imath)}{\widehat{\sigma}^2 - h_0 + h^T \widehat{H}^{-1} h}}.$$

Put $y = g^{-1}(\delta/\widehat{\sigma})$, then $G(y) = \frac{\delta}{\widehat{\sigma}}$, thus, taking derivatives on both sides,

$$\frac{dG(y)}{dy} \cdot \frac{\partial y}{\partial \widehat{\sigma}} = -\frac{\delta}{\widehat{\sigma}^2},$$

from what follows

$$\frac{\partial y}{\partial \widehat{\sigma}} = -\frac{\delta}{\widehat{\sigma}^2} \cdot \frac{1}{\Phi(G^{-1}(\delta/\widehat{\sigma}))}.$$

Substituting in the first expression, we have

$$\frac{\partial g(\widehat{\sigma},\delta)}{\partial \widehat{\sigma}} = -G^{-1}(\delta/\widehat{\sigma}) + \frac{\delta}{\widehat{\sigma}} \cdot \frac{1}{\Phi(G^{-1}(\delta/\widehat{\sigma}))} - \widehat{\sigma}\sqrt{\frac{(\widehat{\mu}-R\imath)^T \widehat{H}^{-1}(\widehat{\mu}-R\imath)}{\widehat{\sigma}^2 - h_0 + h^T \widehat{H}^{-1} h}}\,.$$

Hence

$$\frac{\partial^2 g(\widehat{\sigma},\delta)}{\partial \widehat{\sigma}^2} = \frac{\delta^2 \phi(G^{-1}(\delta/\widehat{\sigma}))}{\widehat{\sigma}^3 \Phi^3(G^{-1}(\delta/\widehat{\sigma}))} + $$
$$+ \frac{(h_0 - h^T \widehat{H}^{-1} h)\sqrt{(\widehat{\mu}-R\imath)^T \widehat{H}^{-1}(\widehat{\mu}-R\imath)}}{(\widehat{\sigma}^2 - h_0 + h^T \widehat{H}^{-1} h)^{\frac{3}{2}}} \geq 0\,.$$

Consequently, the function $g(\widehat{\sigma},\delta)$ is convex in $\widehat{\sigma}$.

For proving $\lim_{\widehat{\sigma}\to+\infty} g(\widehat{\sigma},\delta) = +\infty$, it is enough to prove

$$\lim_{\widehat{\sigma}\to+\infty} \frac{\mu_0 - R - \widehat{\sigma} G^{-1}(\delta/\widehat{\sigma})}{\sqrt{(\widehat{\sigma}^2 - h_0 + h^T \widehat{H}^{-1} h)(\widehat{\mu}-R\imath)^T \widehat{H}^{-1}(\widehat{\mu}-R\imath)}} = +\infty\,.$$

Now, this latter statement is true as

$$\lim_{\widehat{\sigma}\to+\infty} \frac{\mu_0 - R - \widehat{\sigma} G^{-1}(\delta/\widehat{\sigma})}{\sqrt{(\widehat{\sigma}^2 - h_0 + h^T \widehat{H}^{-1} h)(\widehat{\mu}-R\imath)^T \widehat{H}^{-1}(\widehat{\mu}-R\imath)}}$$
$$= \lim_{\widehat{\sigma}\to+\infty} \frac{-\widehat{\sigma} G^{-1}(\delta/\widehat{\sigma})}{\sqrt{(\widehat{\mu}-R\imath)^T \widehat{H}^{-1}(\widehat{\mu}-R\imath)}} = +\infty\,.$$

Thus, the function $g(\widehat{\sigma},\delta)$ is coercive. □

References

[1] Bawa, V.S., Optimal Rules for Ordering Uncertain Prospects, *Journal of Financial Economics*, 2(1), 95–121, 1975.

[2] Bawa, V.S., Admissible Portfolios for all Individuals, *Journal of Finance*, 41, 1169–1183, 1976.

[3] Bawa, V.S. and Lindenberg, E.B., Capital Market Equilibrium in a Mean-Lower Partial Moment Framework, *Journal of Financial Economics*, 5, 189–200, 1977.

[4] Bey, R.P., Estimation the Optimal Stochastic Dominance Efficient Set with a MeanSemivariance Algorithm, *Journal of Financial and Quantitative Analysis*, 14(5), 1059–1070, 1979.

[5] Fishburn, P.C., Mean-Risk Analysis with Risk Associated with Below-Target Returns, *American Economics Review*, 67(2), 116–126, 1977.

[6] Hadar, J. and Russell, W.R., Rules of Ordering Uncertain Prospects, *American Economics Review*, 59, 25–34, 1969.

[7] Harlow, W.V., Asset Allocation in a Downside-Risk Framework, *Financial Analysis Journal*, 28, 28–40, September/October, 1991.

[8] Leland, H.E., Beyond Mean-Variance: Performance Measurement in a Nonsymmetrical World, *Financial Analysts Journal*, 55(1), 27–36, January/February, 1999.

[9] Hogan, W.W., Warren, J.M., Computation of the Efficient Boundary in the E-S Portfolio Selection Model, *Journal of Financial and Quantitative Analysis*, 7(4), 1881–1896, 1972.

[10] Hogan, W.W., Warren, J.M., Towards the Development of an Equilibrium Capital Market Model Based on Semivariance, *Journal of Financial and Quantitative Analysis*, 7(4), 1–11, 1974.

[11] Gotoh, J.Y. and Konno, H., Third Degree Stochastic Dominance and Mean-Risk Analysis, *Management Science*, 46(2), 289–301,2000.

[12] Levy, H., Stochastic Dominance and Expected Utility: Survey and Analysis, *Management Science*, 38(4), 555–593, 1992.

[13] Markowitz, H.M., Portfolio Selection, *Journal of Finance*, 7(1), 77–91, 1952.

[14] Markowitz, H.M., Portfolio Selection: Efficient Diversification of Investment, John Willey, New York, 1959.

[15] Markowitz, H.M., Foundations of Portfolio Theory, *Journal of Finance*, 46(2), 469–477, 1991.

[16] Nantell, T.J. and Price, B., An Analytical Comparison of Variance and Semivariance Capital Market Theories, *Journal of Financial and Quantitative Analysis*, 14(2), 221–242, 1979.

[17] Nawrocki, D.N., A Brief History of Downside Risk Measures, *Journal of Investing*, 8(3), 9–25, 1999.

[18] Ogryczak, W. and Ruszczynski, A., From Stochastic Dominance to Mean-Risk Models: Semideviations as Risk Measures, *European Journal of Operational Research*, 116, 33–50, 1999.

[19] Whitmore, G.A., Third Degree Stochastic Dominance, *American Economics Review*, 60, 457–459, 1970.

Part II

Contributed Talks

9

Modelling electricity prices by the potential jump-diffusion

Svetlana Borovkova and Ferry Jaya Permana

[1] Delft University of Technology, Mekelweg 4, 2628 CD Delft, The Netherlands
`s.a.borovkova@ewi.tudelft.nl`
[2] Delft University of Technology, Mekelweg 4, 2628 CD Delft, The Netherlands
`f.j.permana@ewi.tudelft.nl`

Summary. In liberalized electricity markets prices exhibit features, such as price spikes, rarely seen in other commodity markets. Models for electricity spot price, such as mean-reverting jump-diffusions and regime-switching models are only partially successful in modelling price spikes. In this paper we introduce a new approach to electricity price modelling: a potential function jump-diffusion model, which allows for a continuously varying mean-reversion rate and provides a flexible way to model price spikes.

We analyze electricity spot prices from three major European power exchanges: Amsterdam Power Exchange, UK Power Exchange and European Power Exchange (Germany). The potential function jump-diffusion model is applied to the historical spot prices from these exchanges, and its performance is compared to that of the mean-reverting jump-diffusion. The potential function approach is able to capture price spike behavior and overall characteristics of the data remarkably well, and generally better than traditional mean-reverting models. This approach allows for a continuum of different reversion rates, and hence provides a richer model structure and significantly extends the regime-switching model of Huisman and Mahieu [13].

9.1 Introduction

The recent liberalization of European electricity markets offers benefits to both providers and consumers. It also presents new modelling, pricing and risk management challenges to researchers and practitioners involved in energy markets.

Electricity is a flow commodity of strategic importance, characterized by non-storability, inelasticity of demand and large-scale long-term investments. These factors lead to price features rarely seen in other commodity markets: complex seasonal patterns, periods of huge volatility and price spikes. These characteristics are unsuited to the traditional modelling and pricing methods of financial analysis. Risks in electricity markets are dramatically more pronounced: prices can increase by more than 1000% within hours, and return to their normal levels shortly afterwards – a behavior usually referred to as *spikes*.

Modelling the evolution of the spot price is the core of energy market research. The model for the electricity spot price suggested by Lucia and Schwartz [17] is a mean-reverting diffusion (with seasonally varying mean). This model does not incorporate spikes. However, correctly modelling spikes remains the most difficult part of electricity price modelling, because after an upward jump, the price has to be forced back to normal levels. Jump-diffusions, sometimes suggested for modelling prices of financial assets, are now routinely used by many authors for modelling electricity prices (Deng [9], Knittel and Roberts [16], Escribano, Pena and Villaplana [11], Barone-Adesi and Gigli [4], Cartea and Figueroa [7]). In such models, Poisson jumps are added to a mean-reverting diffusion process. Mean-reverting jump-diffusions provide for spike behavior, since the mean-reversion forces the price to return to its normal level. However, a high rate of mean reversion is needed to force the price back after a large upward jump. Hence, the degree of mean-reversion outside spikes can be grossly misspecified.

So far, the only model successfully separating mean reversion and spikes is the Markov regime-switching model, introduced by Huisman and Mahieu [13]. Their model postulates that the price can be in one of three regimes: normal mean-reversion, spike regime and return from spike to a normal price level. Transitions from one regime to another are governed by a discrete time Markov process, with transition probabilities estimated from historical data (as well as the model parameters in each regime). A similar approach is taken by Geman and Roncoroni [12], where regimes are defined in terms of the price exceeding a certain pre-specified threshold. These regime-switching models allow for a richer structure than the mean reversion with jumps. However, these model did not find widespread acceptance by practitioners, who are accustomed to diffusion-like models, preferring them to regime-switching models.

Here we introduce a new approach for modelling electricity spot prices, based on a recently developed potential field model (Borovkova et al. [5]). This involves a jump-diffusion with drift given by a potential function derived from the process' invariant distribution. The potential forces the price to return to its average level (driven by a seasonal component) after an upward jump. The potential function model is close in spirit to a mean-reverting jump diffusion, with a more general drift term. It is also a generalization of the regime-switching model of Huisman and Mahieu [13], which allows the rate of mean-reversion to be different in each regime. In the potential function model, the rate of reversion to an average price level is a continuous function of the distance from this level, specified by the model builder. This allows for a richer model structure, while still retaining the jump-diffusion framework. In fact, the mean-reverting jump-diffusion features in our model as a special case.

The paper is organized as follows. In the next section we review some known models for electricity spot prices; and in Section 3 we introduce the potential function model. Section 4 is devoted to applications to data from power markets. We perform an extensive analysis of electricity spot prices from three major European markets: The Netherlands, UK and Germany; and test the potential function model on the historical prices from these exchanges. We compare it with the mean-reverting jump-diffusion model in terms of matching characteristics of historical data and simulating new price paths.

9.2 Stochastic Models for Electricity Spot Prices

9.2.1 Overview of Available Models

As mentioned in the introduction, electricity spot prices are characterized by complex seasonal patterns, periods of high volatility and price spikes. All these features can be seen in Figure 9.1, which shows daily electricity spot prices[3] on three major European electricity exchanges: Amsterdam Power Exchange (APX), UK Power Exchange (UKPX, London) and European Energy Exchange (EEX, Leipzig, Germany). On the y-axis is the price in Euros per megawatt/hour, and on the x-axis – the number of calendar days since the first day in the analyzed dataset (Jan. 1 2001 for APX, Apr. 1 2001 for UKPX and Jan. 1 2002 for EEX). Note that the price on the UK power exchange is quoted in Euros and not in British pounds – UKPX is managed by the Amsterdam Power Exchange (APX) and prices are quoted in Euros to ensure better coordination with other European power exchanges.

Price spikes dominate the graphs, but seasonal (yearly and weekly) behavior is also visible. Generally, the average price level is higher in winter and summer than during mid-seasons, and higher on weekdays than on weekends.

The standard approach is to model the spot price $P(t)$ or, more often, its logarithm $Y(t) = \ln P(t)$ on each day t as the sum of the deterministic seasonal component $s(t)$ and a stochastic process $X(t)$:

$$Y(t) = \ln P(t) = s(t) + X(t)\,. \tag{9.1}$$

The seasonal component $s(t)$ usually consists of two periodic functions: with the periods of 1 year and 1 week (sometimes a half-yearly periodic component is added as well).

The specification (9.1) for the (log) price is used by most researchers and practitioners working in energy markets (e.g. Lucia and Schwartz [17], Sorensen [21], Huisman and Mahieu [13], Pilipovic [20] and many others). The particular specification of the stochastic process $X(t)$ leads to different models. For example, in [20], $X(t)$ is modelled as the stationary mean-reverting (Ornstein-Uhlenbeck) process:

$$dX(t) = -\alpha X(t)dt + \sigma dW(t)\,, \tag{9.2}$$

[3] We use the so-called *baseload price* – the daily average of the prices for hourly deliveries.

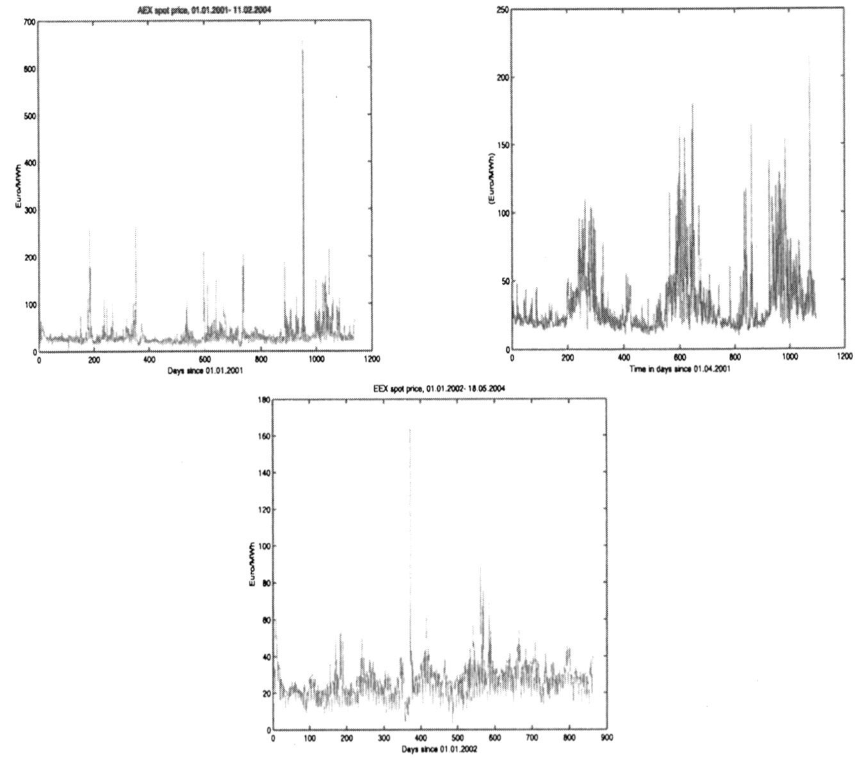

Fig. 9.1. APX (01.01-02.04), UKPX (04.01-03.04) and EEX (01.02-05.04) electricity spot prices

where α is the mean-reversion rate, σ is the volatility and $W(t)$ is the standard Brownian motion.

Lucia and Schwartz [17] model $X(t)$ by the sum of two stochastic processes: a short-term mean reverting component and a long-term equilibrium price level, which follows a geometric Brownian Motion:

$$dX(t) = [-\alpha X(t)dt + \sigma dW_1(t)] + [\mu dt + \eta dW_2(t)], \tag{9.3}$$

with μ being the growth rate and η – the second volatility parameter.

In mean-reverting jump-diffusion models (studied in the articles mentioned in the introduction), $X(t)$ is decomposed into a mean-reverting diffusion and a Poisson jump component:

$$dX(t) = -\alpha X(t)dt + \sigma dW(t) + Jdz(t), \tag{9.4}$$

where $z(t)$ is the Poisson process with intensity λ and J is a random variable representing random jump size.

The regime-switching model by Huisman and Mahieu [13] incorporates a regime-switching mechanism in the specification for $X(t)$ (however, it does not take yearly seasonality into account). Three regimes are defined: "normal" mean-reversion, spike initiation and spike reversal regimes. The transition from one regime to another is given by Markov transition probability matrix, which reflects the fact that the spike initiation regime is usually followed by spike reversal. The mean-reversion rate in the spike reversal regime is assumed to be much higher than the mean-reversion rate in the "normal" regime.

Mean-reverting diffusion models of the type (9.2), (9.3) are unrealistic in that they do not accommodate spikes. The main drawback of mean-reverting jump-diffusion models (9.4) is that the mean-reversion rate is constant, the underlying assumption being that all the shocks affecting the price die out at the same rate, regardless of the shock size. This is in conflict with economic intuition, which suggests that large shocks should die out more rapidly due to powerful forces of supply and demand, while small shocks should slowly revert to the previous price level, due to adjustment of production costs. This is also an empirically observed phenomenon: in most markets, the electricity price returns towards the mean level more rapidly from greater excursions, e.g. an upward jump.

Hence, for the electricity price it is reasonable to assume a non-constant mean-reversion rate, e.g. one that itself depends on the distance from the mean level. The regime-switching approach of Huisman and Mahieu [13] allows for a non-constant mean reversion rate by specifying a number of regimes, each with its own mean-reversion rate (which is still constant within each regime). However, a more general way of modelling this, is to allow the mean-reversion rate to be a *continuous function* of the distance to the mean price level. For instance, the price return can be a quadratic or exponential function of this distance. By this, we would allow for a *continium* of mean-reverting regimes. This can be incorporated into the model by extending the class of possible drift forms. In the next section we shall explore this idea, by specifying the drift term by means of a *potential function.*

9.2.2 Model Estimation Issues

Estimating the parameters of jump-diffusion models of the type (9.4) is non-trivial, and the difficulty in estimation often prevents their implementation in applications. This difficulty originates from the problem of determining whether movements in the underlying process are part of the continuous dynamics, or whether they are part of the jump dynamics.

Several estimation methods have been proposed and studied in the literature for jump-diffusion processes: the (quasi) maximum likelihood (ML) method, the generalized method of moments (GMM) and variations of each. In the ML method, the density of the jump-diffusion returns is derived analytically, by means of characteristic functions or Fourier transforms, or approximated by a mixture of normal distributions (see Beckers [2], Ball and Torous [3]). In the generalized method of moments, the first few moments are derived analytically[4], and the sample moments are set equal to the corresponding population moments. Then the resulting equations are solved for the unknown parameters, providing the parameter estimates.

The maximum likelihood method is efficient and hence more attractive than the generalized method of moments. For many models, GMM has been shown to have the lower overall estimation efficiency, when compared to maximum likelihood-based methods (Zhou [22]). The main attraction of the GMM method is its general applicability; even when the expression for the density function is unknown, moment conditions may still be available in analytical form for many models of practical interest.

However, both ML and GMM methods are known to grossly overestimate the jump frequency and underestimate the mean jump size (Jiang [14], Huisman and Mahieu [13] and other authors point this out). For example, when we applied the maximum likelihood method of Ball and Torous to historical power prices from the Amsterdam Power Exchange (APX), we found that more than 40% of price moves were identified as jumps (while power market practitioners believe the jump frequency to be around 10%). Moreover, such estimation procedures do not explicitly separate jumps from the diffusion component, so graphical methods of jump distribution analysis (histograms, QQ-plots) are inapplicable.

If there were a way of disentangling the jumps from the diffusion component, then the jumps could be analyzed and the jump parameters more accurately estimated. In this paper we shall follow an approach of *jump filtering.* We separate price moves that we classify as jumps by applying a simple sequential filtering procedure, and estimate separately the parameters of the diffusion and the jump components. However, it must be stressed that identifying the jumps without any restrictions is always an ad-hoc task by the researcher, possibly introducing sample selection error to the model.

There is an inherent weakness in all estimation procedures for processes given by a general stochastic differential equation and observed at discrete time intervals. In all these procedures, the theoretical stationary distribution of the process is matched to the empirical distribution of the observed log-returns. However, the exact distribution of the log-returns, given some time discretization step, is not the same as the stationary distribution of the process, since the discretization of the continuous time stochastic differential equation by e.g. the Euler scheme involves higher order terms. For some specific models (such as Geometric Brownian Motion) the exact distribution of the log-returns can be derived, but for most of general models this distribution is not known. Then one can only hope that, for a sufficiently small time discretization step, the process' stationary distribution and the log-returns distribution are close. This is a reasonable assumption and it is supported by

[4] Ait-Sahalia [1] points out the importance of matching higher (up to 4th) order moments

the results on weak convergence of numerical solutions for SDE (Millstein [19], Kloeden and Platen [15]).

A recently introduced approach by Eberlein and Stahl [10] allows exact fitting of the returns generated by the model process to the empirically observed returns. This approach is based on Lévy processes. It is particularly well suited to power prices, whose distributional characteristics can be modelled by very flexible generalized hyperbolic distributions, which arise from Lévy processes.

9.3 Potential Function Approach

The potential function approach was introduced in Borovkova et al. [5] for modelling commodity prices that exhibit multiple attraction regions (such as oil prices). The potential function model is a diffusion model with the drift specified in terms of the potential function. The potential function is closely related to the invariant distribution of the process, hence this approach can encompass multimodality in price distributions. Here we shall apply the potential function approach for a different purpose, namely for modelling a varying rate of mean-reversion in the jump-diffusion context.

The original potential function model postulates that the (log) price process $(X(t))_t$ evolves according to the stochastic differential equation:

$$dX(t) = -U'(X(t))dt + \sigma dW(t)\,, \tag{9.5}$$

where $U : \mathbb{R} \longrightarrow \mathbb{R}$ is a twice continuously differentiable function, such that $U(x) \longrightarrow \infty$ as $|x| \longrightarrow \infty$, and

$$\int_{-\infty}^{\infty} \exp\left(-2U(x)/\sigma^2\right) dx < \infty\,.$$

These conditions assure that the invariant distribution of the process $(X(t))_t$ is the *Gibbs distribution* with density

$$\pi_\sigma(x) = \exp\left(-\frac{2U(x)}{\sigma^2}\right) \tag{9.6}$$

(for proof see e.g. Matkovsky and Schuss [18]).

The relationship (9.6) means that there is one-to-one correspondence between the invariant distribution of the process and the diffusion's drift, given by the potential. So characteristics such as multimodality of the price distribution can be incorporated into the model by the proper choice of the potential function.

The potential (together with the volatility σ) can be estimated from historical data by first estimating

$$G_\sigma(x) = \frac{2}{\sigma^2}U(x) = -\ln(\pi_\sigma(x))$$

by

$$\widehat{G}_\sigma(x) = -\ln(\widehat{\pi}(x)),$$

where $\widehat{\pi}$ is some estimate of the observations' marginal density (e.g. a kernel estimator or a histogram smoothed by a polynomial or a sum of Gaussian densities).

The volatility σ can then be estimated by discretizing the equation (9.5) by Euler scheme with $\Delta t = 1$ day and noting that $\sigma^2/2$ is the linear regression coefficient (without intercept) of the increments $(X_{i+1} - X_i)$ on $(\widehat{G}_\sigma(X_i))$.

The model (9.5) describes a continuous time process. If we observe time series $(X_i)_{i\in\mathbf{N}}$, then we can consider it as a realization of this continuous time process observed at discrete time points. The observations (X_i) come from the distribution with density π_σ given by (9.6).

For a small time interval Δt, the discretization of the equation (9.5)

$$\Delta X_t = X_{t+\Delta t} - X_t = -\nabla U(X_t)\Delta t + \sigma \Delta W_t,$$

gives an Euler scheme for the numerical solution to the diffusion equation (9.5). If we assume that the observation interval is small enough, the approximate model motivated by the above Euler scheme for the observed time series can be given by

$$\bar{X}_{i+1} = \bar{X}_i - \nabla U(\bar{X}_i)h + \varepsilon_i\,. \tag{9.7}$$

Here h is the time interval between observations (for power prices $h = 1$ day) and the (ε_i) are the increments of the process σW_t over the intervals, so they are independent normally distributed random vectors with independent components having mean 0 and variance $h\sigma^2$.

If the underlying process evolves according to (9.5), the exact evolution equation for the time series of observations at discrete intervals is *not* given by (9.7), but by an equation with higher order terms. Conversely, if the exact evolution of the process is given by (9.7), the density of its invariant distribution is no longer π_σ. However, from the results on weak convergence of numerical solutions for SDE (Millstein (1988), Kloeden and Platen (1995)) it follows that, under some differentiability conditions on U, for decreasing h, the invariant distribution of $\bar{X}_i$ converges weakly to the invariant distribution of X_t, i.e., to the Gibbs distribution with the density π_σ. So for a relatively short observation interval, we hope that the invariant distribution is not too far away from the Gibbs distribution. However, as noted in the previous section, matching the empirical returns distribution to the stationary distribution of the underlying process remains here a weak point, just as is the case for all models starting with a stochastic differential equation with general diffusion coefficients.

For more detail on the estimation of the potential model and its applications to commodity prices see Borovkova et al. [5].

In the spirit of mean reverting jump-diffusion models discussed in the previous section, we model the stochastic component of the log-price $X(t)$ by

$$dX(t) = -U'(X(t))dt + \sigma dW(t) + Jdz(t), \tag{9.8}$$

where $U(x)$ is a properly chosen potential function. In the absence of jumps, the process is the diffusion with drift given by a potential function. Note that the mean-reversion is incorporated into (9.8) as a special case, by taking $U(x)$ a quadratic function.

The invariant marginal distribution of the process $X(t)$ evolving as in (9.8) is no longer the Gibbs distribution (9.6) (due to the presence of jumps), so the direct estimation of U from the empirical distribution of observations on $X(t)$ is no longer possible. One way to proceed is to assume a certain parametric form for U. For example, a simple way to model a state-dependent reversion rate, is to assume that the potential function is a polynomial of some fixed (and even) degree, higher than 2 (the degree 2 corresponds to the constant mean-reversion rate). This also assures that the conditions on the potential are satisfied. Then all the parameters of the model (9.8) can be simultaneously estimated from the observed de-seasoned log-prices $(X_i)_i$ by the method of maximum likelihood or the generalized method of moments, in the spirit of Ait-Sahalia [1]. However, also in this case the jump parameters (frequency and mean jump size) may be significantly misspecified.

Here we follow a different route: we apply a sequential jump filtering procedure and estimate the diffusion-related parameters (i.e. the potential function and the volatility σ) from the filtered series, by matching the process' stationary distribution to the empirically observed one, using the relationship (9.6). The jump intensity and jump size parameters are estimated from the series of observed jumps.

9.4 Model Estimation and Application to Historical Prices

We shall work everywhere with log-prices; for APX, UKPX and EEX the daily log-prices (in log(Euro/MW/hour)) are shown in Figure 9.2, for the same periods as in Figure 9.1. The yearly seasonal pattern is much more pronounced for the UK prices, while spikes are much more prominent for the APX prices.

We apply the mean-reversion (with and without jumps) and the potential jump-diffusion models with seasonal mean to the log-prices shown in Figure 9.2. Often all parameters (of the seasonal as well as stochastic components) of a model are estimated simultaneously, by e.g. the maximum likelihood or the Kalman filter. Here we shall adopt a different procedure: a stepwise estimation of the model's components, typically used in traditional time series analysis.

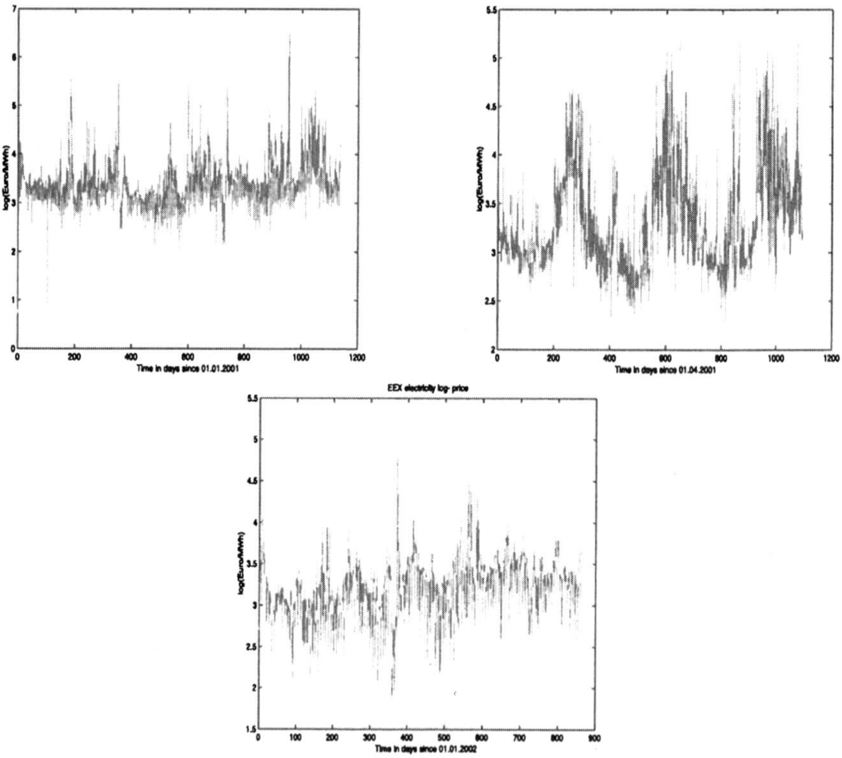

Fig. 9.2. APX, UKPX and EEX log-prices

9.4.1 Filtering Spikes and Estimating Seasonalities

Let $(Y_i)_{i=1}^n$ denote the historical series of observed log-prices. First, we estimate and extract from it deterministic trends and seasonalities. At this stage, we filter out price spikes so as not to disturb estimates of the seasonal components. For this we use the following sequential filtering procedure.

At the first step, we consider as jumps those price moves outside $\widehat{\mu} \pm 2\widehat{\sigma}$ prediction intervals, with $\widehat{\mu}$ and $\widehat{\sigma}$ given by the 30-days moving average and 30-days historical volatility of the price moves (30 day window corresponds to approx. one calendar month). After removing these jumps, we repeat this procedure with the re-estimated 30-days volatility. We continue until no new jumps can be identified. For estimating the seasonal component, the jumps are replaced in the original series by their cutoff values.

A similar jump-removal filtering procedure is also used in Clewlow and Strickland [8]; there, the overall sample variance of price moves is computed and the threshold for identifying jumps is iteratively lowered until the jump removal procedure converges. Our procedure is more data-driven and sequential in nature, however the principle is similar.

For the purposes of electricity price modelling, *upward* jumps (price moves) are relevant, since they reflect external extremal events and indicate initiation of spikes. Downward price moves follow the upward ones and usually do not arise separately in case of electricity. Hence, we shall concentrate our analysis on the upward price moves. These moves of the log-price identified as jumps are shown in Figure 9.3 (in log(Euros/Mw/hour)); on the x-axis is again the number of calendar days since the beginning of the dataset.

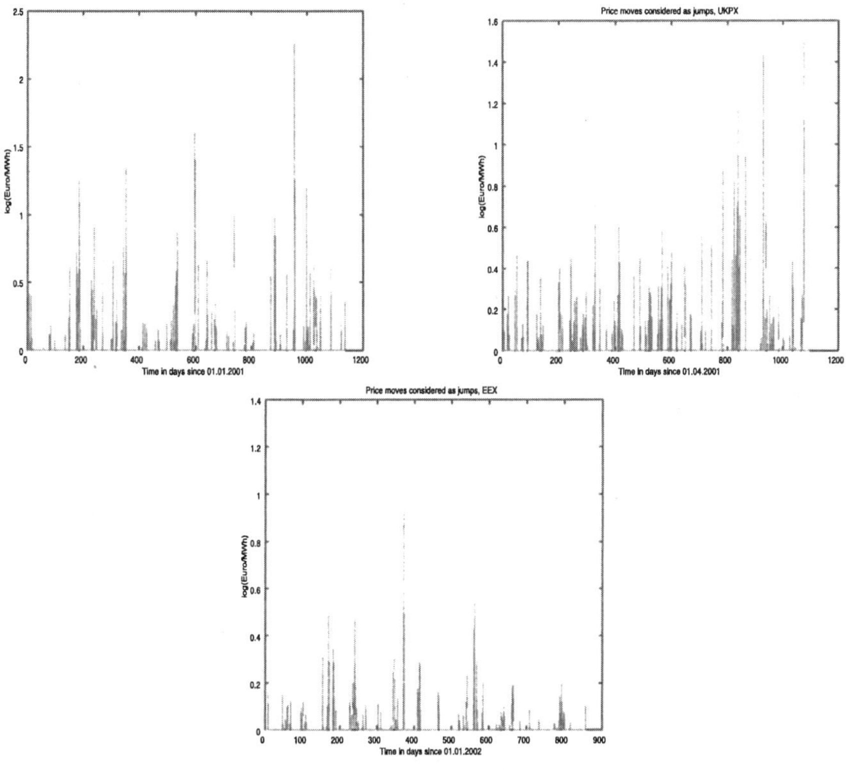

Fig. 9.3. Upward log-price moves considered as jumps

The main advantage of the jump-filtering procedure just described is that we explicitly obtain a series of observed upward price jumps $(J_i)_i$. We then can estimate the frequency of Poisson jump component and the parameters of the jump size distribution directly from the series $(J_i)_i$, and apply to it graphical methods of analysis, such as histograms and QQ-plots.

Descriptive analysis of inter-arrival jump times showed that for all markets the assumption of a Poisson process of jump arrivals is reasonable (i.e. the inter-arrival jump times have approximately exponential distribution). The estimated jump arrival rates are 10% for APX, 13% for UKPX and 10% for

EEX. For comparison: the maximum likelihood estimation procedure of Ball and Torous (1983) produced the estimates for jump frequency in the order of 40% for all three markets, which is clearly a misspecification.

It is often proposed to use the lognormal distribution for the jump size. For APX market we fitted the lognormal distribution to the distribution of the jump size. The resulting fit is shown as the histogram and QQ-plot on the leftmost graphs in Figures 9.4 and 9.5. For UKPX and EEX markets, the exponential distribution gives a much better fit, as the corresponding QQ-plots show. If neither lognormal nor exponential distribution seem suitable, a bootstrap method can be used for simulations, i.e. re-sampling the jump sizes from their empirical distribution.

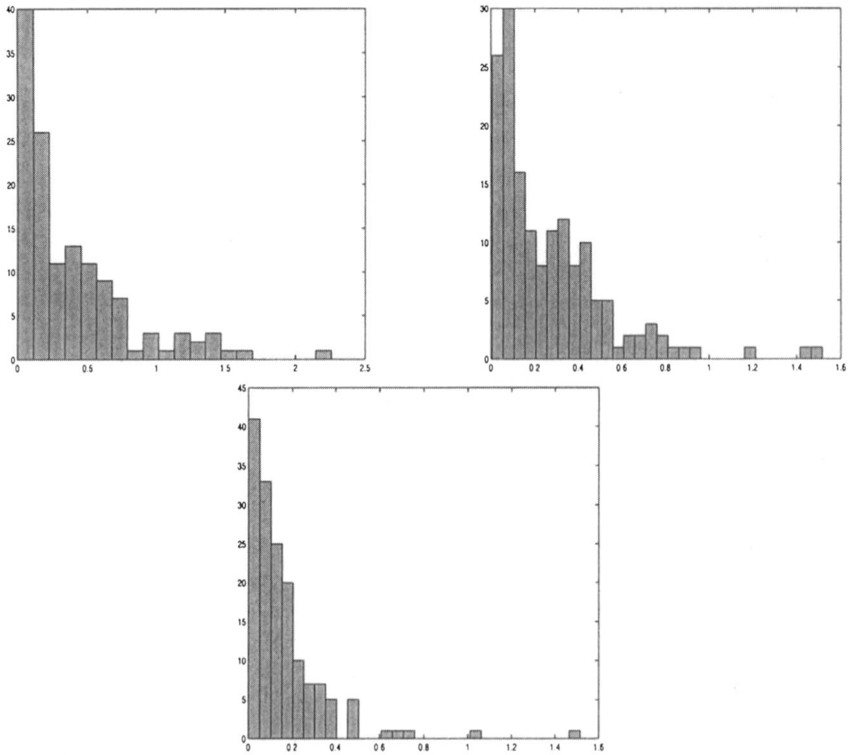

Fig. 9.4. Histograms of observed jump sizes, APX, UKPX and EEX

Table 9.1 summarizes the distributional characteristics of jumps for the three markets.

The above jump filtering procedure is robust to different choices of the time window used to determine the mean and standard deviation. However, increasing the number of standard deviations used for the cutoff leads to a

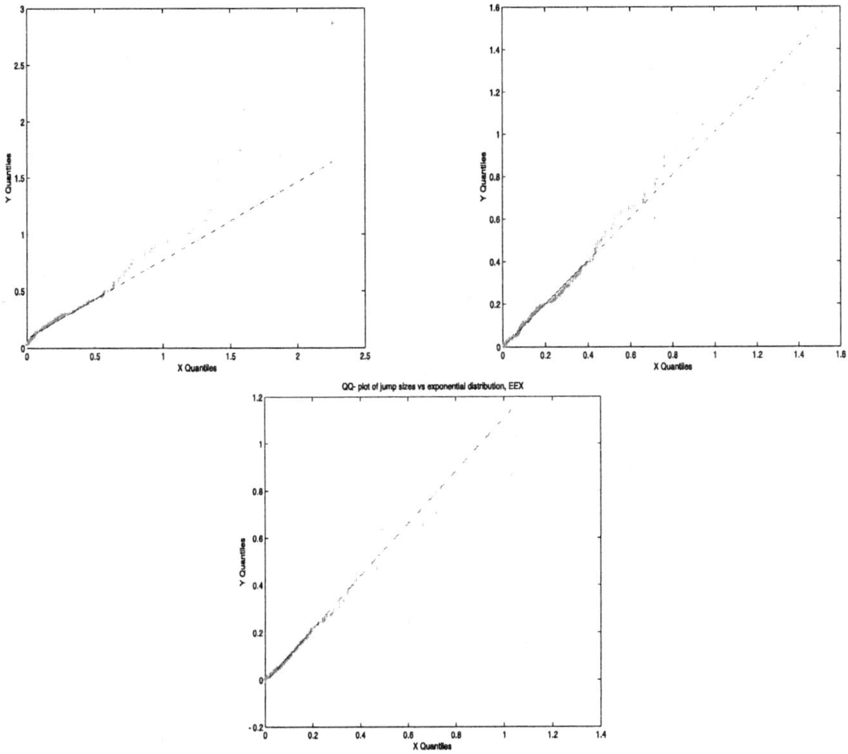

Fig. 9.5. QQ-plots of jump sizes, APX (vs lognormal), UKPX (vs exponential) and EEX (vs exponential)

Jump parameter	APX	UKPX	EEX
frequency	0.10	0.13	0.10
mean jump size	0.33	0.27	0.14
standard deviation	0.24	0.26	0.15

Table 9.1. Distributional characteristics of jumps

lower number of price moves identified as jumps. To check this, we performed jump filtering for APX market with window lengths 60 and 90 days (approx. 2 and 3 months) and with three standard deviations instead of two. Table 9.2 presents the estimated jump frequency and mean jump size for various combinations of the window length (the first number in parenthesis) and the number of standard deviations (the second number in parenthesis). The numbers for different window lengths are close, if the number of standard deviations is kept the same. Increasing this number from 2 to 3 results in an estimated jump frequency of 7% instead of 10%. Hence, the factor 2 is more in line with market participants' perceptions that 10% is representative spike frequency in electricity markets.

	(30,2)	(60,2)	(90,2)	(30,3)	(60,3)	(90,3)
frequency	0.10	0.10	0.10	0.07	0.07	0.07
mean jump size	0.33	0.37	0.35	0.36	0.40	0.39

Table 9.2. Jump characteristics vs. window length and number of standard deviations

After filtering out jumps, we remove the slowly varying yearly trend from each series, estimated by the 12-months moving averages of the log prices. These turned out to be not significantly different from constants for all three markets: 3.4, 3.3 and 3.2 for respectively APX, UKPX and EEX. After subtracting these averages, the yearly seasonal component is estimated by the least-squares fitting of a trigonometric function

$$f(t) = \sum_{k=1}^{2} (A_k \sin(2\pi kt) + B_k \cos(2\pi kt)), \tag{9.9}$$

where choosing $k = 1, 2$ captures yearly as well as half-yearly periodicity. The estimated functions $f(t)$, together with average weekly log-prices, are shown in Figure 9.6 for the three markets (vs the number of weeks since the beginning of the dataset). Note that the summer peak is more pronounced for EEX, less for UKPX and is absent for APX.

The price premium corresponding to a particular day of the week is estimated by the average deviation (over the entire historical dataset) of a particular weekday price from the weekly averages. The estimated daily log-price premium is shown in Figure 9.7. Note that the weekend premium is negative, indicating lower power prices during weekends than the average price level.

The weekly periodic function $w(t)$ is then modelled by a collection of dummy variables

$$w(t) = \sum_{k=1}^{7} \beta_{t,k} DP_k , \tag{9.10}$$

where $\{DP_k, \ k = 1, ..., 7\}$ is the collection of the estimated daily premiums and $\beta_{t,1} = 1$ if the day t is Monday and 0 otherwise, $\beta_{t,2} = 1$ if t is Tuesday and 0 otherwise, and so on.

The estimated yearly periodic function $f(t)$ and the daily premium $w(t)$ are then subtracted from the original log-price series to obtain de-trended and de-seasoned series $(X_i)_{i=1}^n$, shown in Figure 9.8. Note that spikes are present in these series: they have been filtered out only while estimating seasonalities. The series in Figure 9.8 can be considered as realizations of the stochastic component $X(t)$ from eq. (9.1).

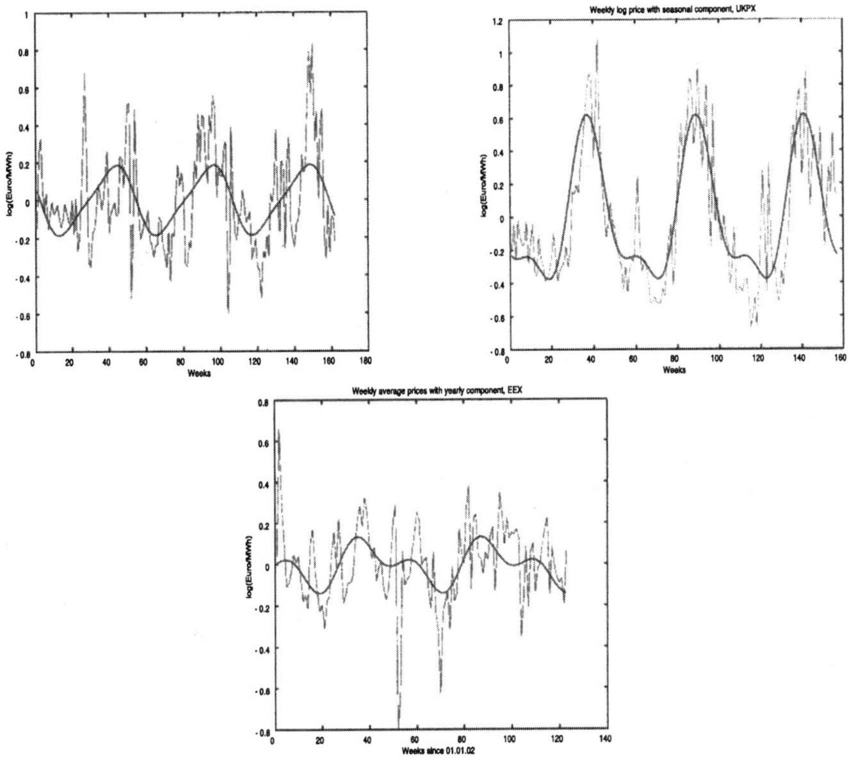

Fig. 9.6. De-trended weekly log-prices with yearly components, APX, UKPX and EEX

9.4.2 Modelling the Stochastic Component

We model the stochastic component of the log-price $X(t)$ in three ways: mean-reverting diffusion, mean-reverting jump-diffusion and potential function jump-diffusion:

$$\text{Model I:} \qquad dX(t) = \alpha(\mu - X(t))dt + \sigma dW(t)\,, \tag{9.11}$$

$$\text{Model II:} \qquad dX(t) = \alpha(\mu - X(t))dt + \sigma dW(t) + Jdz(t)\,, \tag{9.12}$$

$$\text{Model III:} \qquad dX(t) = -U'(X(t))dt + \sigma dW(t) + Jdz(t)\,, \tag{9.13}$$

where $W(t)$ is the standard Brownian motion, σ is the volatility of the diffusion component (assumed constant). The jump component is characterized by $z(t)$: a Poisson process with intensity λ (i.e. $dz(t) = 1$ with probability λdt and $dz(t) = 0$ with probability $1 - \lambda dt$) and by a random jump size J with some specified distribution.

Model I is fitted to the series $(X_i)_i$ (shown in Figure 9.8) by discretizing eq. (9.11) and estimating the parameters by the method of maximum likelihood.

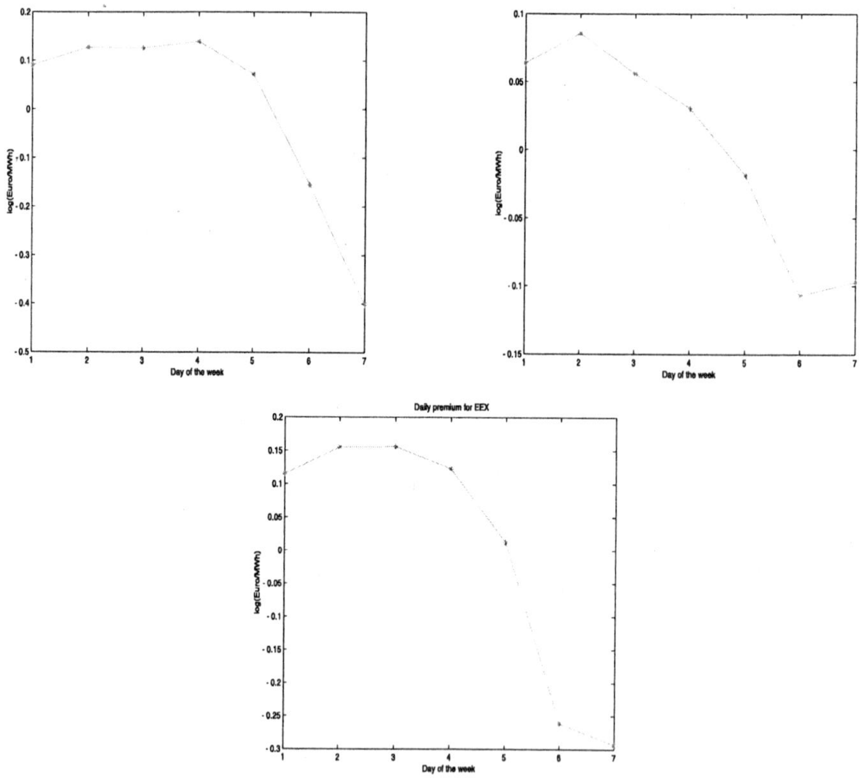

Fig. 9.7. Estimated daily premiums, APX, UKPX and EEX

To estimate Models II and III, we apply the jump-filtering procedure described above to the series in Figure 9.8. We then estimate the parameters of diffusion components from the *filtered series* $(\widetilde{X}_i)_i = (X_i - J_i)_i$, and the parameters of the jump component (frequency of Poisson jumps and the parameters of their size distribution) – from the series of observed jumps $(J_i)_i$. This has a number of advantages: we estimate the rate of mean reversion under "normal" circumstances, reducing scope for its misspecification. Moreover, distributional assumptions regarding jump size and frequency were verified by analyzing histograms and QQ-plots.

Table 9.3 presents the estimated mean-reversion and diffusion parameters for Models I and II, estimated by the method of maximum likelihood from the series $(X_i)_i$ (Model I) or $(\widetilde{X}_i)_i$ (Model II).

Note that the rate of mean reversion is in all cases higher for Model I, although the differences are not very large. This is because for estimating Model II we excluded all the upward price moves, eliminating by this the upward bias in the mean-reversion rate estimate in Model I. The volatility

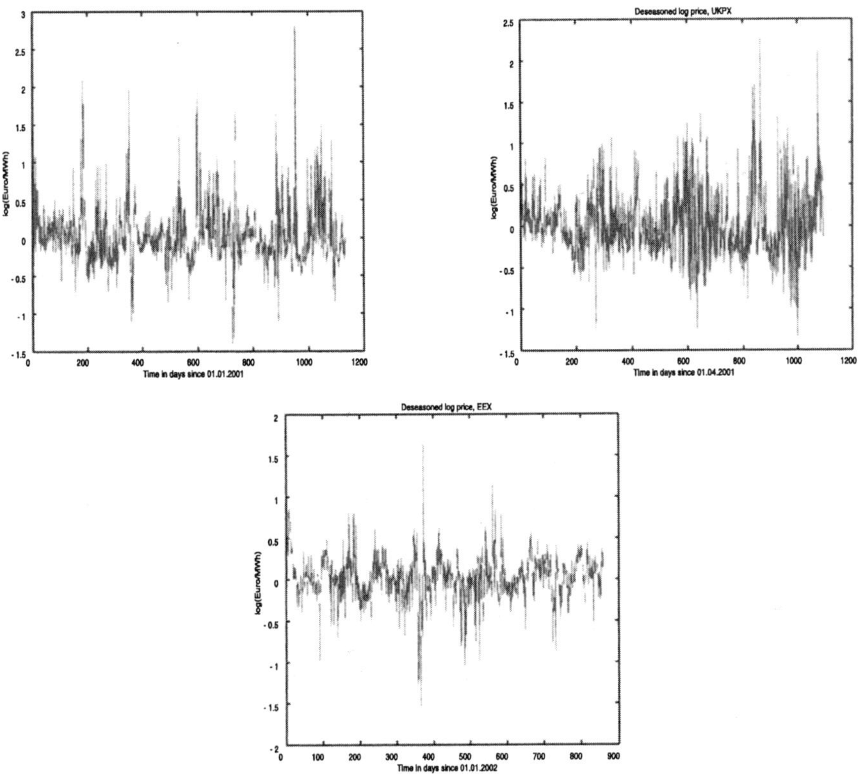

Fig. 9.8. De-seasoned and de-trended log prices, APX, UKPX and EEX

	APX		UKPX		EEX	
Parameter	Model 1	Model 2	Model 1	Model 2	Model 1	Model 2
μ	0.04 (0.01)	0.00 (0.01)	0.03 (0.01)	0.00 (0.01)	0.02 (0.01)	0.00 (0.01)
α	0.41 (0.02)	0.37 (0.02)	0.54 (0.03)	0.51 (0.03)	0.49 (0.03)	0.46 (0.03)
σ	0.48 (0.04)	0.25 (0.03)	0.59 (0.05)	0.31 (0.03)	0.39 (0.04)	0.22 (0.03)

Table 9.3. Estimated mean-reversion parameters for log prices (standard deviation of the estimates is given between brackets)

estimates are much higher (almost double) for Model I than for Model II. This is because in Model I jumps are considered as a part of the diffusion, and their contribution is included in the overall volatility estimate.

All estimates are expressed as daily quantities, so the yearly rates of mean reversion for e.g. APX are 150 (Model I) and 135 (Model II). This means that the average reversion time of the process to a mean level is just over 2.5 days.

All the obtained estimates are similar to those reported by other studies (e.g. Huisman and Mahieu [13], Carnero et al. [6]).

To estimate Model III (the potential function model), we fit a 4th degree polynomial to the log-inverse of the histograms of the filtered series $(\widetilde{X}_i)_i$. The resulting fit is shown in Figure 9.9 for the three markets. The estimated potential is plotted vs. the deviation of the log-price from the systematic seasonal component (on the x-axis). Note that the mean-reverting model would amount to fitting a parabola instead.

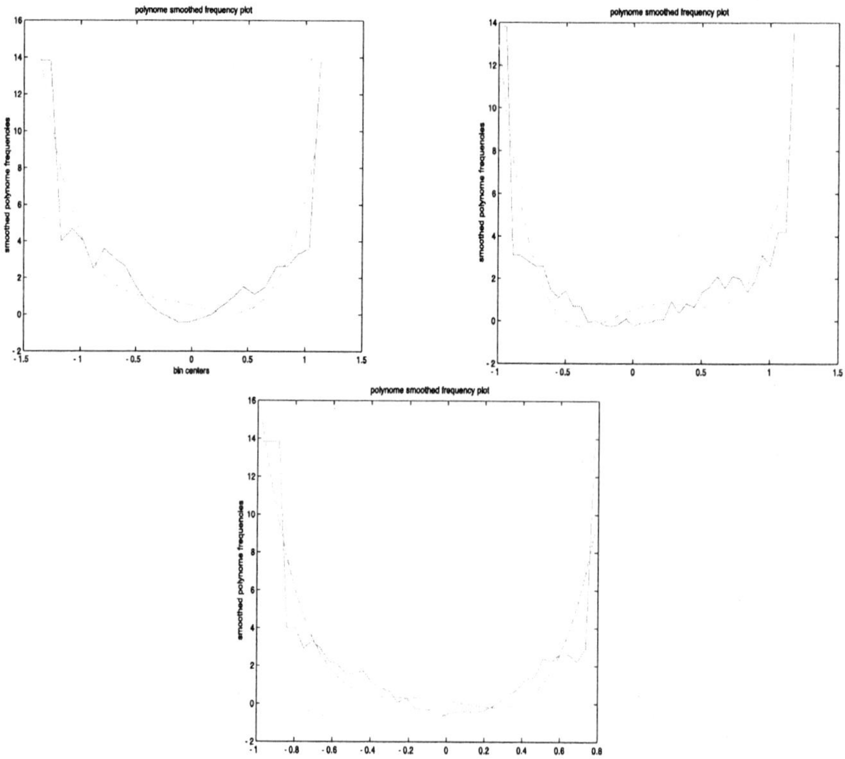

Fig. 9.9. Fitted 4th degree polynomial potential, APX, UKPX and EEX

In the absence of jumps, the process is attracted to the minima of the potential, with the force given by the minus of the potential's derivative. The diffusion term with volatility σ makes sure that the process fluctuates between those minima instead of staying in one of them. When a jump arrives, it places the process further away from the minima, in a region of the potential's steep slope. So after a jump, the process is reverted back with much greater force, given by the higher potential's derivative.

The typical shape of the potential's derivative in our model (as a function of the log-price deviation from the seasonal component) is shown in Figure 9.10. It is low in the region of "normal" price levels and high further away from it. Note that the usual mean-reversion corresponds to approximating the graphs in Figure 9.10 by a straight line.

In a certain sense, the potential model is an extension of the regime-switching model, with a "continuum" of regimes and "continuum" of reversion rates, rather than a number of regimes (three at most) each with its own mean-reversion rate.

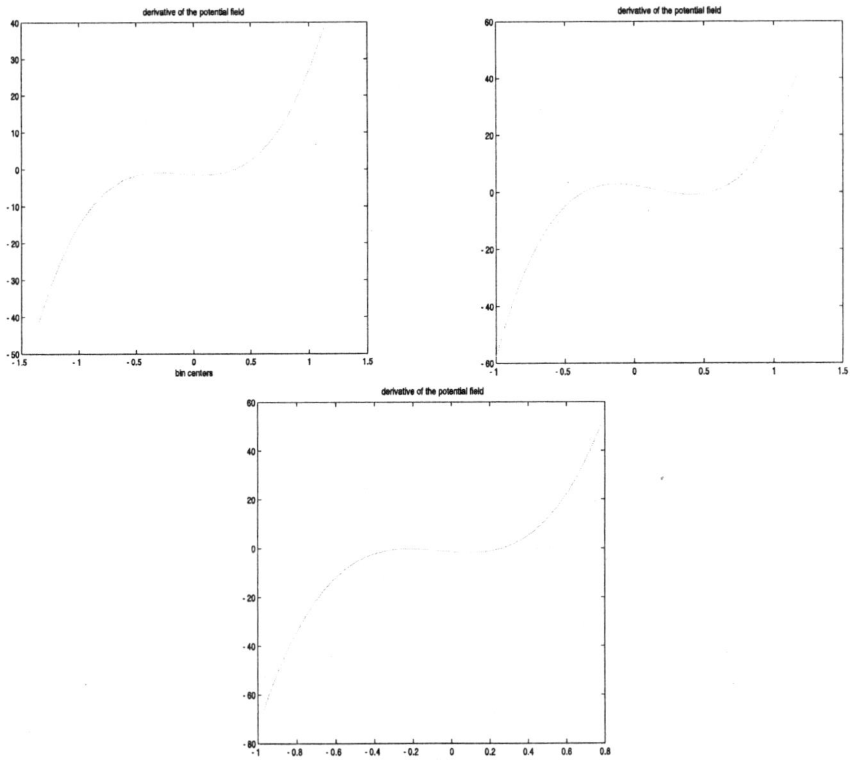

Fig. 9.10. Rate of reversion (derivative of the potential function), APX, UKPX and EEX

The estimated volatilities σ in the potential function model are 0.33, 0.36 and 0.24 for respectively APX, UKPX and EEX. Note that these volatilities are slightly higher than those from Model II (mean-reverting jump-diffusion). This is because in the potential model, price evolution in "normal" regions is mostly driven by the Gaussian noise and not by the drift term. On the contrary, in the mean-reverting model, the influence of the drift term is the same across all price regions.

One of the main purposes of the spot price modelling is to provide a mechanism for simulating price paths (e.g. for scenario simulation, risk management applications, planning of energy industry investments). We simulated log-price paths for the three markets using all three models (shown in Figures 9.11, 9.12 and 9.13).

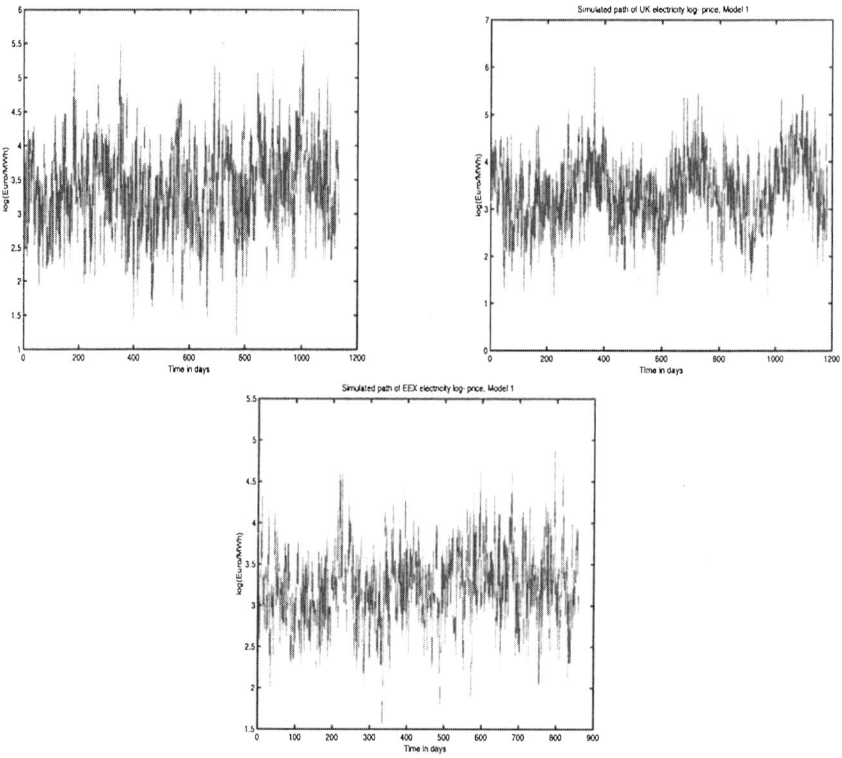

Fig. 9.11. Simulated log-price paths, Model I, APX, UKPX and EEX

The simulation using Model II clearly captures price behavior better than Model I. However, distributional characteristics of the simulated and the original log prices do not match so well (especially skewness and kurtosis), as Table 9.4 shows. Model III captures the characteristic features of the prices (such as spike behavior and general volatility levels) remarkably well. This is also confirmed by comparing distributional characteristics of the observed and the simulated prices, shown in Table 9.4. Note that Model III also matches skewness and kurtosis of the data much better than Models I and II.

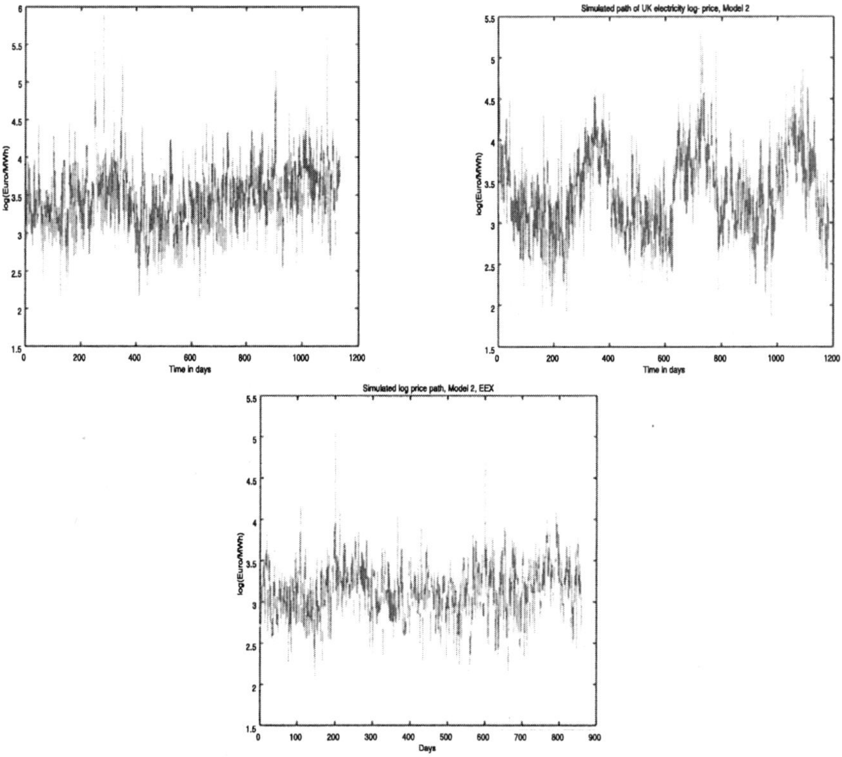

Fig. 9.12. Simulated log-price paths, Model II, APX, UKPX and EEX

	APX				UKPX				EEX			
	0	I	II	III	0	I	II	III	0	I	II	III
mean	3.40	3.48	3.37	3.43	3.33	3.40	3.36	3.33	3.20	3.16	3.18	3.20
st.dev	0.53	0.63	0.48	0.47	0.55	0.79	0.51	0.54	0.39	0.53	0.36	0.39
skewness	1.32	-0.11	0.50	0.78	1.01	0.06	0.28	0.39	-0.23	0.01	0.05	-0.33
kurtosis	6.84	3.18	5.06	6.81	3.66	2.83	2.86	3.15	4.14	2.85	3.15	3.71

Table 9.4. Distributional characteristics of the original (0) and simulated log-prices (with Models I, II and III)

Generally, the potential model allows for a much richer model structure than mean-reversion or regime-switching models. It allows for a state-dependent reversion rate and gives the model builder a wider range of possible specifications of spike behavior. Moreover, the potential model also is capable of capturing finer distributional characteristics of the price in "normal" price regions, such as multimodality.

The central question when fitting the potential model in the jump-diffusion setting is the extrapolation of the potential into the regions of greater price

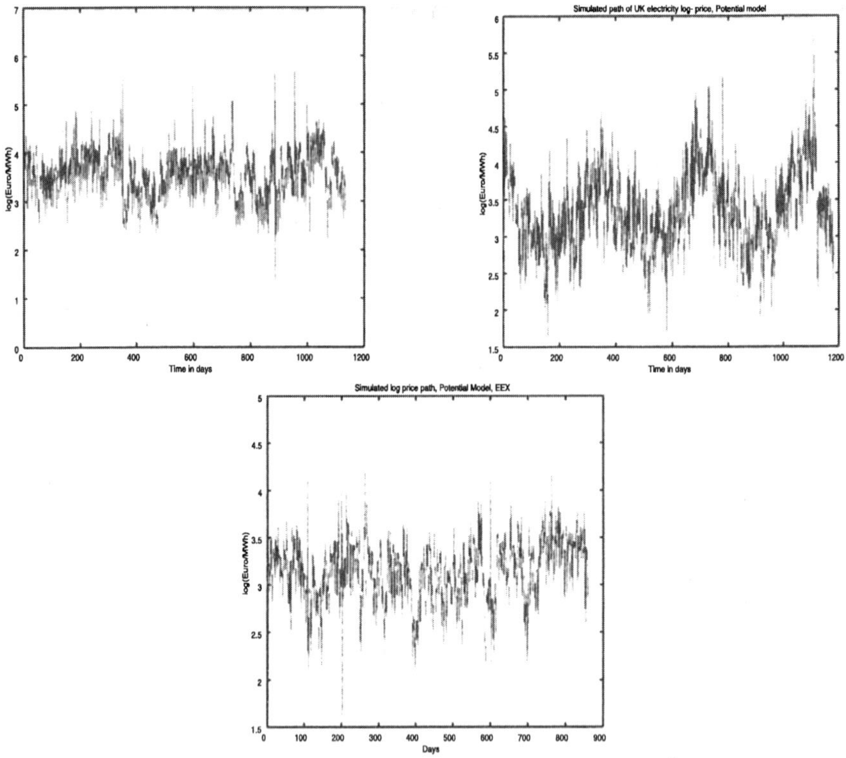

Fig. 9.13. Simulated log-price paths, Model III: potential, APX, UKPX and EEX

moves (i.e. jumps), where not many observations are available. Here we assumed that the potential has the form of fourth degree polynomial. In principle, the choice of appropriate form for the potential is a matter left to the model builder. For some datasets, a polynomial reversion rate might be too strong, and other functional forms may be more appropriate. The only matter of concern when choosing such a functional form, is that the conditions on the potential stated in Section 3 are satisfied.

A data-driven procedure for choosing the functional form of the reversion rate may be feasible. One can investigate how the spike size is related to the spike duration, and infer the appropriate functional relationship, which can in turn be incorporated into the potential. This data driven procedure as well as incorporating more sophisticated functional forms for the potential is the subject of current research.

9.5 Conclusions and Future Work

A thorough investigation of the electricity spot prices on three major European markets has shown that seasonalities and spikes are essential features of electricity prices. Mean reverting jump-diffusion models are only partially capable of capturing spike behavior.

The potential function model, introduced here, allows for a much richer model specification and is flexible enough to replicate a wide variety of spike behavior. The potential's functional form controls the rate of reversion to the average price levels. It can be easily estimated from the data, by combining the estimate of the process' invariant density with the spike filtering procedure. Alternatively, the potential can be specified by the model builder to incorporate the required functional relationship between the spike's size and duration.

The application of the potential function model with polynomial potential to the three series of electricity spot prices showed that characteristic features of the price were captured remarkably well, and generally better than in the mean-reverting jump diffusion model. The potential function model, together with an estimated seasonal price component, can be used for price forecasting and generating hypothetical price paths.

A number of important issues, related to electricity spot price modelling in general and the potential jump diffusion in particular, deserve further investigation.

First, it may be feasible (albeit difficult) to investigate the invariant distribution of the potential function jump-diffusion and verify whether it is related to the potential function, as is the case for the potential function diffusion.

A data-driven approach for deriving the functional form of the potential, and especially its extrapolation into the tails of the price distribution, is desirable. It is essential that such functional forms satisfy the conditions on the potential.

A potential function approach is excellently suited to model several (related) price series simultaneously, by replacing a potential function by a potential field in a higher dimensional space. Such a multivariate extension for potential function diffusions is extensively studied in Borovkova et al. [5], and by extending it to jump-diffusions we will be able to apply the multivariate model to several electricity prices, e.g. the ones analyzed here.

An important open problem in electricity price modelling is developing models that incorporate *spike clustering*, often observed in many electricity markets. Existing models assume a Poisson process of jump arrival, with constant or at most time-dependent (e.g. seasonal) arrival rate. Extreme weather conditions, usually persistent for some time, can be taken as exogenous variables affecting the jump arrival rate, as can be other economic, operational and environmental variables. Alternatively, time series models (e.g. autoregression) for jump arrival rates and jump sizes can be used.

Finally, incorporating time-varying or stochastic volatility into jump-diffusion models is the next step towards more realistic models for electricity prices.

Acknowledgements

The authors wish to thank two anonymous referees for their helpful comments.

References

[1] Y. Ait-Sahalia (2004). Disentangling diffusion from jumps. *Journal of Finance*, forthcoming.

[2] S. Beckers (1981). A note on estimating the parameters of jump-diffusion model of stock returns. *The Journal of Financial and Quantitative Analysis*, 16, 127-140.

[3] C.A. Ball, W.N. Torous (1983). A simplified jump process for common stock returns. *The Journal of Financial and Quantitative Analysis*, 18, 53-65.

[4] G. Barone-Adesi, A. Gigli (2002). Electricity derivatives. *Working paper, NCCR*, Universita della Svizzeria Italiana.

[5] S.A. Borovkova, H.G. Dehling, J. Renkema, H. Tulleken (2003). A potential-field approach to financial time series modelling. *Computational Economics*, 22, 139-161.

[6] M.A. Carnero, S.J. Koopman, M. Ooms (2003). Time series modelling of daily spot prices in European electricity markets. *Preprint.*

[7] A. Cartea, M.G. Figueroa (2004). Pricing in electricity markets: a mean reverting jump-diffusion model with seasonality. *Preprint.*

[8] L. Clewlow, C. Strickland (2000). *Energy Derivatives: Pricing and Risk Management.* Lacima Publications, London.

[9] S. Deng (2000). Pricing electricity derivatives under alternative stochastic spot price models. In: *Proceedings of the 33rd Hawaii International Conference on System Sciences.*

[10] E. Eberlein, G. Stahl (2003). Both sides of the fence: a statistical and regulatory view of electricity risk. *Energy and Power Risk Management*, 8(6), 32-36.

[11] A. Escribano, J.I. Pena, P. Villaplana (2002). Modelling electricity prices: international evidence. *Working paper 02-07, Economic Series 08*, Universidad Carlos III de Madrid.

[12] H. Geman, A. Roncoroni (2003). A class of marked point processes for modelling electricity prices. *Working paper DR 03004*, ESSEC.

[13] R. Huisman, R. Mahieu (2001). Regime jumps in electricity prices. *Energy and Power Risk Management*, 2001, Risk Publications.

[14] G.J. Jiang (1999). Estimation of Jump-Diffusion Processes based on Indirect Inference - with applications to currency exchange rate models. In: *Issues in Computational Economics and Finance*, S.Holly and S. Greenblatt (eds.), Amsterdam: Elsevier.

[15] Kloeden, P.E., Platen, E., 1995. Numerical solutions of stochastic differential equations.

[16] Knittel, C.R., Roberts, M. (2005). Financial Models of Deregulated Electricity Prices: An Application to the California Market, *Energy Economics*, forthcoming.

[17] J. Lucia, E. Schwartz (2002). Electricity Prices and Power Derivatives: Evidence from the Nordic Power Exchange, *Review of Derivatives Research*, 5 (1), 5-50.

[18] B.J. Matkovsky, Z. Schuss (1981). Eigenvalues of the Fokker-Planck operator and the approach to equilibrium for diffusions in potential fields. *SIAM Journal on Applied Mathematics*, 40, 242-254.

[19] Millstein, G.N., 1988. Numerical integration of stochastic differential equations. Kluwer Academic Publishers.

[20] D. Pilipovic (1998). *Valuing and Managing Energy Derivatives.* McGrawHill Publishers.

[21] C. Sorensen (2002). Modeling seasonality in agricultural commodity futures. *Journal of Futures Markets*, 22 (5), 393-426.

[22] H. Zhou (2000). A study of the finite sample properties of EMM, GMM, QMLE and MLE for a square-root interest rate diffusion model. *Federal Reserve Board Discussion Paper* No. 2000-45.

10

Finite dimensional Markovian realizations for forward price term structure models

Raquel M. Gaspar *

Department of Finance, Stockholm School of Economics, Dept. Finance, P.O. Box 6501, 113 83 Stockholm, Sweden
raquel.gaspar@hhs.se

Summary. In this paper we study a fairly general Wiener driven model for the term structure of forward prices.

The model, under a fixed martingale measure, Q, is described by using two infinite dimensional stochastic differential equations (SDEs). The first system is a standard HJM model for (forward) interest rates, driven by a multidimensional Wiener process W. The second system is an infinite SDE for the term structure of forward prices on some specified underlying asset driven by the same W.

We are primarily interested in the *forward prices*. However, since for any fixed maturity T, the forward price process is a martingale under the T-forward neutral measure, the zero coupon bond volatilities will enter into the drift part of the SDE for these forward prices. The interest rate system is, thus, needed as input into the forward price system.

Given this setup we use the Lie algebra methodology of Björk *et al.* to investigate under what conditions on the volatility structure of the forward prices and/or interest rates, the inherently (doubly) infinite dimensional SDE for forward prices can be realized by a finite dimensional Markovian state space model.

10.1 Introduction

In this paper we study forward price models and, in particular, we want to understand when the inherently infinite forward price process can be realized by means of a (Markovian) finite state space model.

From a theoretical point of view, term structures of forward prices are more complex objects than term structures of interest rates or term structures of futures prices. The extra complexity results from the fact that forward prices, with maturity T, are martingales under the T-forward measure, Q^T.

* I thank my supervisor Tomas Björk for helpful comments and constant motivation. I am also indebted to the participants of the Stochastic Finance 2004 conference for their comments, and in particular to Thorsten Schmidt and an anonymous referee for their suggestions on how to improve the earlier version of this paper.

Under such *T-forward measures*, zero-coupon bonds (with maturity T) are numeraires which implies that zero-coupon bond prices volatilities will enter into the dynamics of forward prices. Consequently, in general, term structures of forward prices cannot be studied in isolation, they must be studied under some interest rate setting and a needed input to forward price term structures is the term structure of interest rates.

We model the dynamics directly under the risk-neutral measure Q and our forward price model is described by using two infinite dimensional stochastic differential equations (SDEs), one defining the interest rate setting, another defining the forward contract setting. For the interest rate setting we consider a standard HJM model for (forward) interest rates, driven by a multidimensional Wiener process W. For the forward contract setting we use the Q^T-martingale property of forward prices and the bond prices dynamics induced by the interest rate setting to get a second infinite SDE for the term structure of forward prices, on some specified underlying asset. Without loss of generality we consider that the Wiener process W is the same for both SDEs.

The theoretical literature on term structures of forward prices is not big and has mainly focused on understanding under what conditions, on the dynamics of the state space variables (which are *assumed* to be finite), the term structure is of an *a priori given* specific functional form. Included in this traditional approach are the studies on affine and quadratic term structures of forward prices (see [11] for a recent study integrating these two types of term structures and references).

In this paper we choose a fundamentally different approach. We do *not* assume that the state space model is finite, nor that the term structure of forward prices is of a given specific function form. Instead, we try to understand under what conditions, in terms of the volatility of forward prices and interest rates, we can have a finite dimensional realization (FDR) of forward prices term structure models.

This more systematic way of thinking about term structures was proposed by Björk and Christensen [4] and Björk and Svensson [6], and a more geometric way of thinking about FDR of term structures, was then introduced. The main technical tool of these studies is the Frobenius Theorem, and the main result is that there exists a FDR if and only if the Lie algebra generated by the drift and diffusion terms, of the underlying infinite dimensional (Stratonovich) SDE, is finite dimensional. Filipović and Teichman [10] and [9] increased the applicability of the geometric approach by showing how the theory can be extended to much more general settings than initially considered. Finally, Björk and Landén [5] addressed the question of the actual construction of finite-dimensional realizations, making this geometrical analysis interesting also from an application point of view.

The main area of application of these ideas has been (forward) interest rate term structures, which was the object of study in all the above mentioned papers (for a review study on the geometry of interest rate models see also [2]). More recently this geometric machinery has also been applied to study futures prices term structures (see [3]). As far as our knowledge goes, this techniques have not yet been applied to study forward prices (or any other Q^T-martingales). In the present paper we, thus, take this next natural step.

The main contributions of this paper are as follows.

- We adapt the geometrical analysis of term structures to the case of doubly infinite systems.
- We obtain necessary and sufficient conditions for the existence of a FDR of forward rate term structure models.
- Given that such conditions are satisfied, we derive the dynamics of the underlying finite state space variable.

The organization of the paper is the following.

In Section 10.2 we present the basic setup, derive the doubly infinite SDE that will be the object of study and present the main questions to be answered. Section 10.3 briefly reviews the basic geometrical concepts behind the method of analysis. Sections 10.4, 10.5 and 10.6 are devoted to the actual study of forward price models answering the proposed questions. Section 10.7 resumes our main conclusions and discusses the applicability of the results.

10.2 Setup

The main goal of this study is the study of *forward prices* in a general stochastic interest rate setting.

We, thus, consider a financial market living on a filtered probability space $\left\{\Omega,\ \mathcal{F},\ Q,\ \{\mathcal{F}_t\}_{t\geq 0}\right\}$ carrying an m-dimensional Wiener process W. For reasons that will soon become clear, the main assets we consider are forward contracts (written on some given underlying asset under consideration) and zero-coupon bonds[2].

Let $f_0(t,T)$ denote the forward prices at time t of a forward contract maturing at time T, and $p_0(t,T)$ denote the price at time t of a zero-coupon bond maturing at time T.

Besides the trivial boundary conditions

$$
\begin{aligned}
f_0(T,T) &= S(T)\,, \\
p_0(t,T) &= 1\,,
\end{aligned}
$$

where S is the price process of the underlying asset to the forward contract, arbitrage arguments yields

$$
p_0(t,T) = E_t^Q\left[e^{-\int_t^T R(s)ds}\right], \tag{10.1}
$$

[2] For a textbook discussion of forward contracts and zero-coupon bonds see for instance [1].

$$f_0(t,T) = E_t^T \left[f_0(T,T) \right], \tag{10.2}$$

where R is the short rate of interest and $E_t^Q[\cdot]$, $E_t^T[\cdot]$ denote, respectively, expectations, conditional on $\mathcal{F}_t$, under the martingale measure Q and under the *forward* martingale measure Q^T.

It is also well-known that under *deterministic* interest rate settings, or complete orthogonality between the underlying and the interest rate random sources, forward prices are the same as futures prices (a similar but conceptually easier contract)[3].

In this study we focus on *forward prices* and, in particular, we are interested in analyzing the settings where they are not equivalent to futures prices. Our analysis will, therefore, assume *stochastic interest rates* and that there are at least some *common radom sources* driving *both* the interest rates *and* the underlying to the forward contract.

Thus, in our context, a **forward price model** is only fully defined once we have specified both forward prices dynamics and interest rates dynamics under a same measure (which we choose to be Q) and assumed that these two dynamics are, at least partially, driven by common elements of our multidimensional Wiener process.

Before we present in detail our setting, we start by reparameterizing our variables. A more suitable parameterization for our purposes is the so called Musiela parameterization ([7] and [13]). Under the Musiela parameterization, forward prices and bond prices are given in terms of t and x, where x denotes time **to** maturity, in contrast to T which defined time **of** maturity. Therefore, we will use

$$f(t,x) = f_0(t,t+x) \qquad \text{and} \qquad p(t,x) = p_0(t,t+x)\,. \tag{10.3}$$

10.2.1 The Interest Rate Curve

We consider a standard HJM model for the (forward) interest rates, driven by a multidimensional Wiener process W. Using the Musiela parameterization the dynamics for the interest rates, under Q, are given by[4]

$$dr(t,x) = \left\{ \frac{\partial}{\partial x} r(t,x) + \sigma(t,x) \int_0^x \sigma^*(t,s) ds \right\} dt + \sigma(t,x) dW_t\,, \tag{10.4}$$

where $\sigma(t,x)$ is a given adapted process in $\mathbb{R}^m$ and $*$ denotes transpose.

From the relation between (forward) interest rates and bond prices, we can derive the bond price Q-dynamics.

Lemma 1. *Assume the (forward) interest rates dynamics in (10.4). Then the dynamics of the zero-coupon bond prices, using the Musiela parameterization, is given by*

[3] For a futures price definition, see for instance [1] or [11].

[4] For a textbook treatment of HJM models and the Musiela parameterization for such models, see [1].

$$dp(t,x) = \{R(t) - r(t,x)\}\, p(t,x)dt + p(t,x)v(t,x)dW_t$$

where R is the short interest rate[5] *and the bond prices' volatility, v, is obtained from the (forward) interest rate volatilities as*

$$v(t,x) = -\int_0^x \sigma(t,s)ds \tag{10.5}$$

and hence also adapted.

Proof:Recall the standard relation between (forward) interest rates and bond prices

$$p(t,x) = e^{-\int_0^x r(t,s)ds}.$$

Let us set $y(t,x) = -\int_0^x r(t,s)ds$. Applying the Itô lemma we get

$$\begin{aligned}
dy(t,x) &= -\int_0^x dr(t,s)ds \\
&= -\int_0^x \left[\left(\frac{\partial}{\partial s} r(t,s) + \sigma(t,s)\int_0^s \sigma^*(t,u)du\right)dt + \sigma(t,s)dW_t\right] ds \\
&= -\int_0^x \frac{\partial}{\partial s} r(t,s)dsdt - \int_0 \sigma(t,s)\int_0^s \sigma^*(t,u)dudsdt \\
&\quad \underbrace{-\int_0^x \sigma(t,s)ds}_{v(t,x)}\, dW_t \\
&= \left[\underbrace{r(t,0)}_{R(t)} - r(t,x) - \int_0 \sigma(t,s)\int_0^s \sigma^*(t,u)duds\right] dt + v(t,x)dW_t\,.
\end{aligned}$$

The result follows from $dp(t,x) = p(t,x)dy(t,x) + \frac{1}{2}p(t,x)\left[dy(t,x)\right]^2$ and by notting that

$$\sigma(t,x)\int_0^x \sigma^*(t,u)du = \frac{1}{2}2\sigma(t,x)\int_0^x \sigma^*(t,u)du\,,$$

$$\int_0^x \sigma(t,s)\int_0^x \sigma^*(t,u)duds = \frac{1}{2}\left(\int_0^x \sigma^*(t,u)du\right)^2.$$

□

10.2.2 The Forward Price Curve

Since the forward prices are Q^T-martingales (recall (10.2)), we assume Q^T-dynamics of the form

[5] Recall that in the Musiela parametrization the short rate of interest is $R(t) = r(t,0)$.

$$df_0(t,T) = f_0(t,T)\gamma_0(t,T)dW_t^T\,, \tag{10.6}$$

where we also take γ_0 to be a given adapted process.

Here we use the fact that martingales have zero drift. Note however that, by choosing to model the forward price dynamics as in (10.6), forward prices with different maturities T are modeled under a different martingale measures Q^T.

Reparameterizing using $f(t,x) = f_0(t,t+x)$ give us

$$df(t,x) = \left\{\frac{\partial}{\partial x} f(t,x)\right\} dt + f(t,x)\gamma(t,x)dW_t^T\,,$$

where $T = t + x$ and $\gamma(t,x) = \gamma_0(t,T)$.

It will also simplify matters if we work with the logarithm of forward prices instead of the forward prices themselves. Thus setting

$$q(t,x) = \ln f(t,x)\,, \tag{10.7}$$

we have

$$dq(t,x) = \left\{\frac{\partial}{\partial x} q(t,x) - \frac{1}{2}\|\gamma(t,x)\|^2\right\} dt + \gamma(t,x)dW_t^T\,. \tag{10.8}$$

Note that analyzing the logarithm of forward prices is equivalent to analyzing the forward prices themselves, as we can always use (10.7) to transfer any results on the logarithm of forward prices into results on forward prices.

Finally, to obtain the dynamics of (the logarithm) of forward prices, under the risk-neutral martingale measure, Q, we use the change of numeraire technique (introduced in [12]). Denoting by L the Radon-Nikodym derivative,

$$L(t) = \frac{dQ}{dQ^T} \qquad \text{on } \mathcal{F}_t, \quad 0 \leq t \leq T\,,$$

and recalling the money account B is the numeraire under Q, we have in Musiela parameterization

$$L(t) = p(0,T)\frac{B(t)}{p(t,x)}\,,$$

where $x = T - t$.

Thus, the dynamics of our likelihood process L are given by

$$dL(t) = L(t)\left\{-v(t,x)\right\} dW_t^T$$

i.e., the Girsanov kernel, for the transition from Q^T to Q, is the symmetric of the volatility of zero-coupon bond price with maturity T or, equivalently, with time to maturity $x = T - t$.

Using the above Girsanov kernel, we can easily obtain the (logarithm of) forward prices Q-dynamics from (10.8), that is

$$dq(t,x) = \left\{ \frac{\partial}{\partial x} q(t,x) - \frac{1}{2} \|\gamma(t,x)\|^2 - \gamma(t,x)v^*(t,x) \right\} dt + \gamma(t,x)dW_t \quad (10.9)$$

where, w.l.o.g., we can take W to be multidimensional and the same as in (10.4).

Taking a geometrically oriented interpretation of equations (10.4) and (10.9), we can see each of these equations as infinite dimensional objects. The main infinite dimensional object under study in this paper is (the logarithm) of the **forward price curve**, i.e., the curve $x \to q(t,x)$. This object, however, for general adapted processed σ and γ may depend on, the **interest rate curve**, i.e., the curve $x \to r(t,x)$, another infinite dimensional object.

In principle, both adapted processes σ and γ could depend on q and r. It seems, however, unrealistic to assume that a forward price on a specific underlying (be it the price of a stock, or any other asset) should influence the interest rate volatility.

The opposite is true for forward prices. As mentioned before, these prices are only interesting to study in *stochastic* interest rate settings. This tell us that, maybe, it is realistic that the forward price volatility depends on the interest rates' curve.

With this basic intuition in mind we set some more structure on the volatility processes σ and γ.

Assumption 1. *The adapted processes $\gamma(t,x)$, and $\sigma(t,x)$ have the following functional form in terms of r and q*

$$\gamma(t,x) = \gamma(q_t, r_t, x)\,, \quad (10.10)$$

$$\sigma(t,x) = \sigma(r_t, x)\,, \quad (10.11)$$

where, with a slight abuse of notation, the r.h.s. occurrence of γ and σ denotes deterministic mappings

$$\gamma : \mathcal{H}_q \times \mathcal{H}_r \times \mathbb{R}_+ \to \mathbb{R}^m\,,$$
$$\sigma : \mathcal{H}_r \times \mathbb{R}_+ \to \mathbb{R}^m\,,$$

where $\mathcal{H}_q$ and $\mathcal{H}_r$ are special Hilbert spaces of functions where forward price curves and interest rate curves live, respectively [6].

Note that by imposing Assumption 1, σ does not depend on q_t (the logarithm of the forward price process) and that we restrict ourselves to the study of time homogenous models. Extensions to non-homogeneous models have been considered in [3] and turn out to be straightforward generalization of the homogeneous results.

From now on we will use the short-hand notation $q_t = q(t,x)$, $r_t = r(t,x)$ where we suppress the x-dependence. This shorter notation will be helpful

[6] For details on the construction of the Hilbert spaces $\mathcal{H}_q$ and $\mathcal{H}_r$ we refer to [6], [9] and [10].

when the expressions get messy and is more intuitive from a geometrical point of view.

Using the short-hand notation and Assumption 1, we can rewrite equations (10.4)-(10.9) as

$$dq_t = \left\{ \mathbf{F}q_t - \frac{1}{2}\|\gamma(q_t, r_t)\|^2 - \gamma(q_t, r_t)v^*(r_t) \right\} dt + \gamma(q_t, r_t)dW_t \,, \tag{10.12}$$

$$dr_t = \{\mathbf{F}r_t - \sigma(r_t)v^*(r_t)\} \, dt + \sigma(r_t)dW_t \,, \tag{10.13}$$

where

$$\mathbf{F} = \frac{\partial}{\partial x}$$

and we can interpret the entire system as an object

$$\widehat{q} = (q_t \ , \ r_t)^* \in \mathcal{H}_q \times \mathcal{H}_r \,.$$

As we can see from equations (10.12)-(10.13), the interest rates equation (10.13) does not depend on the forward prices equation (10.12), so the interest rates curve r exist and can be studied in isolation. For a survey study on the geometry of interest rate models see [2]. In contrast to this, the (logarithm of the) forward price equation (10.12), is linked to the interest rate equation (10.13) through $\gamma(q, r)$ and/or $v(r)$[7]. This means, that in general, to study forward prices we will have to study the entire system (10.12)-(10.13).

In the following analysis we will refer to **forward price equation** when referring only to (10.12), to **interest rate equation** when referring only to (10.13), and to **forward price system** when referring to the entire system (10.12)-(10.13).

We can now formulate our main problems.

10.2.3 Main Problems

Problem 1: Under what conditions we have Markovian forward prices?

Problem 2: Is it possible to have a finite realization for the forward prices equation (10.12) but not for interest rates equation (10.13)?

Problem 3: When can the inherently infinite forward price system (10.12)-(10.13) be realized by means of a finite dimensional state space model?

Problem 4: In the cases when a finite dimensional realization (FDR) exists, can we determine the finite dimensional state space model?

The next section introduces the method of analysis.

[7] Given the definition of the bond price volatility, v, in (10.5), if $\sigma(t, x) = \sigma(r_t, x)$ then also $v(t, x) = v(r_t, x)$.

10.3 Method and Basic Geometric Concepts

In this section we describe the general method we will use to attack the presented problems.

The method relies on geometric results from differential geometry and was firstly applied to finance in [4] and [6]. In this section, we adapt the framework of [4] and [6] to our doubly-infinite system case.

To be able to apply the concepts and intuitions of ordinary differential geometry to (stochastic) Itô calculus, we need to rewrite the analysis in terms of Stratonovich integrals instead of Itô integrals.

Definition 1. *For given semi martingales X and Y, the* ***Stratonovich integral*** *of X with respect to Y, $\int_0^t X_s \circ dY_s$, is defined as*

$$\int_0^t X_s \circ dY_s = \int_0^t X_s dY_s + \frac{1}{2}\langle X, Y\rangle_t . \tag{10.14}$$

where the first term on the r.h.s. is the Itô integral and we can define the quadratic variation process $\langle X, Y\rangle$ can be computed via

$$d\langle X, Y\rangle = dX_t dY_t$$

with the usual multiplication rules: $dW \cdot dt = dt \cdot dt = 0$, $dW \cdot dW = dt$.

The Stratonovich formulation is geometrically more convenient because the Itô formula, in Stratonovich calculus, takes the form of the standard *chain rule* in ordinary calculus.

Lemma 2. *Assume that a function $F(t, y)$ is smooth. Then we have*

$$dF(t, Y_t) = \frac{\partial F}{\partial t}(t, Y_t)dt + \frac{\partial F}{\partial y}(t, Y_t) \circ dY_t . \tag{10.15}$$

Let us begin by specifying exactly what we mean with a finite dimensional realization for the forward prices generated by volatilities.

Given the volatility mappings $\gamma : \mathcal{H}_q \times \mathcal{H}_r \times \mathbb{R}_+ \to \mathbb{R}^m$ and $\sigma : \mathcal{H}_r \times \mathbb{R}_+ \to \mathbb{R}^m$ from Assumption 1, the forward prices equation will, in general, depend on the interest rate curve. Thus, our main object of study will be a Stratonovich forward price *system* of the following form:

$$d\widehat{q}_t = \begin{bmatrix} \mu^q(q_t, r_t) \\ \mu^r(r_t) \end{bmatrix} dt + \begin{bmatrix} \gamma(q_t, r_t) \\ \sigma(r_t) \end{bmatrix} \circ dW_t , \tag{10.16}$$

where $\widehat{q} = \begin{bmatrix} q \\ r \end{bmatrix} \in \mathcal{H}_q \times \mathcal{H}_r$.

In special cases, the forward price dynamics may be independent of the interest rate curve, then our object of study is the Stratonovich forward price *equation*,

$$dq_t = \mu^q(q_t)dt + \gamma(q_t) \circ dW_t\,. \tag{10.17}$$

where $q \in \mathcal{H}_q$, and we say that the (logarithm of) forward prices is **Markovian.**

When referring to the **forward price model** we refer to either (10.16) or (10.17), depending on the circumstances.

In this study, we will consider these two possibilities.

Definition 2. *We say that the doubly-infinite SDE (10.16) has a (local) d-dimensional realization at* $\widehat{q}^0 = \left(q^0\ ,\ r^0\right)^*$, *if there exists a point* $z_0 \in \mathbb{R}^d$, *smooth vector fields* $\widehat{a}, \widehat{b}_1 \cdots, \widehat{b}_m$ *on some open subset* $\widehat{\mathcal{Z}}$ *of* $\mathbb{R}^d$ *and a smooth (sub manifold) map* $\widehat{G}\ :\ \widehat{\mathcal{Z}} \to \mathcal{H}_q \times \mathcal{H}_r$, *such that* $\widehat{q} = (q\ ,\ r)^*$ *has a local representation*

$$\widehat{q}_t = \widehat{G}(Z_t) \qquad i.e., \qquad \begin{bmatrix} q_t \\ r_t \end{bmatrix} = \begin{bmatrix} \widehat{G}^q(Z_t) \\ \widehat{G}^r(Z_t) \end{bmatrix} \qquad a.s.$$

where Z is the strong solution of the d-dimensional Stratonovich SDE

$$\begin{cases} dZ_t = \widehat{a}(Z_t) + \widehat{b}(Z_t) \circ dW_t \\ Z_0 = z_0 \end{cases} \tag{10.18}$$

and where W is the same as in (10.16).

Likewise, we say that the SDE (10.17) has a (local) n-dimensional realization at q^0, *if there exist* $z_o \in \mathbb{R}^n$, *smooth vector fields* $a, b_1 \cdots, b_m$ *on some open subset* $\mathcal{Z}$ *of* $\mathbb{R}^n$ *and a smooth (sub manifold) map* $G\ :\ \mathcal{Z} \to \mathcal{H}_q$, *such that q has a local representation*

$$q_t = G(Z_t) \quad a.s.$$

where Z is the strong solution of the d-dimensional Stratonovich SDE

$$\begin{cases} dZ_t = a(Z_t) + b(Z_t) \circ dW_t\,, \\ Z_0 = z_0\,, \end{cases} \tag{10.19}$$

where W is the same as in (10.17).

If the SDE under analysis, (10.16) or (10.17), has a finite dimensional realization (FDR), we say that our forward rate model admits a FDR.

The method of studying existence and construction of FDR for forward price models, relies on some basic concepts from infinite dimensional differential geometry, which we now introduce.

10.3.1 Basic Geometric Concepts

The presentation of the needed geometric concepts follows [6]. These basic concepts will be presented for a general real Hilbert Space Y and we denote by y an element of Y. In practice, the Hilbert space under analysis will be either $\mathcal{H}_q$ when studying Markovian forward prices or $\mathcal{H}_q \times \mathcal{H}_r$ when dealing with the entire forward price system.

Consider a real Hilbert space Y. By an n-dimensional **distribution** we mean a mapping F, which to each $y \in Y$ associates an n-dimensional subspace $F(y) \subseteq Y$. A mapping (vector field) $f : Y \to Y$, is said to **lie in** F if $f(y) \in F(y)$ for every $y \in Y$. A collection $f_1, \ldots, f_n$ of vector fields lying in F **generates** (or spans) F if $\text{span}\{f_1(y), \ldots, f_n(y)\} = F(y)$ for every $y \in Y$, where "span" denotes the linear hull over the real field. The distribution is **smooth** if, for every $y \in Y$, there exist smooth vector fields $f_1, \ldots, f_n$ spanning F. A vector field is smooth if it belongs to C^∞. If F and G are distributions and $G(y) \subseteq F(y)$ for all y we say that F **contains** G, and we write $G \subseteq F$. The **dimension** of a distribution F is defined pointwise as $\dim F(y)$.

Let f and g be smooth vector fields on Y. Their **Lie bracket** is the vector field $[f, g]$, defined by

$$[f, g] = f'g - g'f,$$

where f' denotes the Frechet derivative of f at y, and similarly for g'. We will sometimes write $f'[g]$ instead of $f'g$ to emphasize that the Frechet derivative is operating on g. A distribution F is called **involutive** if for all smooth vector fields f and g lying in F on Y, their lie bracket also lies in F, i.e.

$$f, g \in F \qquad \Rightarrow \qquad [f, g] \in F$$

for all $y \in Y$.

We are now ready to define the concept of a Lie algebra which will play a central role in what follows.

Definition 3. *Let F be a smooth distribution on Y. The* **Lie algebra** *generated by F, denoted by $\{F\}_{LA}$ or by $\mathcal{L}\{F\}$, is defined as the minimal (under inclusion) involutive distribution containing F.*

If, for example, the distribution F is spanned by the vector fields $f_1, \ldots, f_n$ then, to construct the Lie algebra $\{f_1, \ldots, f_n\}_{LA}$, you simply form all possible brackets, and brackets of brackets, etc. of the fields $f_1, \ldots, f_n$, and adjoin these to the original distribution until the dimension of the distribution is no longer increased.

When one tries to compute a concrete Lie algebra the following observations are often very useful. Taken together, they basically say that, when computing a Lie algebra, you are allowed to perform Gaussian elimination.

Lemma 3. *Take the vector fields $f_1, \ldots, f_k$ as given. It then holds that the Lie algebra $\{f_1, \ldots, f_k\}_{LA}$ remains unchanged under the following operations.*

- *The vector field f_i may be replaced by αf_i, where α is any smooth nonzero scalar field.*

- *The vector field f_i may be replaced by*

$$f_i + \sum_{j \neq i} \alpha_j f_j ,$$

where $\alpha_1, \ldots, \alpha_k$ are any smooth scalar fields.

Let F be a distribution and let $\varphi : V \to W$ be a diffeomorphism between the open subsetd V and W on Y. Then we can define a new distribution $\varphi_\star F$ on Y by

$$(\varphi_\star F)(\varphi) := \varphi' F.$$

We now define an useful operator on our Hilbert space Y.

Definition 4. *Let f be a smooth vector field on Y, and let y be a fixed point in Y. Consider the ODE*

$$\begin{cases} \dfrac{dy_t}{dt} = f(y_t) , \\ y_0 = y . \end{cases}$$

We denote the solution y_t as $e^{ft}y$.

Finally, and for future reference, we define a particular type of functions – the quasi-exponential functions – that will turn out useful.

Definition 5. *A **quasi-exponential** (or QE) function is by definition any function of the form*

$$f(x) = \sum_u e^{\lambda_u x} + \sum_j e^{\alpha_j x} \left[p_j(x) \cos(w_j x) + q_j(x) \sin(w_j x) \right]$$

where λ_u,α_j,w_j are real numbers, whereas p_j and q_j are real polynomials.

Important properties of QE functions are given in Lemma 4.

Lemma 4. *The following holds for quasi-exponential functions*

- *A function is QE if and only if it is a component of the solution of a vector valued linear ODE with constant coefficients.*
- *A function is QE if and only if it can be written as $f(x) = ce^{Ax}b$. Where c is a row vector, A is a square matrix and b is a column vector.*
- *If f is QE, then f' is QE.*
- *If f is QE, then its primitive function is QE.*
- *If f and g are QE, then fg is QE.*

10.3.2 Main Results from the Literature

We can now adapt two important theorems from [6] to our forward price problem. The first theorem gives us the general necessary and sufficient conditions for existence of a FDR.

Theorem C.1 (Björk and Svensson). *Consider the SDE in (10.16) and denote $\gamma_1, \cdots, \gamma_m$ and $\sigma_1, \cdots, \sigma_m$ the elements of γ and σ, respectively. Assume that the dimension of the Lie algebra*

$$\left\{ \begin{bmatrix} \mu^q \\ \mu^r \end{bmatrix}, \begin{bmatrix} \gamma_1 \\ \sigma_1 \end{bmatrix}, \dots, \begin{bmatrix} \gamma_m \\ \sigma_m \end{bmatrix} \right\}_{LA}$$

is constant near the initial point $\widehat{q}^0 = \left(q^0 \ , \ r^0\right)^* \in \mathcal{H}_q \times \mathcal{H}_r$.

Then (10.16) possesses an FDR if and only if

$$\dim \left\{ \begin{bmatrix} \mu^q \\ \mu^r \end{bmatrix}, \begin{bmatrix} \gamma_1 \\ \sigma_1 \end{bmatrix}, \dots, \begin{bmatrix} \gamma_m \\ \sigma_m \end{bmatrix} \right\}_{LA} < \infty$$

in a neighborhood of $\widehat{q}^0$.

Likewise, for Markovian forward prices we consider the SDE (10.17), and assume that the dimension of the lie algebra of $\{\mu, \gamma_1, \dots, \gamma_m\}_{LA}$ is constant near $q^0 \in \mathcal{H}_q$. Then (10.17) possesses and FDR if and only if

$$\dim \{\mu, \gamma_1, \dots, \gamma_m\}_{LA} < \infty .$$

Remark 1. To shorten notation we will sometimes use $\left\{ \begin{bmatrix} \mu^q \\ \mu^r \end{bmatrix}, \begin{bmatrix} \gamma \\ \sigma \end{bmatrix} \right\}_{LA}$ instead of $\left\{ \begin{bmatrix} \mu^q \\ \mu^r \end{bmatrix}, \begin{bmatrix} \gamma_1 \\ \sigma_1 \end{bmatrix}, \dots, \begin{bmatrix} \gamma_m \\ \sigma_m \end{bmatrix} \right\}_{LA}$ and $\{\mu, \gamma\}_{LA}$ instead of $\{\mu, \gamma_1, \dots, \gamma_m\}_{LA}$.

The second theorem gives us a parameterization of the curves produced by the forward price model and is a crucial step to the understanding of the construction algorithm.

Theorem C.2 (Björk and Svensson). *Assume that the Lie algebra*

$$\left\{ \begin{bmatrix} \mu^q \\ \mu^r \end{bmatrix}, \begin{bmatrix} \gamma \\ \sigma \end{bmatrix} \right\}_{LA}$$

is spanned by the smooth vector fields $\widehat{f}_1, \dots, \widehat{f}_d$ in $\mathcal{H}_q \times \mathcal{H}_r$.
Then, for the initial point $\widehat{q}^0 = \left(q^0 \ , \ r^0\right)^*$, all forward price and interest rate curves produced by the model will belong to the manifold $\widehat{\mathcal{G}} \in \mathcal{H}_q \times \mathcal{H}_r$, which can be parameterized as $\widehat{\mathcal{G}} = (m)[\widehat{G}]$, where

$$\widehat{G}(z_1, \dots, z_d) = e^{\widehat{f}_d z_d} \dots e^{\widehat{f}_1 z_1} \begin{bmatrix} q^0 \\ r^0 \end{bmatrix}, \tag{10.20}$$

and where the operator $e^{\hat{f}_i z_i}$ *is given in Definition 4.*

Likewise, in the case of Markovian forward prices, and assuming that the Lie algebra $\{\mu, \gamma\}_{LA}$ *is spanned by the smooth vector fields* $f_1, \ldots, f_d$ *in* $\mathcal{H}_q$*. Then, for the initial point* q^0*, all forward price curves produced by the model will belong to the manifold* $\mathcal{G} \in \mathcal{H}$*, which can be parameterized as* $\mathcal{G} = (m)[G]$*, where*

$$G(z_1, \ldots, z_d) = e^{f_d z_d} \ldots e^{f_1 z_1} q^0 \tag{10.21}$$

and where the operator $e^{f_i z_i}$ *is given in Definition 4.*

The manifolds $\widehat{\mathcal{G}}$ and $\mathcal{G}$ in the above theorem are obviously invariant under the forward price model dynamics. Therefore, they will be referred to as the **invariant manifolds** in the sequel. $\widehat{G}$ and G are, thus, local parameterizations of the invariant manifolds $\widehat{\mathcal{G}}$ and $\mathcal{G}$, respectively.

The construction algorithm (Björk and Landén) introduced in [5] is based on idea that, if we are in the case when the forward price system generated by the volatilities $\gamma : \mathcal{H}_q \times \mathcal{H}_r \times \mathbb{R}_+ \to \mathbb{R}^m$ and $\sigma : \mathcal{H}_r \times \mathbb{R}_+ \to \mathbb{R}^m$ admits a FDR, we have

$$\begin{bmatrix} q \\ r \end{bmatrix} = \widehat{G}(Z)$$

and for

$$dZ = \widehat{a}(Z)dt + \widehat{b}(Z) \circ dW_t\,,$$

it must hold that

$$\widehat{G}_\star \widehat{a} = \begin{bmatrix} \mu^q \\ \mu^r \end{bmatrix} \qquad \text{and} \qquad \widehat{G}_\star \widehat{b} = \begin{bmatrix} \gamma \\ \sigma \end{bmatrix}. \tag{10.22}$$

Equivalently for Markovian forward prices, if forward price model generated by $\gamma : \mathcal{H}_q \times R_+ \to R^m$ admits a FDR, then we have

$$q = G(Z)$$

and or

$$dZ = a(Z)dt + b(Z) \circ dW_t\,,$$

it must hold that,

$$G_\star a = \mu \qquad \text{and} \qquad G_\star b = \gamma\,. \tag{10.23}$$

Since we can compute $\widehat{G}$ and G from (10.20) and (10.21), we can solve the system (10.22) for $\widehat{a}$ and $\widehat{b}$, or the system (10.23) for a and b. We note that the equations in (10.22) and (10.23) do not necessarily have unique solutions, but it is enough to find *one* solution.

Note also that by solving (10.22) or (10.23), we obtain the Stratonovich dynamics of our FDR. The Itô dynamics, (which, in general, looks nicer) can be easily obtained using (10.15).

10.3.3 Strategy of Analysis

In the next few sections we will address our four main problems.

In Section 10.4, we answer our *problem one* characterizing the settings when forward prices are Markovian. As it turns out, during this analysis, we will also be able to give a partial answer to *problem two.*

In Section 10.5, we study existence and construction of FDR for Markovian forward prices.

In Section 10.6, we deal with the cases when forward prices are **not** Markovian, studying existence and construction of FDR for the entire forward price system. Studying the entire system we are able to give a complete answer to *question two.*

In sections 10.5 and 10.6, *problems three and four* we will be answered following the scheme.

- Choose a number of vector fields $f_0, f_1, \cdots$ that spans the Lie-algebra we are interested on. For that purpose Lemma 3 is useful to help simplifying the vector fields.
- Conclude under what conditions our Lie-algebra is of finite dimension in view of Theorem C.1.
- Assuming that those conditions hold, compute a local parameterization of the invariant manifold using Theorem C.2.
- Given that parameterization, solve a system of equations of the type (10.22) or (10.23) to obtain the finite state variables dynamics.

10.4 On the Existence of Markovian Forward Prices

Having described the setup and the general method, we now start our analysis.

Recall that our main object of study is the forward price system

$$\begin{cases} dq_t = \left\{ \mathbf{F}q_t - \frac{1}{2}\|\gamma(q_t, r_t)\|^2 - \gamma(q_t, r_t)v^*(r_t) \right\} dt + \gamma(q_t, r_t)dW_t\,, \\ dr_t = \{\mathbf{F}r_t - \sigma(r_t)v^*(r_t)\}\, dt + \sigma(r_t)dW_t\,, \end{cases} \tag{10.24}$$

where $\mathbf{F} = \dfrac{\partial}{\partial x}$ and $v(x, r) = -\displaystyle\int_0^x \sigma(s, r)ds.$

Before we go on, and to exclude patholigical cases from the analysis, we need to impose a regularity condition on forward price models.

Assumption 2. *If $\gamma_i(q_t, r_t) \neq 0$ for some $i \in \{1, \cdots, m\}$, then the following regularity condition holds:*

$$\frac{1}{2}\|\gamma(q_t, r_t)\|^2 + \gamma(q_t, r_t)v^*(r_t) \neq 0.$$

Given Assumption 2 and by mere inspection of (10.24), we see that the answer to our *first problem* – on whether forward prices can be studied without considering the interest rate equation – is *yes* if and only if the terms $\gamma(q_t, r_t)$ and $\gamma(q_t, r_t)v^*(r_t)$ do not depend on r_t.

Remark 2. The (logarithm of the) forward price equation is Markovian if and only if the mappings $\gamma : \mathcal{H}_q \times \mathcal{H}_r \times \mathbb{R}_+ \to \mathbb{R}^m$ and $\gamma v^* : \mathcal{H}_q \times \mathcal{H}_r \times \mathbb{R}_+ \to \mathbb{R}^m$ are constant w.r.t. r.

The first condition – that γ cannot depend on r – is quite straightforward, but let us take a moment to understand what "γv^* constant w.r.t. to r" really means.

Given that γ is not dependent on r, does this mean that v must also be independent of r? The answer to this question is trivially *no*, when we take into consideration the fact that both γ and v are multidimensional. The exercise of explaining this answer, however, helps to establish crucial notation.

Recall that our m- dimensional Wiener process W drives both forward prices and interest rates, and that a multidimensional Wiener process can be seen as a vector of scalar independent Wiener processes. With this in mind, it is possible to understand that, depending on the applications, we may face all the following situations.

- The scalar Wiener processes driving interest rates are orthogonal to the scalar Wiener processes driving the forward prices;
- The scalar Wiener processes driving interest rates and forward prices are the same;
- A part of the scalar Wiener processes driving the interest rates also drives the forward prices (or vice versa);
- Interest rates and forward prices are partially driven by orthogonal scalar Wiener processes and partial driven by the same Wiener processes.

Without loss of generality, we can reorder the scalar Wiener processes inside a multidimensional Wiener process. Assumption 3 bellow, give us the reordering we will assume for our multidimensional Wiener process W.

Assumption 3. *The Q-Wiener process, W, driving both the forward prices and the interest rates is m-dimensional, and the same for both processes. Furthermore, we suppose that W has been reordered as*

$$W = \begin{bmatrix} W^A \\ W^B \\ W^C \end{bmatrix},$$

where W^A, W^B and W^C are, possibly multidimensional, Wiener process such that

- *W^A drives only the forward prices q,*
- *W^B drives only the interest rates r,*

- *W^C drives both forward prices q and interest rates r.*

Finally we establish that $i \in A$ means "W_i is a element of W^A", and similarly for $i \in B$ and $i \in C$.

Assumption 3 has obvious implications for the matrices γ and σ which become then of the following form.

$$\gamma = \begin{bmatrix}\gamma_A \; 0 \; \gamma_C\end{bmatrix} ,$$
$$\sigma = \begin{bmatrix}0 \; \sigma_B \; \sigma_C\end{bmatrix} ,$$

thus, using $v(x, r) = -\int_0^x \sigma(s, r)ds$, we have

$$v = \begin{bmatrix}0 \; v_B \; v_C\end{bmatrix},$$

and

$$\gamma v^* = \begin{bmatrix}\gamma_A \; 0 \; \gamma_C\end{bmatrix} \begin{bmatrix}0 \\ v_B^* \\ v_C^*\end{bmatrix} = \gamma_C v_C^* .$$

From this we see that requiring γ and γv^* independent of r, is nothing but requiring that, γ_A, γ_C and v_C do not depend on r.

The important point here is that *no condition* is imposed on σ_B.

We can now restate Remark 2, using the notation introduced by Assumption 3.

Lemma 5. *Suppose that Assumptions 2 and 3 holds. The (logarithm of) forward prices will be **Markovian** if and only if the volatility mappings γ_A, γ_C and σ_C are constant w.r.t. r. No condition is imposed on σ_B.*

Proof:If γ_A, γ_C and σ_C are constant w.r.t. r, so are $\|\gamma_A\|$, $\|\gamma_C\|$, v_C and $\gamma_C v_C$. The dynamics in of q in (10.24) does not depend on r and forward prices are, thus, Markovian.

To prove the "only if" part we show that dependence of r by γ_A, γ_C or σ_C suffices, under the regularity conditions of Assumption 2, to guarantee non-Markovian forward prices. Suppose, first, that γ_A depends on r. Then $\|\gamma_A\|^2$ also depends on r making the forward prices non-Markovian. Suppose now that γ_C depends on r, then $\|\gamma_C\|^2$ and $\gamma_C v_c^*$ also depend on r. Assumption 2 guarantees that there is no full cancelation and the forward prices are non-Markovian. Finally suppose that σ_C depends on r, then v_c depends on r (since the integral is w.r.t. the variable s and we know $\sigma_C \neq 0$). Since v_c depends on r so does $\gamma_C v_c^*$ and the forward prices are non-Markovian. □

Having established conditions for the forward prices being Markovian, we can go on and try to answer our *second problem* – on whether there exist models which admit a FDR for forward prices but not for interest rates. It turns out that, our unrestricted σ_B for Markovian forward prices, together with general results from the previous literature on interest rates FDR, allows us to give a partial answer already now.

From the previous literature on FDR of **interest rates** we know that only some particular functions $\sigma : \mathcal{H}_r \times \mathbb{R}_+ \to \mathbb{R}^m$ will generate interest rate models that admit a FDR. Concretely, it is shown in [10] that **every** component σ_i must be a weighted sum of quasi-exponential deterministic functions of x, weighted by scalar fields in $\mathcal{H}_r$.

Hence, the fact that σ_B is not driving the forward price equation and can be of *any* form for Markovian forward price models, tell us that, existence of FDR for Markovian forward prices is, in some sense, independent from existence of FDR for interest rates. This leads us to an early answer to our *second question.*

Remark 3. As long as there are FDR for Markovian forward prices, there exist forward price models that allow for a FDR for forward prices but **not** for interest rates.

10.5 Markovian Forward Prices

We now focus on the task of studying FDR for the forward price equation, in the special case where we have *Markovian* forward prices. Thus, in this section the following assumption holds (recall Lemma 5).

Assumption 4. *Consider Assumption 1 and 3 We, assume that the mappings* γ_A, γ_C *and* σ_C *are of the following special forms,*

$$\gamma_A(q,r,x) = \gamma_A(q,x)\,, \qquad \gamma_C(q,r,x) = \gamma_C(q,x)\,, \qquad \sigma_C(r,x) = \sigma_C(x).$$

Note that the specific functional form of σ_C implies we have deterministic σ_C-volatilities and we can interpret σ_i for $i \in C$ as **constant** vector fields in $\mathcal{H}_q$.

Given Assumption 4, the Q-dynamics of (the logarithm) of forward prices can be written as

$$\begin{aligned} dq_t = &\left\{ \mathbf{F}q_t - \frac{1}{2}\left[\|\gamma_A(q_t)\|^2 + \|\gamma_C(q_t)\|^2\right] - \gamma_C(q_t)v_C{}^* \right\} dt + \\ &+ \left[\gamma_A(q_t)\; 0\; \gamma_C(q_t)\right] dW_t\,. \end{aligned} \tag{10.25}$$

Now, we rewrite equation (10.26) in Stratonovich form and obtain

$$\begin{aligned} dq_t = &\left\{ \mathbf{F}q_t - \frac{1}{2}\left[\|\gamma_A(q_t)\|^2 + \|\gamma_C(q_t)\|^2\right] - \gamma_C(q_t)v_C{}^* \right\} dt - \\ &- \frac{1}{2}\sum_{i\in A,C} d\langle \gamma_i(q_t), W_t^i\rangle + \left[\gamma_A(q_t),\, 0,\, \gamma_C(q_t)\right] \circ dW_t\,. \end{aligned}$$

To compute the Stratonovich correction we use the infinite dimensional Itô formula (see [8]) to obtain

$$d\gamma_i(q_t) = (\cdots)\, dt + \gamma_i'(q_t)\gamma_i(q_t)dW_t^i\,, \qquad i \in A, C$$

and, thus

$$d\langle \gamma_i(q_t), W_t^i\rangle = \gamma_i'(q_t)\gamma_i(q_t)dt\,, \qquad i \in A, C.$$

Given the above computations we can write the Stratonovich dynamics of q as

$$\begin{aligned} dq_t = \Big\{ &\mathbf{F}q_t - \frac{1}{2}\left[\|\gamma_A(q_t)\|^2 + \|\gamma_C(q_t)\|^2\right] - \gamma_C v_C{}^* - \\ &-\frac{1}{2}\left[\gamma_A'(q_t)\gamma_A(q_t) + \gamma_C'(q_t)\gamma_C(q_t)\right]\Big\} dt + \left[\gamma_A(q_t)\ 0\ \gamma_C(q_t)\right] \circ dW_t\,, \end{aligned} \tag{10.26}$$

where γ_A' and γ_B' denotes the Frechet derivative. The terms $\gamma_A'(q_t)\gamma_A(q_t)$ and $\gamma_C'(q_t)\gamma_C(q_t)$ should be interpreted as follows,

$$\gamma_A'(q_t)\gamma_A(q_t) = \sum_{i\in A}\gamma_i'(q_t)\gamma_i(q_t)\,, \qquad \gamma_C'(q_t)\gamma_C(q_t) = \sum_{i\in C}\gamma_i'(q_t)\gamma_i(q_t)\,.$$

We start by studying the two easier cases:

(i) the case when γ (i.e, γ_A and γ_C) is also deterministic (σ_C is deterministic by Assumption 4), and
(ii) the case when γ is not deterministic, but has deterministic direction.

10.5.1 Deterministic Volatility

We first consider the case when the functions γ_A and γ_C do not depend on q, so they have the special form

$$\gamma_i(q, x) = \gamma_i(x)\,, \qquad i \in A, C. \tag{10.27}$$

γ_i for $i \in A, C$ are, thus, **constant** vector fields in $\mathcal{H}_q$.

Recall from Assumption 4 that $\sigma_C(r, x) = \sigma_C(x)$.

In this case, the Stratonovich correction term is zero, and equation (10.27) becomes

$$dq_t(x) = \mu(q_t, x)dt + \gamma(x) \circ dW_t\,, \tag{10.28}$$

where

$$\mu(q, x) = \mathbf{F}q - \frac{1}{2}\left[\|\gamma_A(x)\|^2 + \|\gamma_C(x)\|^2\right] - \gamma_C(x)v_C{}^*(x) \tag{10.29}$$

$$\gamma(x) = \left[\gamma_A(x)\ 0\ \gamma_C(x)\right]. \tag{10.30}$$

Since this is a simple case, we choose to include all computations behind the results in the main text to exemplify the technique. In the next sections, when dealing with more complex cases, most of the computations will instead be presented in the appendix, leaving to the main text the intuition behind the results and their discussion.

Existence of a FDR

From Theorem C.1 we know that a FDR exists if and only if

$$\dim \{\mu, \gamma_i; \quad i \in A, C\}_{LA} < \infty .$$

We, thus, need to compute the Lie-algebra, $\{\mu, \gamma\}_{LA}$. Computing the Lie brackets we have, for each i

$$\begin{aligned} [\mu, \gamma_i] &= \mathbf{F}\gamma_i =: f_{1i} , \\ [\gamma_i, f_{1i}] &= 0 , \\ [\mu, f_{1i}] &= \mathbf{F} f_{1i} = \mathbf{F}^2\gamma_i =: f_{2i} , \\ &\vdots \end{aligned}$$

It follows that

$$\{\mu, \gamma\}_{LA} = \text{span}\left\{\mu, \ \mathbf{F}^k\gamma_i \ ; \qquad k = 0, 1 \cdots \quad i \in A, C\right\} . \tag{10.31}$$

Obviously, if a FDR exists, there must exist an n_i for each i, such that

$$\mathbf{F}^{n_i+1}\gamma_i = \sum_{k=0}^{n_i} c_{i,k}\mathbf{F}^k\gamma_i , \tag{10.32}$$

where the $c_{i,k}$ are real numbers.

Proposition 1 tell us under what conditions we will have $\dim \{\mu, \gamma\}_{LA} < \infty$.

Proposition 1. *The (logarithm of the) forward price equation (10.28) admits a finite dimensional realization (FDR) if and only if each component of γ is quasi-exponential (QE). No functional restriction is imposed on the deterministic function σ_C, so in particular, σ_C* does not *have to be a QE function, it can be any deterministic function.*

Proof: Recall from Lemma 4 that γ_i solves the ODE (10.32) if and only if it is a QE function. □

Note that, for Markovian forward prices, the interest rate volatility plays no role in determining existence of FDR. The only restriction on interest rate volatility is that σ_C is deterministic, but that is a result of the Markovian property, not an added requirement imposed to guarantee existence. One other way to see this is to note that only γ shows up in (10.31). As we will soon see, this is specific to the totally deterministic setting.

Remark 4. In the simple deterministic setting, where $\gamma : \mathbb{R}_+ \to \mathbb{R}^m$, Markovian realizations of forward prices are generated only by the volatility of forward prices γ.

In the next example, and to stress this point, we check existence of FDR in a simple model without even specifying the deterministic function σ_C.

Example 1. Assume that forward prices are driven by a one-dimensional Wiener-process W which also drives the interest rates[8]. Furthermore, assume that the interest rate volatility associated to W is deterministic and that we have for the forward price volatility,

$$\gamma(x) = \gamma_C(x) = \alpha e^{-ax}$$

where $\alpha, a \in \mathbb{R}$.

In this case, we have $A \cup B = \emptyset$, $C = \{1\}$ and

$$\mathbf{F}\gamma(x) = -a\alpha e^{-ax} \qquad \Rightarrow \qquad n_1 = 0\,, \quad c_{1,0} = -a\,.$$

Hence,

$$\{\mu, \gamma\}_{LA} = \text{span}\,\{\mu, \gamma\}\,,$$

and the Lie-algebra $\{\mu, \gamma\}_{LA}$ has dimension two. Consequently, there exist a FDR for forward prices in this case. We will get back to this example in the construction part.

Finally, we want to make a remark on the exact dimension of the lie-algebra.

Remark 5. It follows from (10.32) that

$$\dim\{\mu, \gamma\}_{LA} = \dim\left\{\mu, \mathbf{F}^k\gamma_i\ ;\ k = 0, \cdots, n_i,\ i \in A, C\right\} \le 1 + \sum_{i=1}^{m} n_i\,. \tag{10.33}$$

The "$\le$" in (10.33) just reminds us that, given the possibility of Gaussian elimination, there may exist some cancelation effects.

To a better understanding of the above remark, we take the following example.

Example 2. Suppose that

$$\gamma_1(x) = e^{-bx}\,, \qquad \gamma_2(x) = xe^{-bx}.$$

Thus, $n_1 = 0$, $n_2 = 1$, and $\dim\{\mu, \gamma_1, \gamma_2, \mathbf{F}\gamma_2\} \le 4$.

However since $\mathbf{F}\gamma_2 = \gamma_1 - b\gamma_2$,

$$\text{span}\,\{\mu, \gamma_1, \gamma_2, \mathbf{F}\gamma_2\} = \text{span}\,\{\mu,\ \gamma_1,\ \gamma_2\}\,.$$

Hence, in this case, we actually have $\dim\{\mu,\ \gamma\}_{LA} = 3$.

[8] This does not exclude the possibility of more Wiener-processes driving only the interest rates.

Construction of FDR

We now go on to the construction of FDR, in the totally deterministic volatility setup.

First, we obtain a parameterization G of the invariant manifold $\mathcal{G}$. In this case we have that

$$\{\mu,\gamma\}_{LA} = \text{span}\left\{\mu, \mathbf{F}^k\gamma_i\ ; \quad k = 0, 1 \cdots, n_i \quad i \in A, C\right\},$$

and we recall γ_i solves the ODE (10.32).

Using Theorem C.2 we obtain G by computing the operators $\exp\{\mu z_0\}$ and $\exp\left\{\mathbf{F}^k\gamma_i z_{i,k}\right\}$. In order to get $\exp\{\mu z_0\}\, q_0$ we solve

$$\begin{cases} \dfrac{dy_t}{dt} = \mu(y_t, x) = \mathbf{F}y_t + \mathcal{D}(x)\,, \\ y_0 = q_0\,, \end{cases}$$

for

$$\mathcal{D}(x) = -\frac{1}{2}\sum_{i\in A,C}\gamma_i^2(x) - \sum_{i\in C}\gamma_i(x)v_i(x)\,. \tag{10.34}$$

Hence by Definition 4, we have [9]

$$\begin{aligned} e^{\mu t}q_0(x) &= e^{\mathbf{F}t}q_0(x) + \int_0^t e^{\mathbf{F}(t-s)}\mathcal{D}(x)ds \\ &= q_0(x+t) + \int_0^t \mathcal{D}(x+t-s)ds\,. \end{aligned}$$

To obtain the remaining operators we solve

$$\begin{cases} \dfrac{dy_t}{dt} = \mathbf{F}^k\gamma_i\,, \\ y_0 = y\,. \end{cases}$$

Because γ does not depend on t, the solution is

$$e^{\mathbf{F}^k\gamma_i(x)t}y = y + \mathbf{F}^k\gamma_i(x)t\,.$$

It follows that

---

9 From the context, it is clear that $e^{\mathbf{F}t} : \mathcal{H}_q \to \mathcal{H}_q$. From the usual series expansion of the exponential function we have, $e^{\mathbf{F}t}f = \sum_{n=0}^{\infty}\frac{t^n}{n!}\mathbf{F}^n f$. In our case, $\mathbf{F}^n = \frac{\partial^n}{\partial x^n}$, so we have $\left[e^{\mathbf{F}t}f\right](x) = \sum_{n=0}^{\infty}\frac{t^n}{n!}\frac{\partial^n f}{\partial x^n}(x)$, which is a Taylor expansion of f around the point x, so for analytic f we have $\left[e^{\mathbf{F}t}f\right](x) = f(x+t)$.

$$G(z_0, z_{i,k}; i \in A, C \quad k = 0, 1, \cdots, n_i) \tag{10.35}$$

$$= \prod_{i \in A,C; k=0,\dots,n_i} \left(e^{\mathbf{F}^k \gamma_i(x) z_{i,k}} \right) e^{\mu(q,x) z_0} q_0$$

$$= q_0(x + z_0) + \int_0^{z_0} \mathcal{D}(x + z_0 - s) ds + \sum_{i \in A,C} \sum_{k=0}^{n_i} \mathbf{F}^k \gamma_i(x) z_{i,k} . \tag{10.36}$$

Note that the volatility of interest rates show up (through the terms v_i for $i \in C$) only in the deterministic term $\mathcal{D}$ defined in (10.34).

We are now interested in finding a set of factors Z such that

$$q_t = G(Z_t) ,$$

while Z is given by a strong solution to the SDE

$$\begin{cases} dZ_t = a(Z_t)dt + b(Z_t) \circ dW_t , \\ Z_0 = z_0 . \end{cases}$$

For that we need to find a solution to

$$G_\star a = \mu , \qquad G_\star b^i = \gamma_i , \qquad i \in A, C .$$

Simple computations yields

$$G'(z_0, z_{j,k} \; ; j \in A, C \quad k = 0, 1, \cdots, n_j) \begin{pmatrix} h_0 \\ h_{1,0} \\ h_{1,1} \\ \vdots \\ h_{m,n_m} \end{pmatrix} (x)$$

$$= \left[\frac{\partial q_0}{\partial z_0}(x + z_0) + \mathcal{D}(x) + \int_0^{z_0} \frac{\partial \mathcal{D}}{\partial z_0}(x + z_0 - s) ds \right] h_0 +$$

$$+ \sum_{i \in A,C} \sum_{k=0}^{n_j} \mathbf{F}^k \gamma_i(x) h_{i,k} .$$

We can now use the fact that $q = G(Z)$ and that γ satisfies the ODE (10.32) to get

$$
\begin{aligned}
\mu(q_t, x) &= \mathbf{F}q_t + \mathcal{D}(x) \\
&= \frac{\partial}{\partial x} q_0(x+z_0) + \int_0^{z_0} \frac{\partial}{\partial x}\mathcal{D}(x+z_0-s)ds + \sum_{j\in A,C}\sum_{k=0}^{n_j} \mathbf{F}^{k+1}\gamma_j(x) z_{j,k} + \mathcal{D}(x) \\
&= \frac{\partial}{\partial x} q_0(x+z_0) + \int_0^{z_0} \frac{\partial}{\partial x}\mathcal{D}(x+z_0-s)ds + \mathcal{D}(x) + \sum_{j\in A,C}\sum_{k=1}^{n_j} \mathbf{F}^{k}\gamma_j(x) z_{j,k-1} + \\
&\quad + \sum_{j\in A,C} \mathbf{F}^{n_j+1}\gamma_j(x) z_{j,n_j} \\
&= \frac{\partial}{\partial x} q_0(x+z_0) + \int_0^{z_0} \frac{\partial}{\partial x}\mathcal{D}(x+z_0-s)ds + \mathcal{D}(x) + \sum_{j\in A,C}\sum_{k=1}^{n_j} \mathbf{F}^{k}\gamma_j(x) z_{j,k-1} \\
&\quad + \sum_{j\in A,C}\sum_{k=0}^{n_j} c_{j,k}\mathbf{F}^{k}\gamma_j(x) z_{j,n_j} .
\end{aligned}
$$

Thus, from $G_\star a = \mu$, we get

$$
\begin{aligned}
&\left[\frac{\partial q_0}{\partial z_0}(x+z_0) + \mathcal{D}(x) + \int_0^{z_0} \frac{\partial \mathcal{D}}{\partial z_0}(x+z_0-s)ds\right] a_0 + \sum_{j\in A,C}\sum_{k=0}^{n_j} \mathbf{F}^{k}\gamma_j(x) a_{j,k} \\
&= \frac{\partial}{\partial x} q_0(x+z_0) + \int_0^{z_0} \frac{\partial}{\partial x}\mathcal{D}(x+z_0-s)ds + \mathcal{D}(x) + \sum_{j\in A,C}\sum_{k=1}^{n_j} \mathbf{F}^{k}\gamma_j(x) z_{j,k-1} \\
&\quad + \sum_{j\in A,C}\sum_{k=0}^{n_j} c_{j,k}\mathbf{F}^{k}\gamma_j(x) z_{j,n_j}
\end{aligned}
$$

and, since the above expression should hold for all x, while a is not allowed to depend on x, it is possible to identify the following expressions for a

$$
\begin{aligned}
a_0 &= 1 , \\
a_{j,0} &= c_{j,0} z_{j,n_j} , && j \in A, C , \\
a_{j,k} &= c_{j,k} z_{j,n_j} + z_{j,k-1} , && j \in A, C , \quad k = 1, \cdots , n_j .
\end{aligned}
$$

Likewise, from $G_\star b^i = \gamma_i$ for each $i \in A, C$ we get

$$
\begin{aligned}
&\left[\frac{\partial q_0}{\partial z_0}(x+z_0) + \mathcal{D}(x) + \int_0^{z_0} \frac{\partial \mathcal{D}}{\partial z_0}(x+z_0-s)ds\right] b_0 + \sum_{i\in A,C}\sum_{k=0}^{n_j} \mathbf{F}^{k}\gamma_j(x) b^i_{j,k} \\
&= \gamma_i(x)
\end{aligned}
$$

and by simple identification of terms

$$
\begin{aligned}
b^i_0 &= 0 , && i \in A, C , \\
b^i_{j,0} &= 1 , && i = j \quad i, j \in A, C , \\
b^i_{j,k} &= 0 , && \text{for all other } j \text{ and } k .
\end{aligned}
$$

In this totally deterministic setting, the Stratonovich and the Itô dynamics are equivalent, so we have proved the following result.

Proposition 2. *Given the initial forward price curve q_0, the forward prices system generated by γ_A, γ_C and σ_C has a finite dimensional realization given by*

$$q_t = G(Z_t)\,,$$

where G is defined as in (10.36) and the dynamics of the state space variables Z are given by

$$\begin{cases} dZ_0 = dt\,, & \\ dZ_{j,0} = c_{j,0} Z_{j,n_j} dt + dW_t^j\,, & j \in A, C\,, \\ dZ_{j,k} = \left(c_{j,k} Z_{j,n_j} + Z_{j,k-1}\right) dt\,, & j \in A, C\,,\ k = 1, \cdots, n_j\,. \end{cases}$$

We first take the easiest example: the one-dimensional deterministic constant volatility.

Example 3. Assume that forward prices q are driven by a one-dimensional Wiener process that also drive the interest rates r ($C = \{1\}$). Furthermore, assume that the forward price volatility γ is of the following form

$$\gamma(x) = \gamma_C(x) = \alpha\,,$$

where $\alpha \in \mathbb{R}$.

We leave the (scalar) function $\sigma_C(x)$ (thus $v_C(x)$) unspecified to stress the point that it plays no role determining the dimension of the Lie-algebra or constructing the realization.

Then we know $\mathbf{F}\gamma = 0$, thus $n = 1$, $c_{1,0} = 0$ and the dimension of the $\{\mu, \gamma\}_{LA}$ is two.

The invariant manifold is given by

$$G(z_0, z_{1,0}) = q_0(x + z_0) - \frac{1}{2}\alpha^2 z_0 - \alpha \int_0^{z_0} v_C(x + z_0 - s) ds + \alpha z_{1,0}\,,$$

for some deterministic function v_C.

Using Proposition 2, we have $q_t = G(Z_t)$ for the state variable $Z = \begin{bmatrix} Z_0 \\ Z_{1,0} \end{bmatrix}$ with dynamics given by

$$\begin{cases} dZ_0 = dt\,, \\ dZ_{1,0} = dW_t\,. \end{cases}$$

We now recover again Example 1.

Example 1 (Cont.) Recall that we assumed

$$\gamma(x) = \gamma_C(x) = \alpha e^{-ax}\,,$$

which implies $n_1 = 0$, $c_{1,0} = -a$.

Once again we leave σ_C (thus v_C) as an unspecified deterministic function.

In the previous comments it was explained that the Lie-algebra is of dimension 2, so the invariant manifold can be obtained from (10.36),

$$G(z_0, z_{1,0}) = q_0(x + z_0) + \int_0^{z_0} \mathcal{D}(x + z_0 - s)ds + \gamma(x)z_{1,0}\,.$$

In this case

$$\begin{aligned}\mathcal{D}(x) &= -\frac{1}{2}\gamma_C^2(x) - \gamma_C(x)v_C(x)\\ &= -\frac{1}{2}\alpha^2 e^{-2ax} - \alpha e^{-ax} v_C(x)\,.\end{aligned}$$

Thus,

$$\begin{aligned}G(z_0, z_{1,0}) &= q_0(x + z_0) - \int_0^{z_0} \left(\frac{1}{2}\alpha^2 e^{-2a(x+z_0-s)} + \right.\\ &\quad \left. + \alpha e^{-a(x+z_0-s)} v_C(x + z_0 - s)\right) ds + \alpha e^{-ax} z_{1,0}\\ &= q_0(x + z_0) + \frac{1}{2}\frac{\alpha^2}{2a}\left[e^{-2ax} - e^{-2a(x+z_0)}\right] -\\ &\quad -\alpha \int_0^{z_0} \left(e^{-a(x+z_0-s)} v_C(x + z_0 - s)\right) ds + \alpha e^{-ax} z_{1,0}\end{aligned}$$

and from Proposition 2 it follows that the FDR is given by

$$\begin{cases} dZ_0 = dt\,, \\ dZ_{1,0} = -aZ_{1,0}dt + dW_t\,. \end{cases}$$

10.5.2 Deterministic Direction Volatility

We now deal with the second simplest case, that of having *deterministic direction* forward prices volatilities.

Then, we have the following special functional forms for γ_A and γ_C in Assumption 4.

$$\gamma_i(q, x) = \lambda_i(x)\varphi_i(q)\,, \qquad i \in A, C\,,$$

where λ_i is a deterministic function of x (constant vector field in $\mathcal{H}_q$) and φ_i is a scalar vector field in $\mathcal{H}_q$ (i.e., it does not depend on x and depends only on the current forward price curve).

Omitting the x-dependence,

$$\gamma_i(q) = \lambda_i\varphi_i(q)\,, \qquad i \in A, C\,, \tag{10.37}$$

and for future reference we also define

$$\lambda = \left[\lambda_A, 0, \lambda_C\right]. \tag{10.38}$$

On what concerns interest rate volatilities, we maintain the requirement that σ_C is deterministic (since we are still dealing with Markovian forward prices).

In this particular case, the forward price equation can be rewritten as

$$dq_t = \mu(q_t)dt + \gamma(q_t) \circ dW_t \,, \tag{10.39}$$

for

$$\begin{aligned} \mu(q_t) &= \mathbf{F}q_t - \frac{1}{2}\left[\|\gamma_A(q_t)\|^2 + \|\gamma_C(q_t)\|^2\right] - \gamma_C(q_t)v_C{}^* - \\ &\quad -\frac{1}{2}\left[\gamma'_A(q_t)\gamma_A(q_t) + \gamma'_C(q_t)\gamma_C(q_t)\right] \\ \gamma(q_t) &= \left[\gamma_A(q_t)\ 0\ \gamma_C(q_t)\right], \end{aligned} \tag{10.40}$$

and given the functional form for γ_A and γ_C in (10.37) we have the following Frechet derivatives

$$\begin{aligned} \gamma'_A(q_t)\gamma_A(q_t) &= \sum_{i\in A} \lambda_i \varphi'_i(q_t)\left[\lambda_i \varphi_i(q_t)\right], \\ \gamma'_C(q_t,x)\gamma_C(q_t,x) &= \sum_{i\in C} \lambda_i(x)\varphi'_i(q_t)\left[\lambda_i \varphi_i(q_t)\right]. \end{aligned}$$

We see that μ in (10.40) is much more complex than the one previously studied (compare to (10.29)), so, the task of actually computing the Lie algebra $\mathcal{L} = \{\mu, \gamma\}_{LA}$ will not be as straightforward as before.

Using the specific functional forms of γ_A and γ_C in (10.37) we have

$$\begin{aligned} \mu(q) &= \mathbf{F}q - \frac{1}{2}\sum_{i\in A,C} \underbrace{\lambda_i^2}_{D_i} \underbrace{\varphi_i^2(q)}_{\phi_i(q)} - \sum_{i\in C} \overbrace{\varphi_i(q)}^{\text{scalar field}} \underbrace{\lambda_i v_i}_{V_i} - \\ &\quad -\frac{1}{2}\sum_{i\in A,C} \overbrace{\varphi'_i(q)\left[\lambda_i\right]\varphi_i(q)}^{\text{scalar field}}\lambda_i \\ \gamma_i(q) &= \lambda_i \overbrace{\varphi_i(q)}^{\text{scalar field}} \end{aligned}$$

and, given the possibility of Gaussian elimination (Lemma 3), we see that the Lie algebra is in fact generated by the simpler system of vector fields,

$$\begin{aligned} f_0(q) &:= \mathbf{F}q - \frac{1}{2}\sum_{k\in A,C} D_k\phi_k(q) - \sum_{k\in C}\varphi_k(q)V_k \,, \\ f_{1i}(q) &:= \lambda_i \,, \qquad\qquad i \in A, C \,. \end{aligned}$$

where

$$D_k = \lambda_k^2 , \qquad V_k = \lambda_k v_k .$$

We now start computing Lie brackets and simplifying. For all $i \in A, C$ we have

$$\begin{aligned}[f_0, f_{1i}] &= f_0' f_{1i} - f_{1i}' f_0 \\ &= \mathbf{F}\lambda_i - \frac{1}{2} \sum_{k \in A,C} D_k \underbrace{\phi_k'[\lambda_i]}_{\text{scalar field}} - \sum_{k \in C} V_k \underbrace{\varphi_k'[\lambda_i]}_{\text{scalar field}} \\ &=: f_{2i}\end{aligned}$$

and using this new field we have for all $i, j \in A, C$

$$\begin{aligned}[f_{1i}, f_{2j}] &= f_{1i}' f_{2j} - f_{2j}' f_{1i} \\ &= +\frac{1}{2} \sum_{k \in A,C} D_k \underbrace{\phi_k''[\lambda_j; \lambda_i]}_{\text{scalar field}} + \sum_{k \in C} V_k \underbrace{\varphi_k''[\lambda_j; \lambda_i]}_{\text{scalar field}} \\ &=: f_{3ij} .\end{aligned}$$

We know realize that the our lie algebra is hard to handle (even in the one-dimensional Wiener process case[10]). At this point it seems a good idea to note the following.

Remark 6. The Lie-algebra

$$\mathcal{L} = \{\mu, \gamma\}_{LA} \tag{10.41}$$

is included in the larger Lie-algebra

$$\bar{\mathcal{L}} = \{\mathbf{F},\ \lambda_i,\ D_i,\ V_j\ ;\ i \in A, C\ j \in C\}_{LA} . \tag{10.42}$$

That is

$$\{\mu, \gamma\}_{LA} = \{f_0,\ f_{1i}\ ;\ i \in A, C\}_{LA} \subseteq \{\mathbf{F},\ \lambda_i,\ D_i,\ V_j\ ;\ i \in A, C\ j \in C\}_{LA} .$$

There are three important points to make here.

- The fields in the larger Lie-algebra, $\bar{\mathcal{L}}$, are simpler than those in $\mathcal{L}$. That is, none of the field contains sums.
- From the inclusion $\mathcal{L} \subseteq \bar{\mathcal{L}}$ it is obvious that if $\bar{\mathcal{L}}$ has finite dimension also $\mathcal{L}$ does. So, studying the conditions that guarantee $\bar{\mathcal{L}}$ to have finite dimension, give us, at least, *sufficient* conditions for $\mathcal{L}$ to have also finite dimension.

---

[10] We note once again, that if we consider the Wiener process to be one-dimensional, the only interesting case to consider is when that Wiener process drives both forward prices and interest rates, i.e. it belongs to the W^C set. Otherwise, we fall into the futures case already studied in [3]. So, even in that case, we cannot avoid having two parcels not easy to simplify.

- We conjecture that conditions that guarantee $\bar{\mathcal{L}}$ to have finite dimension are also *necessary* conditions for $\mathcal{L}$ to have also finite dimension. The intuition is that since the fields in $\bar{\mathcal{L}}$ are all contained in the fields of $\mathcal{L}$ (as parcels of various sums), if they are "nasty" enough to make the dimension of $\bar{\mathcal{L}}$ infinite, they should make the fields that contain them in $\mathcal{L}$ even "nastier". We will formalize this intuition below.
- Finally, even if the analysis of $\bar{\mathcal{L}}$ is, in the above sense, equivalent to the analysis of $\mathcal{L}$, in the construction sense, studying $\bar{\mathcal{L}}$ will, in principle, generate finite realizations with state variables of higher dimension. This is obviously the price one has to pay for dealing with easier fields. We call these realizations *non-minimal realizations.* An advantage of *non-minimal* realizations is that they are always possible to obtain (as long as the dimension of $\bar{\mathcal{L}}$ is finite).

The following conjecture formally states the idea behind our third point above (and the sketch of the proof, for the one–dimensional case, can be found in the appendix).

Conjecture 1. Consider $\mathcal{L}$ in (10.41) and $\bar{\mathcal{L}}$ in (10.42). Then the following holds

$$\dim(\mathcal{L}) < \infty \qquad \Leftrightarrow \qquad \dim(\bar{\mathcal{L}}) < \infty \,.$$

These ideas can be applied in a more complex setting. They will be used extensively in Section 10.6, when dealing with the entire forward price system.

We now continue our analysis studying the larger Lie-algebra $\bar{\mathcal{L}}$.

Existence of FDR

As mentioned before, in the current deterministic direction setting, the larger Lie-algebra, $\bar{\mathcal{L}}$, is given by

$$\bar{\mathcal{L}} = \{\mathbf{F},\ \lambda_i,\ D_i,\ V_j\ ;\quad i \in A, C \quad j \in C\}_{LA} \,.$$

thus, the basic fields of the enlarged Lie-algebra are

$$\begin{aligned}
g_0(q) &:= \mathbf{F}q\,, \\
g_{1i}(q) &:= \lambda_i\,, && i \in A, C\,, \\
g_{2i}(q) &:= D_i\,, && i \in A, C = 1, \cdots, m\,, \\
g_{3j}(q) &:= V_j\,, && j \in C\,.
\end{aligned}$$

Computing the Lie-brackets we have, in the first step,

$$\begin{aligned}
[g_0, g_{1i}] &= \mathbf{F}\lambda_i =: g_{4i}\,, \\
[g_0, g_{2i}] &= \mathbf{F}D_i =: g_{5i}\,, \\
[g_0, g_{3j}] &= \mathbf{F}V_j =: g_{6j}\,,
\end{aligned}$$

all remaining combinations of lie-brackets from the fields in (10.42) are zero. Using the new vector fields, we easily see that

$$\begin{aligned}[][g_0, g_{4i}] &= \mathbf{F}^2\lambda_i\,, \\ [g_0, g_{2i}] &= \mathbf{F}^2 D_i\,, \\ [g_0, g_{3j}] &= \mathbf{F}^2 V_j\,,\end{aligned}$$

and again all others lie-brackets are zero.

Continuing with similar iterations, it is easy to check that

$$\bar{\mathcal{L}} = \text{span}\left\{\mathbf{F},\ \mathbf{F}^k\lambda_i,\ \mathbf{F}^k D_i,\ \mathbf{F}^k V_j\ ;\quad i \in A, C\ j \in C\ k = 0, 1, \cdots\right\}.$$

Thus, for $\dim(\bar{\mathcal{L}}) < \infty$ there must exist orders n_j^1, n_j^2 and n_j^3 for which

$$\begin{aligned}\mathbf{F}^{n_j^1+1}\lambda_j &= \sum_{k=0}^{n_j^1} c_{j,k}^1 \mathbf{F}^k\lambda_j\,, \qquad \mathbf{F}^{n_j^2+1}D_j = \sum_{k=0}^{n_j^2} c_{j,k}^2 \mathbf{F}^k D_j\,, \ j \in A, C\,, \\ \mathbf{F}^{n_j^3+1}V_j &= \sum_{k=0}^{n_j^3} c_{j,k}^3 \mathbf{F}^k V_j\,, \qquad\qquad\qquad\qquad\qquad j \in C\,,\end{aligned}$$

where $c_{j,k}^I$ are real constants, $k = 0, \cdots, n_j^I$, $j \in A, C$, and $I = 1, 2, 3$.

Indeed, if (10.43) holds,

$$\dim(\bar{\mathcal{L}}) \leq 1 + \sum_{j \in A,C} (n_j^1 + 1) + (n_j^2 + 1) + \sum_{j \in C} (n_j^3 + 1) < \infty\,. \tag{10.43}$$

We now start a brief side discussion on the exact dimension of $\bar{\mathcal{L}}$. As before, we note that the sign "$\leq$" above results from the possibility of Gaussian elimination across the various derivatives of various different fields. This is particularly relevant for the concrete fields of $\bar{\mathcal{L}}$ since $V_j = \lambda_j v_j$ and thus derivatives of λ_j may help to Gaussian simplify the derivatives of V_j. This point will be made clear in Example 4 bellow.

Unfortunately, these simplifications are instance dependent and, thus, in an abstract way it is impossible to be more exact about the dimension of $\bar{\mathcal{L}}$ than in (10.43). The consequence is that when doing the abstract construction of realizations we cannot take into account case-specific Gaussian eliminations. Therefore, when applying the abstract results to concrete models, we may get unnecessarily large realizations. However, since we are considering non-minimal realizations anyway (because we analyze $\bar{\mathcal{L}}$ instead of $\mathcal{L}$), this does not seem a major disadvantage[11]. With this we conclude the side discussion and go on with the abstract analysis.

Proposition 3 give us the necessary and sufficient conditions that guarantee $\dim(\bar{\mathcal{L}}) < \infty$.

[11] The unsatisfied reader can always, when faced with a concrete situation, use the techniques presented and, whenever possible, derive a smaller realization instead of using the abstract results. In Example 4 below, we use both approaches to exemplify the kind of difference one can expect.

Proposition 3. *The dimension of the Lie-algebra $\bar{\mathcal{L}}$ in (10.42) is finite if and only if each component of λ and σ_C is QE.*

Proof: It follows from Lemma 4 that (10.43) holds if and only if λ_i, D_i for $i \in A, C$ and V_j for $j \in C$ are QE functions. It also follows from the properties of QE functions that, λ_i for $i \in A, C$ and σ_j for $j \in C$ QE, suffices to guarantee this requirement. To check the last statement note that λ_i QE $\Rightarrow D_i = \lambda_i^2$ QE; σ_j QE, $\Rightarrow v_j(x) = -\int_0^x \sigma_j(s)ds$ QE; and finally λ_j and v_j QE $\Rightarrow V_j = \lambda_j v_j$ QE. □

Taking together Propositions 3 and Conjecture 1 we have the following general result for forward prices with deterministic direction volatilities.

Proposition 4. *Assume that Conjecture 1 holds. The (logarithm of) forward price equation (10.39) admits a finite dimensional realization if and only if each component of λ and σ_C are quasi-exponential.*

We note that, in contrast to the deterministic forward price volatilities case, in our present setting of deterministic direction forward price volatilities, existence of FDR imposes requirements on the concrete functional form of the deterministic function σ_C: it must be a QE function. This was to be expected, since this time, σ_C actually drives the forward price equation indirectly trough the fields $V_j = \lambda_j v_j$ for $j \in C$.

The following example gives us one, very simple instance where we have a finite dimensional realization for forward prices, considering deterministic direction volatilities.

Example 4. Suppose that forward prices are driven by a one-dimensional Wiener process ($i = 1$) that also drives interest rates ($i \in C$) and that

$$\gamma(q,x) = \gamma_C(q,x) = \underbrace{\alpha e^{-bx}}_{\lambda(x)} \underbrace{q}_{\varphi(q)}, \qquad \sigma_C(x) = \delta e^{-ax}.$$

for $\alpha, \beta, a, b \in \mathbb{R}$. Then we have

$$\begin{aligned}
v(x) &= -\delta \int_0^x e^{-as} ds = \frac{\delta}{a}\left(e^{-ax} - 1\right), \\
\lambda(x) &= \alpha e^{-bx}, \\
D(x) &= \lambda^2(x) = \alpha^2 e^{-2bx}, \\
V(x) &= \lambda(x)v(x) = \alpha e^{-bx}\frac{\delta}{a}\left(e^{-ax} - 1\right) = \frac{\alpha\delta}{a}\left[e^{-(a+b)x} - e^{-bx}\right],
\end{aligned}$$

and

$$\begin{aligned}
\mathbf{F}\lambda(x) &= -b\alpha e^{-bx} = -b\lambda(x) &&\Rightarrow n_1^1 = 0 \; c_{1,0}^1 = -b \\
\mathbf{F}D(x) &= -2b\alpha^2 e^{-2bx} = -2bD(x) &&\Rightarrow n_1^2 = 0 \; c_{1,0}^2 = -2b
\end{aligned}$$

$$\begin{aligned}
\mathbf{F}V(x) &= \tfrac{\alpha\delta}{a}\left[-(a+b)e^{-(a+b)x} - (-b)^k e^{-bx}\right] \\
&= -bV(x) - \alpha\delta e^{-(a+b)x} \\
\mathbf{F}^2 V(x) &= -b\mathbf{F}V(x) + (a+b)\alpha\delta e^{-(a+b)x} \\
&= -b\mathbf{F}V(x) + (a+b)\alpha\delta\left[-bV(x) - \mathbf{F}V(x)\right] \\
&= -(a+2b)\mathbf{F}V(x) - (a+b)bV(x) \qquad \Rightarrow n_1^3 = 1\ c_{1,0}^3 = -(a+b)b \\
&\qquad\qquad\qquad\qquad\qquad\qquad\qquad\qquad c_{1,1}^3 = -(a+2b)
\end{aligned}$$

so we have

$$\{\mu,\gamma\}_{LA} \subseteq \{\mathbf{F},\lambda,D,V,\mathbf{F}V\}_{LA}\,.$$

Thus, $\dim\{\mu,\gamma\}_{LA} \leq \dim\{\mathbf{F},\lambda,D,V\}_{LA} \leq 5$.

Alternatively, we may note that

$$V(x) = \frac{\alpha\delta}{a}\left[e^{-(a+b)x} - e^{-bx}\right] = \frac{\alpha\delta}{a}e^{-(a+b)x} - \frac{\delta}{a}\underbrace{\alpha e^{-bx}}_{\lambda(x)}$$

thus, using Gaussian elimination we can substitute V by $\widetilde{V}(x) = e^{-(a+b)x}$. And since

$$\mathbf{F}\widetilde{V}(x) = -(a+b)e^{-(a+b)x} = -(a+b)\widetilde{V}(x) \quad\Rightarrow\quad \widetilde{n}_1^3 = 0 \quad \widetilde{c}_{1,0}^3 = -(a+b)$$

and we can compute the exact dimension of $\{\mathbf{F},\lambda,D,V,\mathbf{F}V\}_{LA}$,

$$\dim\{\mathbf{F},\lambda,D,V,\mathbf{F}V\}_{LA} = \dim\left\{\mathbf{F},\lambda,D,\widetilde{V}\right\}_{LA} = 4\,.$$

Obviously, in either way we are able to conclude that the forward prices admit a FDR.

Construction of FDR

We would now like to construct a FDR for forward prices, whenever we know that it exists. Since we are using the larger Lie-algebra $\bar{\mathcal{L}}$ and we cannot use case-specific Gaussian elimination in the general case, we are aiming to get *non-minimal* finite realizations.

As before we would like the derive a parameterization $\bar{G}$ of the invariant manifold $\bar{\mathcal{G}}$ and infer, from the functional form of that parameterization, the dynamics of the state variables.

Given the simplicity of the fields spanning $\bar{\mathcal{L}}$, it is straightforward to compute the operators:

$$e^{\mathbf{F}z_0}q_0 = q_0(x+z_0)\,,$$

$$e^{\mathbf{F}^k\lambda_j z_{j,k}^1}q = q + \mathbf{F}^k\lambda_j z_{j,k}^1\,, \quad e^{\mathbf{F}^k D_j z_{j,k}^2}q = q + \mathbf{F}^k D_j z_{j,k}^2\,, \qquad j\in A,C;\quad k=0,1,\cdots,n_j\,,$$

$$e^{\mathbf{F}^k V_j z_{j,k}^3}q = q + \mathbf{F}^k V_j z_{j,k}^3\,, \qquad j\in C;\quad k=0,1,\cdots,n_j\,.$$

Thus,

$$\bar{G}(z_0, z^1_{j,k}, z^2_{j,k}, z^3_{j,k}) = \prod_{j\in A,C} e^{\mathbf{F}^k \lambda_j(x) z^1_{j,k}} \cdot \prod_{j\in A,C} e^{\mathbf{F}^k D_j(x) z^2_{j,k}} \cdot \prod_{j\in C} e^{\mathbf{F}^k V_j(x) z^3_{j,k}}$$

$$= q_0(x+z_0) + \sum_{j\in A,C} \sum_{k=0}^{n^1_j} \mathbf{F}^k \lambda_j z^1_{j,k} + \sum_{j\in A,C} \sum_{k=0}^{n^2_j} \mathbf{F}^k D_j z^2_{j,k}$$

$$+ \sum_{j\in C} \sum_{k=0}^{n^3_j} \mathbf{F}^k V_j z^3_{j,k} \,. \tag{10.44}$$

Hence, in order to find the dynamics of the state variable Z such that $q = \bar{G}(Z)$, and as in the previous section, we take the Stratonovich Q-dynamics to be given by

$$\begin{cases} dZ_t = \bar{a}(Z_t)dt + \bar{b}(Z_t) \circ dW_t \,, \\ Z_0 = z_0 \,, \end{cases} \tag{10.45}$$

and we solve

$$\bar{G}_\star \bar{a} = \mu \,, \qquad \bar{G}_\star \bar{b} = \gamma \tag{10.46}$$

with μ and γ given in (10.37)-(10.40), and $\bar{G}$ from (10.44), to get a strong solution to the SDE (10.45).

The steps are then the usual ones, but with much more messy computations. Using the functional form of $\bar{G}$ in (10.44), it is possible to compute the Frechet derivatives. Then, from the expression for $q = \bar{G}(Z)$ and equations (10.43), we can find an concrete expression to the term $\mathbf{F}q$ in our μ (equation (10.40)). Identification of term in equations (10.46) allows us to determine the Stratonovich drift and diffusion terms. And finally, to obtain the Itô dynamics we calculate the Stratonovich correction term. Following these steps gives us the result in Proposition 5 (the actual computations of can be found in the appendix).

Proposition 5. *Given the initial forward price curve q_0, the forward prices system generated by γ_A, γ_C as in (10.37) and σ_C deterministic, has a finite dimensional realization given by*

$$q_t = \bar{G}(Z_t) \,,$$

where $\bar{G}$ is defined as in (10.44) and the dynamics of the state space variables Z are given by

$$\begin{cases} dZ_0 = dt \,, & \\ dZ^1_{j,0} = c^1_{j,0} Z^1_{j,n^1_j} dt + \varphi(\bar{G}(Z)) dW^j_t \,, & j \in A, C \,, \\ dZ^1_{j,k} = \left(Z^1_{j,k-1} + c^1_{j,k} Z^1_{j,n^1_j} \right) dt \,, & j \in A, C \;\; k = 1, \cdots, n^1_j \,, \end{cases}$$

$$\begin{cases} dZ^2_{j,0} = \left(c^2_{j,0} Z^2_{j,n^2_j} - \frac{1}{2}\varphi^2_j(\bar{G}(Z))\right) dt\,, & j \in A, C\,, \\ dZ^2_{j,k} = \left(Z^2_{j,k-1} + c^2_{j,k} Z^2_{j,n^2_j}\right) dt\,, & j \in A, C\ k = 1, \cdots, n^2_j\,, \\ dZ^3_{j,0} = \left(c^3_{j,0} Z^3_{j,n^3_j} + \varphi(\bar{G}(zZ))\right) dt\,, & j \in C\,, \\ dZ^3_{j,k} = \left(Z^3_{j,k-1} + c^3_{j,k} Z^3_{j,n^3_j}\right) dt\,, & j \in C\ k = 1, \cdots, n^3_j\,. \end{cases}$$

Example 4 (cont.) Recall that we studied an one-dimensional model where $A \cap B = \emptyset$, $C = \{1\}$ and

$$\gamma(q, x) = \gamma_C(q, x) = \underbrace{\alpha e^{-bx}}_{\lambda(x)} \underbrace{q}_{\varphi(q)}\,, \qquad \sigma_C(x) = \delta e^{-ax}.$$

As in the first part of the example, we will first directly apply the abstract results (in the construction part, that is Proposition 5). Then we derive a smaller realization that can be obtained from the case-specific simpler fields.

We start by directly applying Proposition 5. Recall from previous computations that we had $n^1_1 = 0$, $c^1_{1,0} = -b$, $n^2_1 = 0$, $c^2_{1,0} = -2b$, $n^3_1 = 1$, $c^3_{1,0} = -(a+b)b$ and $c^3_{1,1} = -(a+2b)$ and

$$\begin{aligned} V(x) &= \frac{\alpha\delta}{a}\left[e^{-(a+b)x} - e^{-bx}\right], \\ \mathbf{F}V(x) &= \frac{\alpha\delta}{a}\left[-(a+b)e^{-(a+b)x} - (-b)^k e^{-bx}\right]. \end{aligned}$$

Using this we get from (10.44) the parameterization of the realization to be

$$\begin{aligned} &\bar{G}(z_0, z^1_{1,0}, z^2_{1,0}, z^3_{1,0}, z^3_{1,1}) \\ &= q_0(x + z_0) + \lambda(x) z^1_{1,0} + D(x) z^2_{1,0} + V(x) z^3_{1,0} + \mathbf{F}V(x) z^3_{1,1} \\ &= q_0(x + z_0) + \alpha e^{-bx} z^1_{1,0} + \alpha^2 e^{-2bx} z^2_{1,0} + \frac{\alpha\delta}{a}\left[e^{-(a+b)x} - e^{-bx}\right] z^3_{1,0} + \\ &\quad + \frac{\alpha\delta}{a}\left[-(a+b)e^{-(a+b)x} - (-b)^k e^{-bx}\right] z^3_{1,1}\,. \end{aligned}$$

We now note that in our case $\varphi(q) = q$ and by Proposition 5 it follows that the realization is

$$\begin{cases} dZ_0 = dt\,, \\ dZ^1_{1,0} = -bZ^1_{1,0} dt + \bar{G}(Z) dW_t\,, \\ dZ^2_{1,0} = \left(-2bZ^2_{1,0} - \frac{1}{2}(\bar{G}(Z))^2\right) dt\,, \\ dZ^3_{1,0} = \left(-b(a+b)Z^3_{1,1} + \bar{G}(Z)\right) dt\,, \\ dZ^3_{1,1} = \left(Z^3_{1,0} - (a+2b)Z^3_{1,1}\right) dt\,. \end{cases}$$

Now recall that for this particular model we have $\bar{\mathcal{L}} = \left\{\mathbf{F}, \lambda, D, \widetilde{V}\right\}_{LA}$ with $\widetilde{V}(x) = e^{-(a+b)x}$. Thus another (smaller) parameterization, $\widetilde{G}$ is given by

$$\widetilde{G}(\widetilde{z}_0, \widetilde{z}^1_{1,0}, \widetilde{z}^2_{1,0}, \widetilde{z}^3_{1,0}) = q_0(x + \widetilde{z}_0) + \lambda(x)\widetilde{z}^1_{1,0} + D(x)\widetilde{z}^2_{1,0} + \widetilde{V}(x)\widetilde{z}^3_{1,0}$$
$$= q_0(x + \widetilde{z}_0) + \alpha e^{-bx}\widetilde{z}^1_{1,0} + \alpha^2 e^{-2bx}\widetilde{z}^2_{1,0} + e^{-(a+b)x}\widetilde{z}^3_{1,0} .$$

Proposition 5 cannot be used directly, but we can compare the two parameterizations $\bar{G}$ and $\widetilde{G}$ (of the same invariant manifold) above, to get

$$\widetilde{z}_0 = z_0 \,, \; \widetilde{z}^1_{1,0} = z^1_{1,0} - \frac{\delta}{a} z^3_{1,0} + b\frac{\delta}{a} z^3_{1,1} \,, \; \widetilde{z}^2_{1,0} = z^2_{1,0} \,, \; \widetilde{z}^3_{1,0} = \frac{\alpha\delta}{a} z^3_{1,0} - \frac{\alpha\delta}{a}(a+b) z^3_{1,1} .$$

Finally using Itô and simplifying we have

$$\begin{cases} d\widetilde{Z}_0 = dt \,, \\ d\widetilde{Z}^1_{1,0} = \left(-b\widetilde{Z}^1_{1,0} - \frac{\delta}{a}\widetilde{G}(Z)\right) dt + \widetilde{G}(Z) dW_t \,, \\ d\widetilde{Z}^2_{1,0} = \left(-2b\widetilde{Z}^2_{1,0} - \frac{1}{2}(\widetilde{G}(Z))^2\right) dt \,, \\ d\widetilde{Z}^3_{1,0} = \left(-(a+b)\widetilde{Z}^3_{1,0} + \frac{\alpha\delta}{a}\widetilde{G}(Z)\right) dt . \end{cases}$$

The two realizations are equivalent.

10.5.3 The General Case

In the sections 10.5.1 and 10.5.2, we analyzed existence and construction of FDR of Markovian forward prices, under the specific setting of *deterministic* and *deterministic direction* forward price volatility.

A natural question at this point is: *what is the most general functional form, for the volatilities γ and σ, consistent with FDR of Markovian forward prices?* The answer follows from previous results in [10] and from . In the following proposition we adapt it to the Markovian forward prices case. We state it in the form of a proposition.

Proposition 6. *Suppose Assumption 1 holds. There exist a FDR of Markovian forward prices if and only if*

$$\gamma_j(q, r, x) = \sum_{k=0}^{N^j_\gamma} \varphi^j_k(q)\lambda^j_k(x) \,, \qquad j \in A, C \,,$$

$$\sigma_j(r, x) = \begin{cases} \beta_j(x) & \text{if } \gamma_j(q, r, x) = \gamma_j(x) \,, \\ \delta_j(x) & \text{if } \gamma_j(q, r, x) = \gamma_j(q, x) \,, \end{cases} \qquad j \in C \,,$$

where β_j are unrestricted deterministic functions, δ_j,λ^j_k are QE deterministic functions, and φ^j_k are scalar vector fields in $\mathcal{H}_q$.

From Proposition 6, we see the most general situation can be attained by extending *deterministic direction* forward price volatilities to finite sums of deterministic direction parcels. This represents, of course, a relevant extension in terms of model flexibility, but not in terms of complexity of analysis.

The results from section 10.5.2 extend naturally to this most general case, the computations are exactly the same and, in concrete applications, easy to derive. In abstract terms, however, computation get much messier given the additional indices one must keep track of[12].

We now will consider the case when forward prices are **not** Markovian.

10.6 Non-Markovian Forward Prices

Recall that under Assumption 1 – our basic assumption on the volatility processes for forward prices and interest rates γ and σ, respectively – the (logarithm of the) forward price curve q cannot, in general, be studied without incorporating in the analysis the interest rate curve r (recall (10.12)-(10.13)).

In this section, we want to study the circumstances which were not covered by Section 10.5. In that case our forward price model is a doubly infinite system and we set

$$\widehat{q} = \begin{bmatrix} q \\ r \end{bmatrix} .$$

and $\widehat{q}$ belongs to $\mathcal{H}_q \times \mathcal{H}_r$.

The Itô dynamics of $\widehat{q}$ can, thus, also be written in block matrix notation as

$$d\begin{bmatrix} q_t \\ r_t \end{bmatrix} = \left\{ \mathbf{F} \begin{bmatrix} q_t \\ r_t \end{bmatrix} - \frac{1}{2} \begin{bmatrix} \|\gamma(q_t, r_t)\|^2 \\ 0 \end{bmatrix} - \begin{bmatrix} \gamma(q_t, r_t) \\ \sigma(r_t) \end{bmatrix} v^*(r_t) \right\} dt + \begin{bmatrix} \gamma(q_t, r_t) \\ \sigma(r_t) \end{bmatrix} dW_t ,$$

where, as before, we take W to be an m-dimensional Wiener process, and

$$\mathbf{F} = \frac{\partial}{\partial x} , \qquad v(r, x) = - \int_0^x \sigma(r, s) .$$

In this case the Stratonovich correction term is given by

$$-\frac{1}{2} d\langle \begin{bmatrix} \gamma(q_t, r_t) \\ \sigma(r_t) \end{bmatrix} , W_t \rangle = -\frac{1}{2} \sum_{i=1}^{m} d\langle \begin{bmatrix} \gamma_i(q_t, r_t) \\ \sigma_i(r_t) \end{bmatrix} , W_t^i \rangle .$$

Since we have (from the infinite Itô formula)

$$d\begin{bmatrix} \gamma_i(q_t, r_t) \\ \sigma_i(r_t) \end{bmatrix} = (\cdots)dt + \begin{bmatrix} \gamma'_{iq}(q, r) & \gamma'_{ir}(q, r) \\ 0 & \sigma'_{ir}(r) \end{bmatrix} \begin{bmatrix} \gamma_i(q_t, r_t) \\ \sigma_i(r_t) \end{bmatrix} dW_t ,$$

[12] We do not present the abstract results and derivations, as we believe the reader would spend more time understanding the notation, than extending the results of section 10.5.2 to concrete, slightly more general, applications.

with γ'_q, γ'_r and σ'_r the partial Frechet derivatives.

Then for $i = 1, \cdots, m$ we have

$$d\langle \begin{bmatrix} \gamma_i(q_t, r_t) \\ \sigma_i(r_t) \end{bmatrix}, W_t^i \rangle = \begin{bmatrix} \gamma'_{iq}(q_t, r_t) & \gamma'_{ir}(q_t, r_t) \\ 0 & \sigma'_{ir}(r_t) \end{bmatrix} \begin{bmatrix} \gamma_i(q_t, r_t) \\ \sigma_i(r_t) \end{bmatrix} dt$$

and the Stratonovich dynamics for $\widehat{q}$,

$$d\widehat{q}_t = \mu(q_t, r_t)dt + \begin{bmatrix} \gamma(q_t, r_t) \\ \sigma(r_t) \end{bmatrix} \circ dW_t ,$$

$$\mu(q, r) = \mathbf{F} \begin{bmatrix} q \\ r \end{bmatrix} - \frac{1}{2} \begin{bmatrix} \|\gamma(q, r)\|^2 \\ 0 \end{bmatrix} - \begin{bmatrix} \gamma(q, r) \\ \sigma(r) \end{bmatrix} v^*(r) - \\ - \frac{1}{2} \begin{bmatrix} \gamma'_q(q, r) & \gamma'_r(q, r) \\ 0 & \sigma'_r(r) \end{bmatrix} \begin{bmatrix} \gamma(q, r) \\ \sigma(r) \end{bmatrix} .$$

Given that

$$\gamma(q, r) = \begin{bmatrix} \gamma_A(q, r) \; 0 \; \gamma_C(q, r) \end{bmatrix} , \qquad \sigma(r) = \begin{bmatrix} 0 \; \sigma_B(r) \; \sigma_C(r) \end{bmatrix}$$

we note that

$$\begin{bmatrix} \gamma(q, r) \\ \sigma(r) \end{bmatrix} v(r)^* = \begin{bmatrix} \gamma_A(q, r) & 0 & \gamma_C(q, r) \\ 0 & \sigma_B(r) & \sigma_C(r) \end{bmatrix} \begin{bmatrix} 0 \\ v_B(r)^* \\ v_C(r)^* \end{bmatrix} \\ = \begin{bmatrix} \gamma_C(q, r) v_c^*(r) \\ \sigma_B(r) v_c^*(r) + \sigma_C(r) v_c^*(r) \end{bmatrix}$$

and

$$\begin{bmatrix} \gamma'_q(q, r) & \gamma'_r(q, r) \\ 0 & \sigma'_r(r) \end{bmatrix} \begin{bmatrix} \gamma(q, r) \\ \sigma(r) \end{bmatrix} \\ = \begin{bmatrix} \gamma'_{Aq}(q, r) & \gamma'_{Ar}(q, r) \\ 0 & 0 \end{bmatrix} \begin{bmatrix} \gamma_A(q, r) \\ 0 \end{bmatrix} + \begin{bmatrix} 0 & 0 \\ 0 & \sigma'_{Br} \end{bmatrix} \begin{bmatrix} 0 \\ \sigma_B(r) \end{bmatrix} + \\ + \begin{bmatrix} \gamma_C{}'_q(q, r) & \gamma_C{}'_r(q, r) \\ 0 & \sigma_C{}'_r(r) \end{bmatrix} \begin{bmatrix} \gamma_C(q, r) \\ \sigma_C(r) \end{bmatrix} \\ = \begin{bmatrix} \gamma_A{}'_q(q, r) \gamma_A(q, r) + \gamma_C{}'_q(q, r) \gamma_C(q, r) + \gamma_C{}'_r(q, r) \sigma_C(r) \\ \sigma_B{}'_r(r) \sigma_B(r) + \sigma_C{}'_r(r) \sigma_C(r) \end{bmatrix} ,$$

where

$$\gamma_{A}{}'_{q}(q,r)\gamma_{A}(q,r) = \sum_{i\in A} \gamma_{A}{}_{i}{}'_{q}(q,r)\gamma_{A}{}_{i}(q,r)\,,$$

$$\gamma_{C}{}'_{r}(q,r)\sigma_{C}(r) = \sum_{i\in C} \gamma_{C}{}_{i}{}'_{r}(q,r)\sigma_{C}{}_{i}(r)\,,$$

$$\gamma_{C}{}'_{q}(q,r)\gamma_{C}(q,r) = \sum_{i\in C} \gamma_{C}{}_{i}{}'_{q}(q,r)\gamma_{C}{}_{i}(q,r)\,,$$

$$\sigma_{B}{}'_{r}(r)\sigma_{B}(r) = \sum_{i\in B} \sigma_{B}{}_{i}{}'_{r}(r)\sigma_{B}{}_{i}(r)\,,$$

$$\sigma_{C}{}'_{r}(r)\sigma_{C}(r) = \sum_{i\in C} \sigma_{C}{}_{i}{}'_{r}(r)\sigma_{C}{}_{i}(r)\,.$$

We can, finally, identify our main object of study as the following doubly infinite Stratonovich SDE

$$d\widehat{q_t} = \mu(q_t, r_t)dt + \begin{bmatrix} \gamma(q_t, r_t) \\ \sigma(r_t) \end{bmatrix} \circ dW_t\,,$$

where

$$\begin{aligned}\mu(q,r) = \mathbf{F}\begin{bmatrix} q \\ r \end{bmatrix} - \frac{1}{2}\begin{bmatrix} \|\gamma_A(q,r)\|^2 + \|\gamma_C(q,r)\|^2 \\ 0 \end{bmatrix} - \\ - \begin{bmatrix} \gamma_C(q,r)v_C^*(r) \\ \sigma_B(r)v_B^*(r) + \sigma_B(r)v_B^*(r) \end{bmatrix} - \\ -\frac{1}{2}\begin{bmatrix} \gamma_{A}{}'_{q}(q,r)\gamma_A(q,r) + \gamma_{C}{}'_{q}(q,r)\gamma_C(q,r) + \gamma_{C}{}'_{r}(q,r)\sigma_C(r) \\ \sigma_{B}{}'_{r}(r)\sigma_B(r) + \sigma_{C}{}'_{r}(r)\sigma_C(r) \end{bmatrix}\end{aligned} \tag{10.47}$$

$$\begin{bmatrix} \gamma(q,r) \\ \sigma(r) \end{bmatrix} = \begin{bmatrix} \gamma_A(q,r) & 0 & \gamma_C(q,r) \\ 0 & \sigma_B(r) & \sigma_C(r) \end{bmatrix}.$$

Given the general functional forms of γ_A, γ_C, σ_B and σ_C the study of all possible special cases[13] would be exhausting.

In this section, we take, therefore a more agressive strategy: we consider immediately the scenario where each element in γ and σ have **deterministic direction volatility**.

As before, the situation of deterministic direction volatilities can be extended to the case where each element of γ and σ is a finite sum of deterministic direction parcels, and that is the most general possible scenario consistent with existence of FDR[14]. We omit the analysis of this most general scenario because the results can be easily derived from the ones on deterministic direction volatilities, and, in abstract terms, notation becomes almost untractable.

[13] One particular special case would be to take, say, γ_A an σ_B to have deterministic direction and γ_C, σ_C to be deterministic.

[14] For details on why this is the most general scenario we refer to [10].

The deterministic direction setting we will work with is formally stated by the next assumption.

Assumption 5. *The mappings* $\gamma_i : \mathcal{H}_q \times \mathcal{H}_r \times \mathbb{R}_+ \to \mathbb{R}^m$ *and* $\sigma_i : \times \mathcal{H}_r \times \mathbb{R}_+ \to \mathbb{R}^m$ *are of the following functional form*

$$\begin{aligned} \gamma_i(x,q,r) &= \lambda_i(x)\varphi_i(q,r)\,, && i \in A, C\,, \\ \sigma_i(x,r) &= \beta_i(x)\phi_i(r)\,, && i \in B, C\,, \end{aligned} \tag{10.48}$$

where λ_i, β_i *are deterministic functions of* x *and* φ_i, ϕ_i *are scalar vector fields in* $\mathcal{H}_q \times \mathcal{H}_r$ *(i.e., they do not depend on* x *and depend only on the current forward price and interest rate curves).*

We note that under Assumption 5

$$v_i(x,r) = -\int_0^x \sigma_i(r,s)ds = -\int_0^x \beta_i(s)\phi_i(r)ds = -\phi_i(r)\int_0^x \beta_i(s)ds\,;\ i \in B, C.$$

Defining

$$B_i(x) = \int_0^x \beta_i(s)ds \tag{10.49}$$

we, thus, have

$$v_i(x) = -\phi_i(r)B_i(x)\,, \qquad i \in B, C\,.$$

To check if our forward prices model admits a finite dimensional realization, we need to see if

$$\dim \left\{ \mu, \begin{bmatrix} \gamma(q,r) \\ \sigma(r) \end{bmatrix} \right\}_{LA} < \infty\,.$$

Considering (10.47), under Assumption 5, our basic vector fields can be written as

$$\begin{aligned} \mu(q,r) = \mathbf{F}\begin{bmatrix} q \\ r \end{bmatrix} &- \frac{1}{2}\sum_{i\in A,C} \varphi_i^2(q,r)\begin{bmatrix} \lambda_i^2 \\ 0 \end{bmatrix} + \sum_{i \in C}\varphi_i(q,r)\phi_i(r)\begin{bmatrix} \lambda_i B_i \\ 0 \end{bmatrix} + \\ &+ \sum_{i\in B,C}\phi_i^2(r)\begin{bmatrix} 0 \\ \beta_i B_i \end{bmatrix} - \frac{1}{2}\left\{ \sum_{i\in A,C}\varphi'_{iq}(q,r)\left[\lambda_i\right]\varphi_i(q,r)\begin{bmatrix} \lambda_i \\ 0 \end{bmatrix} + \right. \\ &\left. + \sum_{i\in C}\varphi'_{ir}(q,r)\left[\beta_i\right]\phi_i(r)\begin{bmatrix} \lambda_i \\ 0 \end{bmatrix} + \sum_{i \in B,C}\phi'_{ir}(r)\left[\beta_i\right]\phi_i(r)\begin{bmatrix} 0 \\ \beta_i \end{bmatrix} \right\}, \end{aligned}$$

$$\begin{bmatrix} \gamma_i(q,r) \\ \sigma_i(r) \end{bmatrix} = \begin{cases} \varphi_i(q,r)\begin{bmatrix} \lambda_i \\ 0 \end{bmatrix}, & i \in A\,, \\ \phi_i(r)\begin{bmatrix} 0 \\ \beta_i \end{bmatrix}, & i \in B\,, \\ \varphi_i(q,r)\begin{bmatrix} \lambda_i \\ 0 \end{bmatrix} + \phi_i(r)\begin{bmatrix} 0 \\ \beta_i \end{bmatrix}, & i \in C\,. \end{cases}$$

Following the strategy described in Section 10.3.3, we would now start computing Lie brackets of all possible combinations of these fields and, through Gaussian elimination, hopefully, get to a simple set of generators of our Lie-algebra $\mathcal{L} = \{\mu, \delta\}_{LA}$. Based on properties of these generators we would, also hopefully, be able to understand which γ and σ would guarantee a FDR for forward prices.

The particular complex expression for μ above, and the almost impossibility of Gaussian elimination that results from having to handle two infinite SDE at the same time[15], leads to the conclusion that our best hope is again to study a larger Lie-algebra, $\bar{\mathcal{L}}$, and to choose such a Lie-algebra so that the basic fields would be simple.

The following Lemma give us the desired (simple enough) Lie-algebra $\bar{\mathcal{L}}$.

Lemma 6. *Consider the following set of fields in $\mathcal{H}$.*

$$\begin{aligned}
f^0 &= \begin{bmatrix} \mathbf{F}q \\ \mathbf{F}r \end{bmatrix}, \\
f_i^1 &= \begin{bmatrix} \lambda_i \\ 0 \end{bmatrix}, \quad f_i^2 = \begin{bmatrix} D_i \\ 0 \end{bmatrix}, \quad i \in A, C, \\
f_i^3 &= \begin{bmatrix} V_i \\ 0 \end{bmatrix}, \qquad\qquad\qquad i \in C, \\
f_i^4 &= \begin{bmatrix} 0 \\ \beta_i \end{bmatrix}, \quad f_i^5 = \begin{bmatrix} 0 \\ H_i \end{bmatrix}, \quad i \in B, C,
\end{aligned} \tag{10.50}$$

where λ_i, β_i and B_i are deterministic functions of x and defined as in (10.49) and (10.49) and we further define

$$D_i(x) = \lambda_i^2(x), \qquad V_i(x) = \lambda_i(x)B_i(x), \qquad H_i(x) = \beta_i(x)B_i(x). \tag{10.51}$$

Then the following holds

$$\begin{aligned}
\mathcal{L} &= \{\mu, \delta\}_{LA} \subseteq \bar{\mathcal{L}} \\
&= \left\{ \mathbf{F}, \begin{bmatrix} \lambda_i \\ 0 \end{bmatrix}, \begin{bmatrix} D_i \\ 0 \end{bmatrix}, \begin{bmatrix} V_j \\ 0 \end{bmatrix}, \begin{bmatrix} 0 \\ \beta_n \end{bmatrix}, \begin{bmatrix} 0 \\ H_n \end{bmatrix}; \ i \in A, C, \ j \in C, \ n \in B, C \right\}_{LA}.
\end{aligned}$$

Proof: First, note that

$$\begin{aligned}
\varphi_i(q,r), \quad \varphi_i^2(q,r), \quad \varphi_{iq}'(q,r)\,[\lambda_i]\,\varphi_i(q,r), \quad i \in A, B, \\
\phi_i(r), \quad \phi_i^2(r), \quad \phi_{ir}'(r)\,[\beta_i]\,\phi_i(r), \quad i \in B, C, \\
\varphi_i(q,r)\phi_i(r), \quad i \in C,
\end{aligned}$$

[15] To the usual complexity of dealing with multidimensional cases, there is an additional complexity specific of forward price models that results from the fact that $\varphi_i(q,r) \neq \phi_i(r)$. However even under the unrealistic assumption (since the forward price volatility could not depend on the forward prices) where we would assume $\varphi_i(r) = \phi_i(r)$, the complexity of the Stratonovich correction term would not allow us to obtain simple generators for $\mathcal{L}$.

are scalar fields in $\mathcal{H}_q \times \mathcal{H}_r$. The conclusion now follows using Gaussian Elimination (Lemma 3). □

10.6.1 Existence of FDR

Computing Lie-brackets on the basic fields of $\bar{\mathcal{L}}$ is not hard, and the conclusion on the existence of a FDR becomes a straightforward generalization of the easier setups studied in previous sections. Proposition 7 give us the needed conditions.

Proposition 7. *The lie-algebra $\bar{\mathcal{L}}$ is spanned by*

$$\operatorname{span}\left\{ \mathbf{F}\ ,\ \mathbf{F}^k \begin{bmatrix} \lambda_i \\ 0 \end{bmatrix}, \mathbf{F}^k \begin{bmatrix} D_i \\ 0 \end{bmatrix}, \mathbf{F}^k \begin{bmatrix} V_j \\ 0 \end{bmatrix}, \mathbf{F}^k \begin{bmatrix} 0 \\ \beta_n \end{bmatrix}, \mathbf{F}^k \begin{bmatrix} 0 \\ H_n \end{bmatrix}, \right.$$
$$\left. i \in A, C\,, \quad j \in, \quad n \in B, C\,, \quad k = 0, 1, \cdots \right\}$$

and will have a finite dimension if and only if each component of λ and β is QE.

Moreover, under those conditions also each component of D, V, W in (10.51) are QE and

$$\dim\{\bar{\mathcal{L}}\} \leq 1 + \sum_{j \in A,C} (n_j^1 + n_j^2) + \sum_{j \in C} n_j^3 + \sum_{j \in B,C} (n_j^4 + n_j^5)\ , \tag{10.52}$$

for $n_j^i \in \mathbb{N}$, such that,

$$\begin{aligned}
&\mathbf{F}^{n_j^1+1}\lambda_j = \textstyle\sum_{k=1}^{n_j^1} c_{j,k}^1 \mathbf{F}^k \lambda_j\,, \quad \mathbf{F}^{n_j^2+1} D_j = \sum_{k=1}^{n_j^2} c_{j,k}^2 \mathbf{F}^k D_j\,, \quad j \in A, C\,,\\
&\mathbf{F}^{n_j^3+1} V_j = \textstyle\sum_{k=1}^{n_j^3} c_{j,k}^3 \mathbf{F}^k V_j\,, \qquad\qquad\qquad\qquad\qquad\qquad j \in C\,,\\
&\mathbf{F}^{n_j^4+1}\beta_j = \textstyle\sum_{k=1}^{n_j^4} c_{j,k}^4 \mathbf{F}^k \beta_j\,, \quad \mathbf{F}^{n_j^5+1} H_j = \sum_{k=1}^{n_j^5} c_{j,k}^5 \mathbf{F}^k H_j\,, \quad j \in B, C\,,
\end{aligned} \tag{10.53}$$

hold for some $c_{j,k}^I \in \mathbb{R}$ with $k = 0, 1, \cdots, n_j^I$, $j = 1, \cdots, m$, and $I = 1, \cdots, 5$.

Proof:Given the fields of $\bar{\mathcal{L}}$, we have a FDR if and only if (10.53) hold and it follows that (10.52) holds (the "$\leq$" in (10.52) accounts for possible case-specific Gaussian elimination across terms). Finally, (10.53) can be interpreted as ODEs whose solution are QE functions, thus λ_i, for $i \in A, C$; D_j, V_j, for $j \in C$ and H_k for $k \in B, C$ solve (10.53) if and only if they are QE. It remains to show that requiring λ_i and β_j for $i \in A, C$, $j \in B, C$ is sufficient to guarantee that. Given that

$$D_i(x) = \lambda_i^2(x)\,, \quad V_j(x) = \lambda_j(x) \int_0^b \beta_j(s) ds\,, \quad H_j(x) = \beta_j(x) \int_0^b \beta_j(s) ds\,,$$

the result follows from Lemma 4. □

10.6.2 Construction of FDR

Knowing the conditions for existence of a FDR for forward prices, we can now construct the finite dimensional realization. Proposition 8 gives us a *non-minimal* (since it is based on $\bar{\mathcal{L}}$ and cannot take into account case-specific Gaussian elimination) parameterization $\widehat{G}$ of the invariant manifold $\widehat{\mathcal{G}}$.

Note that our parameterization, $\widehat{q} = \widehat{G}(Z)$, will be of the following block matrix form

$$\begin{bmatrix} q \\ r \end{bmatrix} = \begin{bmatrix} \widehat{G}^q(Z) \\ \widehat{G}^r(Z) \end{bmatrix} .$$

Furthermore, by close inspection of (10.50) we realize that the operator generated by f^0,

$$e^{\mathbf{F} z_0} \begin{bmatrix} q_0 \\ r_0 \end{bmatrix} = \begin{bmatrix} \widetilde{q_0} \\ \widetilde{r_0} \end{bmatrix} ,$$

is the only that will affect both $\widehat{G}^q$ and $\widehat{G}^r$. The remaining operators will only affect one component at the time.

The operators generated by f_i^1, f_i^2 for $i \in A, C$ and f_j^3 for $j \in C$, will only affect $\widehat{G}^q$. So,

$$\begin{aligned} e^{\mathbf{F}^k \lambda_j z_{j,k}^1} q &= q + \lambda_j z_{j,k}^1 , & e^{\mathbf{F}^k D_j z_{j,k}^2} q &= q + D_j^1 z_{j,k}^2 , \; j \in A, C , \\ e^{\mathbf{F}^k V_j z_{j,k}^3} q , &= q + V_j z_{j,k}^3 , & & \qquad j \in C . \end{aligned}$$

On the other hand, $\widehat{G}^r$ will be affected by the operators generated by f_j^4, f_j^5 for $j \in B, C$ and we have

$$e^{\mathbf{F}^k \beta_j z_{j,k}^4} r = r + \beta_j z_{j,k}^4 , \qquad e^{\mathbf{F}^k H_j z_{j,k}^5} r = r + H_j z_{j,k}^5 , \; j \in B, C .$$

Once the parameterization has been derived, we can infer the dynamics of the finite dimensional realization, exactly as before. The actual construction of the realization, though cumbersome (and thus presented in the appendix), follow the same ideas of the constructions in previous sections.

Proposition 8. *Suppose Assumption 5 holds. Given the initial forward price curve* q_0 *and the initial interest rate curve* r_0*, the system generated by forward price and interest rate volatilities defined as in (10.49) has a finite dimensional realization, given by*

$$\begin{bmatrix} q_t \\ r_t \end{bmatrix} = \widehat{G}(Z) ,$$

where $\widehat{G}$ *is defined by*

$$\widehat{G}(z_0, z^1_{j,k}, z^2_{j,k}, z^3_{j,k}, z^4_{j,k}, z^5_{j,k}) \tag{10.54}$$
$$= \begin{bmatrix} \widetilde{q}_0 + \sum_{j\in A,C}\sum_{k=0}^{n^1_j} \mathbf{F}^k \lambda_j z^1_{j,k} + \sum_{j\in A,C}\sum_{k=0}^{n^2_j} \mathbf{F}^k D_j z^2_{j,k} + \sum_{j\in C}\sum_{k=0}^{n^3_j} \mathbf{F}^k V_j z^3_{j,k} \\ \widetilde{r}_0 + \sum_{j\in B,C}\sum_{k=0}^{n^4_j} \mathbf{F}^k \beta_j z^4_{j,k} + \sum_{j\in B,C}\sum_{k=0}^{n^5_j} \mathbf{F}^k H_j z^5_{j,k} \end{bmatrix},$$

for $\widetilde{q}_0(x) = q_0(x + z_0)$ *and* $\widetilde{q}_0(x) = q_0(x + z_0)$.

Moreover, the dynamics of the state space variables Z *are given by*

$$\begin{cases} dZ_0 = dt & \\ dZ^1_{j,0} = c^1_{j,0} Z^1_{j,n^1_j} dt + \varphi_j(\widehat{G}^q(Z), \widehat{G}^r(Z)) dW^j_t & j \in A, C \\ dZ^1_{j,k} = \left\{ Z^1_{j,k-1} + c^1_{j,k} Z^1_{j,n^1_j} \right\} dt & j \in A, C \;\; k = 1, \cdots, n^1_j \\ dZ^2_{j,0} = \left\{ c^2_{j,0} Z^2 j, n^2_j - \frac{1}{2}\varphi^2_j(\widehat{G}^q(Z), \widehat{G}^r(Z)) \right\} dt & j \in A, C \\ dZ^2_{j,k} = \left\{ Z^2_{j,k-1} + c^2_{j,k} Z^2_{jn^2_j} \right\} dt & j \in A, C \;\; k = 1, \cdots, n^2_j \\ dZ^3_{j,0} = \left\{ c^3_{j,0} z^3_{j,n^3_j} + \varphi(\widehat{G}^q(Z), \widehat{G}^r(Z)) \phi(\widehat{G}^r(Z)) \right\} dt & j \in C \\ dZ^3_{j,k} = \left\{ Z^3_{j,k-1} + c^3_{j,k} Z^3_{j,n^3_j} \right\} dt & j \in C \;\; k = 1, \cdots, n^3_j \\ dZ^4_{j,0} = c^4_{j,0} Z^4_{j,n^4_j} dt + \phi_j((\widehat{G}^r(Z)) dW^j & j \in B, C \\ dZ^4_{j,k} = \left\{ c^4_{j,0} Z^4_{j,n^4_j} \right\} dt & j \in B, C \;\; k = 1, \cdots, n^4_j \\ dZ^5_{j,0} = \left\{ c^5_{j,0} Z^5_{j,n^5_j} + \phi^2_j(\widehat{G}^r(Z)) \right\} dt & j \in B, C \\ dZ^5_{j,k} = \left\{ Z^5_{j,k-1} + c^5_{j,k} Z^5_{j,n^5_j} \right\} dt & j \in B, C \;\; k = 1, \cdots, n^5_j \end{cases}$$

The next example may help to understand Proposition 8.

10.6.3 Example

Consider a model with the following volatility matrix

$$\begin{bmatrix} \gamma(q,r) \\ \sigma(r) \end{bmatrix} = \begin{bmatrix} \alpha e^{-bx} q & \rho \\ 0 & \delta e^{-ax}\sqrt{r} \end{bmatrix}.$$

Using the usual notation we have two Wiener processes, one of them of type W^A and another of type W^C. So $A = \{1\}$ and $C = \{2\}$. In this example, since the is only one element of each type, we write for the reader's convinience "$\cdot_A$" instead of "$\cdot_1$" and "$\cdot_C$" instead of "$\cdot_2$" all over.

We have

$$
\begin{aligned}
&\gamma_A(q,r,x) = \lambda_A(x)\varphi_A(q,r) \Rightarrow \lambda_A(x) = \alpha e^{-bx} && \varphi_A(q,r) = q\\
&\gamma_C(q,r,x) = \lambda_C(x)\varphi_C(q,r) \Rightarrow \lambda_C(x) = \rho && \varphi_C(q,r) = 0\\
&\sigma_C(r,x) = \beta_C(x)\phi_C(x) \quad\;\; \Rightarrow \beta_C(x) = \delta e^{-ax} && \phi_C(q) = \sqrt{r}\\
&\qquad\qquad\qquad\qquad\qquad\quad\;\; \Rightarrow v_C(r,x) = -B_C(x)\phi(r) && B_C(x) = -\tfrac{1}{a}\left[e^{-ax}-1\right].
\end{aligned}
$$

Moreover, we have

$$
\begin{aligned}
D_A(x) &= \lambda_A^2(x) = \alpha^2 e^{-2bx}\\
V_C(x) &= \lambda_C(x)B_C(x) = -\frac{\rho}{a}\left[e^{-ax}-1\right]\\
H_C(x) &= \beta_C(x)B_C(x) = -\frac{\delta}{a}\left[e^{-2ax}-e^{-ax}\right].
\end{aligned}
$$

Taking all this into account we easily get

$$
\begin{aligned}
&\mathbf{F}\lambda_A = -b\lambda_A && \Rightarrow n_A^1 = 0 \;\; c_{A,0}^1 = -b\\
&\mathbf{F}\lambda_C = 0 && \Rightarrow n_C^1 = 0 \;\; c_{C,0}^1 = 0\\
&\mathbf{F}D_A = -2bD_A && \Rightarrow n_A^2 = 0 \;\; c_{A,0}^2 = -2b\\
&\mathbf{F}D_C = 0 && \Rightarrow n_C^2 = 0 \;\; c_{C,0}^2 = 0\\
&\mathbf{F}V_C = \rho e^{-ax}, \qquad \mathbf{F}^2 V_C = -a\mathbf{F}V_C && \Rightarrow n_C^3 = 1 \;\; c_{C,0}^3 = 0\\
& && \qquad\qquad\quad\; c_{C,1}^3 = -a\\
&\mathbf{F}\beta_C = -a\beta_C && \Rightarrow n_C^4 = 0 \;\; c_{C,0}^4 = -a\\
&\mathbf{F}H_C = -aH_C + \delta e^{-2ax}, \; \mathbf{F}^2 H_C = -3a\mathbf{F}H_C - 2a^2 H_C && \Rightarrow n_C^5 = 2 \;\; c_{C,0}^5 = -2a^2\\
& && \qquad\qquad\quad\; c_{C,1}^5 = -3a.
\end{aligned}
$$

Given this computations, we see the following fields span $\bar{\mathcal{L}}$

$$
\left\{\mathbf{F}, \begin{bmatrix}\lambda_A\\0\end{bmatrix}, \begin{bmatrix}\lambda_C\\0\end{bmatrix}, \begin{bmatrix}D_A\\0\end{bmatrix}, \begin{bmatrix}D_C\\0\end{bmatrix}, \begin{bmatrix}V_C\\0\end{bmatrix}, \begin{bmatrix}\mathbf{F}V_C\\0\end{bmatrix}, \begin{bmatrix}0\\\beta_C\end{bmatrix}, \begin{bmatrix}0\\H_C\end{bmatrix}, \begin{bmatrix}0\\\mathbf{F}H_C\end{bmatrix}\right\}, \tag{10.55}
$$

thus we know that our forward price model admits a FDR since

$$
\dim(\bar{\mathcal{L}}) \le 10 .
$$

From (10.54) we get the parameterization

$$
\begin{bmatrix}
\widehat{G}^q(x, z_0, z_{A,0}^1, z_{C,0}^1, z_{A,0}^2, z_{C,0}^2, z_{C,0}^3, z_{C,1}^3)\\[1em]
\widehat{G}^r(x, z_0, z_{C,0}^4, z_{C,0}^5, z_{C,1}^5)
\end{bmatrix}
$$

$$
= \begin{bmatrix}
q_0(x+z_0) + \alpha e^{-bx} z_{A,0}^1 + \rho z_{C,0}^1 + \alpha^2 e^{-2bx} z_{A,0}^2 + \rho^2 z_{C,0}^2 - \\
\qquad - \frac{\rho}{a}\left[e^{-ax}-1\right] z_{C,0}^3 + \rho e^{-ax} z_{C,1}^3\\
r_0(x+z_0) + \delta e^{-ax} z_{C,0}^4 - \frac{\delta}{a}\left[e^{-2ax}-e^{-ax}\right] z_{C,0}^5 - \\
\qquad - \frac{\delta}{a}\left[-2ae^{-2ax}+ae^{-ax}\right] z_{C,1}^5
\end{bmatrix}, \tag{10.56}
$$

and from Proposition 8 we get the Z dynamics

$$\begin{cases} dZ_0 &= dt\,, \\ dZ^1_{A,0} = -bZ^1_{A,0}dt + \widehat{G}^q(Z)dW^j_t\,, & dZ^1_{C,0} = \widehat{G}^q(Z)dW^j_t\,, \\ dZ^2_{A,0} = \left(-2bZ^2_{A,1} - \frac{1}{2}\left(\widehat{G}^q(Z)\right)^2\right)dt\,, & dZ^2_{C,0} = -\frac{1}{2}\left(\widehat{G}^q(Z)\right)^2 dt\,, \\ dZ^3_{C,0} = \widehat{G}^q(z)\sqrt{\widehat{G}^r(Z))}dt\,, & dZ^3_{C,1} = \left(Z^3_{C,0} - aZ^3_{C,1}\right)dt\,, \\ dZ^4_{C,0} = -aZ^4_{C,0}dt + \sqrt{\widehat{G}^r(Z)}dW^j\,, & \\ dZ^5_{C,0} = \left(-2a^2 Z^5_{C,1} + \widehat{G}^r(Z)\right)dt\,, & dZ^5_{C,1} = \left(Z^5_{C,0} - 3aZ^5_{C,1}\right)dt\,, \end{cases} \tag{10.57}$$

where $\widehat{G}^q$ and $\widehat{G}^r$ are as in (10.57).

It is obvious however, from both (10.57) and (10.57), that this realization is unnecessarily larger. Using the following change of variables we can find a realization of dimension 7 (which is the minimal possible with the fields in (10.55)),

$$\begin{aligned} \widetilde{Z}_0 &= Z_0\,, \\ \widetilde{Z}^1_{A,0} &= Z^1_{A,0}\,, \\ \widetilde{Z}^2_{A,0} &= Z^2_{A,0}\,, \\ \widetilde{Z}^3_{A,0} &= -\frac{\rho}{a}Z^3_{C,0} + \rho Z^3_{C,1}\,, \\ \widetilde{Z}^1_{C,0} &= \rho Z^1_{C,0} + \rho^2 Z^2_{C,0} + \frac{\rho}{a}Z^3_{C,0}\,, \\ \widetilde{Z}^4_{C,0} &= \delta Z^4_{C,0} + \frac{\delta}{a}Z^5_{C,0} - \delta Z^5_{C,1}\,, \\ \widetilde{Z}^5_{C,0} &= -\frac{\delta}{a}Z^5_{C,0} + 2\delta Z^5_{C,1}\,. \end{aligned}$$

We can then use Itô and (10.57) to derive the dynamics of the new state variables.

10.6.4 FDR of Forward Prices Versus FDR of Interest Rates

We finish this study giving a complete answer to our *problem four* – on whether there it is possible to have forward price model which allows for a FDR for forward prices but not for interest rates[16].

We recall a forward price term structure model consists of the following two infinite SDEs.

[16] Recall the partial answer given in Remark 3.

$$dq_t = \left\{ \mathbf{F} q_t - \frac{1}{2} \|\gamma(q_t, r_t)\|^2 - \gamma(q_t, r_t) v^*(r_t) \right\} dt + \gamma(q_t, r_t) dW_t \,, \quad (10.58)$$

$$dr_t = \{ \mathbf{F} r_t - \sigma(r_t) v^*(r_t) \} dt + \sigma(r_t) dW_t \,. \quad (10.59)$$

Proposition 9. *In forward price term structure models, inexistence of a FDR for the interest rate equation (10.58) and existence of a FDR for the forward price equation (10.59), is possible only if forward prices are Markovian* **and** *the conditions of Proposition 6 hold i.e.,*

$$\begin{aligned} \gamma_j(q, r, x) &= \sum_{k=0}^{N_\gamma^j} \varphi_k^j(q) \lambda_k^j(x) \,, & j \in A, C \,, \\ \sigma_j(r, x) &= \beta_j(x) \,, & j \in B \,, \\ \sigma_j(r, x) &= \begin{cases} \beta_j(x) & \text{if } \gamma_j(q, r, x) = \gamma_j(x) \,, \\ \sum_{k=0}^{N_\sigma^j} \omega_k^j \delta_k^j(x) & \text{if } \gamma_j(q, r, x) = \gamma_j(q, x) \,, \end{cases} & j \in C \,, \end{aligned} \quad (10.60)$$

where ω_k^j are deterministic constants, β_j are unrestricted deterministic functions, δ_j^k,λ_k^j are QE deterministic functions, and φ_k^j are scalar vector fields in $\mathcal{H}_q$.

Proof:If forward prices are *not* Markovian, than either γ or γv^* depend on r. If that is the case, we know from Proposition 8 that the dynamics of the state variables of type Z^1, Z^2, Z^3 (the ones showing up directly on the parameterization $\widehat{G}^q$) depend on $\widehat{G}^r$. That is, the forward price realization depend on the interest rate realization, indirectly, through the dynamics of the factors showing up in $\widehat{G}^q$. Thus, if $r = \widehat{G}^r(Z)$ only holds for infinite Z, the forward prices will also be a function on an infinite state variable and, by definition, do not admit a FDR.

If forward prices are Markovian, we know, from Proposition 6, that forward prices admit a FDR if and only if (10.60) hold. On the other hand, we see (10.60) imposes no restriction on σ_B, so we can choose σ_B *not to be* a weighted finite sum of quasi-exponential functions, weighted by scalar fields in $\mathcal{H}_r$, making existence of FDR for interest rates impossible (for further details on this result from the previous literature see [10]). □

10.7 Conclusions and Applicability

Forward prices are only interesting objects of study in settings where the forward measures Q^T differ from the risk-neutral measure Q. In these settings, the study of forward prices depends on zero-coupon bond price volatilities.

Using the Lie-algebraic approach of Björk *et al.*, we have shown that forward prices term structure models consist of a system of two infinite dimensional SDEs, one describing the dynamics of the forward prices themselves and another characterizing the interest rate setting and where the interest rate equation is an input to the forward price equation.

Despite the apparent non-Markovian nature of forward prices, we were able to show that there exist models for which forward prices are, actually, Markovian and identified necessary and sufficient conditions for this Markovian property to hold. Studying Markovian forward prices we concluded that existence of finite dimensional realization (FDR) for Markovian forward prices is, in some sense, independent of existence of FDR for interest rates.

We studied existence and construction of FDR for Markovian forward prices and derived general conditions for existence of FDR. We considered with special detail the pure deterministic and the deterministic direction volatility special cases. ¿From this analysis, we concluded that some results from previous literature can be extended to the forward price term structure case, but also that forward price term structure models are particularly complex.

The dynamics of forward prices has a specially complex drift under the risk-neutral measure Q. A direct consequence of this complexity is that, as soon as we leave the pure deterministic volatility setting, the best we can hope for is to study *non-minimal* Lie-algebras and to find *non-minimal* FDR. Existence of non-minimal realizations is, of course, sufficient to prove existence of FDR, but is in general not necessary. We showed, however, that given the specificity of the forward price equation drift, and for a specific enlargement of the Lie-algebra, existence of FDR for non-minimal realizations is also necessary for existence of FDR, at least for the one-dimensional case. Then we conjectured that this hold for the higher dimension case.

Even if *non-minimal* Lie-algebras are, in the above sense, satisfactory for existence results, they are not as satisfactory for construction results, since we are bound to find realizations with too many variables. Despite this fact, we exemplified how, given a concrete application, we can use the abstract results to obtain a smaller realization (sometimes the minimal one) simply by using a smart change of variables and Itô's lemma.

For non-Markovian forward prices, we showed that whenever there exist FDR for the forward price equation, the dynamics of the state variable depend on the interest rates. Consequently, term structures of forward prices will always (indirectly) depend on interest rates, and existence of a FDR for the interest rate equation is *necessary* for existence of a FDR for the forward price equation. In order to study non-Markovian forward price term structure models, we handle a system of two infinite dimensional SDEs, thus, computations get quite cumbersome. Still, most results are the expected ones, given the previous literature on FDR of interest rates and the study of the forward price equation in the easier Markovian setting.

In terms of the applicability of the results presented here, it is, first of all, important to stress that the characterization of the conditions that guarantee existence of a FDR for forward price term structure models is crucial, in distinguishing the "good" forward price models from the "bad" forward price models. After all, a term structure model that do not allow for a FDR realization cannot be useful for any practical application. For instance, it is impossible to estimate the dynamics of an infinite state variable.

In addition to this *selection applicability*, perhaps, the most important application results from the actual local parameterization of the term structure. This parameterization can help in understanding what are the needed conditions, on the driving volatility vector fields, that will produce term structures consistent with case-specific realities, helping to design good models. In the present study this *design applicability* of the Lie-algebraic approach was left untouched because it is case dependent, and we have focused on general results.

Finally, let alone forward prices, the results derived here are applicable to study term structures of any Q^T-martingale. For example, swap rates and credit spreads are financial instruments with strong connections to Q^T-martingales.

Appendix: Technical Details and Proofs

Conjecture 1 **Proof**:[sketch] The implication $\Leftarrow$ follows immediately from $\mathcal{L} \subseteq \bar{\mathcal{L}}$.

The implication $\Rightarrow$ is much harder to prove. Here, as an illustration, we consider the one-dimensional Wiener process case. We will show the equivalent result that if $\dim(\bar{\mathcal{L}}) = \infty$ then we must also have $\dim(\mathcal{L}) = \infty$.

In this case

$$\begin{aligned} \mathcal{L} &= \{\mu, \gamma\}_{LA} = \{f_0, f_1\}_{LA} , \\ \bar{\mathcal{L}} &= \{\mathbf{F}, \lambda, D, V\} , \end{aligned}$$

for

$$\begin{aligned} f_0(q) &= \mathbf{F}q - \frac{1}{2} D\phi(q) + \varphi(q) V , \\ f_1(q) &= \lambda , \\ D &= \lambda^2 , \\ V &= \lambda v , \end{aligned}$$

and as usual we take $v(x) = -\int_0^x \sigma(s) ds$.

From Lemma 3 we know that $\dim(\bar{\mathcal{L}}) = \infty$ if and only if at least one of the functions γ and σ is not QE. So, assume that γ and σ are *not* QE functions. Then, also D and V are not QE functions.

Let us now, have a look at the original (smaller) Lie-algebra, $\mathcal{L}$. We will try to see if

$$\mathcal{L} = \{\mu, \gamma\}_{LA} = \{f_0, f_1\}_{LA} < \infty .$$

Computing lie-brackets and simplifying

$$\begin{aligned}
[f_0, f_1]\,(q) &= f_0'(q) f_1(q) - f_1{}'(q) f_0(q) \\
&= \mathbf{F}\lambda - \frac{1}{2} D \underbrace{\phi'(q)[\lambda]}_{\text{scalar field}} - V \underbrace{\varphi'(q)[\lambda]}_{\text{scalar field}} \\
&= f_2(q) .
\end{aligned}$$

Continuing this way we need to compute $[f_0, f_2] = f_0'(q) f_2 - f_2{}' f_0(q)$. The first parcel, however give us

$$\begin{aligned}
f_0'(q) f_2(q) &= \mathbf{F}^2\lambda - \frac{1}{2} (\mathbf{F}D)\, \phi'(q)[\lambda] + (\mathbf{F}V)\, \varphi'(q)[\lambda] \\
&= \mathbf{F}^2\lambda + \mathbf{F}D \,.\, (\text{scalar field}) + \mathbf{F}V \,.\, (\text{scalar field})
\end{aligned}$$

and the second parcel (though involving more messy computations) is of the form

$$f_2'(q) f_0(q) = D \,.\, (\text{scalar filed}) + V \,.\, (\text{scalar field}) .$$

So,

$$\begin{aligned}
[f_0, f_2] &= \mathbf{F}^2\lambda + \mathbf{F}D(\text{scalar field}) + \mathbf{F}V \,.\, (\text{scalar field}) + \\
&\quad + D \,.\, (\text{scalar field}) + V \,.\, (\text{scalar field}) =: f_3 \\
[f_0, f_3] &= \mathbf{F}^3\lambda + \mathbf{F}^2 D(\text{scalar field}) + \mathbf{F}^2 V \,.\, (\text{scalar field}) + \\
&\quad + D \,.\, (\text{scalar field}) + V \,.\, (\text{scalar field}) , \\
&\vdots
\end{aligned}$$

It is now easy to see why the dimension of our $\mathcal{L}$ will also be infinite, in this case. It would only be finite if λ, D and V were QE functions, but this hypothesis is excluded by assumption. □

Proposition 5 Proof: Using $\bar{G}$ from (10.44) we have

$$\bar{G}'(z_0, z^1_{j,k}, z^2_{j,k}, z^3_{j,k}) \begin{bmatrix} h_0 \\ h^1_{1,0} \\ \vdots \\ h^2_{1,0} \\ \vdots \\ h^3_{1,0} \\ \vdots \\ h^3_{m,n_j} \end{bmatrix} = \frac{\partial}{\partial z_0} q(x+z_0)h_0 + \sum_{j\in A,C}\sum_{k=0}^{n^1_j} \mathbf{F}^k \lambda_j h^1_{j,k} +$$

$$+ \sum_{j\in A,C}\sum_{k=0}^{n^2_j} \mathbf{F}^k D_j h^2_{j,k} + \sum_{j\in C}\sum_{k=0}^{n^3_j} \mathbf{F}^k V_j h^3_{j,k} \,.$$

From $q = \bar{G}(Z)$ and once again the functional form of $\bar{G}$ in (10.44) we get

$$\begin{aligned}
\mathbf{F}q &= \mathbf{F}\widetilde{q_0} + \sum_{j\in A,C}\sum_{k=0}^{n^1_j} \mathbf{F}^{k+1}\lambda_j z^1_{j,k} + \sum_{j\in A,C}\sum_{k=0}^{n^2_j} \mathbf{F}^{k+1} D_j z^2_{j,k} + \\
&\quad + \sum_{j\in C}\sum_{k=0}^{n^3_j} \mathbf{F}^{k+1} V_j z^3_{j,k} \\
&= \frac{\partial}{\partial x}\widetilde{q_0} + \sum_{j\in A,C}\left[\sum_{k=1}^{n^1_j} \mathbf{F}^k \lambda_j z^1_{j,k-1} + \mathbf{F}^{n^1_j+1}\lambda_{n_j} z^1_{j,n^1_j}\right] + \\
&\quad + \sum_{j\in A,C}\left[\sum_{k=1}^{n^2_j} \mathbf{F}^k D_j z^2_{j,k-1}\mathbf{F}^{n^2_j+1} D_{n_j} z^2_{j,n_j}\right] + \\
&\quad + \sum_{j\in C}\left[\sum_{k=1}^{n^3_j} \mathbf{F}^k V_j z^3_{j,k-1} + \mathbf{F}^{n^3_j+1} V_{n_j} z^3_{j,n^3_j}\right] \\
&= \frac{\partial}{\partial x}\widetilde{q_0} + \sum_{j\in A,C}\left[\sum_{k=1}^{n^1_j} \mathbf{F}^k \lambda_j z^1_{j,k-1} + \sum_{k=0}^{n^1_j} c^1_{j,k}\lambda_k z^1_{j,n_j}\right] + \\
&\quad + \sum_{j\in A,C}\left[\sum_{k=1}^{n^2_j} \mathbf{F}^k D_j z^2_{j,k-1} + \sum_{k=0}^{n^2_j} c^2_{j,k} D_k z^2_{j,n_j}\right] + \\
&\quad + \sum_{j\in C}\left[\sum_{k=1}^{n^3_j} \mathbf{F}^k V_j z^3_{j,k-1} + \sum_{k=0}^{n^3_j} c^3_{j,k} V_k z^3_{j,n_j}\right]
\end{aligned}$$

where we omitted the x-dependence and used $\widetilde{q_0}(x) = q_0(x + z_0)$.

We can now use the above expression to substitute into our μ (recall (10.40)). Thus, $\bar{G}_\star \bar{a} = \mu$ becomes

$$\frac{\partial}{\partial z_0}\widetilde{q}_0\bar{a}_0 + \sum_{j\in A,C}\sum_{k=0}^{n_j^1}\mathbf{F}^k\lambda_j\bar{a}_{j,k}^1 + \sum_{j\in A,C}\sum_{k=0}^{n_j^2}\mathbf{F}^k D_j\bar{a}_{j,k}^2 + \sum_{j\in C}\sum_{k=0}^{n_j^3}\mathbf{F}^k V_j\bar{a}_{j,k}^3$$

$$= \frac{\partial}{\partial x}\widetilde{q}_0 + \sum_{j\in A,C}\left[\sum_{k=1}^{n_j^1}\mathbf{F}^k\lambda_j z_{j,k-1}^1 + \sum_{k=0}^{n_j^1} c_{j,k}^1\lambda_k z_{j,n_j}^1\right] +$$

$$+ \sum_{j\in A,C}\left[\sum_{k=1}^{n_j^2}\mathbf{F}^k D_j z_{j,k-1}^2 + \sum_{k=0}^{n_j^2} c_{j,k}^2 D_k z_{j,n_j}^2\right] +$$

$$+ \sum_{j\in C}\left[\sum_{k=1}^{n_j^3}\mathbf{F}^k V_j z_{j,k-1}^3 + \sum_{k=0}^{n_j^3} c_{j,k}^3 V_k z_{j,n_j}^3\right] -$$

$$-\frac{1}{2}\sum_{j\in A,C} D_j\varphi_j^2(q) - \frac{1}{2}\sum_{j\in A,C}\varphi_j'(q)\left[\lambda_j\right]\varphi_j(q)\lambda_j\,,$$

and $\bar{G}_\star \bar{b}^i = \gamma_i$ for all $i \in A, C$

$$\frac{\partial}{\partial z_0}\widetilde{q}_0\bar{b}_0^i + \sum_{j\in A,C}\sum_{k=0}^{n_j^1}\mathbf{F}^k\lambda_j\bar{b}_{j,k}^{1i} + \sum_{j\in A,C}\sum_{k=0}^{n_j^2}\mathbf{F}^k D_j\bar{b}_{j,k}^{2i} + \sum_{j\in C}\sum_{k=0}^{n_j^3}\mathbf{F}^k V_j\bar{b}_{j,k}^{3i}$$

$$= \varphi_i(q)\lambda_i\,.$$

Identification of terms, and again use of $q = \bar{G}(Z)$ yields

$$\begin{aligned}
\bar{a}_0 &= 1\,,\\
\bar{a}_{j,0}^1 &= c_{j,0}^1 z_{j,n_j^1}^1 - \tfrac{1}{2}\varphi_j'(\bar{G}(z))\left[\lambda_j\right]\varphi_j(\bar{G}(z))\,,\ j\in A,C\,,\\
\bar{a}_{j,k}^1 &= z_{j,k-1}^1 + c_{j,k}^1 z_{j,n_j^1}^1\,, && j\in A,C\,,\ \ k = 1,2,\cdots n_j^1\,,\\
\bar{a}_{j,0}^2 &= c_{j,0}^2 z^2 j, n_j^2 - \tfrac{1}{2}\varphi_j^2(\bar{G}(z))\,, && j\in A,C\,,\\
\bar{a}_{j,k}^2 &= z_{j,k-1}^2 + c_{j,k}^2 z_{jn_j^2}^2\,, && j\in A,C\,,\ \ k = 1,2,\cdots n_j^2\,,\\
\bar{a}_{j,0}^3 &= c_{j,0}^3 z^3 j, n_j^3 + \varphi(\bar{G}(z))\,, && j\in C\,,\\
\bar{a}_{j,k}^3 &= z_{j,k-1}^3 + c_{j,k}^3 z_{j,n_j^3}^3\,, && j\in C\,,\ \ k = 1,2,\cdots n_j^3\,,\\
\bar{b}_0^i &= 0\,,
\end{aligned}$$

$$\bar{b}_{j,k}^{1i} = \begin{cases}\varphi_j(\bar{G}(z)) & \text{if } \ i = j \ \text{and} \ \ k = 0\\ 0 & \text{if } \ i \neq j \ \text{or} \ \ k = 1,2,\cdots n_j^1\end{cases}\,,\ j\in A,C\,,$$

$$\begin{aligned}
\bar{b}_{j,k}^{2i} &= 0\,, && j\in A,C\,,\ \ k = 1,2,\cdots n_j^2\,,\\
\bar{b}_{j,k}^{3i} &= 0\,, && j\in A,C\,,\ \ k = 1,2,\cdots n_j^3.
\end{aligned}$$

Note that the factors $z^1_{j,0}$ are driven by the scalar wiener process W^j and that all remaining factors have diffusion terms equal to zero. Thus only for $z^1_{j,0}$ the Itô dynamics differ from the Stratonovich dynamics. The above $\bar{a}^1_{j,0}$ is the Stratonovich drift. Given the form of $\bar{a}^1_{j,0}$, the diffusion for $z^1_{j,0}$. We easily get the Itô drift to be simply $\bar{a}^{1\ \text{Itô}}_{j,0} = c^1_{j,0} z^1_{j,n^1_j}$. □

Proposition 8 Proof: We first compute the parameterization $\widehat{G}$. From Theorem C.2 and the special shape of the basic fields in (10.50), we have

$$\begin{bmatrix} \widehat{G}^q(Z) \\ \widehat{G}^r(Z) \end{bmatrix} = \begin{bmatrix} \prod_{j\in A,C} \left(e^{\mathbf{F}^k \lambda_j z^1_{j,k}} e^{\mathbf{F}^k D_j z^2_{j,k}} \right) \prod_{j\in C} \left(e^{\mathbf{F}^k V_j z^3_{j,k}} \right) e^{\mathbf{F} z_0} q_0 \\ \prod_{j\in B,C} \left(e^{\mathbf{F}^k \beta_j z^4_{j,k}} e^{\mathbf{F}^k H_j z^5_{j,k}} \right) e^{\mathbf{F} z_0} r_0 \end{bmatrix} \tag{10.61}$$

$$= \begin{bmatrix} \widetilde{q}_0 + \sum_{j\in A,C} \sum_{k=0}^{n^1_j} \mathbf{F}^k \lambda_j z^1_{j,k} + \sum_{j\in A,C} \sum_{k=0}^{n^2_j} \mathbf{F}^k D_j z^2_{j,k} + \sum_{j\in C} \sum_{k=0}^{n^3_j} \mathbf{F}^k V_j z^3_{j,k} \\ \widetilde{r}_0 + \sum_{j\in B,C} \sum_{k=0}^{n^4_j} \mathbf{F}^k \beta_j z^4_{j,k} + \sum_{j\in B,C} \sum_{k=0}^{n^5_j} \mathbf{F}^k H_j z^5_{j,k} \end{bmatrix}.$$

Using $\widehat{G}$ from (10.61) we have

$$\widehat{G}^{q'}(z_0, z^1_{j,k}, z^2_{j,k}, z^3_{j,k}) \begin{bmatrix} h_0 \\ h^1_{1,0} \\ \vdots \\ h^2_{1,0} \\ \vdots \\ h^3_{m,n^3_j} \end{bmatrix} = \frac{\partial}{\partial z_0} \widetilde{q}_0 h_0 + \sum_{j\in A,C} \sum_{k=0}^{n^1_j} \mathbf{F}^k \lambda_j h^1_{j,k} +$$

$$+ \sum_{j\in A,C} \sum_{k=0}^{n^2_j} \mathbf{F}^k D_j h^2_{j,k} + \sum_{j\in C} \sum_{k=0}^{n^3_j} \mathbf{F}^k V_j h^3_{j,k},$$

and

$$\widehat{G}^{r'}(z_0, z^4_{j,k}, z^5_{j,k}) \begin{bmatrix} h_0 \\ h^4_{1,0} \\ \vdots \\ h^5_{m,n^3_j} \end{bmatrix} = \frac{\partial}{\partial z_0} \widetilde{r}_0 h_0 + \sum_{j\in B,C} \sum_{k=0}^{n^4_j} \mathbf{F}^k \beta_j h^4_{j,k} +$$

$$+ \sum_{j\in A,C} \sum_{k=0}^{n^5_j} \mathbf{F}^k H_j h^2_{j,k},$$

where we omitted the x-dependence and used $\widetilde{q_0}(x) = q_0(x+z_0)$, $\widetilde{r_0}(x) = r_0(x+z_0)$.

From $q = \widehat{G}^q(Z)$, $r = \widehat{G}^r(Z)$ and once again the functional form of $\widehat{G}$ in (10.61) we get

$$
\begin{aligned}
\mathbf{F}q &= \mathbf{F}\widetilde{q_0} + \sum_{j\in A,C}\sum_{k=0}^{n_j^1}\mathbf{F}^{k+1}\lambda_j z_{j,k}^1 + \sum_{j\in A,C}\sum_{k=0}^{n_j^2}\mathbf{F}^{k+1}D_j z_{j,k}^2 + \sum_{j\in C}\sum_{k=0}^{n_j^3}\mathbf{F}^{k+1}V_j z_{j,k}^3 \\
&= \frac{\partial}{\partial x}\widetilde{q_0} + \sum_{j\in A,C}\left[\sum_{k=1}^{n_j^1}\mathbf{F}^k\lambda_j z_{j,k-1}^1 + \mathbf{F}^{n_j^1+1}\lambda_{n_j^1} z_{j,n_j^1}^1\right] + \\
&\quad + \sum_{j\in A,C}\left[\sum_{k=1}^{n_j^2}\mathbf{F}^k D_j z_{j,k-1}^2 \mathbf{F}^{n_j^2+1} D_{n_j^2} z_{j,n_j^2}^2\right] + \\
&\quad + \sum_{j\in C}\left[\sum_{k=1}^{n_j^3}\mathbf{F}^k V_j z_{j,k-1}^3 + \mathbf{F}^{n_j^3+1} V_{n_j^3} z_{j,n_j}^3\right] \\
&= \frac{\partial}{\partial x}\widetilde{q_0} + \sum_{j\in A,C}\left[\sum_{k=1}^{n_j^1}\mathbf{F}^k\lambda_j z_{j,k-1}^1 + \sum_{k=0}^{n_j^1} c_{j,k}^1 \lambda_k z_{j,n_j}^1\right] + \\
&\quad + \sum_{j\in A,C}\left[\sum_{k=1}^{n_j^2}\mathbf{F}^k D_j z_{j,k-1}^2 + \sum_{k=0}^{n_j^2} c_{j,k}^2 D_k z_{j,n_j^2}^2\right] + \\
&\quad + \sum_{j\in C}\left[\sum_{k=1}^{n_j^3}\mathbf{F}^k V_j z_{j,k-1}^3 + \sum_{k=0}^{n_j^3} c_{j,k}^3 V_k z_{j,n_j^3}^3\right],
\end{aligned}
$$

$$
\begin{aligned}
\mathbf{F}r &= \mathbf{F}\widetilde{r}_0 + \sum_{j\in b,C}\sum_{k=0}^{n_j^4}\mathbf{F}^{k+1}\beta_j z_{j,k}^4 + \sum_{j\in B,C}\sum_{k=0}^{n_j^5}\mathbf{F}^{k+1}H_j z_{j,k}^5 \\
&= \frac{\partial}{\partial x}\widetilde{r}_0 + \sum_{j\in B,C}\left[\sum_{k=1}^{n_j^4}\mathbf{F}^k\beta_j z_{j,k-1}^4 + \mathbf{F}^{n_j^4+1}\beta_{n_j^4} z_{j,n_j^4}^4\right] + \\
&\quad + \sum_{j\in B,C}\left[\sum_{k=1}^{n_j^5}\mathbf{F}^k H_j z_{j,k-1}^5 \mathbf{F}^{n_j^5+1} H_{n_j^5} z_{j,n_j^5}^5\right] \\
&= \frac{\partial}{\partial x}\widetilde{r}_0 + \sum_{j\in B,C}\left[\sum_{k=1}^{n_j^4}\mathbf{F}^k\beta_j z_{j,k-1}^4 + \sum_{k=0}^{n_j^4} c_{j,k}^4\beta_k z_{j,n_j^4}^4\right] + \\
&\quad + \sum_{j\in B,C}\left[\sum_{k=1}^{n_j^5}\mathbf{F}^k H_j z_{j,k-1}^5 + \sum_{k=0}^{n_j^5} c_{j,k}^5 H_k z_{j,n_j^5}^5\right].
\end{aligned}
$$

We can now use the above expressions to substitute into our μ, recall we have

$$
\begin{aligned}
\mu(q,r) = \mathbf{F}\begin{bmatrix} q \\ r \end{bmatrix} - \frac{1}{2}\sum_{i\in A,C}\varphi_i^2(q,r)\begin{bmatrix}\lambda_i^2 \\ 0\end{bmatrix} + \sum_{i\in C}\varphi_i(q,r)\phi_i(r)\begin{bmatrix}\lambda_i B_i \\ 0\end{bmatrix} + \\
+ \sum_{i\in B,C}\phi_i^2(r)\begin{bmatrix} 0 \\ \beta_i B_i\end{bmatrix} - \frac{1}{2}\Bigg\{\sum_{i\in A,C}\varphi_{iq}'(q,r)\left[\lambda_i\right]\varphi_i(q,r)\begin{bmatrix}\lambda_i \\ 0\end{bmatrix} + \\
+ \sum_{i\in C}\varphi_{ir}'(q,r)\left[\beta_i\right]\phi_i(r)\begin{bmatrix}\lambda_i \\ 0\end{bmatrix} + \sum_{i\in B,C}\phi_{ir}'(r)\left[\beta_i\right]\phi_i(r)\begin{bmatrix}0 \\ \beta_i\end{bmatrix}\Bigg\},
\end{aligned}
$$

$$
\begin{bmatrix}\gamma_i(q,r) \\ \sigma_i(r)\end{bmatrix} = \begin{cases}
\varphi_i(q,r)\begin{bmatrix}\lambda_i \\ 0\end{bmatrix}, & i \in A, \\
\phi_i(r)\begin{bmatrix}0 \\ \beta_i\end{bmatrix}, & i \in B, \\
\varphi_i(q,r)\begin{bmatrix}\lambda_i \\ 0\end{bmatrix} + \phi_i(r)\begin{bmatrix}0 \\ \beta_i\end{bmatrix}, & i \in C.
\end{cases}
$$

Defining

$$
\mu(q,r) = \begin{bmatrix}\mu^q(q,r) \\ \mu^r(r)\end{bmatrix},
$$

identification of terms in $\widehat{G}_\star^q\widehat{a}^q = \mu^q$, $\widehat{G}_\star^r\widehat{a}^r = \mu^r$, gives us

$$
\begin{aligned}
\widehat{a}_0 &= 1\,,\\
\widehat{a}^1_{j,0} &= \begin{cases} c^1_{j,0} z^1_{j,n^1_j} - \frac{1}{2}\varphi'_j(\widehat{G}^q(z),\widehat{G}^r(z))\,[\lambda_j]\,\varphi_j(\widehat{G}^q(z),\widehat{G}^r(z))\,, & j \in A\,,\\ c^1_{j,0} z^1_{j,n^1_j} - \frac{1}{2}\varphi'_j(\widehat{G}^q(z),\widehat{G}^r(z))\,[\lambda_j]\,\varphi_j(\widehat{G}^q(z),\widehat{G}^r(z)) - \\ \quad -\frac{1}{2}\varphi'_j(\widehat{G}^q(z),\widehat{G}^r(z))\,[\beta_j]\,\phi_j(\widehat{G}^r(z))\,, & j \in C\,,\end{cases}
\end{aligned}
$$

$$
\begin{aligned}
\widehat{a}^1_{j,k} &= z^1_{j,k-1} + c^1_{j,k} z^1_{j,n^1_j}\,, && j \in A, C\,,\ k = 1,2,\cdots n^1_j\,,\\
\widehat{a}^2_{j,0} &= c^2_{j,0} z^2_{j,n^2_j} - \frac{1}{2}\varphi^2_j(\widehat{G}^q(z),\widehat{G}^r(z))\,, && j \in A, C\,,\\
\widehat{a}^2_{j,k} &= z^2_{j,k-1} + c^2_{j,k} z^2_{jn^2_j}\,, && j \in A, C\,,\ k = 1,2,\cdots n^2_j\,,\\
\widehat{a}^3_{j,0} &= c^3_{j,0} z^3_{j,n^3_j} + \varphi(\widehat{G}^q(z),\widehat{G}^r(z))\phi(\widehat{G}^r(z))\,, && j \in C\,,\\
\widehat{a}^3_{j,k} &= z^3_{j,k-1} + c^3_{j,k} z^3_{j,n^3_j}\,, && j \in C\,,\ k = 1,2,\cdots n^3_j\,,\\
\widehat{a}^4_{j,0} &= c^4_{j,0} z^4_{j,n^4_j} - \frac{1}{2}\phi'_j(\widehat{G}^r(z))\,[\beta_j]\,\phi_j(\widehat{G}^r(z))\,, && j \in B, C\,,\\
\widehat{a}^4_{j,k} &= z^4_{j,k-1} + c^4_{j,k} z^4_{j,n^4_j}\,, && j \in B, C\,,\ k = 1,2,\cdots n^4_j\,,\\
\widehat{a}^5_{j,0} &= c^5_{j,0} z^5_{j,n^5_j} + \phi^2(\widehat{G}^r(z))\,, && j \in B, C\,,\\
\widehat{a}^5_{j,k} &= z^5_{j,k-1} + c^5_{j,k} z^5_{j,n^5_j}\,, && j \in B, C\,,\ k = 1,2,\cdots n^5_j\,.
\end{aligned}
$$

To get the diffusion terms we identify terms in $\widehat{G}^q_\star \widehat{b}^{qi} = \gamma_i$ for $i \in A, C$ and $\widehat{G}^r_\star \widehat{b}^{ri} = \sigma_i$ for $i \in B, C$,

$$
\begin{aligned}
\widehat{b}^i_0 &= 0\,,\\
\widehat{b}^{1i}_{j,k} &= \begin{cases} \varphi_j(\widehat{G}^q(z),\widehat{G}^r(z))\,, & \text{if } i = j, \text{ and } k = 0,\\ 0\,, & \text{otherwise}\,,\end{cases} \qquad j \in A, C\,,
\end{aligned}
$$

$$
\begin{aligned}
\widehat{b}^{2i}_{j,k} &= 0\,, && j \in A, C\,,\ k = 1,2,\cdots n^2_j\,,\\
\widehat{b}^{3i}_{j,k} &= 0\,, && j \in A, C\,,\ k = 1,2,\cdots n^3_j\,,\\
\widehat{b}^{4i}_{j,k} &= \begin{cases} \phi_j(\widehat{G}^r(z))\,, & \text{if } i = j \text{ and } k = 0\,,\\ 0\,, & \text{if } i \neq j \text{ or } k = 1,2,\cdots n^4_j\,,\end{cases} && j \in B, C\,,\\
\widehat{b}^{5i}_{j,k} &= 0\,, && j \in B, C\,,\ k = 1,2,\cdots n^5_j\,.
\end{aligned}
$$

Note that this implies that the factors $Z^1_{j,0}$, $Z^4_{j,0}$ are driven by the scalar wiener process W^j so, in particular for $j \in C$ the same Wiener process drives the two variables. All remaining factors have diffusion terms equal to zero. Thus only for $z^1_{j,0}$ and $z^4_{j,0}$, the Itô dynamics differ from the Stratonovich dynamics. We easily get the Itô drifts to be simply

$$\widehat{a}_{j,0}^{1\ \text{Itô}} = c_{j,0}^{1} z_{j,n_j^1}^{1},$$
$$\widehat{a}_{j,0}^{4\ \text{Itô}} = c_{j,0}^{4} z_{j,n_j^4}^{4}.$$

□

References

[1] T. Björk. *Arbitrage Theory in Continuous Time.* Oxford University Press, 2nd edition, 2004.

[2] T. Björk. On the geometry of interest rate models. In *Paris-Princeton Lectures on Mathematical Finance 2003*, volume 1847, pages 133–215. Springer Lecture Notes in Mathematics, 2004.

[3] T. Björk, M. Blix, and C. Landén. A note on the existence of finite dimensional realizations for futures prices. Work in Progress, 2004.

[4] T. Björk and B. Christensen. Interest rate dynamics and consistent forward rate curves. *Mathematical Finance*, 9(4):323–348, 1999.

[5] T. Björk and C. Landén. On the construction of finite dimensional realizations for nonlinear forward rate models. *Finance and Stochastics*, 6(3):303–331, 2002.

[6] T. Björk and L. Svensson. On the existence of finite dimensional realizations for nonlinear forward rate models. *Mathematical Finance*, 11(2):205–243, 2001.

[7] A. Brace and M. Musiela. A multifactor Gauss Markov implementation of Heath, Jarrow, and Morton. *Mathematical Finance*, 4:259–283, 1994.

[8] G. Da Prato and J. Zabzcyk. *Stochastic Equations in Infinite Dimensions.* Cambridge University Press, 1992.

[9] D. Filipović and J. Teichmann. On finite dimensional term structure models. Working paper, Princeton University, 2002.

[10] D. Filipović and J. Teichmann. Existence of invariant manifolds for stochastic equations in infinite dimension. *J. Funct. Anal.*, 197(2):398–432, 2003.

[11] R.M. Gaspar. General quadratic term structures for bond, futures and forward prices. SSE/EFI Working paper Series in Economics and Finance, 559, 2004.

[12] H. Geman, N. El Karoui, and J-C. Rochet. Changes of numéraire, changes of probability measure and option pricing. *Journal of Applied Probability*, 32:443–458, 1995.

[13] M. Musiela. Stochastic PDE:s and term structure models. Preprint, 1993.

11

Good Portfolio Strategies under Transaction Costs: A Renewal Theoretic Approach

Albrecht Irle and Jörn Sass*

[1] Mathematisches Seminar, Universität Kiel, D-24098 Kiel, Germany
irle@math.uni-kiel.de

[2] RICAM, Austrian Academy of Sciences, Altenberger Str. 69, A-4040 Linz, Austria
joern.sass@oeaw.ac.at

Summary. This paper looks at portfolio theory in the presence of costs of transactions. A fundamental result is given in Morton and Pliska (1995)([15]) where renewal theoretic arguments and the theory of optimal stopping are used to derive optimal strategies for maximizing the asymptotic growth rate under purely fixed costs which are proportional to the portfolio value. Our paper is also devoted to maximizing the asymptotic growth rate but here we consider a combination of fixed and proportional costs. Motivated by various structural results in the work on optimal portfolio theory we introduce a class of natural trading strategies which can be described by four parameters, two for the stopping boundaries and two for the new risky fractions (fraction of the wealth invested in the stock). In this class the problem can be simplified by renewal theoretic arguments to treating one period between two trading times, where we then have to start the new risky fraction process according to the invariant distribution. This yields an explicit form for the asymptotic growth rate that can be maximized in these four parameters. The computation of best strategies in this class thus is simple, and we provide various examples. Preliminary considerations based on the fundamental results of Bielecki and Pliska (2000)([5]) and the results of this paper point out that in fact an allover optimal impulse control strategy can be found within this class.

Introduction

Typical transaction costs considered in portfolio theory are constant costs, fixed costs (proportional to the portfolio value), and proportional costs (proportional to the transaction volume). While the latter penalizes the size of the transaction, the first two punish the frequency of trading. So a combination of both types is of interest, as well from a practical as a theoretical point of view. The trading strategies of interest are so called impulse control strategies consisting of a sequence of stopping times at which trading takes place and

* Supported by the Austrian Academy of Sciences

the transactions at those times. Optimal impulse control strategies can be described as solutions of quasi variational inequalities. Usually, determining such solutions leads to very hard problems, and already the numerical evaluation can become difficult. As a remedy one can turn to classes of strategies which can be handled by probabilistic means and which provide good strategies for the problem at hand, hopefully even containing an optimal one.

As we are interested in explicit solutions we shall work within the framework of the Black Scholes model. We consider an investor who faces proportional costs and fixed costs. Our objective is the maximization of the asymptotic growth rate

$$R(x,\pi) = \liminf_{t\to\infty} \frac{1}{t}\, \mathrm{E}\big[\ln X_t \,|\, X_0 = x, \pi_0 = \pi\big], \tag{11.1}$$

where X_t is the value of the portfolio at time $t\eta 0$ and π_0 the initial fraction of X_0 invested in the stock.

Without transaction costs, the problem of maximizing the expected utility was solved by Merton, [14]. For logarithmic and power utility, the optimal trading strategy is given by a constant $\widehat{\eta}$ which is the optimal risky fraction (i.e. the fraction of wealth invested in the stock). This constant fraction is equivalent to the Merton line in the wealth space which consists of the vectors of wealth invested in the bond and wealth invested in the stock. To keep the risky fraction constant involves continuous trading which, under transaction costs, is no longer suitable.

In the last decade, many excellent papers in portfolio theory dealing with transaction costs have appeared. These costs are usually defined in three different ways: proportionally to the volume of trade (proportional cost), proportionally to the portfolio value (fixed cost) or consisting of a constant component and proportional cost (constant plus proportional costs). In a majority of these papers an infinite horizon discounted consumption criterion under logarithmic or power utility is considered. It has turned out that the structure of the solution depends essentially on the type of the transaction costs and only to a lesser amount on the optimization criterion or utility function.

The first type of costs considered were purely proportional costs for which the optimal solution is given by a cone in which it is optimal not to trade at all. This cone corresponds to an interval for the risky fraction. Reaching the boundaries, infinitesimal trading occurs in such a way that the wealth process just stays in the cone. This kind of behaviour was first stated in [13]. A rigorous proof for a discounted consumption criterion can be found in [7] using methods of stochastic control theory and showing that the wealth process is a diffusion, reflected at the boundaries of the cone. [1], [10], [19] prove under weaker assumptions the existence and uniqueness of a viscosity solution for the corresponding HJB equation, and [2], [20] derive under somewhat different proportional costs similar no-transaction regions for the asymptotic growth rate.

To avoid the occurrence of infinitesimal trading at the boundary, it seems reasonable to add a constant component to the transaction fee which punishes very frequent trading. This was done in another line of papers which deal with constant and proportional costs. An investor has now to choose discrete trading times and optimal transactions at these times. Thus methods of optimal impulse control have to be used. [8] and [11] achieve general existence results for finite and infinite horizon and determine optimal strategies for maximizing the utility of terminal wealth for the identity as utility function and for exponential utility. Maximizing the discounted consumption under power utility for an infinite horizon and allowing for continuous consumption without costs, [16] derive quasi-variational HJB inequalities whose solution yields the optimal strategy. The insight, already obtained in [11], is that there is still some no-transaction region, but reaching the boundary transactions will be done in such a way that the wealth process restarts at some curve between boundary and Merton line.

An elegant approach is provided by [15] for purely fixed costs. There it is shown that, for the objective of maximizing the expected asymptotic growth rate (11.1), a factorization of the wealth process into the wealth gained per period is possible which leads under logarithmic utility to an additive representation. Using a reduction to one trading period they proceed by solving an optimal stopping problem for linear costs for the risky fraction process. A very general cost structure is treated in [5] where a set of quasi variational inequalities is derived whose solution yields the optimal trading strategy for (11.1) if the costs contain a constant or fixed component.

The approaches as described above lead to plausible optimal strategies in dependence of the type of transaction costs considered. Numerical solutions and approximations of optimal strategies are frequently difficult, cf. [3], [5], [6], [11], and it is often not obvious how to use these strategies in practice.

Our goal is to carry over the results of [15] to transaction costs which have proportional costs in addition to the fixed costs which is often the case for private investors. To get insight into this situation we introduce a class of natural trading strategies which can be described by four parameters (a, b, α, β), a and b for the stopping boundaries and α, β for the new risky fractions (fraction of the wealth invested in the stock). When the risky fraction process reaches a or b trading occurs in such a way that the new risky fractions are α or β, respectively. Stopping at a corresponds to buying and stopping at b to selling stocks. This class is motivated by the results discussed above. The cone obtained for proportional costs corresponds to an interval for the risky fraction process. The results in [15] say that for fixed costs a constant new risky fractions is optimal and, by the results for combined constant and proportional costs, we have to expect that due to the proportional costs we have in our case two different new risky fractions, one after buying and one after selling.

In this class the problem can be simplified to one period between two trading times by renewal theoretic arguments, where we have to start the new risky fraction process according to the invariant distribution. This yields an explicit functional that has only to be maximized in these four parameters. The optimality in the greater class of impulse control strategies may then be shown afterwards using the explicit structure we derive.

We proceed as follows: after presenting some notation and the solution without transaction costs in Section 11, we introduce the model for the fixed and proportional costs in Section 11, first using controls given by stopping times and the transactions (amount of money invested in the stock) at those times. In Theorem C.1 we then show that a factorization of the wealth process is possible when we express the control strategy in terms of the new risky fractions instead of the transactions.

Since the factors in Theorem C.1 do not depend on the wealth it is very convenient to reformulate the control problem in terms of the risky fractions, leading to the definition of NRF-strategies in Definition 2. The equivalence of both approaches can be shown. In Section 11 we introduce our subclass of natural trading strategies, those trading strategies with constant boundaries (CB-strategies) given by four parameters (a, b, α, β) as described above.

After collecting some results about the risky fraction process without trading, still in Section 11, we approach the problem of maximizing the asymptotic growth rate.

Of course there will be no unique optimal trading strategy, as up to some finite time horizon the strategy could be changed without changing the asymptotic growth rate. Only under strong assumptions it is possible to derive general existence and uniqueness results for this criterion which are applicable in our situation, e.g. in [4] and [12]. Even in models with countable state space there are simple examples that no optimal strategy exists or that no optimal strategy is stationary, see [18].

For our CB-strategies we can obtain a regenerative structure which allows us to reduce the problem with renewal theoretic arguments to one period between two buyings. Using the invariant distribution of the embedded Markov chain of new risky fractions the problem can finally be reduced to one period between two trading times and provides a manageable expression for the asymptotic growth rate. This Theorem C.2 corresponds to [15, Proposition 3.1]. But in our situation we have to integrate over the invariant distribution of the Markov chain of the new risky fractions. Then this expression has to be maximized over the CB-strategies. Due to the explicit representation in Corollary 1 of the asymptotic growth rate depending only on the four parameters (a, b, α, β), this can be done numerically quite easily. In Section 11 we also provide an existence result and some examples. We close in Section 11 with a heuristic approach based on [5] how to verify the optimality in the class of all impulse control strategies.

Trading and Optimization Without Transaction Costs

We will consider one *bond* or *bank account* and one *stock* with price processes $(B_t)_{t\eta 0}$ and $(S_t)_{t\eta 0}$, respectively, which evolve according to the Black Scholes model. Hence the prices are given for *interest rate* $r\eta 0$, *trend* $\mu \in \mathbb{R}$, and *volatility* $\sigma > 0$ by $B_0 = S_0 = 1$ and

$$dB_t = B_t\, r\, dt, \quad dS_t = S_t(\mu\, dt + \sigma\, dW_t)\,,$$

where $(W_t)_{t\eta 0}$ is a standard Brownian motion on a suitable probability space $(\Omega, \mathcal{F}, P)$. Let $(\mathcal{F}_t)_{t\eta 0}$ denote the augmented filtration generated by $(W_t)_{t\eta 0}$.

Without transaction costs the trading of an investor may be described by the *initial capital* $x > 0$ and by the *risky fraction process* $(\eta_t)_{t\eta 0}$, where η_t is the fraction of the total portfolio value (wealth) which the investor chooses to hold in the stock at time t. In this section we call $(\eta_t)_{t\eta 0}$ *admissible*, if it is adapted, measurable, and bounded. Given x and $(\eta_t)_{t\eta 0}$ the *wealth process* $(X_t^\eta)_{t\eta 0}$ is defined self-financing as the continuous solution of

$$dX_t^\eta = (1 - \eta_t) X_t^\eta\, r\, dt + \eta_t\, X_t^\eta \big(\mu\, dt + \sigma\, dW_t\big)\,, \quad X_0^\eta = x\,.$$

Our objective is the maximization of the *asymptotic growth rate* (11.1) over all admissible risky fraction processes $\eta = (\eta_t)_{t\eta 0}$. Using

$$\mathrm{E}[\ln X_t^\eta \mid X_0^\eta = x] = \ln x + \mathrm{E}\left[\int_0^t \Big(r + \eta_s(\mu - r) - \tfrac{1}{2}(\eta_s\, \sigma)^2\Big)\, ds \,\Big|\, X_0^\eta = x\right] \quad (11.2)$$

a simple pointwise maximization yields as optimal solution $\eta_t = \widehat{\eta}$, $t\eta 0$, where

$$\widehat{\eta} = \frac{\mu - r}{\sigma^2} \quad \text{and} \quad \widehat{R} = R^{\widehat{\eta}} = r + \frac{1}{2}\Big(\frac{\mu - r}{\sigma}\Big)^2. \quad (11.3)$$

This constant *optimal risky fraction* $\widehat{\eta}$ corresponds to the well known *Merton line*.

Multiplying the prices by e^{-rt} we see that we may assume from now on $r = 0$ for simpler notation, so $B_t = 1$, $t\eta 0$. Like in (11.3), for general r the solution for (11.1) is then obtained by adding the rate r and using $\mu - r$ instead of μ.

Fixed and Proportional Transaction Costs

Let us now assume that an investor faces transaction fees. With current wealth $x > 0$ there has to be paid for a transaction of volume $\Delta \in \mathbb{R}$ the *transaction costs*

$$\widetilde{c}(x, \Delta) = \delta\, x + \gamma\, |\Delta|\,, \quad (11.4)$$

where $\delta \in (0, 1)$ and $\gamma \in [0, 1 - \delta)$. We call $\delta\, x$ and $\gamma|\Delta|$ *fixed* and *proportional* costs, respectively. Note that the definition of the fixed cost is the same as in [15]. This type of costs may be interpreted as managing costs.

It is convenient to use two processes to describe the evolution of the wealth. We use the wealth process $(X_t)_{t\eta 0}$ and the risky fraction process $(\pi_t)_{t\eta 0}$.

Since $\delta > 0$, the natural class of strategies to consider are impulse control strategies where trading occurs at time points $\tau_0 \leq \tau_1 \leq \ldots$, see [5]. In view of (11.4) and [5, Proposition 4.1] we can restrict to trading times which are separated.

Definition 1.

(i) *An* impulse control strategy $(\tau_n, \Delta_n)_{n\in\mathbb{N}_0}$ *consists of stopping times* $0 = \tau_0 \leq \tau_1 \leq \ldots \leq \infty$ *with respect to* $(\mathcal{F}_t)_{t\eta 0}$, *the* trading times, *which satisfy* $\tau_n \to \infty$ *a.s. and* $\tau_n < \tau_{n+1}$ *on* $\{\tau_n < \infty\}$, *and of* $\mathcal{F}_{\tau_n}$*-measurable,* $\mathbb{R}$*-valued random variables* Δ_n, $n \in \mathbb{N}$, *the* transactions.

(ii) *The wealth process* X *and the risky fraction process* π *controlled by an impulse control strategy* $\widetilde{K} = (\tau_n, \Delta_n)_{n\in\mathbb{N}_0}$ *are introduced for initial values* $x > 0, \pi \in [0,1]$ *by*

$$X_0 = x\,, \quad \pi_0 = \pi\,, \quad V_0 = X_0 - \widetilde{c}(X_0, \Delta_0)\,, \quad \eta_0 = (\pi_0 X_0 + \Delta_0)/V_0\,,$$

and for $n \in \mathbb{N}$ *on* $\{\tau_n < \infty\}$ *by*

$$X_t = \left(1 - \eta_{n-1} + \eta_{n-1} S_t / S_{\tau_{n-1}}\right) V_{n-1}\,, \quad t \in (\tau_{n-1}, \tau_n]\,, \tag{11.5}$$

$$\pi_t = \eta_{n-1} V_{n-1} S_t / (S_{\tau_{n-1}} X_t)\,, \quad t \in (\tau_{n-1}, \tau_n]\,, \tag{11.6}$$

$$V_n = X_{\tau_n} - \widetilde{c}(X_{\tau_n}, \Delta_n)\,, \tag{11.7}$$

$$\eta_n = (\pi_{\tau_n} X_{\tau_n} + \Delta_n)/V_n\,. \tag{11.8}$$

We may use an upper index $\widetilde{K}$ *to indicate that the processes are controlled by* $\widetilde{K}$, *e.g.* $X_t^{\widetilde{K}}$.

(iii) *An impulse control strategy* $\widetilde{K}$ *is* admissible *if* $X_t^{\widetilde{K}} > 0$ *and* $\pi_t^{\widetilde{K}} \in (0,1)$ *for all* $t\eta 0$.

So $(X_t)_{t\eta 0}$ and $(\pi_t)_{t\eta 0}$ are defined left continuous with right-hand limits. We call V_n the *new wealth* and η_n the *new risky fraction.* Because $(1 - \eta_{n-1})V_{n-1}$ is the new value invested in the bond and $\eta_{n-1}V_{n-1}$ the new value invested in the stock at time τ_{n-1} these parts evolve without trading according to the dynamics of the bond and of the stock, respectively, yielding (11.5) and (11.6). At τ_n the new wealth V_n is the wealth before trading minus the transaction costs to be paid, and the new risky fraction η_n is the new amount $\pi_{\tau_n} X_{\tau_n} + \Delta_n$ invested in the stock divided by the new wealth, leading to (11.7), (11.8). Note that (iii) implies that short selling is not allowed.

The investor has always to pay at least the fixed costs when stopping, in particular fees have to be paid at $\tau_0 = 0$ even if $\Delta_0 = 0$. But the initial payment does not matter for our objective, the maximization of

$$R^{\widetilde{K}}(x, \pi) = \liminf_{t\to\infty} \frac{1}{t} \operatorname{E}\left[\ln X_t^{\widetilde{K}} \mid X_0 = x, \pi_0 = \pi\right], \quad x > 0, \pi \in (0,1)\,, \tag{11.9}$$

over all admissible impulse control strategies $\widetilde{K}$. We like to find an optimal strategy $\widetilde{K}^*$ for which $R^* = \sup\{R^{\widetilde{K}} : \widetilde{K} \text{ admissible }\} = R^{\widetilde{K}^*}$. It is obvious that an optimal impulse control strategy may not exist, e.g. if $\widehat{\eta} \notin (0,1)$. So we would like to compare with the following degenerate strategies.

Lemma 1. *Let R^0, R^1, and $\widetilde{R}(\pi)$ denote the asymptotic growth rates for the strategies given by $\tau_1 = \infty$ and $\eta_0 = 0$, $\eta_0 = 1$, and $\eta_0 = \pi \in (0,1)$, respectively. Then*

$$R^0 = 0\,, \quad R^1 = \mu - \sigma^2/2\,, \quad \text{and} \quad \widetilde{R} \leq \max\{0, \mu - \sigma^2/2\}\,.$$

Proof: For R^0 and R^1 the result is obvious since the corresponding wealth processes are given by $X_t^0 = B_t$ and $X_t^1 = S_t$. For $\widetilde{R}(\pi)$ the wealth is given by $\widetilde{X}_t = HB_t + GS_t$, where $G = (1-\pi)V_0$ and $H = \pi V_0$ are the constant number of stocks and bonds the investor holds for $t > 0$. So,

$$\begin{aligned} \mathrm{E}\ln \widetilde{X}_t &\leq \mathrm{E}\ln(2\max\{H, G\,S_t\}) \\ &= \ln 2 + \mathrm{E}\max\{\ln H, \ln G + (\mu - \sigma^2/2)t + \sigma W_t\}\,. \end{aligned}$$

Therefore $\widetilde{R} = \lim_{t\to\infty} \frac{1}{t}\,\mathrm{E}\ln X_t \leq \max\{0, \mu - \sigma^2/2\}$. □

So we only have to analyze admissible strategies and compare the result with $\max\{0, \mu - \sigma^2/2\}$.

The following theorem provides a factorization of the wealth process and hence is basic for the subsequent analysis; compare [5, Theorem 3.6].

Theorem C.1. *Let $(\tau_n, \Delta_n)_{n\in\mathbb{N}}$ be an admissible impulse control strategy. Then for $t > 0$*

$$X_t = X_0\,\frac{1-\delta + A_0\gamma\,\pi_0}{1 + A_0\gamma\,\eta_0}\,\frac{1-\eta_{M_t}}{1-\pi_t}\prod_{n=1}^{M_t}\left(\frac{1-\eta_{n-1}}{1-\pi_{\tau_n}}\,\frac{1-\delta+A_n\gamma\,\pi_{\tau_n}}{1+A_n\gamma\,\eta_n}\right),$$

where $M_t = \sup\{n \in \mathbb{N}_0 : \tau_n < t\}$ and $A_n = \mathrm{sign}((1-\delta)\eta_n - \pi_n)$, $n \in \mathbb{N}$.

Proof: Eliminating V_n in (11.7), (11.8), using (11.4), and solving for Δ_n yields

$$\Delta_n = \frac{(1-\delta)\eta_n - \pi_n}{1 + \gamma\,\eta_n\,\mathrm{sign}(\Delta_n)}\,X_{\tau_n}\,.$$

Admissibility requires $\eta_n \in (0,1)$. So the denominator is strictly positive, hence

$$\mathrm{sign}(\Delta_n) = \mathrm{sign}((1-\delta)\eta_n - \pi_n) = A_n\,.$$

Therefore,

$$\widetilde{c}(X_{\tau_n}, \Delta_n) = \frac{\delta + \gamma\,A_n(\eta_n - \pi_{\tau_n})}{1 + \gamma\,A_n\eta_n}\,X_{\tau_n}$$

and substituting in (11.7) yields

$$V_n = \frac{1-\delta + A_n \gamma\, \pi_{\tau_n}}{1 + A_n \gamma\, \eta_n}\, X_{\tau_n}\,. \tag{11.10}$$

From (11.5), (11.6) we have $(1-\pi_t)X_t = (1-\eta_n)V_n$ for all $t \in (0,\infty) \cap (\tau_n, \tau_{n+1}]$, hence

$$V_n = \frac{1-\delta + A_n \gamma\, \pi_{\tau_n}}{1 + A_n \gamma\, \eta_n}\, \frac{1-\eta_{n-1}}{1-\pi_{\tau_n}}\, V_{n-1}$$

on $\{\tau_n < \infty\}$, $n\eta 1$. By induction we obtain

$$V_{M_t} = V_0 \prod_{n=1}^{M_t} \left(\frac{1-\eta_{n-1}}{1-\pi_{\tau_n}}\, \frac{1-\delta + A_n \gamma\, \pi_{\tau_n}}{1 + A_n \gamma\, \eta_n} \right),$$

hence (iii) follows from (11.10) for $n = 0$ and from $(1-\pi_t)X_t = (1-\eta_{M_t})V_{M_t}$. □

Remark 1. In Theorem C.1 the first factor under the product sign corresponds to the gain/loss due to the stock holdings in $(\tau_{n-1}, \tau_n]$ and the second factor to the transaction costs paid at τ_n.

So it seems convenient to reformulate the control problem in terms of the new risky fractions since the factors of X_t/X_0 in Theorem C.1 depend only on the risky fractions. Hence it is suitable to reformulate the control problem in such a way that the control is given by choosing the risky fraction η_n instead of the transaction Δ_n. That this can be done in a well defined way is guaranteed by the following lemma.

Lemma 2.

(i) *An impulse control strategy* $\widetilde{K} = (\tau_n, \Delta_n)_{n\in\mathbb{N}_0}$ *is* admissible *if and only if* $\Delta_n \in \mathcal{D}(X^{\widetilde{K}}_{\tau_n}, \pi^{\widetilde{K}}_{\tau_n})$, $n \in \mathbb{N}_0$, *where*

$$\mathcal{D}(x,\pi) = \{\Delta \in \mathbb{R} \,:\, \pi\, x + \Delta > 0, x - \widetilde{c}(x,\Delta) - D > 0, x - \widetilde{c}(x,\Delta) > 0\}.$$

Further $\mathcal{D}(x,\pi) \neq \emptyset$ *for all* $x > 0$, $\pi \in (0,1)$.

(ii) $f_\eta(x,\pi,\cdot) : \mathcal{D}(x,\pi) \to (0,1), \Delta \mapsto \dfrac{\pi\, x + \Delta}{x - \delta\, x - \gamma\,|\Delta|}$ *is a bijection for all* $x > 0, \pi \in (0,1)$.

(iii) $\eta^{\widetilde{K}}_n = f_\eta(X^{\widetilde{K}}_{\tau_n}, \pi^{\widetilde{K}}_{\tau_n}, \Delta_n)$ *for any admissible impulse control strategy* $\widetilde{K}$.

Proof: (i) follows directly from Definition 1.
(ii) Since $1-\delta-\gamma\,\pi > 0$ and $x-\delta\, x-\gamma\,|\Delta| > 0$, we have for all $x > 0, \pi \in (0,1)$, $\Delta \in \mathcal{D}(x,\pi)$, $\Delta \neq 0$,

$$\frac{\partial}{\partial \Delta} f_\eta(x,\pi,\Delta) = \frac{1-\delta + a\,\gamma\,\pi}{(x-\delta\,x - a\,\gamma\,\Delta)^2}\, x\eta \frac{1-\delta - \gamma\,\pi}{(x-\delta\,x - a\,\gamma\,\Delta)^2}\, x > 0\,,$$

where $a = \text{sign}(\Delta)$. So f_η is strictly increasing and continuous on $\mathcal{D}(x,\pi)$. Computing

$$\mathcal{D}(x,\pi) = \begin{cases} (-\pi\, x,\ (1-\pi-\delta)x/(1-\gamma)), & \text{if } 1-\pi-\delta \le 0\,, \\ (-\pi\, x,\ (1-\pi-\delta)x/(1+\gamma)), & \text{if } 1-\pi-\delta > 0\,. \end{cases}$$

and evaluating f_η at the boundary points yields $f_\eta(x,\pi,\mathcal{D}(x,\pi)) = (0,1)$. (iii) follows from (11.7), (11.8). □

So we have a one-to-one correspondence between admissible impulse control strategies and NRF-strategies defined below.

Definition 2.

(i) *A* new risky fraction impulse control strategy (NRF-strategy) $(\tau_n,\eta_n)_{n\in\mathbb{N}_0}$ *consists of stopping times* $0=\tau_0\le\tau_1\le\ldots\le\infty$ *with respect to* $(\mathcal{F}_t)_{t\eta 0}$*, which satisfy* $\tau_n\to\infty$ *a.s. and* $\tau_n<\tau_{n+1}$ *on* $\{\tau_n<\infty\}$*, and of* $\mathcal{F}_{\tau_n}$*-measurable random variables* η_n *with values in* $(0,1)$*. We call* τ_n *the* nth trading time *and* η_n *the* new risky fraction *at* τ_n*.* $\mathcal{K}$ *will denote the class of NRF-strategies.*

(ii) *Given an NRF-strategy and initial risky fraction* π_0 *and new risky fraction* η_0 *in* $(0,1)$*, the* risky fraction process $(\pi_t)_{t\eta 0}$ *is defined by*

$$\pi_t = \frac{\eta_n\, S_t/S_{\tau_n}}{1-\eta_n+\eta_n\, S_t/S_{\tau_n}} \qquad \textit{for all} \quad t\in(0,\infty)\cap(\tau_n,\tau_{n+1}], \quad n\in\mathbb{N}_0.$$

By

$$A_n = \text{sign}((1-\delta)\eta_n - \pi_{\tau_n}) \qquad \textit{on} \qquad \{\tau_n<\infty\}\,, \quad n\in\mathbb{N}_0\,,$$

we introduce the type of trading*. The wealth process* X^K *for an NRF-strategy* $K=(\tau_n,\eta_n)_{n\in\mathbb{N}}$ *is defined by the representation in Theorem C.1.*

(iii) *The (one stage)* gain function g *is*

$$g:(0,1)^2\to\mathbb{R},\ g(\pi,\eta) = \ln\frac{1-\eta}{1-\pi} + \ln\frac{1-\delta+\text{sign}((1-\delta)\eta-\pi)\gamma\,\pi}{1+\text{sign}((1-\delta)\eta-\pi)\gamma\,\eta}\,.$$

In Definition 2 (ii) the type of trading A_n is only introduced to simplify the notation, cf. Theorem C.1 Note that $A_n=-1,0,+1$ correspond to selling, holding, buying stocks. Using the gain function g and rearranging the factors in Theorem C.1 we can express the wealth process by

$$\ln X_t = \ln X_0 + \ln\frac{1-\delta+A_0\gamma\,\pi_0}{1+A_0\gamma\,\eta_0} + \sum_{n=1}^{M_t} g(\pi_{\tau_n},\eta_n) - \ln(1-\pi_t)\,.$$

This yields for $P_\pi = P(\,\cdot\,|\eta_0=\pi)$

$$R^K(\pi) = \liminf_{t\to\infty} \frac{1}{t}\, \mathrm{E}_\pi \left[\sum_{n=1}^{M_t} g(\pi_{\tau_n}, \eta_n) - \ln(1-\pi_t) \right]. \tag{11.11}$$

By Lemma 2 there is a one-to-one correspondence between admissible impulse control strategies and NRF-strategies, hence $R^* = \sup\{R^K : K \in \mathcal{K}\}$. So we are looking for a $K^* \in \mathcal{K}$ for which $R^* = R^{K^*}$.

Advantages of the reformulation in terms of η are the easier admissibility conditions, the direct use of the representation in Theorem C.1 leading to the simpler representation (11.11) instead of (11.9), and that we hence only have to control $(\pi_t)_{t\eta 0}$.

Control Strategies with Constant Boundaries

We shall now introduce our class of good trading strategies given by only four parameters, a, b to describe the boundaries of the continuation region (no trading region) and α, β for the new risky fractions when reaching these boundaries.

Definition 3. *An NRF-strategy* $K = (\tau_n, \eta_n)_{n\in\mathbb{N}_0}$ *is called a* trading strategy with constant boundaries *(*CB-strategy*), if there exist* a, b, α, β *such that* $0 < a < b < 1$ *and* $a < \alpha \le \beta \le (1-\delta)b$, *the inter trading times are given by stopping times with constant boundaries* a, b, *i.e.*

$$\tau_n = \inf\{t > \tau_{n-1} : \pi_t \notin (a,b)\}, \quad n \in \mathbb{N},$$

and the control variable η_n *satisfies* $\eta_0 \in (a,b)$ *and for* $n \in \mathbb{N}$

$$\eta_n = \varphi(\pi_{\tau_n}), \qquad \text{where} \qquad \varphi(b) = \beta,\ \varphi(a) = \alpha.$$

A CB-strategy will be denoted by (a,b,α,β). *By* $\mathcal{K}_{CB}$ *we denote the class of CB-strategies and by* $R^*_{CB} = \sup\{R^K : K \in \mathcal{K}_{CB}\}$ *the optimal asymptotic growth rate in this class.*

Between two trading times no trading occurs. In such a period the risky fraction process evolves like the processes in Definition 4, see also [15]. In Lemma 3 we gather some results about this process for further reference.

Definition 4. *For the fraction* $\eta_0 \in (0,1)$ *after trading and subtracting of transaction costs in* $t = 0$ *the* risky fraction process without trading $(\pi^o_t)_{t\eta 0}$ *is defined by*

$$\pi^o_t = \frac{\eta_0\, S_t}{1 - \eta_0 + \eta_0\, S_t}.$$

Furthermore we denote for given $(a,b,\alpha,\beta) \in \mathcal{K}_{CB}$

$$\tau^o = \inf\{t\eta 0 : \pi^o_t \notin (a,b)\}.$$

Lemma 3. *Suppose* $(a, b, \alpha, \beta) \in \mathcal{K}_{CB}$. *For* $\pi \in [a, b]$,

$$P_\pi(\pi^o_{\tau^o} = b) = \frac{h_0(\pi) - h_0(a)}{h_0(b) - h_0(a)},$$

and

$$\mathrm{E}_\pi \tau^o = \frac{h_0(\pi) - h_0(a)}{h_0(b) - h_0(a)}(h_1(b) - h_1(a)) + h_1(a) - h_1(\pi),$$

where in the case $\widehat{\eta} \in (0,1) \setminus \{\frac{1}{2}\}$

$$h_0(x) = -\left(\frac{1-x}{x}\right)^{2\widehat{\eta}-1}, \quad h_1(x) = -\frac{2 \ln \frac{1-x}{x}}{\sigma^2(2\widehat{\eta}-1)},$$

and in the case $\widehat{\eta} = \frac{1}{2}$

$$h_0(x) = \ln \frac{1-x}{x}, \quad h_1(x) = \frac{1}{\sigma^2}\left(\ln \frac{1-x}{x}\right)^2.$$

Proof: The risky fraction process without trading $(\pi^o_t)_{t\eta 0}$ is a diffusion in $(0,1)$ with dynamics

$$d\pi^o_t = \pi^o_t(1-\pi^o_t)\Big((\mu - \pi^o_t\sigma^2)dt + \sigma\, dW_t\Big),$$

which can be seen from applying Itô's rule to $(\pi^o_t)_{t\eta 0}$. The generator L_π of $(\pi^o_t)_{t\eta 0}$ is given on $\mathcal{C}^2_c(0,1)$, the two times continuously differentiable functions on $(0,1)$ with compact support, by

$$L_\pi h(x) = (1-x)x(\mu - \sigma^2 x)h'(x) + \tfrac{1}{2}(1-x)^2 x^2 \sigma^2 h''(x).$$

Solutions of $L_\pi h_0 = 0$ and $L_\pi h_1 = 1$ on $(0,1)$ are, separately for the two cases $\widehat{\eta} \in (0,1) \setminus \{\frac{1}{2}\}$ and $\widehat{\eta} = \frac{1}{2}$, those h_0, h_1 given in the proposition. By Doob's Optional Sampling Theorem, see e.g. [17, Theorem II.3.2], the result follows with truncation from the fact that $(h(\pi^o_t) - \int_0^t L_\pi h(\pi^o_s)ds)_{t\eta 0}$ is a martingale for all $h \in \mathcal{C}^2_c(0,1)$, see [17, Proposition VII.1.6]. □

The Regenerative Structure

Suppose that a CB-strategy $K = (a, b, \alpha, \beta) \in \mathcal{K}_{CB}$ is given and τ^o is the corresponding stopping time with constant boundaries a, b for $(\pi^o_t)_{t\eta 0}$ as defined in Section 11. We always use $P_\pi = P(\,\cdot\,|\eta_0 = \pi)$. Since

$$|\ln(1-\pi_t)| \le |\ln(1-b)| \quad \text{and} \quad |g(\pi_{\tau_n}, \eta_n)| \le \max\{|g(b,\beta)|, |g(a,\alpha)|\}$$

the representation (11.11) yields

$$R^K(\pi) = \liminf_{t\to\infty} \frac{1}{t}\, \mathrm{E}_\pi \sum_{n=1}^{\bar{M}_t} g(\pi_{\tau_n}, \eta_n), \quad \pi \in (0,1), \tag{11.12}$$

where from now on $\bar{M}_t = \sup\{n \in \mathbb{N}_0 : \tau_n \le t\}$.

Lemma 4.

(i) $P_\pi(\tau_n - \tau_{n-1} \in \cdot \,|\, \mathcal{F}_{\tau_{n-1}}) = P_{\eta_{n-1}}(\tau^o \in \cdot)$.
(ii) $P_\pi(\pi_{\tau_n} \in \cdot \,|\, \mathcal{F}_{\tau_n}) = P_{\eta_{n-1}}(\pi^o_{\tau^o} \in \cdot)$. *Thus* $\mathrm{E}_\pi[g(\pi_{\tau_n}, \eta_n)|\mathcal{F}_{\tau_{n-1}}] = \mathrm{E}_{\eta_{n-1}} g(\pi^o_\tau, \varphi(\pi^o_\tau))$.
(iii) K *is an NRF-strategy satisfying* $\tau_n > \tau_{n-1}$, $\mathrm{E}_\pi \tau_n < \infty$ *for all* $n \in \mathbb{N}$, $\pi \in (0,1)$.

Proof: (i) and (ii) are immediate from the strong Markov property of the underlying Brownian motion. (iii) follows from Lemma 3 □

Proposition 1. *The process* $(\pi_{\tau_n})_{n\in\mathbb{N}}$ *is a homogeneous Markov chain with state space* $\{a,b\}$, *initial distribution* $P_{\eta_0}(\pi^o_{\tau^o} \in \cdot)$. *The transition probabilities* $p_{x,y} = P_\pi(\pi_{\tau_n} = y|\pi_{\tau_{n-1}} = x) = P_{\varphi(x)}(\pi^o_{\tau^o} = y)$ *satisfy* $p_{x,y} > 0$, $x, y \in \{a,b\}$. *The invariant distribution* $(p, 1-p)^\top$ *of* $(\pi_{\tau_n})_{n\in\mathbb{N}}$ *is given by*

$$p = \frac{p_{b,a}}{p_{a,b} + p_{b,a}}.$$

Proof: The Markov property follows from the strong Markov property of $(W_t)_{t\eta 0}$, the adaptedness of $(\pi_{\tau_n})_{n\in\mathbb{N}}$, and $\eta_n = \varphi(\pi_{\tau_n})$. The transition probabilities can be computed from Lemma 3, which shows that they are strictly positive. With the notation in the proposition, the invariant distribution is given by

$$(p,\, 1-p)\begin{pmatrix} p_{a,a} & p_{a,b} \\ p_{b,a} & p_{b,b} \end{pmatrix} = \begin{pmatrix} p \\ 1-p \end{pmatrix}.$$

□

Definition 5. *The measure* ν *on* $\{\alpha, \beta\}$ *given by*

$$\nu(\alpha) = 1 - \nu(\beta) = p, \quad \text{if} \quad \alpha < \beta,$$

p *as in Proposition 1, and* $\nu \equiv 1$, *if* $\alpha = \beta$, *is called the* invariant distribution of K. *The probability measure corresponding to* ν *as initial distribution is*

$$P_\nu = p\,P_\alpha + (1-p)P_\beta.$$

The following lemma contains the technical part of the proof of Proposition 2.

Lemma 5. *Define* $N_1 = \inf\{n \in \mathbb{N} : \pi_{\tau_n} = a\}$. *Suppose that* $(Z_n)_{n\in\mathbb{N}}$ *is real valued, adapted to* $(\mathcal{F}_{\tau_n})_{n\in\mathbb{N}}$ *and, for all* $\pi \in (0,1)$, $n \in \mathbb{N}$, *satisfies*

$$\mathrm{E}_\pi|Z_n| < \infty \quad \text{and} \quad P_\pi(Z_n \in \cdot \,|\, \mathcal{F}_{\tau_{n-1}}) = P_{\eta_{n-1}}(Z \in \cdot),$$

where Z *is distributed like* Z_1. *Then for all* $\pi \in (0,1)$

$$\mathrm{E}_\pi\Big|\sum_{n=1}^{N_1} Z_n\Big| < \infty \quad \text{and} \quad \mathrm{E}_\alpha \sum_{n=1}^{N_1} Z_n = \frac{1}{p}\,\mathrm{E}_\nu Z,$$

where $\nu = (p, 1-p)^\top$ *is the invariant distribution of* K.

Proof: A computation like below substituting $P_\pi(\pi^o_{\tau^o} = y)$ for $p_{a,y}$ yields $\mathrm{E}_\pi|\sum_{n=1}^{N_1} Z_n| < \infty$. By Lemma 4 and Proposition 1

$$\begin{aligned}
\mathrm{E}_\alpha \sum_{k=1}^{N_1} Z_k &= \sum_{n=1}^{\infty} \mathrm{E}_\alpha \Big[\sum_{k=1}^{n} Z_k \,\Big|\, N_1 = n \Big] P_\alpha(N_1 = n) \\
&= p_{a,a} \mathrm{E}_\alpha [Z | \pi^o_{\tau^o} = a] \\
&\quad + \sum_{n=2}^{\infty} p_{a,b} (p_{b,b})^{n-2} p_{b,a} \Big(\mathrm{E}_\alpha [Z | \pi^o_{\tau^o} = b] + (n-2) \mathrm{E}_\beta [Z | \pi^o_{\tau^o} = b] + \\
&\quad + \mathrm{E}_\beta [Z | \pi^o_{\tau^o} = a] \Big) \\
&= \mathrm{E}_\alpha [Z | \pi^o_{\tau^o} = a] \, p_{a,a} + \mathrm{E}_\alpha [Z | \pi^o_{\tau^o} = b] \, p_{a,b} \\
&\quad + p_{a,b} / p_{b,a} \, (\mathrm{E}_\beta [Z | \pi^o_{\tau^o} = a] \, p_{b,a} + \mathrm{E}_\beta [Z | \pi^o_{\tau^o} = b] \, p_{b,b}) \\
&= \mathrm{E}_\alpha Z + p_{a,b} / p_{b,a} \, \mathrm{E}_\beta Z \\
&= \tfrac{1}{p} \, (p \, \mathrm{E}_\alpha Z + (1-p) \mathrm{E}_\beta Z) = \tfrac{1}{p} \, \mathrm{E}_\nu Z \, .
\end{aligned}$$

□

Proposition 2. *Define* $N_0 = 0$,

$$N_k = \inf\{ n > N_{k-1} \, : \, \pi_{\tau_n} = a \} \, , \quad k \in \mathbb{N} \, ,$$

and

$$Y_k = \sum_{n=N_{k-1}+1}^{N_k} g(\pi_{\tau_n}, \eta_n) \, , \quad k \in \mathbb{N} \, .$$

Then

(i) $(Y_k, \tau_{N_k} - \tau_{N_{k-1}})_{k \in \mathbb{N}}$ *are i.i.d. random variables under* P_α.

(ii) $\mathrm{E}_\pi N_1$, $\mathrm{E}_\pi \tau_{N_1}$, $\mathrm{E}_\pi |Y_1| < \infty$ *for* $\pi \in (0,1)$ *and*

$$p \, \mathrm{E}_\alpha \tau_{N_1} = \mathrm{E}_\nu \tau^o \, , \qquad p \, \mathrm{E}_\alpha Y_1 = \mathrm{E}_\nu \, g(\pi^o_{\tau^o}, \varphi(\pi^o_{\tau^o})) \, .$$

Proof: The strong Markov property of $(W_t)_{t \eta 0}$ yields (i), since $\eta_{N_k} = \alpha$ for all $k \in \mathbb{N}_0$. (ii) follows from Lemma 5 using $Z_n = 1$ to obtain the result for N_1, $Z_n = \tau_n - \tau_{n-1}$ for τ_{N_1}, and $Z_n = g(\pi_{\tau_n}, \eta_n)$ for Y_1. □

The regenerative structure in Proposition 2 (i) enables us now, to simplify the determination of the asymptotic growth rate to one period between two buyings. Using (ii) reduces this to one trading period starting with the invariant distribution.

Theorem C.2. *The asymptotic growth rate of K is*

$$R^K = \frac{\mathrm{E}_\nu\, g(\pi^o_{\tau^o}, \varphi(\pi^o_{\tau^o}))}{\mathrm{E}_\nu \tau^o},$$

where ν is the invariant distribution of K. In particular R^K is independent of the initial conditions.

Proof: Let N_k, $k \in \mathbb{N}_0$, and Y_n, $n \in \mathbb{N}$ be given as in Proposition 2. For $N(t) = \sup\{k \in \mathbb{N}_0 : \tau_{N_k} \le t\}$ and $\bar{M}_t = \sup\{n \in \mathbb{N}_0 : \tau_n \le t\}$, we have $N_{N(t)} \le M_t$. Due to the integrability of Y_1, its independence of Y_k, $k > 1$, and to $\eta_{N_1} = \varphi(\pi_{\tau_{N_1}}) = \alpha$, it follows from (11.12) for all $\pi \in (0,1)$

$$R^K(\pi) = \liminf_{t\to\infty} \frac{1}{t}\Bigg(\mathrm{E}_\pi Y_1 - \mathrm{E}_\alpha Y_1 + \mathrm{E}_\alpha\Bigg[\sum_{k=1}^{N(t)+1} Y_k\Bigg]$$
$$-\mathrm{E}_\pi Y_{N(t)+1} + \mathrm{E}_\pi\Bigg[\sum_{n=N_{N(t)}+1}^{\bar{M}(t)} g(\pi_{\tau_n}, \eta_n)\Bigg]\Bigg).$$

We will number the terms to which we apply the lim inf by '1' to '5'. Term 1 and term 2 are bounded by Proposition 2, and the boundedness of term 5 can be shown as in the proof of Lemma 5. For term 4 we refer to the proof of [18, Theorem 3.16], in which the convergence $\frac{1}{t}\mathrm{E}_\pi Y_{N(t)+1} \to 0$ is proved in exactly our situation using the renewal theorem. So only term 3 remains,

$$R^K(\pi) = \liminf_{t\to\infty} \frac{1}{t}\, \mathrm{E}_\alpha\Bigg[\sum_{k=1}^{N(t)+1} Y_k\Bigg].$$

Proposition 2 says that $(\tau_{N_k} - \tau_{N_{k-1}})_{k\in\mathbb{N}}$ constitutes under P_α a sequence of integrable i.i.d. random variables. Therefore, we are in the setting of renewal theory, where $N(t)$, $t\eta 0$, counts the renewals up to time t. An elementary result gives us $\mathrm{E}\, N(t) < \infty$ for all $t\eta 0$, e.g. [18, Proposition 3.2]. Moreover according to Proposition 2, with respect to P_α, $(Y_k)_{k\in\mathbb{N}}$ is a sequence of integrable i.i.d. random variables and $N(t)+1$ is a stopping time for $(N_k)_{k\in\mathbb{N}}$. Thus Wald's identity yields

$$\mathrm{E}_\alpha\Bigg[\sum_{k=1}^{N(t)+1} Y_k\Bigg] = \mathrm{E}_\alpha\big[N(t)+1\big]\,\mathrm{E}_\alpha Y_1.$$

Application of the elementary renewal equation to term 4 finally shows

$$\frac{1}{t}\,\mathrm{E}_\alpha\Bigg[\sum_{k=1}^{N(t)+1} Y_k\Bigg] = \Bigg(\frac{\mathrm{E}_\alpha\big[N(t)\big]}{t} + \frac{1}{t}\Bigg)\mathrm{E}_\alpha Y_1 \longrightarrow \frac{\mathrm{E}_\alpha Y_1}{\mathrm{E}_\alpha \tau_{N_1}} \quad (t\to\infty).$$

This implies the independence of the initial value π. The statement follows now simply by Proposition 2 (ii):

$$R^K = \frac{\mathrm{E}_\alpha Y_1}{\mathrm{E}_\alpha \tau_{N_1}} = \frac{\frac{1}{p}\,\mathrm{E}_\nu\, g(\pi^o_{\tau^o}, \varphi(\pi^o_{\tau^o}))}{\frac{1}{p}\,\mathrm{E}_\nu \tau^o} = \frac{\mathrm{E}_\nu\, g(\pi^o_{\tau^o}, \varphi(\pi^o_{\tau^o}))}{\mathrm{E}_\nu \tau^o}\,.$$

□

Remark 2. We introduced the wealth simply as the sum of the positions in bond and stock. If we instead evaluate the portfolio at time T by its utility after selling the stocks, this would correspond to a factor $(1-\delta-\gamma\,\pi_T)$. Since for a CB-strategy $\mathrm{E}_\pi \ln(1-\delta-\gamma\,\pi_T)$ is bounded we would get the same asymptotic growth rate as in our case and hence the same optimal CB-strategy.

Existence of Optimal CB-Strategies

For a CB-strategy $K=(a,b,\alpha,\beta)$ Corollary 1 shows the dependence of the asymptotic growth rate on the four parameters explicitly. Based on this representation Theorem C.3 provides an existence result in the class $\mathcal{K}_{CB}$.

Corollary 1. *For $K=(a,b,\alpha,\beta)\in\mathcal{K}_{CB}$,*

$$R^K = \frac{p\,g(a,\alpha)+(1-p)g(b,\beta)}{p(h_1(a)-h_1(\alpha))+(1-p)(h_1(b)-h_1(\beta))}\,,$$

where

$$p=\frac{h_0(b)-h_0(\beta)}{h_0(\alpha)-h_0(a)+h_0(b)-h_0(\beta)}$$

and h_0, h_1 were defined in Lemma 3

Proof: From the definition of the invariant distribution,

$$\mathrm{E}_\nu g(\pi^o_{\tau^o},\varphi(\pi^o_{\tau^o})) = p\,g(a,\alpha)+(1-p)g(b,\beta)\,.$$

So Theorem C.2, Lemma 3 and Proposition 1 yield the representation. □

Theorem C.3. *If there exists a CB-strategy K satisfying $R^K>\max\{0,\mu-\sigma^2/2\}$ then a CB-strategy K^* exists which is optimal in $\mathcal{K}_{CB}$, i.e. $R^{K^*}=R^*_{CB}$.*

Proof: Suppose $\mu-\sigma^2/2>0$. Then $2\widehat{\eta}-1=2(\mu-\sigma^2/2)/\sigma^2>0$. Using Lemma 3 the denominator in Corollary 1 becomes

$$\frac{1}{\mu-\sigma^2/2}\left(p\ln\frac{a(1-\alpha)}{(1-a)\alpha}+(1-p)\ln\frac{b(1-\beta)}{(1-b)\beta}\right), \qquad (11.13)$$

where p is given as in Corollary 1 using $h_0(\pi)=-((1-\pi)/\pi)^{2\widehat{\eta}-1}$. Note that

$$\lim_{b\nearrow 1} p = \frac{-h_0(\beta)}{h_0(\alpha)-h_0(a)-h_0(\beta)} \in (0,1)\,, \quad \lim_{a\searrow 0} p = 0\,.$$

Thus, using (11.13) and keeping the other parameters constant,

$$\lim_{b \nearrow 1} R^{(a,b,\alpha,\beta)} = \mu - \frac{\sigma^2}{2},$$
$$\lim_{a \searrow 0} R^{(a,b,\alpha,\beta)} = \frac{g(b,\beta)}{\ln \frac{b(1-\beta)}{(1-b)\beta}} \left(\mu - \frac{\sigma^2}{2}\right) < \mu - \frac{\sigma^2}{2},$$
$$\lim_{\alpha \searrow a} R^{(a,b,\alpha,\beta)} = -\infty .$$

The latter is due to $\delta > 0$. It implies $R^{(a,b,\alpha,\beta)} < 0$ for $b \to a$. Therefore for all $\varepsilon > 0$ the set

$$\mathcal{K}_\varepsilon = \{(a,b,\alpha,\beta) \in \mathcal{K}_{CB} \; : \; R^{(a,b,\alpha,\beta)} \eta \mu - s^2/2 + \varepsilon\}$$

is closed in $\mathbb{R}^4$, hence compact. If there exists $K = (a,b,\alpha,\beta) \in \mathcal{K}_{CB}$ satisfying $R^K > \mu - \sigma^2/2$, then $\mathcal{K}_\varepsilon$ will be non-empty for some $\varepsilon_0 > 0$ and due to compactness, an optimal solution exists in $\mathcal{K}_\varepsilon$ since R is continuous.

In the cases $\mu - \sigma^2/2 < 0$ and $\mu - \sigma^2/2 = 0$ the arguments are similar, only that in the latter symmetric case we have to use those h_0, h_1 given for $\widehat{\eta} = \frac{1}{2}$ in Lemma 3. □

Remark 3. Let us not allow short selling or borrowing. Then in the case $\mathcal{K}_\varepsilon = \emptyset$ for all $\varepsilon > 0$ it will be optimal to invest only in the stock if $\mu - \sigma^2/2\eta 0$ and to invest only in the money market if $\mu - \sigma^2/2 \leq 0$, see Lemma 1. In this sense, always a strategy exists which is optimal in $\mathcal{K}_{CB} \cup \{K^0, K^1\}$, where K^0, K^1 correspond to these extreme strategies.

Example 1. For parameters $\mu = 0.096$, $\sigma = 0.4$, $r = 0$ we obtain from (11.3) $\widehat{\eta} = 0.6000$ and $\widehat{R} = 0.0288$ if no transaction costs have to be paid. For costs $\delta = 0.01\%$, $\gamma = 0.3\%$ we obtain the optimal values

$$\widehat{a} = 0.4876,\ \widehat{b} = 0.7081, \qquad \widehat{\alpha} = 0.5680,$$
$$\widehat{\beta} = 0.6338,\ R^*_{CB} = 0.0284, \quad \mathrm{E}_\nu \tau^o = 1.2501.$$

In Figure 11.1 we analyze the dependency on the choice of the boundaries. In Figure 11.1 (i) we compute $R^{(a,b,\widehat{\alpha},\widehat{\beta})}$ on $0 < a < \widehat{\alpha}$, $\widehat{\beta}/(1-\delta) \leq b < 1$, and in (ii) $R^{(\widehat{a},\widehat{b},\alpha,\beta)}$ on $\widehat{a} < \alpha \leq \widehat{\eta}$, $\widehat{\eta} \leq \beta \leq (1-\delta)\widehat{b}$. The fat lines correspond to the optimal parameters, so the optimum lies at their intersection.

In Figure 11.2 (i) we plot the optimal boundaries depending on γ for constant $\delta = 0.01\%$, where the case $\gamma = 0$ is the setting of [15], and in (ii) depending on δ for constant $\gamma = 0.3\%$. Note that in the latter case we cannot start at $\delta = 0$ since the case of purely proportional costs is not covered by our model.

Example 2. In Example 1 we chose parameters μ and σ such that $\widehat{\eta}$, the optimal fraction without costs, lies in $(0,1)$ far away from the boundaries. If we use instead $\mu = 0.159$ we get

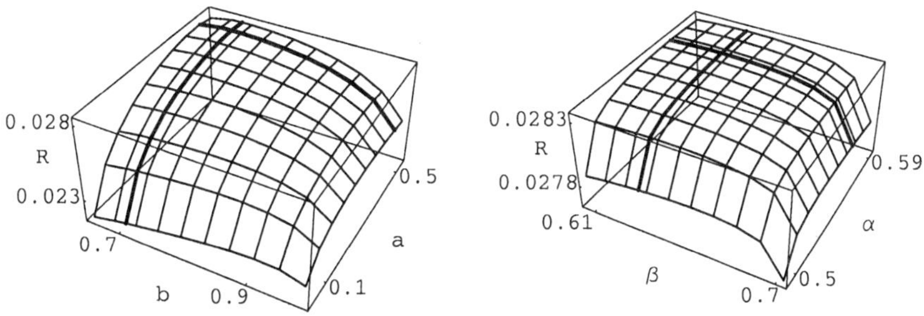

Fig. 11.1. Dependency of R on the choice of (i) b, a and (ii) on β, α

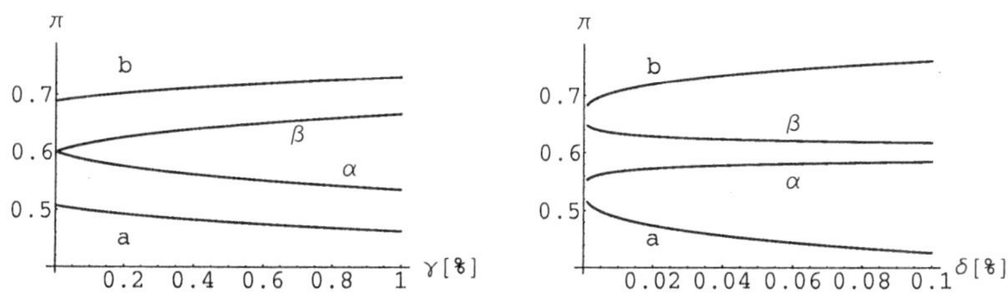

Fig. 11.2. Dependency of the trading regions on the (i) proportional costs γ, (ii) fixed costs δ

$$\widehat{\eta} = 0.993750, \quad \widehat{R} = 0.079003, \quad R^1 = 0.079000.$$

So $\widehat{R}$ is only slightly larger than $R^1 = \mu - \sigma^2/2$. Thus it can be expected that under high enough transactions costs no optimal solution in $\mathcal{K}_{CB}$ exists, i.e. $R^K < R^1$ for all CB-strategies K. And that is exactly what is happening as can be shown numerically by evaluating $R^K = R^{(a,b,\alpha,\beta)}$ for all possible b while choosing the other parameters optimally.

Numerically it can be shown that, in Example 2, the left-hand derivative when b approaches 1 is strictly positive while it is negative in Example 1 as we can see from Figure 11.1 (i). Since we know from the proof of Theorem C.3 that R^K converges to R^1 when b approaches 1, the sign of the left hand-derivative at 1 seems to be a good criterion for the existence of an optimal strategy in $\mathcal{K}_{CB}$. If it is negative we know from Theorem C.3 that one exists. But unfortunately the dependency on μ, or equivalently on $\widehat{\eta}$, is not monotone, hence an exact criterion is difficult to obtain.

Optimality in the Class of Impulse Control Strategies

We give a short outline how to verify the optimality of an optimal CB-strategy in the class of impulse control strategies, using the results of [5]. The rigorous and lengthy treatment will be deferred to a forthcoming publication.

The conditions (C1)–(C4) in [5] are satisfied and [5, Proposition 4.2] can be reformulated for our cost structure. Similar as in Section 6 of that paper, the quasi variational inequalities, which we have to solve, can – using the generator L_π defined in the proof of Lemma 3 – be reduced to

$$\begin{aligned} &(I)\ \ L_\pi\psi(\pi) + (\mu - \sigma^2/2\,\pi)\pi - \lambda \le 0, \quad \pi \in (0,1)\,, \\ &(II)\ \ \sup_{\eta\in[0,1]} \Big\{c(\eta,\pi) + \psi(\eta)\Big\} - \psi(\pi) \le 0\,, \quad \pi \in [0,1], \\ &\qquad (I)\times(II) = 0\,, \quad \pi \in (0,1)\,, \end{aligned}$$

where for $\eta \in (0,1)$, $\pi \in [0,1]$

$$c(\eta,\pi) = \ln(1-\delta+\mathrm{sign}((1-\delta)\eta-\pi)\gamma\,\pi) - \ln(1+\mathrm{sign}((1-\delta)\eta-\pi)\gamma\,\eta)\,.$$

If $r \neq 0$ we have to substitute in (I) $\mu - r$ for μ and $\lambda - r$ for λ.

Theorem C.4. *Suppose $K = (a,b,\alpha,\beta) \in \mathcal{K}_{CB}$ and that we can find a solution ψ for (I) with $\lambda = R^K$ satisfying $(II) = 0$ at a, b and $(II) < 0$ on (a,b), and that we can find an extension $\widehat{\psi}$ of $\psi|_{(a,b)}$ such that the supremum in (II) is attained at α if $\pi \in [0,a]$ and at β if $\pi \in [b,1]$. If $\widehat{\psi}$ lies in the Sobolev space $H^2(0,1)$ and the corresponding conditions in [5, Theorem 5.1] are satisfied, then K is an optimal impulse control strategy, i.e. $R^* = R^K$.*

If we have found a CB-strategy $K^* = (a,b,\alpha,\beta)$ that is optimal in $\mathcal{K}_{CB}$ with value R^{K^*}, we can proceed as follows to verify that K^* is optimal in $\mathcal{K}$.

(1) A solution of $(I) = 0$ is given by (use $\lambda = R^{K^*}$)

$$\psi_C(\pi) = C\,h_0(\pi) + R^{K^*}h_1(\pi) + \ln(1-\pi)\,, \quad C \in \mathbb{R}\,.$$

(2) Choose C^* such that for ψ_{C^*} we have equality in (II) at a and b.
(3) Define ψ as

$$\psi(\pi) = \begin{cases} \psi_{C^*}(\alpha) + c(\alpha,\pi)\,, & \pi \in [0,a)\,, \\ \psi_{C^*}(\pi) = C^*h_0(\pi) + R^{K^*}h_1(\pi) + \ln(1-\pi)\,, & \pi \in [a,b]\,, \\ \psi_{C^*}(\beta) + c(\beta,\pi)\,, & \pi \in (b,1]\,. \end{cases} \tag{11.14}$$

(4) Show (numerically) $(II) < 0$ on (a,b), that the supremum in (II) is attained at β if $\pi \in [b,1]$ and at α if $\pi \in [0,a]$, and that ψ is differentiable at a and b.
(5) Show (numerically) that $(I) \le 0$ on $(0,1)$.
(6) Show that the corresponding conditions in [5, Theorem 5.1] are satisfied.

Then ψ is bounded, lies in $H^2(0,1)$, and $(I) \times (II) = 0$ holds on $(0,1)$ by the definition of ψ. So K^* is optimal in the class of impulse control strategies.

Example 3. For the parameters of Example 1 an optimal ψ is given by (11.14), where we use the constants R^*_{CB}, $\widehat{a}$, $\widehat{b}$, $\widehat{\alpha}$, $\widehat{\beta}$ obtained in Example 1 and h_0, h_1 are given in Lemma 3 using the $\widehat{\eta} \neq \frac{1}{2}$ from Example 1. As described in the algorithm above, we can compute $C^* = -6.355669$ and the smooth pasting properties can be verified numerically. In Figure 11.3 we plot (i), left-

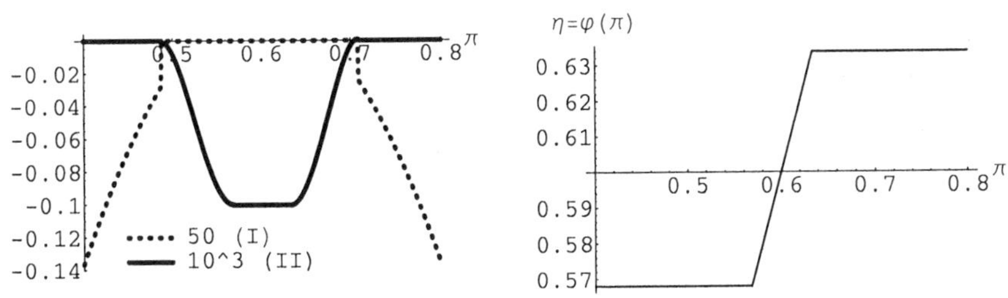

Fig. 11.3. Example 3: Verification of the optimality of the CB-strategy in Example 1.

hand, the functional in (I) as dotted curve and the supremum in (II) as solid curve, and (ii), right-hand, the corresponding maximizer. Note that in (i) the region where the solid curve is negative corresponds to the no-trading region, the boundary points where it first touches the axis are $\widehat{a}$, $\widehat{b}$, and that in the trading regions the maximizers in (ii) are given by $\widehat{\alpha}$ and $\widehat{\beta}$.

Remark 4. Comparing with a direct solution of the quasi variational inequalities our approach has the numerical advantage that due to the explicit structure of the asymptotic growth rate as given in Corollary 1 we only have to maximize in four parameters. This is much more stable than solving the quasi variational inequalities since this would involve finding the parameters C and λ, the latter corresponding to the optimal growth rate, which is as a function of the parameters a, b, α, β very flat around the optimal parameters, cf. Figure 11.1.

Remark 5. Our model contains as special case the purely fixed costs ($\gamma = 0$), but not the case of purely proportional costs ($\delta = 0$), cf. Figure 11.2. Following a heuristic approach like e.g. in [20] it can be shown that the HJB equation for the maximization of the asymptotic growth rate for purely proportional costs is of the form

$$\max\{L_\pi \psi(\pi) + (\mu - \tfrac{\sigma^2}{2}\,\pi)\pi - \lambda, \psi'(\pi) - \tfrac{\gamma}{1+\gamma\,\pi}, -\psi'(\pi) - \tfrac{\gamma}{1-\gamma\,\pi}\} = 0\,,$$

where equality holds in the first component if no trading is optimal, in the second component if buying is optimal, and in the third component if selling is optimal. Note that e.g. [1], [7], [10], [19] use a different optimization criterion and [2], [20] use different cost structures and hence they obtain slightly different HJB equations.

Further note that we can obtain these inequalities from (I), (II) by taking the limit $\delta \to 0$ in (II) and taking the derivatives with respect to π in the two cases that we sell and that we buy, observing that ψ' is decreasing in the no-trading region. Numerically the solutions converge. So the case of purely proportional costs can be seen as a limiting case of our model. To make these arguments rigorous, including the convergence of the optimal strategies, will be the objective of a future publication.

References

[1] M. Akian, J.L. Menaldi, A. Sulem (1996). *On an investment-consumption model with transaction costs.* SIAM Journal on Control and Optimization 34, 329–364.

[2] M. Akian, A. Sulem, M.I. Taksar (2001). *Dynamic optimization of a long-term growth rate for a portfolio with transaction costs and logarithmic utility.* Mathematical Finance 11, 153–188.

[3] C. Atkinson, P. Wilmott (1995). *Portfolio management with transaction costs: An asymptotic analysis of the Morton and Pliska model.* Mathematical Finance 5, 357–368.

[4] R.N. Bhattacharya, M. Majumdar (1989). *Controlled semi-Markov models under long-run average rewards.* Journal of Statistical Planning and Inference 22, 223–242.

[5] T.R. Bielecki, S.R. Pliska (2000). *Risk sensitive asset management with transaction costs.* Finance and Stochastics 4, 1–33.

[6] J.-P. Chancelier, B. Øksendal, A. Sulem (2002). *Combined stochastic control and optimal stopping, with application to portfolio optimization under fixed transaction costs.* Proceedings of the Steklov Institute of Mathematics 2 (237), 140–163.

[7] M.H.A. Davis, A.R. Norman (1990). *Portfolio selection with transaction costs.* Mathematics of Operations Research 15, 676–713.

[8] J.E. Eastham, K. J. Hastings (1988). *Optimal impulse control of portfolios.* Mathematics of Operations Research 13, 588–605.

[9] K. JaneVcek, S. Shreve (2004). *Asymptotic analysis for optimal investment and consumption with transaction costs.* Finance and Stochastics 8, 181–206.

[10] Y. Kabanov, C. Klüppelberg (2004). *A geometric approach to portfolio optimization in models with transaction costs.* Finance and Stochastics 8, 207–227.

[11] R. Korn (1998). *Portfolio optimisation with strictly positive transaction costs and impulse control.* Finance and Stochastics 2, 85–114.

[12] S.A. Lippmann (1971). *Maximal average-reward policies for semi-Markov decision processes with arbitrary state and action space.* The Annals of Mathematical Statistics 42, 1717–1726.

[13] M. J. P. Magill, M. Constantinides (1976). *Portfolio selection with transaction costs.* Journal of Economic Theory 13, 245–263.

[14] R.C. Merton (1969). *Lifetime portfolio selection under uncerttainty: The continuous-time case.* Review of Economics and Statistics 51, 247–257.

[15] A.J. Morton, S.R. Pliska (1995). *Optimal portfolio management with fixed transaction costs.* Mathematical Finance 5, 337–356.

[16] B. Øksendal, A. Sulem (2002). *Optimal consumption and portfolio with both fixed and proportional transaction costs.* SIAM Journal of Control and Optimization 40, 1765–1790.

[17] D. Revuz, M. Yor (1991). *Continuous Martingales and Brownian Motion.* Springer, Berlin.

[18] S.M. Ross (1970). *Applied Probability Models with Optimization Applications.* Holden-Day, San Francisco.

[19] S.E. Shreve, H.M. Soner (1994). *Optimal investment and consumption with transaction costs.* Annals of Applied Probability 4, 609–692.

[20] M. Taksar, M.J. Klass, D. Assaf (1988). *A diffusion model for optimal portfolio selection in the presence of brokerage fees.* Mathematics of Operations Research 13, 277–294.

Power and Multipower Variation: inference for high frequency data

Jeannette H.C. Woerner

Institut für Mathematische Stochastik, Universität Göttingen, Maschmühlenweg 8-10, D-37073 Göttingen, Germany
woerner@math.uni-goettingen.de

Summary. In the framework of stochastic volatility models there is a wide range of applications of power, bipower and multipower variation, i.e. the sum of appropriately scaled absolute values of log-returns and neighbouring log-returns raised to a certain power. Given high frequency data we can use the concept of power and multipower variation in the context of model selection, namely to determine if the underlying process possesses a jump component, as well as estimating the integrated volatility both in classical and Lévy type stochastic volatility models.

In this paper we will focus on bipower and multipower variation for classical stochastic volatility models. These concepts provide more robustness against jump components for estimators needed for pricing classical variance or volatility swaps. Furthermore, a combination of power and multipower variation can be used to separate the continuous and the jump part of the quadratic variation and hence gives insight in determining whether the classical purely continuous stochastic volatility model is appropriate.

Introduction

In the last years the concept of power variation, i.e. taking

$$\sum_i (t_i - t_{i-1})^\gamma |X_{t_i} - X_{t_{i-1}}|^p \text{ as } \max_i |t_i - t_{i-1}| \to 0\,,$$

where X_t denotes the log-price process, or more recently the concept of k-power variation, i.e. taking

$$\sum_i \prod_{j=i}^{i+k-1} (t_j - t_{j-1})^{\gamma_{j-i+1}} |X_{t_j} - X_{t_{j-1}}|^{p_{j-i+1}}\,,$$

became popular as a powerful tool for analyzing high frequency data. The starting point was made when the link between the mathematical concept of quadratic variation and integrated volatility was established. Integrated volatility is not only useful as a measure for the level of volatility, but also explicitly needed for pricing variance and volatility swaps, cf. Howison et al. [17]. These financial instruments became increasingly attractive to investors, since they avoid direct exposure to underlying assets, but make it possible to hedge volatility risk. Barndorff-Nielsen and Shephard [5, 6] established consistency and distributional results for power variation estimates of the p-th integrated volatility in classical diffusion and stable stochastic volatility models respectively, whereas Woerner [27] considered quite general Lévy type stochastic volatility models.

However, in contrast to this theory, empirical studies (e.g. Andersen and Bollerslev [1, 2], Ding, Granger, and Engle [13], Granger and Ding [15], and Granger and Sin [14]) have shown that estimates in the classical stochastic volatility models seem to perform better for $p = 1$, than for $p = 2$. Woerner [26, 28] provided a theoretical explanation for this empirical finding by analyzing the power variation estimates in the presence of a jump component. Namely, when jumps are present in the mean process of a classical stochastic volatility model or as an additive component, the integrated volatility can never be estimated consistently for $p = 2$ by power variation estimates, but depending on the activity of the jump component it can be done for $p = 1$. Hence the empirical finding, that $p = 2$ does not lead to good results, suggests that the data either possesses a noise component, which causes jumps in the data, or the classical stochastic volatility model is not appropriate but should include jumps.

With the concept of bipower and multipower variation it is however possible to overcome this problem and estimate the squared integrated volatility consistently also in the presence of jumps, which is important for pricing variance swaps. The concept of bipower variation in the presence of a finite activity jump component was introduced by Barndorff-Nielsen and Shephard [8] and applied to testing for large jumps by Barndorff-Nielsen and Shephard [7].

In this paper we focus on extending their result to processes with a more general, possibly correlated, mean process and general semimartingale jump components, which especially include the popular Lévy processes such as generalized hyperbolic ones, NIG and CGMY processes. Hence our result provides robust, consistent estimators, allowing for a feasible distributional theory, for the squared integrated volatility within a large class of models, including the BNS model with leverage (Barndorff-Nielsen and Shephard model, i.e. the volatility process follows an Ornstein-Uhlenbeck type process). Furthermore, our result combined with the result for power variation can be applied in the context of model selection to determine jump components. This provides a different approach to the one based on non-normed power variation which was suggested in Woerner [26].

Models and Notation

The concept of power variation in a mathematical framework was introduced in the context of studying the path behaviour of stochastic processes in the 1960ties, cf. Berman [11], Hudson and Tucker [19] for additive processes or Lepingle [21] for semimartingales. Assume that we are given a stochastic process X on some finite time interval $[0,t]$. Let n be a positive integer and denote by $S_n = \{0 = t_{n,0}, t_{n,1}, \cdots, t_{n,n} = t\}$ a partition of $[0,t]$, such that $0 < t_{n,1} < t_{n,2} < \cdots < t_{n,n}$ and $\max_{1 \le k \le n}\{t_{n,k} - t_{n,k-1}\} \to 0$ as $n \to \infty$. Now the p-th power variation is defined to be

$$\sum_{i=1}^{n} |X_{t_{n,i}} - X_{t_{n,i-1}}|^p = V_p(X, S_n)\,.$$

We are interested in the limit as $n \to \infty$, hence the setting of high frequency data. Well established are for convergence in probability the cases for $p = 1$, where finiteness of the limit means that the process has bounded variation, and $p = 2$, called quadratic variation, which is finite for all semimartingale processes. However, in the stochastic volatility setting, i.e. for the moment assuming that our process is of the form $\int_0^t \sigma_s dB_s$, only the case $p = 2$ leads to a non-trivial limit. Obviously, for $p > 2$ the limit is zero and for $p < 2$ the limit is infinity. Whereas for jump processes we do not have this distinction, for $p \ge 2$ the p-th power variation converges to the p-th power of the jumps. This makes it possible to analyze, with the concept of power variation, if the data is derived from a purely continuous process or a process possessing jumps (cf. Woerner [26]).

Returning to the classical stochastic volatility setting, an extension of the concept of power variation is to introduce an appropriate norming sequence, as it was done in Barndorff-Nielsen and Shephard [6], which allows to find non-trivial limits even in the cases where the non-normed power variation limit would be zero or infinity. Let us introduce the following notation for the normed p-th power variation

$$\sum_{i=1}^{n} \Delta_{n,i}^{\gamma} |X_{t_{n,i}} - X_{t_{n,i-1}}|^p = V_p(X, S_n, \Delta_n^{\gamma})\,,$$

where $\gamma \in I\!R$ and $t_{n,i} - t_{n,i-1} = \Delta_{n,i}$ denotes the distance between the i-th and the $i-1$-th time-point. When we have equally spaced observations, $\Delta_{n,i}$ is independent of i and the normed power variation reduces to $\Delta_n^{\gamma} V_p(X, S_n)$. As long as we stick to a purely continuous stochastic volatility model this extension enables us to estimate the p-th integrated volatility for all p, cf. Barndorff-Nielsen and Shephard [6]. Whereas, when we have a jump component in the mean process, such as for example in the BNS model with leverage, in the case $p = 2$ it is not possible to reach the power integrated volatility, though it is the most popular for practitioners.

A further extension of power variation is the concept of normed bipower or multipower variation, which makes it possible to include also the case for the squared or even higher order integrated volatility. This also provides another possibility to distinguish between purely continuous and processes with jumps by comparing the power and multipower variation. Let us introduce the following notation for the normed $r+s$-th bipower variation

$$\sum_{i=1}^{n-1} \Delta_{n,i+1}^{\gamma}|X_{t_{n,i+1}} - X_{t_{n,i}}|^{r} \Delta_{n,i}^{\delta}|X_{t_{n,i}} - X_{t_{n,i-1}}|^{s} = V_{r,s}(X, S_n, \Delta_n^{\gamma}, \Delta_n^{\delta}),$$

where $\gamma, \delta \geq 0$. Analogously for the $\sum_{i=1}^{k} p_i$-th k-power variation

$$\begin{aligned}&\sum_{i=1}^{n-k+1} \prod_{j=i}^{i+k-1} (t_j - t_{j-1})^{\gamma_{j-i+1}} |X_{t_j} - X_{t_{j-1}}|^{p_{j-i+1}} \\ &= V_{p_1,\cdots,p_k}(X, S_n, \Delta_n^{\gamma_1}, \cdots, \Delta_n^{\gamma_k}).\end{aligned}$$

When we have equally spaced observations, the normed $\sum_{i=1}^{k} p_i$-th k-power variation reduces to $\Delta_n^{\gamma_1+\cdots+\gamma_k} V_{p_1,\cdots,p_k}(X, S_n)$.

The concept of bipower and multipower variation was introduced by Barndorff-Nielsen and Shephard [8] for jump components with only finite activity. In this paper we focus on an extension of their result to general jump processes, which makes it possible to get results for the BNS model with leverage.

Let us now briefly review the stochastic processes which we will need in the following. We start with a general semimartingale process X, which is widely used in finance. For an overview both under financial and theoretical aspects see Shiryaev [24]. In its canonical representation a semimartingale may be written as

$$X = X_0 + B(h) + X^c + h * (\mu - \nu) + (x - h(x)) * \mu,$$

or for short with the predictable characteristic triplet $(B(h), \langle X^c \rangle, \nu)$, where X^c denotes the continuous local martingale component, $B(h)$ is predictable of bounded variation and h is a truncation function, behaving like x around the origin. Furthermore, $\mu((0,t] \times A; \omega) = \sum(I_A(J(X_s)), 0 < s \leq t)$, where $J(X_s) = X_s - X_{s-}$ and $A \in \mathcal{B}(\mathbb{R} - \{0\})$, is a random measure, the jump measure, and ν denotes its compensator, satisfying $(x^2 \wedge 1) * \nu \in \mathcal{A}_{loc}$, i.e. the process $(\int_{(0,t]\times \mathbb{R}} (x^2 \wedge 1) d\nu)_{t \geq 0}$ is locally integrable. Semimartingale models include the well-established continuous diffusions, jump-diffusions, stochastic volatility models, as well as Lévy processes.

Lévy processes are a special class of semimartingales where we have independent and stationary increments. They are given by the characteristic function via the Lévy-Khinchin formula

$$E[e^{iuX_t}] = \exp\{t(i\alpha u - \frac{\sigma^2 u^2}{2} + \int (e^{iux} - 1 - iuh(x))\nu(dx))\},$$

where α denotes the drift, σ^2 the Gaussian part and ν the Lévy measure. Hence σ^2 and α determine the continuous part and the Lévy measure the frequency and size of jumps. If $\int(1 \wedge |x|)\nu(dx) < \infty$ the process has bounded variation, if $\int \nu(dx) < \infty$ the process jumps only finitely many times in any finite time-interval, called finite activity, it is a compound Poisson process. Furthermore, the support of ν determines the size and direction of jumps. A popular example in finance are subordinators, where the the support of the Lévy measure is restricted to the positive half line, hence the process does not have negative jumps and the process is of bounded variation in addition. For more details see Sato [22].

A measure for the activity of the jump component of a semimartingale is the generalized Blumenthal-Getoor index,

$$\beta = \inf\{\delta > 0 : (|x|^\delta \wedge 1) * \nu \in \mathcal{A}_{loc}\},$$

where $\mathcal{A}_{loc}$ is the class of locally integrable processes. This index β also determines, that for $p > \beta$ the sum of the p-th power of jumps will be finite. Note that if we are in the framework of Lévy processes, being an element of a locally integrable process reduces to finiteness of the integral of $|x|^\delta \wedge 1$ with respect to the Lévy measure, since it is deterministic, cf. Blumenthal and Getoor [12].

Let us now introduce the stochastic volatility models. In the Black and Scholes framework the logarithm of an asset price X_t is modelled as a geometric Brownian motion or as the solution of the stochastic differential equation

$$dX_t = (\mu + \beta\sigma^2)dt + \sigma dB_t\,,$$

where μ, β and σ are constants. One possibility of overcoming the problems of the Black-Scholes framework and capturing the empirical facts of excess kurtosis, skewness, fat tails and volatility smile, is to introduce a random spot volatility process σ leading to the simplest case of a stochastic volatility model. Now the logarithm of an asset price X_t is modelled as the solution to the following diffusion equation

$$dX_t = (\mu + \beta\sigma_t^2)dt + \sigma_t dB_t\,, \tag{12.1}$$

where σ is assumed to satisfy a second stochastic differential equation. Transforming (12.1) to an integrated form leads to

$$X_t = \mu t + \beta \int_0^t \sigma_s^2 ds + \int_0^t \sigma_s dB_s\,,$$

up to an additive constant, or in a more general formulation

$$X_t = \alpha_t + \int_0^t \sigma_s dB_s\,,$$

where α is some stochastic process. Traditionally, it is assumed that the mean process α is of locally bounded variation and Lipschitz-continuous.

The main differences between the various stochastic volatility models lie in the stochastic differential equation the spot volatility process is assumed to satisfy. We will recall different examples, which our estimating results can be applied to.

Example 1. Assume that the price process can be described by the following diffusion equation

$$dX_t = \mu X_t dt + \sigma_t X_t dB_t \,.$$

Then Hull and White [20] model σ^2 by a geometric Brownian motion, i.e.

$$d\sigma_t^2 = \alpha\sigma_t^2 dt + \chi\sigma_t^2 dW_t \,,$$

where W is a Brownian motion independent of B.

Scott [23] and Stein and Stein [25] model σ by an Ornstein-Uhlenbeck process, i.e.

$$d\sigma_t = -\delta(\sigma_t - \theta)dt + kdW_t \,,$$

where again W is a Brownian motion independent of B.

Barndorff-Nielsen and Shephard [4] model σ^2 by an Ornstein-Uhlenbeck type process of the form

$$d\sigma_t^2 = -\lambda\sigma_t^2 dt + dZ_{\lambda t} \,.$$

Here Z is a subordinator without drift independent of the Brownian motion B. The time scale λt is chosen to ensure that the marginal law of σ^2 is not affected by the choice of λ. Note that though σ^2 exhibits jumps, X is still continuous.

A possibility of including leverage in this model is to add a jump component of the form of the driving subordinator to the price process. The BNS model including leverage, as it was suggested by Barndorff-Nielsen and Shephard [4] is given by

$$\begin{aligned} dX_t &= \{\mu + \beta\sigma_t^2\}dt + \sigma_t dB_t + \rho d\bar{Z}_{\lambda t} \\ d\sigma_t^2 &= -\lambda\sigma_t^2 dt + dZ_{\lambda t} \,, \end{aligned}$$

where $\bar{Z}_t = Z_t - E(Z_t)$.

Robustness of multipower variation

In the context of stochastic volatility models in practice neither the structure of the underlying spot volatility process is known nor the process is observed continuously. This makes it difficult to infer the volatility as it is necessary for option pricing and hedging.

The concept of power variation provides results under quite mild regularity assumptions on the spot volatility and the mean process, but not in the case $p \geq 2$, when jumps are present. With bipower variation estimates, we can consistently estimate all the cases we can deal with power variation and besides also the cases $p < 4$ with jumps. Hence we get a larger class of easily computable estimators of the p-th integrated volatility for $p < 4$.

The following theorem is an extension of Barndorff-Nielsen and Shephard [8] to a larger class of jump components. Our result makes it possible to include a large part of the class of general purely discontinuous semimartingales as jump component and also more general continuous components which both may be correlated to other components of the model.

Theorem C.1. *Let $X_t = Z_t + Y_t + \int_0^t \sigma_s dB_s$. Suppose that $Y = Y^{(1)} + Y^{(2)}$, where $Y_t^{(1)} = \int_0^t a_s ds$ is locally Lipschitz-continuous and independent of B. $Y^{(2)}$ is continuous and satisfies for $p > 0$ as $n \to \infty$*

$$\Delta_n^{1-\frac{p}{2}} V_p(Y^{(2)}, S_n) \xrightarrow{p} 0 \,.$$

Let Z be a semimartingale with Blumenthal-Getoor index β which satisfies either
a) $\langle Z^c \rangle = 0$, if $1 \leq \beta < 2$, or
b) $\langle Z^c \rangle = 0$ and $B(h) + (x-h) * \nu = 0$, if $\beta < 1$.
Furthermore, assume that the volatility process σ^2 is independent of the Brownian motion B and a.s. locally Riemann-integrable. Then we obtain for $r, s > 0$ with $\max(r,s) < 2$

$$\mu_r^{-1} \mu_s^{-1} \Delta_n^{1-\frac{r+s}{2}} V_{r,s}(X, S_n) \xrightarrow{p} \int_0^t \sigma_u^{r+s} du \,,$$

as $n \to \infty$, where $\mu_p = E(|u|^p) = \frac{2^{p/2}\Gamma(\frac{p+1}{2})}{\Gamma(1/2)}$ and $u \sim N(0,1)$.

Proof: We wish to show that in probability as $n \to \infty$

$$\Delta_n^{1-\frac{r+s}{2}} \sum_{j=1}^{n-1} |V_j + Z_j|^r |V_{j+1} + Z_{j+1}|^s \to \mu_r \mu_s \int_0^t \sigma_u^{r+s} du \,, \tag{12.2}$$

where $Z_j = Z_{j\Delta_n} - Z_{(j-1)\Delta_n}$ and $V_j = Y_{j\Delta_n} - Y_{(j-1)\Delta_n} + \int_{(j-1)\Delta_n}^{j\Delta_n} \sigma_s dB_s$. Let us assume $Y^{(2)} = 0$ for simplicity first.

Our proof uses a similar technique as Woerner [28, Theorem 3.1]. Since for exponents less than or equal to 1 our proof relies on the triangular inequality[1] and for the exponent bigger than 1 on Minkowski's inequality[2] we have to look at three different cases.

First look at $0 < r, s \leq 1$. Using the triangular inequality we obtain

[1] $\sum_{i=1}^n |a_i + b_i|^p \leq \sum_{i=1}^n |a_i|^p + \sum_{i=1}^n |b_i|^p$, $p \leq 1$
[2] $(\sum_{i=1}^n |a_i + b_i|^p)^{1/p} \leq (\sum_{i=1}^n |a_i|^p)^{1/p} + (\sum_{i=1}^n |b_i|^p)^{1/p}$, $p > 1$

$$\sum_{j=1}^{n-1} |V_j + Z_j|^r |V_{j+1} + Z_{j+1}|^s - \sum_{j=1}^{n-1} |V_j|^r |V_{j+1}|^s$$
$$\leq \sum_{j=1}^{n-1} |V_j|^r |Z_{j+1}|^s + \sum_{j=1}^{n-1} |Z_j|^r |V_{j+1}|^s + \sum_{j=1}^{n-1} |Z_j|^r |Z_{j+1}|^s ,$$

$$\sum_{j=1}^{n-1} |V_j|^r |V_{j+1}|^s - \sum_{j=1}^{n-1} |V_j + Z_j|^r |V_{j+1} + Z_{j+1}|^s$$
$$\leq \sum_{j=1}^{n-1} |V_j|^r |V_{j+1} + Z_{j+1} - Z_{j+1}|^s - \sum_{j=1}^{n-1} |V_j + Z_j|^r |V_{j+1} + Z_{j+1}|^s$$
$$\leq \sum_{j=1}^{n-1} |V_j|^r |V_{j+1} + Z_{j+1}|^s + \sum_{j=1}^{n-1} |V_j|^r |Z_{j+1}|^s -$$
$$- \sum_{j=1}^{n-1} |V_j + Z_j|^r |V_{j+1} + Z_{j+1}|^s$$
$$\leq \sum_{j=1}^{n-1} |Z_j|^r |V_{j+1} + Z_{j+1}|^s + \sum_{j=1}^{n-1} |V_j|^r |Z_{j+1}|^s$$
$$\leq \sum_{j=1}^{n-1} |Z_j|^r |V_{j+1}|^s + \sum_{j=1}^{n-1} |Z_j|^r |Z_{j+1}|^s + \sum_{j=1}^{n-1} |V_j|^r |Z_{j+1}|^s .$$

This yields

$$\left| \sum_{j=1}^{n-1} |V_j + Z_j|^r |V_{j+1} + Z_{j+1}|^s - \sum_{j=1}^{n-1} |V_j|^r |V_{j+1}|^s \right|$$
$$\leq \sum_{j=1}^{n-1} |V_j|^r |Z_{j+1}|^s + \sum_{j=1}^{n-1} |Z_j|^r |V_{j+1}|^s + \sum_{j=1}^{n-1} |Z_j|^r |Z_{j+1}|^s$$

and we obtain

$$P\left(\left|\Delta_n^{1-\frac{r+s}{2}}\sum_{j=1}^{n-1}|V_j+Z_j|^r|V_{j+1}+Z_{j+1}|^s-\mu_r\mu_s\int_0^t\sigma_u^{r+s}du\right|>\lambda\right)$$

$$\leq P\left(\left|\Delta_n^{1-\frac{r+s}{2}}\sum_{j=1}^{n-1}|V_j+Z_j|^r|V_{j+1}+Z_{j+1}|^s-\right.\right.$$

$$\left.\left.-\Delta_n^{1-\frac{r+s}{2}}\sum_{j=1}^{n-1}|V_j|^r|V_{j+1}|^s\right|>\lambda/2\right)+$$

$$+P\left(\left|\Delta_n^{1-\frac{r+s}{2}}\sum_{j=1}^{n-1}|V_j|^r|V_{j+1}|^s-\mu_r\mu_s\int_0^t\sigma_u^{r+s}du\right|>\lambda/2\right)$$

$$\leq P\left(\left|\Delta_n^{1-\frac{r+s}{2}}\sum_{j=1}^{n-1}|V_j|^r|Z_{j+1}|^s\right|>\lambda/6\right)+ \tag{12.3}$$

$$+P\left(\left|\Delta_n^{1-\frac{r+s}{2}}\sum_{j=1}^{n-1}|Z_j|^r|V_{j+1}|^s\right|>\lambda/6\right)+ \tag{12.4}$$

$$+P\left(\left|\Delta_n^{1-\frac{r+s}{2}}\sum_{j=1}^{n-1}|Z_j|^r|Z_{j+1}|^s\right|>\lambda/6\right)+ \tag{12.5}$$

$$+P\left(\left|\Delta_n^{1-\frac{r+s}{2}}\sum_{j=1}^{n-1}|V_j|^r|V_{j+1}|^s-\mu_r\mu_s\int_0^t\sigma_u^{r+s}du\right|>\lambda/2\right). \tag{12.6}$$

(12.6) tends to zero as $n\to\infty$ by Barndorff-Nielsen and Shephard [8, Theorem 5]. Hence it remains to show that also (12.3), (12.4) and (12.5) tend to zero as $n\to\infty$.

However, let us first look at the other parameter constellations for r and s. W.l.o.g. assume that $r<s$, $0<r\leq 1$, $s>1$. Using the triangular inequality and Minkowski's inequality we obtain

$$\left(\sum_{j=1}^{n-1}|V_j+Z_j|^r|V_{j+1}+Z_{j+1}|^s\right)^{1/s}$$

$$\leq\left(\sum_{j=1}^{n-1}\left||V_j+Z_j|^{r/s}V_{j+1}+|V_j+Z_j|^{r/s}Z_{j+1}\right|^s\right)^{1/s}$$

$$\leq\left(\sum_{j=1}^{n-1}|V_j+Z_j|^r|V_{j+1}|^s\right)^{1/s}+\left(\sum_{j=1}^{n-1}|V_j+Z_j|^r|Z_{j+1}|^s\right)^{1/s}$$

$$\leq\left(\sum_{j=1}^{n-1}|V_j|^r|V_{j+1}|^s\right)^{1/s}+\left(\sum_{j=1}^{n-1}|Z_j|^r|V_{j+1}|^s\right)^{1/s}+$$

$$+\left(\sum_{j=1}^{n-1}|V_j|^r|Z_{j+1}|^s\right)^{1/s}+\left(\sum_{j=1}^{n-1}|Z_j|^r|Z_{j+1}|^s\right)^{1/s},$$

$$\left(\sum_{j=1}^{n-1}|V_j|^r|V_{j+1}|^s\right)^{1/s} =$$
$$= \left(\sum_{j=1}^{n-1}|V_j + Z_j - Z_j|^r|V_{j+1} + Z_{j+1} - Z_{j+1}|^s\right)^{1/s}$$
$$\leq \left(\sum_{j=1}^{n-1}\left(|V_j + Z_j|^r + |Z_j|^r\right)|V_{j+1} + Z_{j+1}|^s\right)^{1/s} + \left(\sum_{j=1}^{n-1}|V_j|^r|Z_{j+1}|^s\right)^{1/s}$$
$$\leq \left(\sum_{j=1}^{n-1}|V_j + Z_j|^r|V_{j+1} + Z_{j+1}|^s\right)^{1/s} + \left(\sum_{j=1}^{n-1}|Z_j|^r|V_{j+1}|^s\right)^{1/s} +$$
$$+ \left(\sum_{j=1}^{n-1}|Z_j|^r|Z_{j+1}|^s\right)^{1/s} + \left(\sum_{j=1}^{n-1}|V_j|^r|Z_{j+1}|^s\right)^{1/s}.$$

This yields

$$\left|\left(\sum_{j=1}^{n-1}|V_j + Z_j|^r|V_{j+1} + Z_{j+1}|^s\right)^{1/s} - \left(\sum_{j=1}^{n-1}|V_j|^r|V_{j+1}|^s\right)^{1/s}\right|$$
$$\leq \left(\sum_{j=1}^{n-1}|V_j|^r|Z_{j+1}|^s\right)^{1/s} + \left(\sum_{j=1}^{n-1}|Z_j|^r|V_{j+1}|^s\right)^{1/s} + \left(\sum_{j=1}^{n-1}|Z_j|^r|Z_{j+1}|^s\right)^{1/s}.$$

For $r < s$ with $r, s > 1$ we can proceed similarly. For the steps where we used the triangular inequality before, we simply have to perform another transformation to use Minkowski's inequality instead and the triangular inequality for the exponent $r/s < 1$. Let us look at one case for illustration

$$\left(\sum_{j=1}^{n-1}|V_j + Z_j|^r|V_{j+1}|^s\right)^{1/s}$$
$$\leq \left[\left(\sum_{j=1}^{n-1}\left||V_{j+1}|^{s/r}V_j + |V_{j+1}|^{s/r}Z_j\right|^r\right)^{1/r}\right]^{r/s}$$
$$\leq \left[\left(\sum_{j=1}^{n-1}|V_j|^r|V_{j+1}|^s\right)^{1/r} + \left(\sum_{j=1}^{n-1}|Z_j|^r|V_{j+1}|^s\right)^{1/r}\right]^{r/s}$$
$$\leq \left(\sum_{j=1}^{n-1}|V_j|^r|V_{j+1}|^s\right)^{1/s} + \left(\sum_{j=1}^{n-1}|Z_j|^r|V_{j+1}|^s\right)^{1/s}.$$

We obtain similarly as for the previous case

$$
\begin{aligned}
&P\left(\left|\left(\Delta_n^{1-\frac{r+s}{2}}\sum_{j=1}^{n-1}|V_j+Z_j|^r|V_{j+1}+Z_{j+1}|^s\right)^{1/s}-\right.\right.\\
&\quad\left.\left.-\left(\mu_r\mu_s\int_0^t\sigma_u^{r+s}du\right)^{1/s}\right|>\lambda\right)\\
&\leq P\left(\left|\left(\Delta_n^{1-\frac{r+s}{2}}\sum_{j=1}^{n-1}|V_j|^r|Z_{j+1}|^s\right)^{1/s}\right|>\lambda/6\right)+\\
&\quad+P\left(\left|\left(\Delta_n^{1-\frac{r+s}{2}}\sum_{j=1}^{n-1}|Z_j|^r|V_{j+1}|^s\right)^{1/s}\right|>\lambda/6\right)+\\
&\quad+P\left(\left|\left(\Delta_n^{1-\frac{r+s}{2}}\sum_{j=1}^{n-1}|Z_j|^r|Z_{j+1}|^s\right)^{1/s}\right|>\lambda/6\right)+\\
&\quad+P\left(\left|\left(\Delta_n^{1-\frac{r+s}{2}}\sum_{j=1}^{n-1}|V_j|^r|V_{j+1}|^s\right)^{1/s}-(\mu_r\mu_s\int_0^t\sigma_u^{r+s}du)^{1/s}\right|>\lambda/2\right).
\end{aligned}
$$

This expression tends to zero under the same conditions as for the other case. Hence it remains to show under which conditions on r and s, we obtain in probability

$$
\lim_{n\to\infty}\Delta_n^{1-\frac{r+s}{2}}\sum_{j=1}^{n-1}|V_j|^r|Z_{j+1}|^s=0\,, \tag{12.7}
$$

$$
\lim_{n\to\infty}\Delta_n^{1-\frac{r+s}{2}}\sum_{j=1}^{n-1}|Z_j|^r|Z_{j+1}|^s=0\,. \tag{12.8}
$$

We use the same technique as is Woerner [26], namely splitting the process Z in a component with finitely many jumps bigger than epsilon, for which we can argue similarly as in Barndorff-Nielsen and Shephard [8], and a process with infinitely many small jumps. We define $I_j(\epsilon)$ to be one if there are no jumps of absolute value bigger than ϵ in the j-th time interval and zero otherwise. Furthermore, denote

$$
Z_t^\epsilon=Z_t-\sum(J(Z_s):|J(Z_s)|>\epsilon,0<s\leq t)\,,
$$

where $J(Z_s)=Z_s-Z_{s-}$, hence denote the jumps of Z. It is clear that $|Z_j|^sI_j(\epsilon)\leq|Z_j^\epsilon|^s$.

Now we can proceed with the proof of (12.7) using Hölder's inequality with $1/p+1/q=1$ and $1/a+1/b=1$ and obtain in probability

$$\lim_{n\to\infty} \Delta_n^{1-\frac{r+s}{2}} \sum_{j=1}^{n-1} |V_j|^r |Z_{j+1}|^s$$

$$= \lim_{\epsilon\to 0}\lim_{n\to\infty} \Delta_n^{1-\frac{r+s}{2}} \sum_{j=1}^{n-1} |V_j|^r |Z_{j+1}|^s I_{j+1}(\epsilon) +$$

$$+ \lim_{\epsilon\to 0}\lim_{n\to\infty} \Delta_n^{1-\frac{r+s}{2}} \sum_{j=1}^{n-1} |V_j|^r |Z_{j+1}|^s (1 - I_{j+1}(\epsilon))$$

$$\leq \left(\lim_{\epsilon\to 0}\lim_{n\to\infty} \left(\Delta_n^{1-\frac{pr}{2}} \sum_{j=1}^{n-1} |V_j|^{pr} \right)^{1/p} \right) \times$$

$$\times \left(\lim_{\epsilon\to 0}\lim_{n\to\infty} \left(\Delta_n^{1-\frac{qs}{2}} \sum_{j=1}^{n-1} |Z_{j+1} I_{j+1}(\epsilon)|^{qs} \right)^{1/q} \right) + \tag{12.9}$$

$$+ \left(\lim_{\epsilon\to 0}\lim_{n\to\infty} \left(\Delta_n^{1-\frac{ar}{2}} \sum_{j=1}^{n-1} |V_j|^{ar} \right)^{1/a} \right) \times$$

$$\times \left(\lim_{\epsilon\to 0}\lim_{n\to\infty} \left(\Delta_n^{1-\frac{bs}{2}} \sum_{j=1}^{n-1} |Z_{j+1}(1 - I_{j+1}(\epsilon))|^{bs} \right)^{1/b} \right) \tag{12.10}$$

$$= 0\,.$$

In fact for the first and the third term we know by Barndorff-Nielsen and Shephard [6, Theorem 1] that in probability it tends to $(\mu_{pr} \int_0^t \sigma_u^{pr} du)^{1/p}$ and $(\mu_{ar} \int_0^t \sigma_u^{ar} du)^{1/a}$ respectively. Note that Condition 1 of Theorem 1 is only needed for the CLT not for the consistency result. Hence it remains to show that the second and the fourth term tend to zero in probability.

For the second term we can establish as in Woerner [26, Theorem 3.1] and Hudson and Mason [18] that for sufficiently small Δ_n, $\sup_j |Z_j^\epsilon| \leq 2\epsilon$ and $\Delta_n \leq 2\epsilon$. We fix η, $\epsilon < \eta/4$ and take $c \in (\beta, qs)$

$$\limsup_{n\to\infty} \Delta_n^{1-\frac{qs}{2}} \sum_{j=1}^{n-1} |Z_{j+1}|^{qs} I_{j+1}(\epsilon)$$

$$\leq \limsup_{n\to\infty} \Delta_n^{1-\frac{qs}{2}} \sum_{j=1}^{n-1} |Z_{j+1}^\epsilon|^{qs}$$

$$\leq (2\epsilon)^{1-\frac{qs}{2}+sq-c} K_c \left[\left(\sum |J(Z_s)|^c : |J(Z_s)| < \eta, 0 < s \leq t \right) + \right.$$

$$\left. + \left(\limsup_{n\to\infty} \sum_{j=1}^{n-1} |Z_{j+1}^\eta|^c \right) \right], \tag{12.11}$$

with some constant K_c independent of ϵ. Both sums are finite by the conditions on Z. Namely under condition a) finiteness in (12.11) is ensured by

Lepingle [21], whereas under condition b) finiteness is ensured by Hudson and Mason [18]. Note that Hudson and Mason [18] originally formulated their result for additive processes, but their arguments can also be applied to general semimartingales satisfying condition b). Hence we can let $\epsilon \to 0$ and obtain the desired result provided $\beta < qs$ and $\beta < 1 + (qs/2)$. This does not impose a restriction, since we can always take q sufficiently large and p accordingly small.

For the fourth term we know that the sum is finite, since we only have finitely many jumps, hence the term tends to zero as $bs < 2$. Since from the Hölder condition we have $b > 1$ this implies $s < 2$. As we have to satisfy the same for r and s exchanged we get the condition

$$\max(r, s) < 2\,.$$

Finally it remains to prove (12.8)

$$\lim_{n\to\infty} \Delta_n^{1-\frac{r+s}{2}} \sum_{j=1}^{n-1} |Z_j|^r |Z_{j+1}|^s$$

$$= \lim_{n\to\infty} \Delta_n^{1-\frac{r+s}{2}} \sum_{j=1}^{n-1} |Z_j|^r I_j(\epsilon) |Z_{j+1}|^s I_{j+1}(\epsilon) \tag{12.12}$$

$$+ \lim_{n\to\infty} \Delta_n^{1-\frac{r+s}{2}} \sum_{j=1}^{n-1} |Z_j|^r (1 - I_j(\epsilon)) |Z_{j+1}|^s I_{j+1}(\epsilon) \tag{12.13}$$

$$+ \lim_{n\to\infty} \Delta_n^{1-\frac{r+s}{2}} \sum_{j=1}^{n-1} |Z_j|^r I_j(\epsilon) |Z_{j+1}|^s (1 - I_{j+1}(\epsilon)) \tag{12.14}$$

$$+ \lim_{n\to\infty} \Delta_n^{1-\frac{r+s}{2}} \sum_{j=1}^{n-1} |Z_j|^r (1 - I_j(\epsilon)) |Z_{j+1}|^s (1 - I_{j+1}(\epsilon)) \tag{12.15}$$

$$= 0\,.$$

(12.12), (12.13) and (12.14) can be shown as above. For (12.12) we can use Hölder inequality for $1/a + 1/b + 1/c = 1$

$$\Delta_n^{1-\frac{r+s}{2}} \sum_{j=1}^{n-1} |Z_j|^r I_j(\epsilon) |Z_{j+1}|^s I_{j+1}(\epsilon)$$

$$= \left(\sum_{j=1}^{n-1} \Delta_n \right)^{1/a} \left(\Delta_n^{1-\frac{br}{2}} \sum_{j=1}^{n-1} |Z_j|^{br} I_j(\epsilon) \right)^{1/b} \times$$

$$\times \left(\Delta_n^{1-\frac{cs}{2}} \sum_{j=1}^{n-1} |Z_{j+1}|^{cs} I_{j+1}(\epsilon) \right)^{1/c}.$$

This does not impose any further restriction on r, s since we can always take b and c sufficiently large by choosing a close to one. For (12.15) we can use the same argument as in Barndorff-Nielsen and Shephard [8, Theorem 6], namely that if we take the partition fine enough there are no contiguous jumps, as there are only finitely many ones. This implies that also (12.15) is equal to zero, which yields the desired result for $Y^{(2)} = 0$. If we now repeat the same argument for $V_j + Z_j$ and $Y_j^{(2)}$ we obtain the desired result under the condition stated for $Y^{(2)}$, which completes our proof. □

Let us discuss the conditions on the processes Z, σ and Y first.

Condition b) can be replaced by a simpler one, if Z is an additive process with $\beta < 1$. In this case we can take the truncation function $h = 0$ and replace b) by $\sigma^2 = \mu = 0$, where σ^2 denotes the Gaussian part and μ the drift of the additive process.

By Barndorff-Nielsen et.al. [3] the condition that σ and Y_1 are independent of B may be replaced by assuming that the volatility can be written in the following form

$$\sigma_t = \sigma_0 + \int_0^t b_s ds + \int_0^t c_s dW_s + \int_0^t d_s dB_s + \int_0^t \int_E h \circ w(s,x)(\mu - \nu)(ds, dx) \\ + \int_0^t \int_E (w - h \circ w(s,x))\mu(ds, dx) ,$$

where W is a Brownian motion independent of B and μ a Poisson random measure, for more details of the notation see Barndorff-Nielsen et.al. [3].

As we know from Woerner [27] the condition on $Y^{(2)}$ is satisfied if it is Hölder continuous of the order $1/2 + \gamma$, $\gamma > 0$, which implies that we can for example add a fractional Brownian motion with Hurst exponent $H \in (1/2, 1]$.

Next we want to discuss the choice of the exponents. The conditions on r and s provide us with more potential for robust estimation than the power variation method. In contrast to the power variation, where we can estimate the p-th integrated volatility robustly only for $p < 2$, we can now estimate it for $p < 4$. This means that we can include the case $p = 2$ as needed for the variance swap. However, we can still not estimate the integral of the fourth power of the volatility as needed for the asymptotic theory of the quadratic variation.

This problem can be solved by passing to multipower variation:

Theorem C.2. *Let* $X_t = Z_t + Y_t + \int_0^t \sigma_s dB_s$. *Suppose that* Y, Z *and* σ *satisfy the same conditions as in Theorem C.1. For all* $1 \leq i \leq k$, $k \geq 2$ *suppose* $r_i > 0$ *with* $\max_i r_i < 2$, *then we obtain as* $n \to \infty$

$$\prod_{i=1}^k \mu_{r_i}^{-1} \Delta_n^{1 - \frac{\sum_{i=1}^k r_i}{2}} V_{r_1, \cdots, r_k}(X, S_n) \xrightarrow{p} \int_0^t \sigma_u^{\sum_{i=1}^k r_i} du .$$

Proof: The proof is analogous to the bipower case. We simply get contributions from n permutations of the V and Z terms which can be treated in the same way using the generalized Hölder inequality. □

We can see that with the k-power variation we can estimate the p-th integrated volatility consistently for $p < 2k$.

Example 2. Let us look at a very general model of the form

$$X_t = \sum_{i=1}^{k} Y_t^{(i)} + \sum_{j=1}^{l} Z_t^{(j)} + \int_0^t \sigma_s dB_s \,,$$

where the $Y^{(i)}$ are Hölder continuous processes of the order $1/2+\gamma$ and $Z^{(j)}$ are pure jump processes which all may be correlated with all other components of the model. We only need that the Blumenthal-Getoor index is less than two, hence we can include the whole class of generalized hyperbolic processes, stable processes and CGMY processes with $\alpha < 2$ and $Y < 2$ respectively. Now we look at three most interesting cases from the point of view of option pricing:

- Estimation of $\int_0^t |\sigma_s| ds$:
  We need the absolute value for pricing volatility swaps. We obtain a consistent estimator by using the p-th power variation with $p = 1$.

- Estimation of $\int_0^t \sigma_s^2 ds$:
  We need the squared integrated volatility for pricing variance swaps, as well as for the asymptotic variance for the absolutes values. By power variation we do not get a robust estimator, but if we move on to bipower variation estimates we get a consistent estimator.

- Estimation of $\int_0^t \sigma_s^4 ds$:
  We need this quantity for the distributional theory of the squared integrated volatility. Here even the bipower variation estimate is not sufficient, but with the tripower variation estimate we get a robust and consistent estimator.

Besides of providing consistent estimators for the integrated volatility, Theorem C.1. implies that we can separate the continuous and the jump part of the quadratic variation, especially if Z is a purely discontinuous additive process without drift and $\beta < 2$. This fact can be used to analyze if we have a purely continuous process or some jump component. This was studied by Barndorff-Nielsen and Shephard [8] for the finite activity jump case and applied to testing in Barndorff-Nielsen and Shephard [7].

Example 3 (BNS model including leverage). We look again at the BNS model including leverage

$$\begin{aligned} dX_t &= \{\mu + \beta\sigma_t^2\}dt + \sigma_t dB_t + \rho d\bar{Z}_{\lambda t} \\ d\sigma_t^2 &= -\lambda\sigma_t^2 dt + dZ_{\lambda t} \,, \end{aligned}$$

where $\bar{Z}_t = Z_t - E(Z_t)$ and it is assumed that the Lévy process Z is independent of the Brownian motion B. The continuous part of the mean process is independent of B and Lipschitz continuous. Hence for the equally spaced setting we obtain

$$
\begin{aligned}
&V_2(X,S_n)\mu_2^{-1} \xrightarrow{p} \int_0^t \sigma_s^2 ds + \sum (J(Z_s)^2; 0 < s \le t) \\
&V_{r,s}(X,S_n)\mu_r^{-1}\mu_s^{-1} \xrightarrow{p} \int_0^t \sigma_s^2 ds \\
&V_2(X,S_n)\mu_2^{-1} - V_{r,s}(X,S_n)\mu_r^{-1}\mu_s^{-1} \xrightarrow{p} \sum (J(Z_s)^2; 0 < s \le t)\,,
\end{aligned}
$$

as long as $r + s = 2$. Hence this provides results even when Z has as much activity as a hyperbolic Lévy motion or a Normal Inverse Gaussian process which have both $\beta = 1$.

Distributional Theory

In the previous section we studied consistency and robustness of the multipower variation estimators. Now we want to look at the distributional theory, which makes it possible to construct tests and calculate confidence regions. As for the power variation estimates the distributional theory unfortunately does not hold in the same generality as the consistency results. Especially we need stronger conditions on the volatility process and on the jump component, where we can only allow for jumps with moderate activity, namely Blumenthal-Getoor index less than one. Our result is an extension of Barndorff-Nielsen and Shephard [8] to a larger class of semimartingale jump components. Together with M. Winkel they are currently also working on extensions of their results to Lévy jumps with different methods as presented here (cf. Barndorff-Nielsen et.al. [10], Barndorff-Nielsen and Shephard [9]).

Theorem C.3. *Let $X_t = Z_t + Y_t + \int_0^t \sigma_s dB_s$. Suppose that $Y_t = \int_0^t a_s ds$ is locally Lipschitz-continuous, additionally Y is independent of B and Z is a semimartingale with Blumenthal-Getoor index $\beta < 1$ which satisfies $\langle Z^c \rangle = 0$, $B(h) + (x - h) * \nu = 0$. Assume that $\sigma_t > 0$ is independent of B_t, locally Riemann integrable, pathwise bounded away from zero and has the property that for some $\gamma > 0$ and $n \to \infty$*

$$
\sqrt{\Delta_n} \sum_{j=1}^n |\sigma^\gamma(\eta_{n,j}) - \sigma^\gamma(\chi_{n,j})| \xrightarrow{p} 0 \qquad (12.16)
$$

for any $\chi_{n,j}$ and $\eta_{n,j}$ such that

$$
0 \le \chi_{n,1} \le \eta_{n,1} \le t_{n,1} \le \chi_{n,2} \le \eta_{n,2} \le t_{n,2} \cdots \le \chi_{n,n} \le \eta_{n,n} \le t\,.
$$

Then we obtain for $r, s > 0$ *with* $\max(r, s) < 1$ *and* $r + s > \beta/(2 - \beta)$,

$$\frac{\Delta_n^{1-\frac{r+s}{2}} V_{r,s}(X, S_n) - \mu_r \mu_s \int_0^t \sigma_u^{r+s} du}{\Delta_n^{1/2} \sqrt{C \int_0^t \sigma_u^{2r+2s} du}} \xrightarrow{\mathcal{D}} N(0,1)\,,$$

as $n \to \infty$, *where* $\mu_p = E(|u|^p)$ *with* $u \sim N(0,1)$ *and* $C = \mu_{2r}\mu_{2s} + 2\mu_r\mu_s\mu_{r+s} - 3\mu_r^2\mu_s^2$.

Proof: The idea is to use the result for the model with no jump component, which we know from Barndorff-Nielsen and Shephard [8] for $r = s$ or from Barndorff-Nielsen et.al. [3] for the general case, where the asymptotic normality can be deduced straight forward for the independent case from the stable convergence towards an integral w.r.t. a Brownian motion. We use the same notation as in Theorem C.1,

$$\begin{aligned} &\frac{\Delta_n^{1-\frac{r+s}{2}} V_{r,s}(X, S_n) - \mu_r \mu_s \int_0^t \sigma_u^{r+s} du}{\Delta_n^{1/2} \sqrt{C \int_0^t \sigma_u^{2r+2s} du}} \\ &\quad = \frac{\Delta_n^{1-\frac{r+s}{2}} V_{r,s}(X, S_n) - \Delta_n^{1-\frac{r+s}{2}} V_{r,s}(V, S_n)}{\Delta_n^{1/2} \sqrt{C \int_0^t \sigma_u^{2r+2s} du}} + \\ &\quad\quad + \frac{\Delta_n^{1-\frac{r+s}{2}} V_{r,s}(V, S_n) - \mu_r \mu_s \int_0^t \sigma_u^{r+s} du}{\Delta_n^{1/2} \sqrt{C \int_0^t \sigma_u^{2r+2s} du}}\,. \end{aligned}$$

For the last term we know that it converges to $N(0,1)$, hence we obtain the desired result if the first summand tends to zero in probability. For the numerator we can proceed as in Theorem C.1, hence together with the norming sequence we have to show that in probability,

$$\lim_{n\to\infty} \Delta_n^{\frac{1}{2}-\frac{r+s}{2}} \sum_{j=1}^{n-1} |V_j|^r |Z_{j+1}|^s = 0\,, \tag{12.17}$$

$$\lim_{n\to\infty} \Delta_n^{\frac{1}{2}-\frac{r+s}{2}} \sum_{j=1}^{n-1} |Z_j|^r |Z_{j+1}|^s = 0\,, \tag{12.18}$$

which leads similarly as in Theorem C.1 to

$$
\begin{aligned}
&\lim_{n\to\infty} \Delta_n^{\frac{1}{2}-\frac{r+s}{2}} \sum_{j=1}^{n-1} |V_j|^r |Z_{j+1}|^s \\
&\leq \left(\lim_{\epsilon\to 0} \lim_{n\to\infty} \left(\Delta_n^{1-\frac{pr}{2}} \sum_{j=1}^{n-1} |V_j|^{pr} \right)^{1/p} \right) \times \\
&\quad \times \left(\lim_{\epsilon\to 0} \lim_{n\to\infty} \left(\Delta_n^{\frac{1}{2}-\frac{q}{2p}-\frac{qs}{2}} \sum_{j=1}^{n-1} |Z_{j+1} I_{j+1}(\epsilon)|^{qs} \right)^{1/q} \right) + \\
&\quad + \left(\lim_{\epsilon\to 0} \lim_{n\to\infty} \left(\Delta_n^{1-\frac{ar}{2}} \sum_{j=1}^{n-1} |V_j|^{ar} \right)^{1/a} \right) \times \\
&\quad \times \left(\lim_{\epsilon\to 0} \lim_{n\to\infty} \left(\Delta_n^{\frac{1}{2}-\frac{b}{2a}-\frac{bs}{2}} \sum_{j=1}^{n-1} |Z_{j+1}(1 - I_{j+1}(\epsilon))|^{bs} \right)^{1/b} \right) \\
&= 0\,.
\end{aligned}
$$

The first and the third term tend to a finite limit as in Theorem C.1. For the fourth term to tend to zero we need

$$\frac{1}{2} - \frac{b}{2a} - \frac{bs}{2} > 0\,.$$

This leads to the restriction $s < 1$, hence $\max(r, s) < 1$. For the second term we use the same method as in (12.11), this implies that we need

$$\frac{1}{2} - \frac{q}{2p} - \frac{qs}{2} + sq - \beta > 0\,.$$

Examining the possible ranges of q and s we obtain that $\beta < 1$. For (12.18) we can proceed as in Theorem C.1. The part with both components having small jumps leads to an additional restriction on r and s, namely we have to satisfy simultaneously

$$
\begin{aligned}
1 - b\frac{s + \frac{1}{q}}{2} + bs - \beta &> 0 \\
bs &> \beta \\
1 - a\frac{r + \frac{1}{p}}{2} + ar - \beta &> 0 \\
ar &> \beta\,,
\end{aligned}
$$

with $1/p + 1/q = 1$ and $1/a + 1/b = 1$. This leads to

$$
\begin{aligned}
s &> \frac{\beta}{q(2-\beta)} \\
r &> \frac{\beta}{p(2-\beta)}\,,
\end{aligned}
$$

hence $r + s > \beta/(2 - \beta)$.

This yields the desired result under the stated conditions on β, r and s. □

As for the power variation estimator we have to impose stronger conditions for the distributional theory on the different components on the model than for consistency. From Barndorff-Nielsen and Shephard [6] we know that the condition (12.16) on σ is satisfied for the BNS model and for the special case $r = s = 1$ it can be dropped (cf. Barndorff-Nielsen and Shephard [7]). By Barndorff-Nielsen et.al. [3] the condition on the volatility process may be replaced by

$$\sigma_t = \sigma_0 + \int_0^t b_s ds + \int_0^t c_s dW_s + \int_0^t d_s dB_s + \int_0^t \int_E h \circ w(s,x)(\mu - \nu)(ds, dx)$$
$$+ \int_0^t \int_E (w - h \circ w(s,x))\mu(ds, dx),$$

which is weaker than (12.16) and allows us to skip the condition that σ and Y are independent of B. Of course, then we do not get asymptotic normality, but stable convergence to an integral w.r.t. a Brownian motion as in Barndorff-Nielsen et.al. [3]. This implies that our result can also be applied to the Heston model, cf. Heston [16].

Furthermore, the activity of the jump component must not be too large. But anyway we can include compound Poisson processes, Gamma processes, inverse Gaussian processes, stable processes with $\alpha < 1$ and CGMY processes with $Y < 1$. However, as we have the restriction $r + s < 2$ we do not get a distributional result for the squared integrated volatility case by bipower variation. As for the consistency we can generalize the result to multipower variation.

Theorem C.4. *Let $X_t = Z_t + Y_t + \int_0^t \sigma_s dB_s$. Suppose that Y, Z and σ satisfy the same conditions as in Theorem 3.1. For all $1 \leq i \leq k$, $k \geq 2$ and $r_i > 0$ with $\max_i r_i < 1$ and $\sum_{i=1}^k r_i > \beta/(2-\beta)$, we obtain as $n \to \infty$*

$$\frac{\Delta_n^{1-\frac{\sum_{i=1}^k r_i}{2}} V_{r_1,\cdots,r_k}(X, S_n) - \prod_{i=1}^k \mu_{r_i} \int_0^t \sigma_u^{\sum_{i=1}^k r_i} du}{\Delta_n^{1/2} \sqrt{C \int_0^t \sigma_u^{2\sum_{i=1}^k r_i} du}} \xrightarrow{\mathcal{D}} N(0,1),$$

as $n \to \infty$. Here $\mu_p = E(|u|^p)$ with $u \sim N(0,1)$ and

$$C = \prod_{l=1}^k \mu_{2r_l} - (2k-1) \prod_{l=1}^k \mu_{r_l}^2 +$$
$$+2 \sum_{i=1}^{k-1} \prod_{l=1}^i \mu_{r_l} \prod_{l=k-i+1}^k \mu_{r_l} \times \prod_{l=1}^{k-i} \mu_{r_l + r_{l+i}}.$$

Proof: The proof is analogous to the previous one relying on the generalized Hölder inequality as in Theorem C.2. □

Now we can see that we can get a distributional theory for the squared integrated volatility by using tripower variation.

So far, our results of Theorem C.3 and C.4 are not feasible, since in general we do not know the variance, but it can be made feasible by plugging in a consistent and robust multipower estimator as we have deduced in the previous section.

Example 4. Let us look at a general model of the form

$$X_t = Y_t + \sum_{j=1}^{l} Z_t^{(j)} + \int_0^t \sigma_s dB_s \,,$$

where the Y is Lipschitz-continuous and independent of B. The $Z^{(j)}$ are pure jump processes with Blumenthal-Getoor index less than one, which all may be correlated with all other components of the model. Hence we can have compound Poisson processes, Gamma processes, inverse Gaussian processes, stable processes with $\alpha < 1$, CGMY processes with $Y < 1$. Now we look at the two most interesting cases:

- Estimation of $\int_0^t |\sigma_s| ds$:
  We need the absolute value for pricing volatility swaps. We get a consistent and robust estimator and the corresponding distributional theory by using bipower variation as an estimator for $\int_0^t |\sigma_s| ds$ and for the variance term.

- Estimation of $\int_0^t \sigma_s^2 ds$:
  We need the squared integrated volatility for pricing variance swaps. We get a consistent and robust estimator and the corresponding distributional theory by using tripower variation as an estimator for $\int_0^t \sigma_s^2 ds$ and for the variance term.

Acknowledgement

I would like to thank Prof. Dr. O.E. Barndorff-Nielsen for helpful discussions, Prof. Dr. N. Shephard for drawing my attention to the concept of power and bipower variation and both for making their latest research available to me. Furthermore, I would like to thank Prof. Dr. J. Jacod, two referees and the associate editor for helpful comments.

References

[1] T. G. Andersen and T. Bollerslev, Intraday periodicity and volatility persistence in financial markets, *Journal of Empirical Finance*, 4:115–158, 1997.

[2] T. G. Andersen and T. Bollerslev, Answering the skeptics: yes, standard volatility models do provide accurate forecasts, *International Economics Review*, 39:885–905, 1998.

[3] O. E. Barndorff-Nielsen and E. Graversen and J. Jacod and M. Podolskij and N. Shephard, A central limit theorem for realised power and bipower variation of continuous semimartingales, Technical report, 2004a. To appear in *From Stochastic Analysis to Mathematical Finance*, Festschrift for Albert Shiryaev.

[4] O. E. Barndorff-Nielsen and N. Shephard, Non-Gaussian Ornstein-Uhlenbeck-based models and some of their uses in financial economics (with discussion), *Journal of the Royal Statistical Society, Series B*, 63:167–241, 2001.

[5] O. E. Barndorff-Nielsen and N. Shephard, Power variation and time change. Technical report, www.levyprocess.org, 2002. Forthcoming in Theory of Probability and its Applications.

[6] O. E. Barndorff-Nielsen and N. Shephard, Realised power variation and stochastic volatility models, *Bernoulli*, 9:243–265, 2003a.

[7] O. E. Barndorff-Nielsen and N. Shephard, Econometrics of testing for jumps in financial econometrics using bipower variation, Technical report, www.levyprocess.org, 2003b.

[8] O. E. Barndorff-Nielsen and N. Shephard, Power and bipower variation with stochastic volatility and jumps, to appear in Journal of Financial Econometrics, 2004.

[9] O. E. Barndorff-Nielsen and N. Shephard, Multipower variation and stochastic volatility, Working paper, 2004c.

[10] O. E. Barndorff-Nielsen and N. Shephard and M. Winkel, Limit Theorems for multipower variation in the presence of jumps, Working paper, 2004b.

[11] S. M. Berman, Sign-invariant random variables and stochastic processes with sign invariant increments, *Trans. Amer. Math. Soc*, 119:216–243, 1965.

[12] R. M. Blumenthal and R. K. Getoor, Sample functions of stochastic processes with stationary independent increments, *J. Math. Mech.*, 10:493–516, 1961.

[13] Z. Ding and C. W. J. Granger and R.F. Engle, A long memory property of stock market returns and a new model, *Journal of Empirical Finance*, 1:83–106, 1993.

[14] C. J. W. Granger and C.-T. Sin, Modelling the absolute returns of different stock indices: exploring the forecastability of an alternative measure of risk, Working paper, Department of Economics, University of California at San Diego, 1999.

[15] C. W. J. Granger and Z. Ding, Some properties of absolute returns, an alternative measure of risk, *Annals d'Economie et de Statistique*, 40:67–91, 1995.

[16] S. Heston, A closed form solution for options with stochastic volatility with applications to bond and currency options, *Review of Financial Studies*, 6:327–343, 1993.

[17] S. Howison and A. Rafailidis and H. Rasmussen, On the pricing and hedging of volatility derivatives, Working Paper, OCIAM, University of Oxford, 2003.

[18] W. N. Hudson and J. D. Mason, Variational sums for additive processes, *Proc. Amer. Math. Soc*, 55:395–399, 1976.

[19] W. N. Hudson and H. G. Tucker, Limit theorem for variational sums, *Trans. Amer. Math. Soc*, 191:405–426, 1974.

[20] D. Hull and A. White, The pricing of options on assets with stochastic volatilities, *Journal of Finance*, 42:281–300, 1987.

[21] D. Lepingle, La variation d'ordre p des semi-martingales, *Z. Wahrscheinlichkeitstheorie und Verw. Gebiete*, 36:295-316, 1976.

[22] K. Sato, *Lévy Processes and Infinitely Divisible Distributions*, Cambridge University Press, Cambridge, 1999.

[23] L. Scott, Option pricing when the variance changes randomly: theory, estimation and an application, *Journal of Financial and Quantitative Analysis*, 22:419-438, 1987.

[24] A. N. Shiryaev, *Essentials of Stochastic Finance: Facts, Models and Theory*, World Scientific, Singapore, 1999.

[25] E. Stein and C. Stein, Stock price distributions with stochastic volatility: an analytic approach, *Review of Financial Studies*, 4:727–752, 1991.

[26] J. H. C. Woerner, Variational Sums and Power variation: a unifying approach to model selection and estimating in semimartingale models, *Statistics & Decisions*, 21:47-68, 2003a.

[27] J. H. C. Woerner, Purely discontinuous Lévy Processes and Power Variation: inference for integrated volatility and the scale parameter, 2003-MF-07, Working Paper Series in Mathematical Finance, University of Oxford, 2003b.

[28] J. H. C. Woerner, Estimation of Integrated Volatility in Stochastic Volatility Models, *Appl. Stochastic Models Bus. Ind.*, 21:27–44, 2005.